Photoemission Studies of High-Temperature Superconductors

This book describes the current status of photoelectron spectroscopic techniques, both theoretical and experimental, that have been applied to the study of the cuprate ("high-temperature") superconductors, together with the results derived from such measurements.

The techniques described include angle-resolved photoelectron spectroscopy of valence electrons, core-level spectra (XPS), and some special variations, such as resonance photoemission. Attention is paid to the difficulties in interpretation of such spectra and to the problems in obtaining good sample surfaces and high resolution. Some comparison with results from other experimental techniques is made. The authors also outline expected future developments in the techniques.

This book will be of great interest to graduate students and researchers in physics, chemistry and materials science with an interest in high-temperature superconductors.

T0235155

CAMBRIDGE STUDIES IN LOW TEMPERATURE PHYSICS

EDITORS

A.M. Goldman
Tate Laboratory of Physics, University of Minnesota
P.V.E. McClintock
Department of Physics, University of Lancaster
M. Springford
Department of Physics, University of Bristol

Cambridge Studies in Low Temperature Physics is an international series which contains books at the graduate text level and above on all aspects of low temperature physics. This includes the study of condensed state helium and hydrogen, condensed matter physics studied at low temperatures, superconductivity and superconducting materials and their applications.

Photoemission Studies of High-Temperature Superconductors

David W. Lynch and Clifford G. Olson

Iowa State University

CAMBRIDGE UNIVERSITY PRESS
Cambridge, New York, Melbourne, Madrid, Cape Town, Singapore, São Paulo

Cambridge University Press
The Edinburgh Building, Cambridge CB2 2RU, UK

Published in the United States of America by Cambridge University Press, New York

www.cambridge.org
Information on this title: www.cambridge.org/9780521551892

First published 1999
This digitally printed first paperback version 2005

A catalogue record for this publication is available from the British Library

Library of Congress Cataloguing in Publication data

Lynch, David W.
Photoemission studies of high-temperature superconductors / David
W. Lynch and Clifford G. Olson.
p.   cm.
Includes bibliographical references and index.
ISBN 0 521 55189 7
1. Copper oxide superconductors. 2. High temperature
superconductors. 3. Photoemission. 4. Photoelectron spectroscopy.
I. Olson, Clifford G., 1942–   . II. Title.
QC611.98.C64L96 1999
537.6′235–dc21   98–7169   CIP

ISBN-13 978-0-521-55189-2 hardback
ISBN-10 0-521-55189-7 hardback

ISBN-13 978-0-521-01949-1 paperback
ISBN-10 0-521-01949-4 paperback

Contents

Preface

There are several reasons this book was written. One was to provide a survey of all of the information obtained to date on cuprates by photoelectron spectroscopy, a technique that has been one of the more productive techniques for providing information on the electronic structure of cuprates. Thus the book serves as a review of research, but such reviews soon become dated in a rapidly moving field. Another reason was to provide a textbook, albeit of a limited nature, for persons entering the field of photoelectron spectroscopy. This aspect of the book should be useful for a longer time. We have limited our discussion to only the techniques used in photoemission on cuprates, excluding other applications of photoelectron spectroscopy, but, in fact, not much has been omitted. Finally, we hope that experienced theorists and experimentalists from fields other than photoelectron spectroscopy will learn something of the difficulties with the photoemission experiments and problems with the interpretation of the data. Photoemission data are widely quoted and we hope to provide a better understanding of the phenemona and the experimental difficulties for those who use such data in the future. In the past few years there has been much activity in the study of complex oxides, e.g., manganates, nickelates, ruthenates, often by photoelectron spectroscopy. We hope those whose interests lie with these materials rather than cuprates will find much of value in this book.

The level of writing is such that a graduate student in physics with a one-year introductory course in solid-state physics, e.g., from the texts by Ashcroft and Mermin or Kittel, should have little trouble with most of the book. Chapter 3 contains the notation of second quantization, but those sections using such notation only outline the approach. They were not intended to equip a reader to carry out calculations in many-body physics. Persons in other fields, e.g., materials science and chemistry, should be able to read the book with little more difficulty. Numerous references will guide readers to the literature, both in theoretical and experimental aspects of photoelectron spectroscopy.

There has been some controversy among the practitioners of photoelectron spectroscopy on cuprates. Certainly many of the early measurements were made on poor samples, but there are more recent disagreements in data taken on single crystals. Disagreements in interpretation also have occurred. We have tried to take a balanced view in our presentation of experimental results and their interpretation. Controversies usually can be settled by further measurements.

There is still no widely accepted theory for superconductivity in the cuprates, nor even for the description of their electronic structure in the normal state. However, data from photoelectron spectroscopy are used frequently for comparison with calculations. We have made some comparisons with calculated spectra, but we have not tried to cover thoroughly the vast theoretical literature. We have not referenced all of the theoretical work in which a spectral function for a model cuprate has been calculated. To do so would have increased significantly the size of the volume. Some of the theoretical work will become dated rather rapidly once a final description for the electronic structure of the cuprates is at hand, but it seems premature to select right now which work that will be.

Many book prefaces comment on how difficult it is to stop writing. In the present case it was not so hard to stop if one could stay away from the library after a certain date and be careful about which mail one opened. As a result of such discipline, this book covers the literature quite thoroughly, except for conference proceedings, until the end of 1996. Most of the pertinent publications appearing in the *Physical Review, Physical Review Letters*, and *Physica C* in the first half of 1997 have also been cited, but not those in other journals.

Notation is always a problem. We have used generally accepted symbols for most quantities. The price paid for this is having the same symbol used for several different quantities. An example is that α is used for the absorption coefficient, the fine-structure constant, and a critical exponent in the description of x-ray absorption edges. However, they appear in different parts of the book, never together. We believe such multiple use of some symbols will not cause any difficulty. Almost all symbols have been defined upon first use, and some that reappear many pages after their first appearance, have been defined again upon reappearance. Some common symbols for physical quantities have not been defined, e.g., $\hbar$ and k_B.

Several people have provided assistance in writing this book. We have received bibliographies, reprints, and preprints from Z.-X. Shen, Juan Carlos

Campuzano, and Takashi Takahashi. Bruce Harmon read much of the first half of the manuscript and Mike Norman read a near-final version of the entire manuscript. Both made many helpful comments. Needless to say any errors or omissions in this book should not be attributed to anyone but its authors. Finally, Rebecca Shivvers saved the LaTeX manuscript from being unprintable after it was attacked by a bug in the word processing program originally used.

This book was written with partial support from the Ames Laboratory which is operated by Iowa State University for the U.S. Department of Energy under contract No. 74405-ENG-82.

David W. Lynch
Ames

Chapter 1

Introduction

Between 1911 and 1986, superconductivity was strictly a low-temperature phenomenon. The highest critical temperature T_c for any superconductor was 23.2 K for Nb_3Ge. For any possible applications, the only useful refrigerants were liquid helium and liquid hydrogen. For much of this period, an understanding of the microscopic origin of superconductivity was lacking. Bardeen, Cooper, and Schrieffer published the BCS theory in 1957, and the more general Eliashberg theory soon followed. Theoretical predictions of the highest T_c one could hope for were not very useful for finding new materials with T_c above 23.2 K. In 1986, Bednorz and Müller discovered that La_2CuO_4 went superconducting when Ba substituted for some of the La. Later doping studies with Ca, Sr, and Ba showed that in this system T_c reached 38 K. Soon thereafter, Wu et al. found that suitably doped $YBa_2Cu_3O_7$ had a T_c of 92 K, well above the boiling point of liquid nitrogen. In the intervening years several other classes of high-temperature superconductors were found. All of these had several properties in common. All contained one or more planes of Cu and O atoms per unit cell, all had structures which were related to the cubic perovskite structure, and all were related to a "parent compound" which was antiferromagnetic and insulating. Doping this parent compound (in one of several ways) produced a metal which was superconducting and, for some optimum doping, T_c could be very high, usually above the boiling point of liquid nitrogen. For some doping ranges the material exhibited semiconducting behavior in the normal state, but was still superconducting at lower temperatures. Considering various doping schemes, over 300 cuprates have been studied in some fashion. Fewer than this have been studied by any type of photoelectron spectroscopy, and only a handful to date have proven amenable to study by angle-resolved photemission.

After the discovery of the cuprate superconductors, two other classes of materials were discovered which would have attracted great attention as high-temperature superconductors were it not for the previous existence of the cuprate-based materials. $BaBiO_3$ is a perovskite-structured insulator. Obviously it contains no Cu and it is not magnetic. It also contains no two-dimensional planes of Ba or Bi and O atoms. It can be doped to produce a metal, and an optimum doping results in T_c values higher than that of Nb_3Ge. $Ba_{1-x}K_xBiO_3$ has a peak T_c of 30 K for $x = 0.4$. The fullerene or buckyball molecule, C_{60}, forms an insulating crystal with several crystal structures. C_{60} can be intercalated with alkali metals, e.g., K. At a composition of K_3C_{60} the material is metallic with a minimum in resistivity as a function of the C : K ratio, and this metal is superconducting. The highest T_c achieved to date is 33 K using two alkali metals to make Cs_2RbC_{60}.

In order to understand superconductivity at a microscopic level in a given material one needs an understanding of the electronic structure of the material, a microscopic mechanism for the coupling of electrons into pairs, and a scheme for forming a coherent condensate from a large number of such pairs. The BCS theory with extensions by Eliashberg and Bogoliubov provided the last of these. The BCS model used the electron–phonon interaction to form the Cooper pairs. The electronic structure was assumed to be simple, that of a gas of Landau quasi-particles, with no band structure effects. Thus in this model all metals should become superconducting at some, possibly very low, temperature. The electronic band structure and the actual phonon spectra were introduced later, but finding increases in T_c by examining electronic structures and phonon spectra to select new materials is far from a science. There were no predictions of a high T_c for a cuprate. Recently Emery and Kivelson reintroduced a picture in which T_c is limited not by the breaking of quasiparticle pairs, but by the loss of long-range phase order. Pairs may then exist above T_c. They suggested this picture is most likely to be valid for underdoped cuprates which are superconductors with very low superfluid densities. Although there were a number of results from several types of experiment on $YBa_2Cu_3O_{7-x}$ that could be explained by the occurrence of a pseudogap, such a pseudogap, presumably the same one, was found in recent photoemission studies of $Bi_2Sr_2CaCu_2O_{8-x}$. Since this pseudogap closely resembled in magnitude and symmetry the gap found in the superconducting phase, it may be attributed to pair formation above T_c.

At the time of this writing the data base of the *High-T_c Update* newsletter at the Ames Laboratory has over 60 500 entries. There are at least two journals whose entire contents are devoted to superconductivity, mostly in cuprates in the past few years, *Journal of Superconductivity* and *Physica C*. The mechanism for superconductivity in the cuprates is still not known, or if one of the several mechanisms already proposed is correct, it is not yet widely accepted as correct. It is not yet known how best to describe the normal state of the cuprates. Are they Fermi liquids or not, and if not, how are they different? Should we approach the electronic structure of the normal state with conventional band theory, modified to take better account of correlation effects or is a highly correlated many-body model a better, or even necessary, starting point? What is the excitation spectrum in the superconducting state? Is there evidence for pairing above T_c? Can cuprate superconductivity be described by the BCS model with phonon coupling or some other coupling mechanism giving the Cooper pairs? What is the symmetry of the pairs? What is so special about Cu and O? Is the correct model for cuprate superconductivity the same for electron doping as for hole doping? Does superconductivity in doped $BaBiO_3$ have any connection with superconductivity in the cuprates? What is the role of the newly discovered stripe phases? Photoelectron spectroscopy can address these questions and provide insight which can aid in answering them, but it rarely gives a direct answer. It is an excited-state spectroscopy, and the connection between the observed spectrum and the ground-state properties is usually not a direct one, especially when the non-interacting- or independent-electron model is not accurate. Even for simple metals like Na, photoelectron spectra show departures from predictions of the independent-particle model. Often the reasons for such departures are known and spectra that agree with experiment can be calculated. Comparable understanding of the photoelectron spectra of high-T_c superconductors is still lacking.

Photoelectron spectroscopy (often called photoemission spectroscopy, although the photon is not emitted) has been employed increasingly since about 1970 to obtain electronic structure information. In the most common experimental setup a beam of monochromatic radiation illuminates the surface of a sample in vacuum and the kinetic energy spectrum of the emitted electrons is measured. The electrons analyzed may be most of those emitted (angle-integrated mode) or only those in a small solid angle that may be varied (angle-resolved mode). Using ultraviolet or soft x-ray photons, angle-integrated photoelectron spectra bear a resemblance to a density of states, and by varying the photon energy some information about the nature of the electronic wave functions can be gained. If the analysis of the photoelectron kinetic energies is done with angle resolution as well, and the sample is a single crystal, electron momentum information becomes available and one can map the energy bands. One result of this type of study is an experimental Fermi surface. With soft x-ray photons one can also excite core levels. Their binding energies, line shapes, and the occurrence of satellite lines give information on local bonding. All these spectroscopies have given considerable detailed information on a wide variety of materials. However, this information is not obtained readily. Because all the electrons in a solid may respond to the photohole, the spectrum may deviate from that expected from a one-electron model, e.g., band structure, by a small or large amount. A well-known example is the photoelectron spectrum of the 4f electrons in Ce. There is approximately one 4f electron per atom in Ce metal, but the photoelectron spectrum has two peaks reliably associated with the 4f initial state. A rather elaborate many-electron model is needed to understand this result, but with it one learns a great deal about the ground state of Ce metal. Similar, but less obvious, problems arise with so-called band mapping.

Naturally, in the early days of the cuprate superconductors many scientists with photoelectron spectrometers made measurements on the best samples of the time. These were sintered ceramics containing many small crystallites of the desired material, and perhaps small amounts of another phase or of unreacted starting materials. X-ray diffraction was used to select samples that were primarily of one phase. Fresh surfaces were obtained by fracturing or scraping *in situ*. Early efforts were centered on questions of Cu valence and on which atoms the holes produced by doping resided. Data were sample dependent, and rarely did the spectra show the Fermi edge characteristic of a metal, even though the samples exhibited bulk metallic conductivity and were superconducting upon cooling. The fracture may have taken place on grain boundaries where a minority phase may have precipitated, and the spectra may have been dominated by this minority phase. The fresh surfaces may have reacted with molecules in the ambient, or impurity atoms may have diffused from the bulk so rapidly that they reached equilibrium before photoelectron spectroscopic studies of surface stability were begun. A common problem in the early data was the occurrence of a satellite peak around 8–12 eV in the valence-band spectra, a peak now known not to be intrinsic to clean surfaces of those high-T_c materials frequently studied.

As single crystals became available in sizes suitable for ultraviolet and soft x-ray photoelectron spectroscopy, the quality of the data improved rapidly. Samples freshly cleaved in ultrahigh vacuum at low temperature were stable for many hours at low temperature. The photoelectron spectra showed the expected Fermi edge, and angle-resolved photoemission mapped a few bands, including the important ones near the Fermi level. Improvements in instrumental resolution and in crystal quality led to much more informative data, especially on $Bi_2Sr_2CaCu_2O_8$ and, to a lesser extent, on $YBa_2Cu_3O_7$. In the former, the Fermi surface was mapped, bands just below it were mapped, photohole lifetimes were determined, although at energies too far below the Fermi energy to allow theoretical work to be applicable, and the opening of a gap below T_c was observed. The anisotropy of the gap in the basal plane was revealed later when yet better crystals became available. The effects of electron doping on the electronic structure were also studied in Nd_2CuO_4.

The superconducting gap or order parameter in cuprate superconductors is about 25 meV or less, depending on crystallographic direction. It was measured by photoemission spectroscopy using photons with an energy of about 20 eV, which eject photoelectrons of about 15 eV kinetic energy. In the first such measurement the photon bandpass was about 20 meV and the electron energy analyzer resolution also 20 meV. Additional energy broadening can be introduced by the combined effect of band dispersion and finite angle resolution. Neglecting the latter, the overall resolution was about 28 meV, so the accuracy of the gap measurement was not very great. It was first determined from the difference in two spectral edges about 20 meV apart, each measured with 28 meV resolution. Normally one would try to measure a 20 meV excitation using a probe with an energy of the order of 20 meV, i.e., by an infrared absorption or reflection measurement, or a tunneling study. Low-energy electron energy-loss spectroscopy and Raman spectroscopy would also be promising because of the high resolution in such experiments, although the probing particle is more energetic. Such measurements were, in fact, carried out, both on early samples and crystals of increasing quality. The infrared measurements showed a number of gap-like features, but continuing work has cast doubt on the early interpretation, and it not clear yet that gap information can be extracted. The early tunneling and electron energy-loss spectra similarly were controversial. The photoelectron studies on $Bi_2Sr_2CaCu_2O_8$ in 1989–90 gave the first widely accepted data on the electronic structure of a high-T_c material. At the time of this writing, there are comparable data only on one other family of high-T_c materials, $YBa_2Cu_3O_7$, but a gap in the superconducting phase was not measured reliably by photoelectron spectroscopy for many years. Recent (1997) results from two groups show the presence of a gap and suggest its symmetry. Crystals, or at least the cleaved surfaces of crystals, of Tl- and Hg-based high-T_c cuprates are not yet of good enough quality to have yielded useful photoelectron spectra.

One of the results of the application of photoelectron spectroscopy to the cuprates is that we have learned that we may not yet know how to interpret such spectra. Insulators or large-band-gap semiconductors had not been studied much, partly because of the effects of surface charging during the measurements and partly because metals and semiconductors simply attracted more interest. Thus the role of correlation effects on the spectra, especially valence-band spectra, were not sufficiently studied. Since the surge of interest in the cuprates began, other oxides have become important to study, in particular, the manganates. These oxides resemble the cuprates in many ways. Although they are not superconducting, they exhibit very large magnetoresistance effects, the so-called "giant" and colossal" magnetoresistance.

Photoelectron spectroscopy can yield information useful in technology as well as fundamental electronic structure information. The loss of oxygen to the vacuum and its recovery from overlayers, the reactions with adsorbed water vapor and with overlayers of metals, potentially useful as electrical contacts, have been studied. Multilayer films containing alternate layers of high T_c-super-conductors have been produced and the interfaces studied by photoelectron spectroscopy.

In the following chapters, we describe photoelectron spectroscopy and its instrumentation, and results from studies on high-T_c superconductors, preceded by some examples of photoelectron spectroscopy on some simpler systems. Photoelectron spectroscopy is described in considerable detail, for in trying to gain a better understanding of the electronic structure of high-T_c superconductors by photoelectron spectroscopy it is important to know what approximations are being made in its interpretation and what aspects of the interpretation of spectra are controversial or even inadequately studied. An example of the latter is the background of "inelastically scattered" electrons. We begin the description of photoemission with the three-step model which is known not to be suitable in most cases. We do this because it is a way to introduce many concepts that are themselves of considerable validity. An example is the treatment of inelastic scattering. We follow this with a description of the correct one-step model. Although it is correct, it is difficult to apply in many cases without approximations. For example, it should be used to calculate the inelastic scattering effects in the photoelectron spectrum, but this is often done in an approximate or empirical way, taking ideas from the three-step model. The chapter on photoemission is not intended to give compete coverage of the subject, for we include only techniques and phenomena that have been used on high-temperature superconductors, omitting any discussion of gas-phase photoemission, adsorbates, overlayers, and several other topics. We discuss instrumentation primarily as it impacts on issues of resolution and other limitations on the spectra. It is important to know why improvements in instrumentation are not easy to achieve. Since it makes little sense to study the results of photoelectron spectroscopy in isolation, we describe briefly several other related electron and photon spectroscopies that give similar or complementary information, although the discussion will be in less detail than that for photoelectron spectroscopy.

Before discussing the photoelectron spectroscopy of high-temperature superconductors we describe our current understanding of the photoelectron spectroscopy and the electronic structure of some simpler materials: Na, Cu, Ni, NiO, Cu_2O, and CuO. NiO and CuO are not really simple, but they bear some resemblance to cuprates. Our understanding of the differences between the measured photoelectron spectra and those expected on the basis of the independent-electron model are quite well understood for these materials, while they are not yet as well understood for the cuprates. We precede all these with a survey of the properties of the high-T_c superconductors that are relevant for photoemission studies. These include the crystal structure and the calculated electronic structures, but almost none of the other physical properties, for which many reviews exist. The central part of the book, Chapters 6–9, is an exposition of the studies of the electronic structure of the two classes of high-T_c materials mentioned above as the only ones for which adequate single crystals have been available. There has been much more work here than one might at first think, for there have been numerous studies of the effects of doping on the electronic structure, as well as the changes seen in cooling below T_c. Moreover, within one class there are several compounds with different numbers of CuO_2 planes per unit cell. Our emphasis is on valence-band studies carried out by angle-resolved photoelectron spectroscopy, but we also discuss less exhaustively the results from core-level spectroscopy. Connections are made with the results of other spectroscopies.

There has been an enormous amount of theoretical work on cuprates, ranging from discussions of speculative ideas to extensive numerical calculations. There is not yet widespread agreement on the type of normal state that exists in the cuprates nor on the pairing mechanism. Any theoretical model must account for the photoelectron spectra as well as experimental results by many other techniques. Disagreement with photoelectron spectra may not mean the model is wrong, for it may well be that we still have things to learn about how to interpret photoelectron spectra in systems in which correlation is important. We mention a number of theoretical papers, but have made no attempt to compare systematically photoelectron spectra from cuprates with several of the microscopic models for which such spectra have been calculated. Nearly all published calculated spectra bear some resemblance to experimental spectra!

Except in this introduction, we have tried to give many references to the literature. However, despite searches of data bases and old-fashioned library searches, we are sure to have missed a number of papers on photoelectron spectroscopy of high-T_c materials. We certainly have missed a number of papers on x-ray induced photoelectron spectroscopy (XPS) on cuprates. We deliberately omitted a number of papers published before 1990, admittedly an arbitrarary date. Many of these papers reported results on polycrystalline samples of unknown or demonstrably poor quality. Although some of the results in these papers may be useful, it is not easy to know which results are comparable with those found on better samples.

A very large number of papers has appeared in conference proceedings. These papers can be found readily in literature searches when the proceedings are published as special issues of journals, but not when they appear as books. Some published conference papers appear again in a journal, not always in an expanded form, and we may not reference both. We have not tried to carry out a thorough referencing of all work on areas not central to this book, e.g., crystal structures or all of the photoelectron spectroscopy ever carried out on one of our examples, Cu. Instead we content ourselves with a few references to provide an entry to the literature. In the two chapters dealing with photoemission theory and experimental techniques, we have given many references to original papers and reviews, but we have not attempted to be encyclopedic, nor have we traced references back to find the first publication on a particular topic. General references on superconductivity are Tinkham (1996), Kuper (1968), de Gennes (1966), Rose-Innes and Rhoderick (1969), Poole *et al.* (1995), Waldram (1996), and Schrieffer (1964), while general references on the cuprates are books by Burns (1992), Cyrot and Pavuna (1992), Lynn (1990), Plakida (1995), a review by Brenig (1995), and a series of books of reviews edited by Ginsberg (1989, 1990, 1992, 1994, 1996). There are also many conference proceedings in book form or as special issues of journals. In addition, there are the journals devoted to superconductivity, e.g., *Physica C* and the *Journal of Superconductivity*. The majority of the papers in them deal with cuprates.

Chapter 2

Structure and electronic structure of cuprates

Photoemission studies generally give information on the electronic structure of a material, not the geometrical structure, although some surface structural information can be obtained by photoelectron spectroscopy. Knowledge of the crystal structures of the high-T_c materials is important in photoemission studies in several ways. First, crystal structures are required to determine the crystal potential for the calculation of the electronic structures. Even if a simple model of the crystal is being used, as in cluster calculations, the model should resemble part of the full crystal structure. Second, angle-resolved photoemission provides information in reciprocal space: energies as a function of angle or wave vector. The Bravais lattice of the crystal determines the Brillouin zone, hence the "space" in which theory and experiment often are compared. Third, the amount and nature of the anisotropy of the crystal structure, although difficult to quantify, are helpful in determining the anisotropy one might expect in physical properties, including photoelectron spectra. Orienting single crystals for angle-resolved photoemission studies also requires some structural knowledge. Fourth, there is an important experimental consideration. Most photoelectron spectroscopy is carried out on clean surfaces produced by cleaving in ultrahigh vacuum. It often is crucial to know what the cleavage surface is. Core-level photoelectron spectroscopy can help identify the atomic character of a cleaved surface, but knowledge of the crystal structure is needed and some idea of interplanar bonding is helpful. Fortunately, a structure determination is usually one of the first studies made on any new material. In the case of the cuprates, however, this has not been straightforward. Not only are the structures sensitive to composition, but as time has passed, increasingly subtle modifications of previously determined structures have been found.

All the cuprates to be discussed have sheets of Cu and O atoms in the ratio of 1:2 in a square, or nearly square, array (Fig. 2.1). The sheets are either planar or slightly puckered. Such sheets may be separated from each other by several sheets containing atoms of oxygen and another metal, or there may be n CuO_2 sheets adjacent to each other, these n sheets being separated from the nearest pair of n similar sheets by several of the layers of metal and oxygen atoms. Some of the high-T_c cuprates have one or more chains of Cu and O atoms in the ratio 1:1 running parallel to one axis of the squares of the CuO_2 planes. This is shown schematically in Fig. 2.2.

In Section 2.1 we consider in some detail the crystal structures of the two "families" of high-T_c materials on which photoelectron studies of the valence band have been successful. There have been extensive studies of the crystal struc-

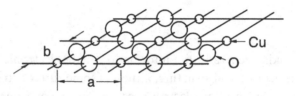

Fig. 2.1   Schematic CuO_2 plane. (Burns, 1992)

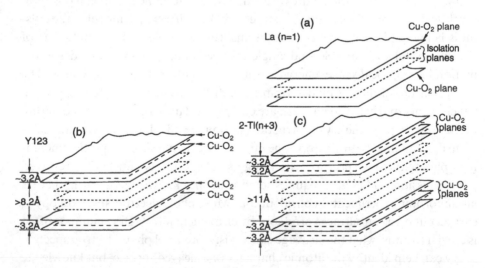

Fig. 2.2   Schematic stacking of CuO_2 planes (solid) with planes of other atoms, isolation planes (short dashes) separating them. (a) Single CuO_2 planes as in La_2CuO_4. (b) "Double" CuO_2 planes as in $YBa_2Cu_3O_7$. (There is an Y plane in between the pair, shown by longer dashes.) (c) "Triple" CuO_2 planes as in $Tl_2Ba_2Ca_2Cu_3O_{10}$. (There is a Ca plane in between each CuO_2 plane, shown by longer dashes.) (Burns, 1992)

tures, and these continue. Not only do the lattice parameters and crystal structures vary as functions of temperature, oxygen content, and cation substitutions,
but different heat treatments could lead to the ordering of the dopants or oxygen
vacancies. More complete descriptions of crystal structures can be found in
Beyers and Shaw (1989), Santoro (1990), Hazen (1990), Chen (1990), Burns
(1992), Shaked *et al.* (1994), and Egami and Billinge (1994).

2.1   Crystal structures

Shaked *et al.* (1994) list 26 different crystal structures for cuprate superconductors. Not all these materials have been studied successfully by photoelectron
spectroscopy, and very few have been studied by angle-resolved photoelectron
spectroscopy. Many of the latter are listed in Table 2.1, although only
$Bi_2Sr_2CaCu_2O_8$ and $YBa_2Cu_3O_7$ have been studied extensively. Not all of the
lattices, space groups, and structures are known precisely because of the occurrence of superlattices. For this reason, sometimes different lattices and space
groups are reported.

Figure 2.3 shows the orthorhombic unit cell of Y123. The unit cell is almost
tetragonal, with the lengths of the lattice vectors **a** and **b** differing by about 2%,
and it is about three times as long in the c- direction as in the a- and b- directions.

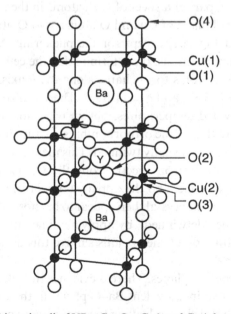

Fig. 2.3   Orthorhombic unit cell of $YBa_2Cu_3O_7$. Cu1 and Cu4 denote inequivalent Cu
sites. O1–O4 denote inequivalent O sites. (Burns, 1992)

Table 2.1   *Space groups and Bravais lattices of some cuprates studied by photoemission*

Compound	"Shorthand" notation	Maximum T_c (K)	Bravais lattice	Space group
$(La_{2-x}Sr_x)CuO_4$	214	38	I-tetragonal	4/mmm
$(La_{2-x}Sr_x)CaCu_2O_6$		60	I-tetragonal	4/mmm
$Bi_2Sr_2CuO_6$	Bi2201	20	I-tetragonal	4/mmm
$Bi_2Sr_2CaCu_2O_8$	Bi2212, BSCCO	85	orthorhombic	mmm
$Bi_2Sr_2Ca_2Cu_3O_{10}$	Bi2223	110	I-tetragonal	4/mmm
YBa_2Cu3O_7	Y123, YBCO	92	orthorhombic	mmm
$YBa_2Cu_4O_8$	Y124	80	A-orthorhombic	mmm
$(Nd,Ce,Sr)_2CuO_4$	NCCO	30	I-tetragonal	4/mmm

(In this and the following, **c** is the longest lattice vector. Some references use the crystallographers' convention that **b** is the longest vector.) The unit cell can be viewed approximately as the stacking along the c-axis of three perovskite-structured cubes, with Cu atoms on the cube corners, O on the face centers, Y at one cube center, and Ba at the other two. Starting at the center of the cell illustrated, there is a single Y atom, part of a sheet of such atoms in the a–b plane. On each side of this is a puckered plane of Cu and O atoms, two O atoms per Cu atom, arranged like those in Fig. 2.1, except for the puckering. Moving outwards, there is a plane of Ba and O atoms, then, terminating the cell, a plane of Cu and O in the ratio 1:1. These atoms form chains along the b-axis. The lattice parameters are $a = 3.82$, $b = 3.88$, and $c = 11.65$ Å. There is also a tetragonal form of Y123, stable at elevated temperatures, but we need not consider it here. Its structure is very close to that of the orthorhombic structure just described.

Single crystals of Y123 are usually twinned. This means that in a macroscopic region sampled by photoelectron spectroscopy the Cu–O chains run in two orthogonal directions, not one, and the apparent anisotropy is reduced. Many macroscopic properties of the crystal then appear to be those of a tetragonal crystal. Such crystals can be "detwinned" by applying uniaxial stress while annealing, but only a small fraction of the crystals survive this detwinning procedure without breaking.

Y123 cleaves between a–b planes. There is evidence that cleaving leaves Ba on the surface. There are two inequivalent Ba–O planes in the unit cell. From Fig. 2.3 we see that if the crystal cleaves leaving the upper Ba–O plane on the surface, the plane directly beneath it is a CuO_2 plane, while if it is the lower Ba–O plane, the plane directly below is the Cu–O "chain" plane. These are not equivalent cleavage planes. Moreover, with just single Ba–O planes, a cleave that leaves a Ba–O plane on the surface has a 50% chance of leaving it on the surface removed upon cleaving rather than on the one remaining in position for subsequent measurements. This leaves a CuO_2 plane on the surface to be measured. Similarly the other possible cleave also has a 50% probability of leaving a Cu–O chain plane on the surface to be measured. This means that approximately half the surface area will be a Ba–O plane, one or the other mentioned above possibly being preferred, and half will be a Cu–O or CuO_2 plane, again, one or the other possibly dominating, depending on bonding considerations and/or defect structures. One Ba–O plane or the other may dominate, each sharing half the surface with the corresponding Cu–O or CuO_2 plane. It is also possible that both Ba–O and Cu–O planes are present on the surface, i.e., there are four types of surface on one cleaved face. We do not yet know exactly which bonds are broken during cleavage, so we cannot yet identify the cleavage planes, other than to say the surface is partly one of the Ba–O planes. Further complicating the issue, there are also reports of "anomalous" cleaves that leave an Y surface for at least part of the surface (Schroeder et al., 1993a,b). Calandra and Manghi (1992) have calculated the electronic structure of Y123 terminated by each of the six possibilities mentioned above, and there are differences that, in principle, could be measured. This has been discussed further by Calandra and Manghi (1994). Bansil et al. (1992) calculated angle-resolved photoemission spectra for all six possible (100) terminating surfaces and compared the results with measurements. They concluded that the cleaved surface is a Ba–O plane with a CuO_2 plane just below. If so, the cleavage is between a Ba–O plane and a Cu–O chain layer. This being so, some cleaves or parts of all cleaves, should exhibit the Cu–O chain layer as the surface. In support of this picture a scanning tunneling microscopy (STM) study of Y123 by Edwards et al. (1992) showed that cleavage, carried out at low temperatures in ultrahigh vacuum, occurred between a Ba–O plane and the layer of Cu–O chains. We conclude that it is possible that not all cleaves leave the same type of surface.

Figure 2.4 shows the orthorhombic unit cell of Bi2212. The dimensions are approximately $a = 5.42$, $b = 5.38$, and $c = 30.72$ Å. This as an extremely tall unit cell. Cleaved surfaces are a–b planes. Photoemission with 10–100 eV photons commonly employed for valence-band studies is very surface sensitive. The mean free path for escape of photoelectrons without inelastic scattering is of the order of 10 Å, which means that for Bi2212 and some other high-T_c materials, the data of primary interest come almost entirely from the top-most unit cell. The structure in Fig. 2.4 is almost tetragonal. (When it is to be considered tetragonal, a different unit cell is used, with a- and b-axes rotated 45° with respect to those in Fig. 2.4.) However, the structure in the figure is not the real structure.

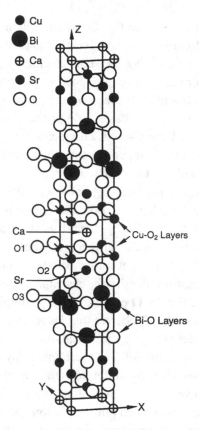

Fig. 2.4 Orthorhombic cell of $Bi_2Sr_2CaCu_2O_8$. O1–O3 denote inequivalent O sites. (Pickett, 1989)

A superlattice with an incommensurate wave vector directed along the b-direction occurs. The b-axis periodicity is about 27 Å. The actual structure is obtained from that of Fig. 2.4 by the displacement of atoms, primarily in the Bi–O planes, but there is also some displacement of all other atoms (Yamamoto *et al.*, 1990). A closer approximation has a unit cell approximately 5 times longer in the b-direction. This superlattice usually has been ignored in calculations of the electronic structure, but its effects have appeared in photoelectron spectra, as will be shown in Chapter 6.

Starting at the middle of the unit cell for $Bi_2Sr_2CaCu_2O_8$ (Fig. 2.4) and moving toward the top or bottom of the cell, we see a Ca layer, a CuO_2 plane, a plane of Sr–O, then two Bi–O planes. Then there is a Sr–O plane, a Cu–O plane and the terminating Ca plane, only one atom of which belongs to the unit cell shown. The calculated charge density between the two adjacent Bi–O planes is very low (Krakauer and Pickett, 1988), leading to easy cleavage between them, and this is believed to be the cleavage plane. Cleaving between either of the two pairs of Bi–O planes in Fig. 2.4 leads to the same sequence of layers below the Bi–O plane, i.e., to equivalent surfaces. In principle, valence-band photoelectron spectroscopy can determine the surface plane, especially if there is just one, if aided by calculations of the spectra expected for different terminations. Quantitative low-energy electron diffraction (LEED) can do the same. Only a limited amount of such work has been done. XPS of core levels can also be used to determine which atoms of the many in a formula unit lie nearest the surface. Scanning tunneling microscopy (STM) and its variants can also be used to study a surface with atomic resolution, but little work has been done on surfaces cleaved at low temperature, the type of surface used for much of the photoelectron spectroscopy carried out to date on the cuprates. However, surfaces cleaved at room temperature show the superlattice present in the bulk, as revealed by LEED (Lindberg *et al.*, 1988) and STM (Shih *et al.*, 1989). The latter work also showed that the surface Bi–O plane was non-metallic.

$YBa_2Cu_4O_8$ has the structure shown in Fig. 2.5. The orthorhombic unit cell is nearly tetragonal, with dimensions of $a \cong b \cong 3.9$ and $c = 27.2$ Å. This structure

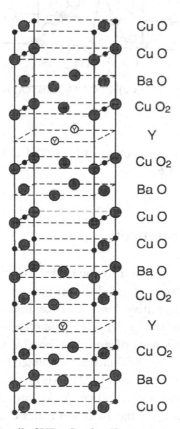

Cu O
Cu O
Ba O
Cu O_2
Y
Cu O_2
Ba O
Cu O
Cu O
Ba O
Cu O_2
Y
Cu O_2
Ba O
Cu O

Fig. 2.5   Orthorhombic unit cell of $YBa_2Cu_4O_8$. (Santoro, 1990)

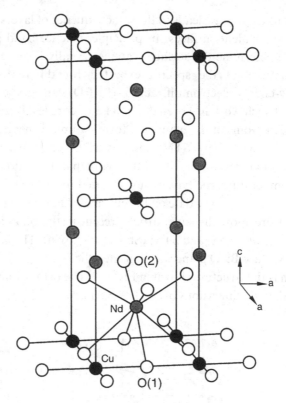

Fig. 2.6   Tetragonal unit cell of $(Nd,Ce)_2CuO_4$. O(1) and O(2) denote inequivalent O sites. (Hazen, 1990)

is close to that of Y123. The structure of $(Nd,Sr)_2CuO_4$, the tetragonal 214T* structure, is shown in Fig. 2.6.

2.2   Electronic structures

The Brillouin zones are determined by the lattices reciprocal to the Bravais lattices. The only Bravais lattices we need to consider are primitive and body-centered tetragonal, and primitive and base-centered orthorhombic. Their Brillouin zones are depicted in Fig. 2.7. The symmetry lines and symmetry points are labeled. If the a- and b-axes of an orthorhombic lattice are of nearly the same length, sometimes a tetragonal lattice is assumed for purposes of electronic structure calculations. An example is Bi2212 whose structure is often described with a base-centered orthorhombic lattice. A close approximation uses a body-centered tetragonal lattice, and the Brillouin zones change accordingly. Since the a- and b-axes of the body-centered tetragonal lattice are at 45° with respect

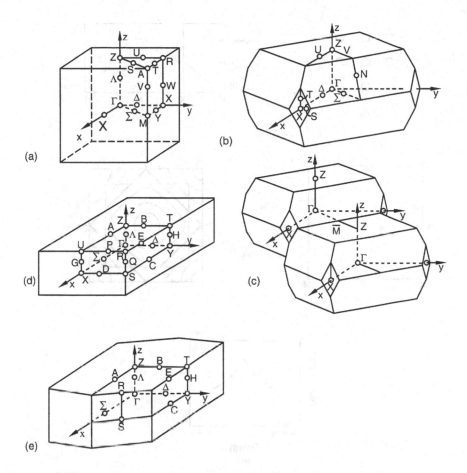

Fig. 2.7 Brillouin zones with symmetry lines and points labeled for (a) primitive tetragonal space lattice, (b) body-centered tetragonal lattice with $c/a > 1$ ((c) shows how neighboring zones contact each other), (d) primitive orthorhombic lattice, and (e) one-face-centered orthorhombic lattice. (Burns, 1992 and Koster, 1955)

to those of the rhombohedral lattice, the k_x- and k_y-axes of the two Brillouin zones also are at 45° to each other. This means that in one lattice, a wave vector along the [100] direction in reciprocal space is parallel to a nearest-neighbor Cu–O bond, while in the other, it is at 45° to the bond, i.e., it bisects the angle between the Cu–O bonds in the CuO_2 sheet.

The electron energy bands from the CuO_2 planes are common to all the cuprate superconductors. Their contribution to the electronic structure should persist qualitatively in complete self-consistent three-dimensional band calculations. Figure 2.8(a) shows a square-planar array of atoms with an s-like atomic wave function on each atom. With small amounts of overlap between nearest neighbors only, the tight-binding band structure of Fig. 2.8(b) results. The Brillouin zone for this lattice is shown in Fig. 2.8(c). Fig. 2.8(d) shows

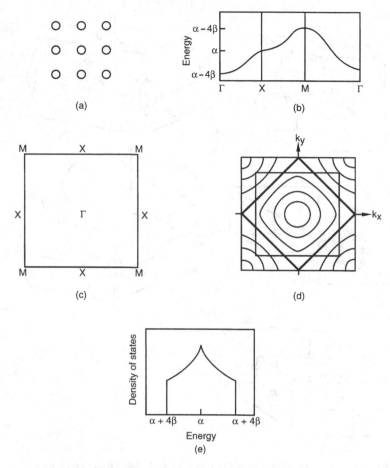

Fig. 2.8 (a) Primitive square lattice. The circles represent s-like wave functions.
(b) Tight-binding bands for s-states on the lattice of part (a). α and β are the usual tight-binding parameters, an average energy and a hopping (or transfer) energy, respectively.
(c) Brillouin zone for the primitive square lattice. (d) Curves of constant energy from the band of part (b). The heavy line is for the energy $E = \alpha$, the Fermi energy for a half-filled band. (e) The density of states for the band of part (b). Adapted from Burns (1992).

surfaces of constant energy, and Fig. 2.8(e) shows the density of states. For exactly half filling, the Fermi surface is the square shown with a heavier line, and the Fermi level falls right at the van Hove critical point in the center of the density of states. Although this model is too simple to represent Cu 3d and O 2p states, it has many features in common with the next approximation.

Figure 2.9(a) has Cu atoms on the lattice points of a square lattice and O atoms in the center of each side, the actual arrangement in a CuO_2 plane. From cluster calculations (see later) we know that most of the states composed of Cu 3d–O 2p hybrids are filled, and the uppermost states, where the Fermi level of cuprates is expected to lie, will be composed of the antibonding hybrids of the

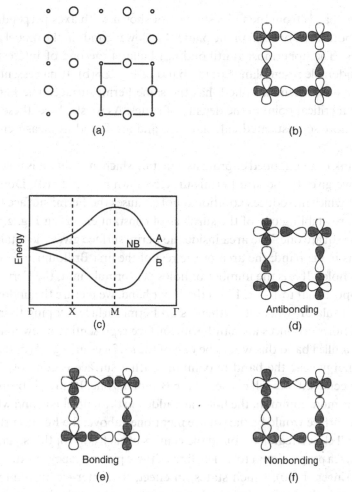

Fig. 2.9 (a) CuO_2 basis on a primitive square lattice. The small circles represent Cu, the large ones, O. (b) Schematic picture of atomic Cu $3d_{x^2-y^2}$ wave functions and those O $2p_x$ and $2p_y$ functions that hybridize with them. The shading represents a changed sign. (c) Tight-binding bands obtained from the atomic functions of part (b). A = antibonding, NB = nonbonding, B = bonding. (d) Phasing of the atomic functions for the state at $\overline{M}$ in the antibonding band. (e) Phasing of the atomic functions for the state at $\overline{M}$ in the bonding band. (f) Phasing of the atomic functions for the state at $\overline{M}$ in the nonbonding band. Part (b) also shows the phases of the atomic functions for one of the degenerate states at Γ. Adapted from Burns (1992).

Cu $3d_{x^2-y^2}$ states and the O $2p_x$ and $2p_y$ states. The wave functions are shown schematically in Fig. 2.9(b). These combine to give the tight-binding bands of Fig. 2.9(c). The bonding and antibonding bands result from the hybridization of the states shown in Fig. 2.9(b) In fact, the relative phasing of the wave functions in Fig. 2.9(b) is that for the three-fold degenerate state at Γ. The relative phases of the atomic states at the $\overline{M}$ point are shown in Figs. 2.9(d–f). There are more nonbonding bands from the O 2p states, not shown, with axes perpendicular to the Cu–O bond axis, but still in the plane. The lower bands in the model are filled in cuprates. The uppermost (antibonding) band is the one of interest, and it bears considerable resemblance to the band of Fig. 2.8(b). It has essentially the same shape and, when half filled, has the same Fermi surface. The Fermi level then lies at a critical point in the density of states. As we shall see, these features persist in more sophisticated calculations, and are found in measurements on the cuprates.

If we think of an undoped cuprate as a metal, which in reality it is not, then the model above gives the square Fermi surface shown in Fig. 2.8(d). Doping with an element which introduces additional holes causes the Fermi surface to shrink somewhat, resembling one of the surfaces of constant energy in Fig. 2.8(d) that is inside the square one. The area inside the Fermi surface gives the total number of electrons in the band; the area of the rest of the first Brillouin zone gives the number of holes. If x is the number of holes per formula unit, the Fermi surface area is proportional to $1 - x$. If, on the other hand, we picture the undoped cuprate as an insulator, which it is, there is no Fermi surface. Doping to introduce additional holes produces a small Fermi surface representing a few holes. If we start from a filled band this would be one of the surfaces in Fig 2.8(d) that represents an energy near the band maximum. Such a surface is composed of four small arcs, centered on each corner of the Brillouin zone. Its area is proportional to x. If, as is more common, the holes are added to the half-filled band whose virtual Fermi surface would be the square mentioned above, the Fermi surface consists of small closed regions around the centers of each edge of this square, again with total area proportional to x. The first of these pictures obeys Luttinger's theorem (Luttinger, 1960), which states, in effect, that inter-electron interactions do not change the volume of the Fermi surface, although they may alter its shape. The second picture is not in accord with Luttinger's theorem. The first picture gives a "large" Fermi surface, the second, a "small" one. Transport property measurements as a function of carrier concentration were best described by the small Fermi surface. As we shall see, photoemission studies of cuprates with hole concentrations varying around the optimal doping which maximizes T_c, i.e., metallic samples, show large Fermi surfaces which grow and shrink qualitatively as expected with doping. This surprised a number of people when it was first discovered. However, when less-doped samples are studied, pieces of the large Fermi surface are missing, and there is a trend toward the small one. Starting with the insulating phase and doping toward the metallic phase has not yet been achieved in photoemission studies.

The square lattice with the CuO_2 basis is found in all cuprates, sometimes planar and sometimes puckered, but the rest of the three-dimensional lattice determines the Brillouin zone. Sometimes the actual unit cell is approximated by another one of higher symmetry, in which case the axes may be rotated 45°, as mentioned above. Since all angle-resolved photoemission studies to date have worked with samples whose surfaces were basal planes, the dispersion of the bands is measured as a function of k_x and k_y. The resultant Fermi surfaces are displayed in a two-dimensional Brillouin zone related to that of Fig. 2.8(c), but sometimes with different symmetry notation because the zone used is a section through the three-dimensional zone, the perpendicular bisecting plane of the k_z-axis of that zone. These zones are shown in Fig. 2.10. The edges of each of the four Brillouin zones shown are parallel to the Cu–O bond directions in the CuO$_2$ planes. The zone for the square lattice is shown in Fig. 2.10(a). The line Γ–X from the center of the zone to the center of the right-hand edge is parallel to k_x. That from the center to the center of the top edge is parallel to k_y, but its terminus is labeled X, for all points labeled X are equivalent by symmetry. Figure 2.10(b) is for a body-centered tetragonal lattice, e.g., that of Nd$_2$CuO$_4$. The line from Γ to G$_1$, if extended, continues across the top, not the central plane, of the next unit cell, passing through the point Z in the center of the top of that zone, before going on to the Γ point of the zone beyond. The point G$_1$ sometimes is labeled $\overline{\text{M}}$. Figure 2.10(c) is for the body-centered tetragonal lattice usually used for the orthorhombic approximation to Bi2212. Figure 2.10(d) is for the orthorhombic lattice, such as that of Y123. In this figure, the square is actually a

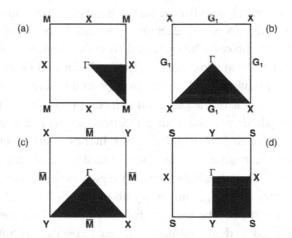

Fig. 2.10 Brillouin zone sections for several cuprate structures. (a) Square-planar lattice of Fig. 2.8(a) and 2.9(a). (b) Body-centered tetragonal space lattice such as that for Nd$_2$CuO$_4$. (The points G$_1$ may be labeled $\overline{\text{M}}$ or not labeled. They are not labeled in Fig. 2.5 (c). See text.) (c) Body-centered tetragonal space lattice as for Bi2212. (d) Primitive orthorhombic lattice as for Y123. The Cu–O bond axes in the CuO$_2$ planes are parallel to the edges of the squares. The shaded regions represent the irreducible "wedges" from which the rest of the zone, Fermi surface, and bands within it may be obtained by symmetry. In the case of the tetragonal lattices, (b) and (c), the corners of the squares should be truncated, as can be seen in Fig. 2.7. Adapted from Shen and Dessau (1995).

rectangle with sides of slightly different length. Note that in parts (a) and (d), the
Γ–X directions are parallel to the Cu–O bonds, while in (b) and (c), they are at
45° to them.

The one-electron band structures have been calculated for many high-T_c
materials. In addition to band structures, one-electron energies as a function of
wave vector, these calculations give wave functions from which the atomic subshell character can be found, electric dipole matrix elements calculated, and
wave function symmetry identified. Modern band structure calculations, no matter how they are carried out, often rely on several assumptions. The first is the
independent-electron approximation; each electron moves in a potential determined by the average charge density of all the other electrons. The wave function
is then a single product of one-electron wave functions. This is the Hartree
approximation. Incorporating the Pauli principle by using a single determinant
of the product wave functions leads to the Hartree–Fock approximation. This
causes the appearance of a non-local exchange potential in the Schrödinger equation. A few calculations are still carried out with the Hartree–Fock equations,
iterating to self-consistency.

Local-density functional theory (Hohenberg and Kohn, 1964; Kohn and
Sham, 1965), which does not require the independent-electron approximation,
is used in most electronic structure calculations to obtain the ground-state energy
and charge density of the system. In practice, wave functions usually are introduced. There are several choices for the forms of the potential and the basis functions. With the same form of exchange potential the results should be
independent of these choices when iterated to self-consistency, and this is the
case. In the local-density approximation (LDA) (Jones and Gunnarsson, 1989;
Pickett, 1989) the exchange and correlation potentials are handled approximately by taking the value of the potential at a point $\mathbf{r}$ as equal to the exchange-
correlation potential for an interacting electron gas with a uniform density
equal to the crystal charge density at the point $\mathbf{r}$. In the density functional formalism the individual eigenvalues can not be assumed to have meaning, but they
are often taken with considerable success to represent states occupied by electrons up to the Fermi level, E_F, and empty above E_F. Similarly, the wave functions can not be assumed to be meaningful. The density functional formalism
has not been extended to deal routinely with excited states, although there has
been recent activity to develop a better theory for excited states. Thinking of an
excited state as an unoccupied state below E_F and an occupied one above E_F,
the corresponding two eigenvalues from the LDA calculation may give a useful
picture for the excitation. Using the difference in eigenvalues as the excitation
energy sometimes is fairly accurate, but it is not correct in general, and it can be
misleading. We give some examples involving simple (elemental) metals in later
chapters. The region of applicability of LDA bands to high-T_c materials is a

subject often discussed. Angle-resolved photoemission has been used to map bands for comparison with LDA band structures. There is some agreement. Some people are surprised by the fact that there is some agreement, and some by the fact that the agreement is not better.

LDA band structures are known to have some deficiencies. When applied to semiconductors and insulators not containing transition metals or rare earths, the calculated band gap is always too small. Self-interaction corrections (Svane and Gunnarsson, 1990) have been used to improve agreement. More important for dealing with cuprates is the fact that the potentials used in the LDA or earlier electronic structure calculations treat electron–electron correlation in an average way. This allows fluctuations to occur in which the local electron density can build up with insufficient cost in repulsive energy. When the electrons are rather uniformly spread throughout the crystal, as with conduction electrons in Al, this deficiency in the LDA method appears not to be important. (However, see the example for Na in Chapter 5.) The extreme limit when the energetics are dominated by the correlation energy occurs when the electrons are well-localized on individual atoms or ions. Then there is an energy cost of U to transfer an electron from one atom or ion to another similar atom or ion. In such cases with an odd number of electrons per atom the bands split into two bands. The lower one is full and the upper one is empty, with a separation of U between the band centers, with each band width much smaller than U. The system is then an insulator, usually treated with the Hubbard Hamiltonian. In metallic systems U is not the atomic Coulomb repulsion, for it is screened by the conduction electrons. Instead of being 10–15eV, it is reduced to 5 eV or so, still a large value. This correlation effect is important in insulators and in magnetic systems. A well-known example of this is NiO. Conventional band theory predicts this material is a metal because the incomplete Ni 3d subshell leads to a Fermi level within a band based on the Ni 3d states. NiO is, however, an antiferromagnetic insulator. This discrepancy is viewed often as an example of the inapplicability of the density functional formalism to systems where correlation is more dominant than the average crystal potential, expressed in terms of the ratio of U to a band width. Since the parent materials of some of the high-T_c superconductors are insulators, even antiferromagnetic insulators, it is hard to believe that the LDA has any relevance for them. Even though the LDA may not be adequate to handle the insulating parent compounds, it may be useful for the doped metallic superconductors. The metallic phase may be qualitatively different from the insulating phase, and a single starting point for describing both may not be possible.

Two improvements to the LDA have been made. One includes the self-interaction correction (Svane and Gunnarsson, 1990). The other incorporates the Coulomb repulsion U into LDA. The so-called LDA+U method was introduced (Anisimov et al., 1991) to take better account of correlation than the usual

LDA calculation. For the orbitals expected to be somewhat localized, hence to have larger correlation effects, U and a corresponding exchange integral J, both screened, are calculated (Gunnarsson et al., 1989), not introduced as ad-hoc parameters. Such orbitals are the 3d orbitals of Cu or Ni. The calculation on NiO by either improved LDA method gives a charge-transfer insulator, not a metal. NiO is believed to be such an insulator (Sawatzky and Allen, 1985; Hüfner, 1994). The observed energy gap is not U for the Ni 3d orbitals, but the smaller energy Δ between the top of the filled O 2p band and the empty upper Ni 3d Hubbard band, based on Ni 3d orbitals. A charge-transfer insulator has a filled band composed of states based on one type of atom or ion and an empty band Δ higher based on states on the other atom or ion (Zaanen et al., 1985). The excitation of an electron across the band gap then involves the transfer of an electron from one atom to its neighbor. NiO has an upper and lower Hubbard band composed of states based on Ni atoms, with another filled band, based on O 2p states, located in the gap between the Hubbard bands. The observed band gap, Δ, is smaller than U, neglecting band widths. The same LDA + U method was applied to $CaCuO_2$, a model parent compound for cuprate high-T_c materials, with success, in that the correct insulating behavior was found, but the behavior upon hole doping, a transition to a metallic state, was not given correctly. LDA + U has also been used to describe the effects of rare earth (R) doping in $R_{1-x}Pr_xBa_2Cu_3O_7$ (Liechtenstein and Mazin, 1995).

There are other improvements that have been made to the LDA to better incorporate electron–electron interactions. The many-electron picture of the electron gas deals with quasiparticles, electrons or holes dressed by a cloud of holes or electrons, respectively. (They may also be dressed by a phonon cloud. This is important for conventional superconductivity.) These form a Fermi liquid. There is a one-to-one correspondence between a quasiparticle and a single-electron state. Quasiparticles have finite lifetimes due to the electron–electron interactions, now screened. The effects of the interactions are incorporated into a complex self-energy $\widetilde{\Sigma}$, the real part of which represents an energy shift from the one-electron eigenvalue and the imaginary part of which is related to the lifetime of the excitation by the uncertainty principle. This concept has been extended to electrons in a periodic potential, and complex self-energies have been calculated for correction of the LDA eigenvalues. Such corrections generally improve agreement with experiment. Because of the coupling to other electrons, the photoexcitation of a quasielectron occurs as a peak of intensity reduced from that of the excitation of the bare electron by a renormalization factor of $Z \leq 1$. This is called the coherent part of the spectrum. The remainder of the oscillator strength, the fraction $Z - 1$, is spread over a range of energies, and is called the incoherent part of the spectrum. The step at the Fermi edge at $T = 0$ is no longer of height unity but of height Z. $Z(\mathbf{k}, E)$ may be interpreted as the fraction of single-particle character in the quasiparticle (excitation) with wave vector $\mathbf{k}$ and energy E (Mahan, 1990), or as the contribution of a single quasiparticle to the spectral density function. The remaining strength, $Z - 1$, arises from the excitation of many quasiparticles. The value of Z for a wave vector on the Fermi surface is equal to the size of the discontinuous drop in the momentum distribution for the wave vector. The momentum distribution,

$$n(\mathbf{k}) = \int_{-\infty}^{\infty} f_{\text{FD}}(E) A(\mathbf{k}, E) \mathrm{d}E,$$

can be measured by angle-resolved photoelectron spectroscopy (Chapter 7). $A(\mathbf{k}, E)$ is the spectral density function, which can be obtained from the self-energy (Chapter 3), and f_{FD} is the Fermi–Dirac distribution function.

Another approach to electronic structure specifically directed at interpreting photoemission and inverse photoemission spectra with the concomitant relaxation of the electrons around the photohole or added electron is to calculate the ground state of the initial N-electron system and the ground state of the final $(N - 1)$- or $(N + 1)$-electron system. Since the hole or added electron in the final state breaks the translational symmetry, the calculation must be done on a cluster or by the use of supercells (Zunger and Freeman, 1977). Supercells are large unit cells, and a photohole or extra electron can be added to each. The cells are repeated periodically so Bloch wave functions can be used. For metallic systems the screening of the extra charge is quite localized if the extra charge is well localized, so the cells do not have to be made very large before interactions between the extra charge in neighboring cells is negligible. This method works well for core photoemission and for photoemission from the 4f states of rare earths (Freeman et al., 1987).

The best way to describe the electronic structure of the metallic cuprates has not yet been demonstrated conclusively. After beginning with either a localized or delocalized picture, each of which is described below, modifications due to the electron–electron correlations usually must be considered. This can lead to (i) the conventional Fermi liquid model outlined above, (ii) unconventional Fermi liquids, in which the energy dependence of the real and imaginary parts of the self energy is different from those of a Fermi liquid, or (iii) yet other types of behavior. These have been discussed by Pickett et al. (1992). Despite a great deal of work in the years since 1992, their discussion still needs little modification.

Returning to the LDA description, roughly speaking, and ignoring degeneracies, each atomic electron state that is considered to be a valence electron state contributes one band that is likely to run below the Fermi level in some part of the Brillouin zone and contribute to the ground-state electronic structure. For $YBa_2Cu_3O_7$ there are 3 Y 4d and 4s electrons, 4 Ba 6s electrons, 33 Cu 3d and 4s electrons, and 28 O 2p electrons. There is one formula unit per unit cell, so we expect something like 34 bands within about ± 10 eV of the Fermi energy. The most important bands, those near the Fermi energy, are derived from the states listed above for Cu and O. The band structure diagrams are accurately called "spaghetti diagrams" in such cases. In fact, if the bands were uniformly dense in the region below E_F, there would not be much sense in measuring angle-resolved photoemission spectra, even if selection rules reduced the number of possible transitions. Too many spectral peaks would overlap and one could not study dispersion. However, often near the Fermi level there are only a few bands, and these have been mapped successfully in several cases.

Electronic structure calculations are rarely carried out to energies far above the Fermi energy, even though these states are the final states in photoemission. There are many reasons for this. They are subject to fewer experimental checks than states at and below the Fermi energy. Angle-resolved inverse photoemission could be considered for testing the accuracy of high-lying bands, but the experimental resolution is not very high, and worse, the lifetime widths of these states are large. There are also calculational problems. The wave function expansions need more terms for good convergence of high-energy states, and there are problems with using the same potential for high-lying states as is used for states near the Fermi energy. Finally, and most important, there is the conceptual problem discussed above; the one-electron states are not system eigenstates, and their energies may not be close to quasiparticle energies, even with self-energy corrections. Thus the LDA bands calculated to date, with a few exceptions, are limited to all the filled valence bands and bands going only a few eV above the Fermi energy.

We present the LDA bands for several high-T_c superconductors in Figs. 2.11 through 2.18. These have been reviewed several times (Hass, 1989; Pickett, 1989; Wang, 1990; Yu and Freeman, 1994) We do not present more-recent LDA + U results because they have not been carried out on as many materials, and they appeared after many of the angle-resolved photoemission studies had been completed and compared with the LDA bands we present.

The LDA electronic structure of orthorhombic Y123 (Krakauer *et al.*, 1988a) is shown in Fig. 2.11 (a)–(c). See also Massidda *et al.* (1987) and Freeman *et al.* (1989). As is easily seen there are many bands in the first 5 eV below the Fermi energy. Those arising primarily from the Cu 3d and O 2p states are emphasized in (b) and (c), where the contributions from the Cu–O chains and CuO$_2$ planes are shown separately. The important bands for superconductivity are those crossing the Fermi level, and in the first 0.5 eV below the Fermi level, the number of bands is not intimidating. Note that just below E_F, nearly all bands have significant Cu–O character, the exception being near the midpoint between Γ and Y. Pickett (1989) discussed these bands in more detail, and compared them with bands calculated by others, especially with regard to differences near the Fermi

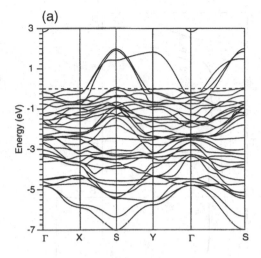

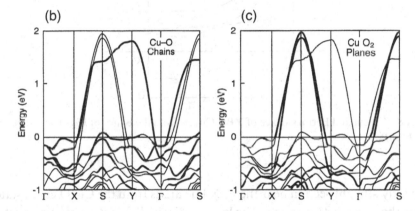

Fig. 2.11 Calculated LDA energy bands for $YBa_2Cu_3O_7$. The horizontal dashed line in (a) denotes the Fermi energy. Figure 2.11 (b) and (c) indicate by the thickness of the lines the weighting of Cu–O states to the bands, separating the contributions of Cu–O chains (b) from those of the CuO_2 planes (c). Note the different energy scales between (a) and (b), (c). (Krakauer *et al.*, 1988a)

level. He also discussed the charge density resulting from the calculation, the density of states, and the Fermi surface. The Fermi surface for Y123 is shown in Figure 2.12 (Pickett *et al.*, 1990). The Γ–Y–S–X plane is one quarter of the section of the Brillouin zone for $k_z = 0$. The two sections of the Fermi surface curving 90° in this plane are from from bands based primarily on states of the atoms on the CuO_2 planes. That there are two such nearby sections of the Fermi surface and two nearly parallel bands in Fig. 2.11 is a consequence of there being two CuO_2 planes in the unit cell. The small interaction between these planes lifts the degeneracy the bands that the two isolated planes would have.

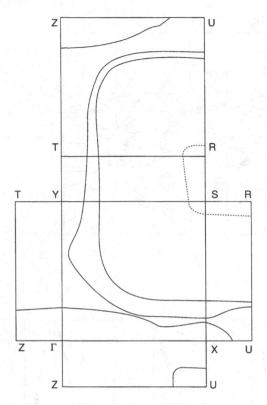

Fig. 2.12 Calculated Fermi surface of $YBa_2Cu_3O_7$ from the bands of Fig. 2.11. (Pickett *et al.*, 1990)

The nearly straight section near the Γ–X line arises from the Cu–O chain states, as does the small section near S. The latter varies in different electronic structure calculations for this material (Pickett, 1989). The T–Z–U–R plane is the top surface of the Brillouin zone normal to k_z. The bands of Y123 show significant dispersion along k_z, making the treatment of Y123 as a two-dimensional material unreliable for some purposes. This dispersion is shown better in Fig. 2.13, the projection of the Fermi surface on the k_x–k_y plane at $k_z = 0$. The shaded regions represent points of the Fermi surface for all values of k_z. Their widths are an indication of the dispersion along k_z.

Figure 2.14 shows the LDA electronic structure of Bi2212, assumed to be body-centered tetragonal (Krakauer and Pickett, 1988a). The bands are equally as dense as in Y123 more than 1 eV below the Fermi level (not shown). Most of the bands near the Fermi level arise primarily from the Cu 3d and O 2p states in the CuO_2 planes. The exception is the band that crosses E_F steeply between Γ and Z. This arises from the Bi 6p- and O 2p-derived bands 2–3 eV above E_F. There is some mixing with the Cu 3d and O 2p states from the CuO_2 plane

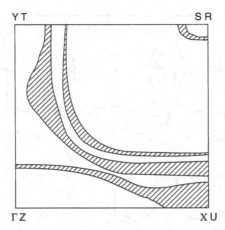

Fig. 2.13 Projection of the Fermi-surface sections of Fig. 2.12 onto the $k_z = 0$ plane. The shaded regions mark where there is a state at the Fermi surface for any k_z. The width of such regions is an indication of dispersion along k_z. (Pickett *et al.*, 1990)

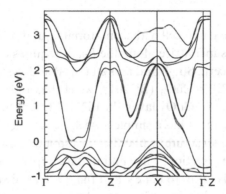

Fig. 2.14 LDA band structure of $Bi_2Sr_2CaCu_2O_8$. The dashed line denotes the Fermi level. The left-hand panel from Γ to Z is in the central k_x–k_y plane of one zone and across the top of surface of the next zone. (See Fig. 2.7.) The right-hand panel from Γ to Z is along the k_z direction, along which there is little dispersion. (Krakauer and Pickett, 1988a)

atoms near the Fermi level. The bands are not plotted for wave vectors along k_z because there is very little dispersion. This material is very anisotropic and can be described as two-dimensional for many purposes. Pickett (1989) provides an extensive discussion of the calculated charge density and density of states. The Fermi surface resulting from the bands of Fig. 2.14 is shown in Fig. 2.15. The splitting of the bands derived primarily from the CuO_2 planes due to their weak interaction is not shown in this figure. The smaller closed surfaces centered halfway between Γ and Z, at a point often labeled $\overline{M}$, arise from the Bi–O bands mentioned above, while the larger closed surface is from the CuO_2 plane states.

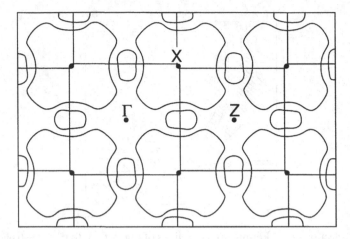

Fig. 2.15   The intersection of the Fermi surface from the bands of Fig. 2.14 with the (100) plane of the Brillouin zone. The Γ and Z points are not in the same plane in one Brillouin zone, but the zones connect as in Fig. 2.7, alternating top and central planes. (Krakauer and Pickett, 1988a)

The use of a tetragonal unit cell with a one-formula-unit basis may be too simple. Formation of the superlattice along the b-axis requires enough atomic displacement to be of concern. Singh and Pickett (1995) addressed this problem for the simpler cuprate, $Bi_2Sr_2CuO_6$ (Bi2201), which has only two CuO_2 layers instead of the two pairs of CuO_2 layers in Bi2212. It still has the Bi–O layers which distort the most, and it has a simple model, a $\sqrt{2} \times 2\sqrt{2}$ superlattice, that closely mimics the true structure, while the corresponding model for Bi2212 is five times longer in one direction. They find that the superlattice causes shifts in the Bi–O-derived bands of up to 0.4 eV and that the Fermi surface is distorted by measurable amounts, shown in Fig. 2.16. They emphasize that for Bi2212 the full orthorhombic unit cell should be used instead of the tetragonal approximation in calculations, and that experimentalists should not use the four-fold symmetry of the tetragonal lattice to reduce the amount of data they take.

The bands of body-centered tetragonal Nd_2CuO_4 are shown in Fig. 2.17 (Massidda *et al.*, 1989). This material is interesting because, when suitably doped with Ce replacing some Nd, it is an n-type superconducting material, whereas all other cuprates are p-type, including Nd_2CuO_4, when oxygen deficient. Once again, the band structure is complicated, but in the first 0.5 eV below E_F, the bands are quite simple and arise from Cu 3d and O 2p states on the atoms of the CuO_2 plane. The rather simple Fermi surface is shown in Fig. 2.18. This Fermi surface is not for Nd_2CuO_4 itself, but for $Nd_{1.85}Ce_{0.15}CuO_4$, a composition which is metallic and superconducting. This was obtained with the rigid band model by raising the Fermi level in Fig. 2.17 by 0.2 eV (Massidda *et al.*, 1989).

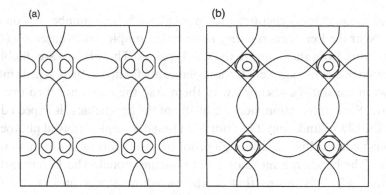

Fig. 2.16 Fermi surface for Bi2201 intersection with the basal plane of the Brillouin zone. (a) Results for the orthorhombic unit cell, more correct for the actual crystal. (b) Fermi surface for an assumed tetragonal unit cell mapped onto the Brillouin zone for the orthorhombic structure. The tetragonal Brillouin zone has edges rotated 45° with respect to those shown. (Singh and Pickett, 1995)

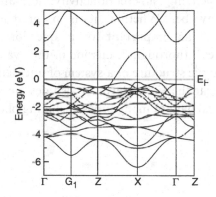

Fig. 2.17 LDA band structure of Nd_2CuO_4. (Massidda *et al.*, 1989)

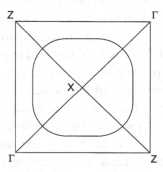

Fig. 2.18 The intersection of the Fermi surface from the bands of Fig. 2.17 with the (100) plane of the Brillouin zone. (Massidda *et al.*, 1989)

The calculated band structures presented above have a number of common features. Near the Fermi energy they have rather simple bands based on Cu 3d and O 2p states from the atoms in the CuO_2 planes. These bands are doubled if there are two planes per unit cell. If these planes are far apart, there is little interaction between the states associated with them, and the bands are then two-fold degenerate. Some interaction, hence a lifting of the degeneracy, is expected and found in the LDA bands, but it has not been found in angle-resolved photoemission experiments on Bi2212, although proof that the splitting is too small to see is lacking. The bands are more or less two-dimensional. The Fermi surfaces show "nesting." This means that in the two-dimensional projection of the Fermi surface a line of states on the Fermi surface is separated from another line on the Fermi surface by a constant or nearly constant wave vector. In three dimensions there might be nearly parallel surfaces, i.e., nearly parallel planes. This leads to a critical point in the density of states at or near the Fermi level. For perfect nesting this critical point is a logarithmic singularity and it occurs in a region of high density of states. This has consequences for many physical properties of the cuprates, including superconductivity, no matter what the microscopic mechanism may be. Whether it is a prerequisite for high-T_c superconductors is still an important open question. A recent review (Markiewicz, 1997) gives a thorough discussion of the "van Hove scenario," a description of the electronic structure of a system of interacting electrons with a van Hove singularity near the Fermi level. It also describes and interrelates a great deal of experimental data of many types, including photoemission data. It is well worth reading.

2.3   Surface states

Band calculations carried out with a periodic potential leave out surface states. In the 1930s surface states were singled out for special study, but now it is possible to include them in a band calculation by the use of supercells. Instead of taking a unit cell of the crystal as the fundamental repeated unit in the periodic potential, one considers several layers of periodic material, usually at least seven, separated by an empty space of at least a few atomic spacings. In the stacking direction the unit cell (supercell) is then taken to be the layers of atoms plus half of the vacuum space on each side. In the other two directions the normal unit cell may be used. The number of layers of atoms may be increased from a small number till the eigenvalues of the bulk states do not depend on the number of layers, and the width of the vacuum region is varied till the eigenvalues of the surface states do not depend on the width.

Such a calculation results in states that can be placed in three classifications. Most states are bulk states. Their energies match those of a three-dimensional calculation and the wave functions are periodic throughout the slab. Surface states have wave functions localized near the surface, and energies located in a gap of the eigenvalue spectrum for the bulk states. Surface-state wave functions decay exponentially into the vacuum and their envelopes decay exponentially into the slab with different decay lengths. They are periodic in the plane of the slab. Surface resonances resemble surface states in their localization near the surface, but their energies lie within a band of bulk states, to which they appear coupled. Because the wave functions of surface states extend into the vacuum, these states are likely to be involved in any surface chemistry. General references on surface states are Davison and Levine (1970), Forstmann, (1978), Garcia-Molinar and Flores (1979), Davison and Stęślicka (1992), and Desjonquères and Spanjaard (1993). If similar supercell calculations are carried out for multilayers, e.g., alternating slabs of two materials, interface states may be found, states which decay exponentially on both sides of the interface and do not disperse with the normal component of wave vector. These states bear considerable resemblance to surface states. The details of the surface states depend on the structure near the surface. If the surface reconstructs or relaxes, the surface states may change considerably, possibly even disappear.

Surface states need not be located below the Fermi level. They may occur at higher energies, but if they are at energies more than the work function above the Fermi energy, they will be degenerate with states outside the crystal, even though they may lie in a gap of the bulk band states. In this case they are probably broadened beyond recognition by rapid decay.

Since photoemission is very surface sensitive, surface states are readily studied by photoelectron spectroscopy. In angle-resolved photoemission they exhibit dispersion for wave vectors in the plane of the surface, but no dispersion with change in the normal component of wave vector. Surface states above the Fermi energy may play a role as final states in inverse photoemission. In many so-called two-dimensional materials, e.g., $TiSe_2$, the bulk states themselves do not disperse much with wave vector normal to the planes of the layers of atoms. In this sense they also resemble surface or interface states. Since the cuprates are layered materials, often appearing almost two-dimensional, their bulk states will have some properties in common with surface states. They may also exhibit true surface states. It may not be important to pursue the differences between surface and bulk states for layered materials, for the differences may be subtle.

Calandra and Manghi (1992) carried out tight-binding calculations on Y123, using different planes as the surface termination. They found surface states existed for most of the terminations considered. These states may, however, become altered, or disappear as surface states, if any surface reconstruction occurs. Hu (1994) and Yang and Hu (1994) have calculated gap states near E_F in the superconducting phase of a model $d_{x^2-y^2}$ superconductor.

Most of the interpretation of the photoelectron spectra of the cuprates has been in terms of bulk states. Surface effects have not been separated and studied as well as they have been for simpler materials such as Cu. For crystals that are nearly two-dimensional, such as the cuprates, bulk states often have some of the characteristics of surface states, making the separation more difficult. Moreover, in the many-body approach such a separation may not be possible in principle because of interference effects between the two in photoelectron spectra. In any case, some reinterpretation of the photoelectron spectra of cuprates may be necessary in the future.

2.4   Cluster models

The band picture makes use of wave functions spread throughout the crystal. Wave vector is a good quantum number. The opposite approach, using localized wave functions, is that used for atoms and molecules. The local Coulomb interactions are then emphasized more. In the case of cuprates, this approach should include both Cu and O atoms, so a cluster is considered as the fundamental unit. A great deal of work on correlation in clusters was in existence before cuprates were discovered, for clusters were used for molecules and for the study of insulating compounds.

It is relatively easy to set up the equivalent of the independent-electron model of the energy levels for a cluster. One starts with the atomic orbitals of the atoms, and, making use of symmetry, hybridizes them. The difficulty occurs in treating accurately the excited states, those with a hole in a previously-filled level, perhaps also with an excited electron on the cluster. This is trivially done in the independent-electron model; electrons are removed from one orbital and placed in another with the orbital energies remaining independent of occupancy. A better treatment of electron correlations is needed for most applications. Atomic multiplet theory has a long history of treating this problem, as has configuration interaction (Shirley, 1978). A second difficulty, usually not appearing in the study of molecules or clusters, but important for the cuprates, is how to treat doping, the addition or removal of less than one electron per cluster formula unit.

For the cuprates the simplest cluster might be two-dimensional, a Cu^{2+} ion at the center of a square of O^{2-} ions. This cluster then has a net charge of $-6e$. In real structures, there may be O ions directly above and below the Cu atom, but at a slightly larger distance than the in-plane Cu–O distance, hence less covalently bonded. When embedded in a three-dimensional crystal the planar square may distort slightly toward a diamond shape, and the apical oxygen atoms may move a little off the z-axis. The Cu^{2+} ion brings nine valence electrons to the cluster. The highest filled and lowest empty levels on the Cu ion are the five-fold degenerate (ignoring spin) 3d levels. These split in the square-planar crystal field into a two-fold level and three non-degenerate levels as shown in Fig. 2.19, where the levels are ordered as they would be for the Cu ion surrounded by the four O^{2-} ions. Each O^{2-} ion has six valence electrons, two each in the p_x, p_y, and p_z orbitals. Some of the p_x and p_y orbitals, those with axes directed toward the Cu site, hybridize more strongly with the Cu 3d orbitals than do the remaining O p orbitals.

Eskes *et al.*, (1990) carried out a calculation on just such a cluster for comparison with CuO, in which the actual geometry is a bit distorted from that of the square cluster. They started with the five Cu 3d orbitals of Fig. 2.19 and five symmetry-adapted linear combinations of the O 2p orbitals, all filled except for one hole. The crystal field splitting from the ionic lattice, shown in Fig. 2.19, was neglected as being smaller than Cu–O hybridization effects. Thus they began with degenerate Cu 3d levels. The three Cu 3d orbitals with the most planar charge density and the O $2p_x$ and $2p_y$ orbitals are shown schematically in Fig. 2.20. The symmetry-adapted O orbitals that contribute to the bonding hybrids are $(1/2)(p_{x1} + p_{y2} - p_{x3} - p_{y4})$, a_1 symmetry, hybridizing with $d_{3z^2-r^2}$, $(1/2)(p_{x1} - p_{y2} - p_{x3} + p_{y4})$, b_1 symmetry, hybridizing with $d_{x^2-y^2}$, and $(1/2)(p_{y1} + p_{x2} - p_{y3} - p_{x4})$, b_2 symmetry, hybridizing with d_{xy}. The remaining two d orbitals, degenerate, with e symmetry, hybridize primarily with O p_z orbitals. The Cu 3d and O 2p orbitals were allowed to hybridize to give two sets of four energy levels, one bonding and one antibonding. The hybridization was carried out with an O 2p–O 2p transfer integral and four Cu 3d–O 2p transfer integrals, not all independent parameters, and the energy Δ_{pd} to transfer an electron to the Cu atom from an O atom. The energy levels are shown in Fig.

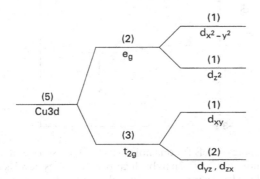

Fig. 2.19 Splitting of the degenerate Cu 3d levels (*left*) in a crystal field of cubic (*center*) and square-planar symmetry (*right*). The numbers indicate degeneracies (not including spin) and the ordering of the levels (sign of the crystal field) is that expected in a cuprate.

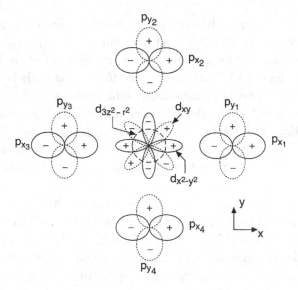

Fig. 2.20 Cu 3d and important O 2p orbitals used in calculations on a $(CuO_4)^{6-}$ square-planar cluster. The O p_z and Cu $3d_{xz}$ and $3d_{yz}$ orbitals are not shown. (Eskes *et al.*, 1990)

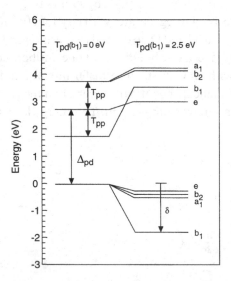

Fig. 2.21 Calculated energy levels for one hole in a $(CuO_4)^{6-}$ square-planar cluster. The left-hand levels are without Cu 3d–O 2p hybridization. The lowest level is that of the Cu 3d orbital. The upper three levels are O 2p-based, all raised by the charge-transfer energy Δ_{pd}, and split by the O–O hybridization matrix element T_{pp}. The right-hand set is with hybridization. All of the latter are non-degenerate except the two-fold degenerate e levels. (Eskes *et al.*, 1990)

2.21. The left-hand side omits Cu–O hybridization, but includes the charge-transfer energy and O–O hybridization. The right-hand side includes the Cu–O hybridization. This is a plot of hole energy levels and the ground state has the hole in an orbital of b_1 ($d_{x^2-y^2}$) symmetry, but the average occupancy of this state is less than unity. The ground-state configurations were a combination of $|d^9\rangle$ and $|d^{10}\underline{L}\rangle$ for each of the four pairs of bonding–antibonding hybrid orbitals. $\underline{L}$ stands for a hole on a ligand orbital. (This use of more than one atomic configuration is called configuration interaction.) Three parameters were varied to produce these levels. The ground state has a hybridization energy δ of -1.77 eV as shown, and an average 3d hole occupation number of 0.67. Calculations of final states in photoemission and inverse photoemission, in which there is one more hole and one less hole, respectively, were also carried out. In photoemission, the additional hole in the final state may be in the valence levels discussed above or in a core level. These will be discussed in Chapter 5.

Eskes et al.(1990) added the two apical oxygen atoms to the cluster just discussed and recalculated the valence-band photoelectron spectrum. The result was very close to that of the square-planar cluster. This suggests that the square cluster is an adequate starting point for discussing the spectroscopy of cuprates. As we shall see in Chapter 5, Eskes et al. were able to use a cluster model for CuO to provide considerable insight into the types of final states in each part of the valence-band photoelectron spectrum. Sawatzky (1990, 1991) has given a good summary of the work on clusters by his group up till 1991.

Pickett (1989) discusses some of the problems with using clusters to calculate ab-initio energy eigenvalues. For example, the intra-atomic Coulomb repulsive energy may be partially screened by electrons not included in the cluster. For the example given above, the Cu atom has the proper number of nearest neighbors, but the O atoms do not have the correct number of Cu atoms around them. This deficiency in the model is partially corrected by the use of effective values of some of the parameters, e.g., transfer integrals. Its effect can also be tested if the corresponding calculations can be done on clusters of different sizes. Such work will be discussed in Chapter 5 in connection with photoemission spectra of CuO. Finally, one can account for the effect of the polarization of the atoms outside the cluster in an empirical fashion, as was done in a cluster calculation on NiO by Janssen and Nieuwpoort (1988).

2.5   Localized models for electronic structure

The cluster calculations just described qualitatively can be placed in several classes of models for localized systems. The LDA band picture has been widely used for metallic systems for many years. Unmodified, it does not give good values for the energy gap in semiconductors and insulators and it fails to predict the insulating nature of transition-metal oxides. Mott suggested that a different viewpoint was necessary for insulators. As the interatomic spacing in a crystal increases in a thought experiment, the width W of the valence band decreases. The energy U to transfer an electron from one site to another site is independent of interatomic spacing if we ignore screening by electrons on other atoms. In a metal with its normal interatomic spacing, $W > U$ and the decrease in kinetic energy from spreading out the wave functions more than compensates for the increase in Coulomb energy occurring when charge from one site overlaps charge on another. At much larger spacings, $W < U$, and for a monovalent material, we have each electron staying on just one site, one electron per site. The material then is an insulator. In more complex systems one concentrates on W and the band structure when $W > U$ is expected, and on U and the on-site correlations when $U > W$ is expected. When W and U are comparable, as for some transition-metal oxides, the situation is less simple and only a few systems have been studied.

Hubbard (1955, 1963) gave this picture quantitative expression with the Hubbard Hamiltonian:

$$H = -t \sum_{i,j,\sigma} a_{i\sigma}^{\dagger} a_{j\sigma} + U \sum_{i} n_{i+} n_{i-}.$$

Here $a_{i\sigma}$ annihilates an electron of spin σ at lattice site i and $a_{i\sigma}^{\dagger}$ is the corresponding creation operator. For cuprates, the sites are the Cu sites and the electrons are assumed to be the 3d electrons. t is the transfer integral or hopping integral, in this case between neighboring Cu sites, and U the Coulomb repulsion between two electrons on the same Cu site. $n_{i\sigma}$ is the number operator, $a_{i\sigma}^{\dagger} a_{i\sigma}$. U and t are positive energies. The first term, the kinetic energy term, by itself would give the tight-binding band structure if it were applied to a lattice, except that the potential used in t contains some Coulomb repulsion from the last term when it is used in tight-binding band structure calculations. (A more elaborate treatment would allow next-nearest-neighbor hopping, also with a different value of t as an additional parameter.) The last term represents the on-site Coulomb repulsion. The addition of this term removes the independent-electron approximation. This model is used frequently for the cuprates and NiO. However, U may not have a single value representing the transfer of one d electron from one transition-metal site to a distant transition-metal site. Its value will depend on the multiplet structure of the transition metal, for the angular distribution of 3d charge is not the same for all members of a multiplet. Materials fitting this description are called Mott–Hubbard insulators.

This is the single-band Hubbard Hamiltonian and it may appear in other forms. For a half-filled band as in the insulating parent cuprate compounds, it is useful to use a representation with hole operators. For cuprates the Hubbard

Hamiltonian may also be extended to the three-band or three-state form by adding two more terms like the first one, each new sum representing $2p_x$ and $2p_y$ states on the O sites, a term hybridizing the O $2p_x$ and $2p_y$ states with the Cu $3d_{x^2-y^2}$ states, and possibly Coulomb repulsion terms for the O sites. Cu 4s electrons usually are not considered directly. They are assumed to have been transferred to the O sites, and back-transfer occurs only by hybridization with 3d states. O $2p_z$ states are omitted as well.

The solutions to this Hamiltonian are many and varied. They depend on the number of electrons, the lattice geometry, and the relative magnitudes of t and U. Closed-form solutions have not been found for a large periodic array of atoms. Exact solutions have been found for small clusters by diagonalizing a large matrix. The smallest meaningful cluster is a square-planar ($Cu^{2+}O_4^{2-}$) cluster with 25 electrons if the O $2p_z$ states are not considered. This represents a cuprate plane with the uppermost band exactly half filled. Then the Hamiltonian must be solved with one or more additional holes to represent hole doping toward the superconducting composition. An additional hole also represents the final state in photoemission.

There are other model Hamiltonians related to the Hubbard Hamiltonian. The t–J model replaces the Coulomb repulsion term in the one-band Hubbard Hamiltonian with a term involving only spin operators, a version of the antiferromagnetic Heisenberg Hamiltonian for spin $\frac{1}{2}$. This may be done if $U \gg t$. There is one electron per site with a large U. Then a perturbation expansion converts the Hubbard Hamiltonian to

$$H = -t \sum_{i,j,\sigma} a_{i\sigma}^\dagger a_{j\sigma} - J \sum_{i,j} \mathbf{S}_i \cdot \mathbf{S}_j,$$

where $J = 4t^2/U$. The spin operators $\mathbf{S}$ are introduced in a way analogous to their appearance in magnetism; the eigenvalues of the second term above are equal, to second order, to the eigenvalues of the corresponding original Hamiltonian.

One problem with the one-band Hubbard Hamiltonian is that the O 2p states appear to be missing. Zhang and Rice (1988) showed that a hole on a Cu site could carry O 2p character with it as it moved on a lattice of Cu sites. They started with a two-band Hubbard Hamiltonian with Cu $3d_{x^2-y^2}$ and O $2p_x$ and $2p_y$ states, hybridization between them, and Coulomb repulsion on the Cu site. U was taken to be much larger than the hybridization, and H was reduced to the t–J form, but with the value of J modified by the presence of the O 2p states. They sought the ground state for a two-dimensional square-planar lattice that was one hole short of being half full, i.e., with one hole per Cu site, plus one extra hole. This hole could be present as the result of doping or a photoemission process. They showed that the lowest state for the hole was a singlet, with the extra hole localized on four O sites around one Cu site. The distribution of the hole over the four sites reduces the Coulomb repulsive energy and the kinetic energy, thereby imparting stability. Moreover, the motion of this this entity, the "Zhang–Rice singlet," through the lattice could be described by a single-band t–J Hamiltonian with an effective hopping integral. (An analog of the Zhang–Rice singlet appears in calculations by the LDA + U method using a supercell with an extra hole (Anisimov *et al.*, 1992).)

It also is possible to include the oxygen sites explicitly in a three-band model with a Hamiltonian

$$H = E_{\rm d} \sum_i d_i^\dagger d_i + E_{\rm p} \sum_j p_j^\dagger p_j - t_{\rm Cu-O} \sum_{i,m} [d_{i\sigma}^\dagger p_{i+m} + ({\rm c.c.})]$$
$$+ (t_{\rm O-O}/2) \sum_{j,m'} [p_j^\dagger p_{j+m} + ({\rm c.c.})] + U_{\rm dd} n_{j+} n_{j-}.$$

Here i denotes the lattice (Cu) sites, j the O sites, m denotes the nearest neighbors of a Cu site, and m' the nearest-neighbor O site to site j. There are two hopping parameters, $t_{\rm Cu-O}$ and $t_{\rm O-O}$. Holes are created on Cu sites by $d^\dagger$ and on O sites by $p^\dagger$. The abbreviation c.c. stands for complex conjugate. The above Hamiltonian can be extended by adding Coulomb interactions for the O sites, $U_{\rm pp}$, and a nearest-neighbor Coulomb interaction $U_{\rm dp}$. This Hamiltonian is more complex than the one-band Hamiltonian and the inclusion of Cu–O hopping leads to additional dispersion and distortions of the Fermi surface as $t_{\rm O-O}/t_{\rm Cu-O}$ increases from zero (Markiewicz, 1997).

The Anderson impurity Hamiltonian (Anderson, 1961) is

$$H = \sum_{k,\sigma} E_k c_{k\sigma}^\dagger c_{k\sigma} + \sum_\sigma E_\sigma d_\sigma^\dagger d_\sigma + (U/2) \sum_\sigma n_\sigma n_{-\sigma}$$
$$+ \sum_{k,\sigma} V_k (c_{k\sigma}^\dagger d_\sigma + d_\sigma^\dagger c_{k\sigma}).$$

The first term represents the conduction electrons. By itself, it would describe the one-electron band structure. The second term represents the electrons localized on a single impurity. This model is widely used in descriptions of magnetic impurities. The creation operator $d_\sigma^\dagger$ creates a 3d electron on the single Cu site, considered as an impurity. The third term is the Coulomb repulsion between electrons on the impurity site, and the fourth describes the hybridization between electrons localized on the impurity and the delocalized conduction electrons. There is just a single impurity site in this model. A periodic array of impurity sites would give an "Anderson lattice," a more difficult problem to study. As for the Hubbard Hamiltonian, there is a rich variety of solutions depending on the relative sizes of the parameters and the number of electrons. Often the lattice geometry is not considered and the wave vector is dropped as a quantum number in the first term. Then V_k in the last term depends only on energy. More commonly it is treated as a constant.

2.6 Doping

Many of the cuprate superconductors are non-stoichiometric. The nearest stoichiometric compounds are insulating; usually they are antiferromagnetic. Upon doping with holes the Néel temperature drops, eventually to zero, then the sample becomes metallic. Upon further increase in hole concentration the metallic sample becomes superconducting, and the transition temperature T_c rises, then falls, with increasing hole concentration. This is shown schematically in the generic phase diagram of Fig. 2.22. Optimal doping refers to the hole concentration for which T_c is a maximum. Underdoped and overdoped samples have hole concentrations lower or higher, respectively, than optimally doped samples. A more sophisticated version of the phase diagram of Fig. 2.22 was given by Emery and Kivelson (1995) starting from rather general concepts.

Doping with holes may be carried out by oxygen addition or by the substitution of a metal atom for one of the metallic constituents of the material, but usually not Cu. The doping metal atom has a valence lower than that of the original atom. Doping with metal atoms can be done only at the time of crystal growth, while oxygen can be added to or removed from a crystal by annealing in suitable atmospheres. The hole concentration need not increase by one for each dopant atom, and the relationship between dopant concentration and hole concentration need not be linear, especially in Bi2212.

Describing doping a crystal with either electrons or holes in the independent-electron picture is rather straightforward. The Fermi level of the insulator lies in

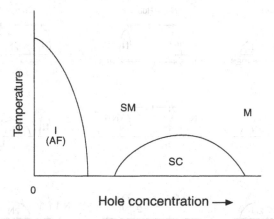

Fig. 2.22 Schematic (and not universal) phase diagram for cuprates. Structural changes have been omitted. The zero of hole concentration is the state with one hole per Cu atom, i.e., the stoichiometry of the parent compound. I(AF) is an antiferromagnetic insulator, SC is the superconducting phase, M is a metal and SM is a "strange metal." The boundary between M and SM will be discussed in Chapter 7. The region at low temperature between the antiferromagnetic insulator and the superconductor is not well characterized.

the gap. Added electrons go to the bottom of the conduction band and the Fermi level moves there. Added holes go to the top of the valence band and the Fermi level follows. This is shown in the left-hand column of Fig. 2.23 for a hypothetical monovalent monatomic semiconductor (Meinders *et al.*, 1993), where an intrinsic semiconductor with $2N$ states in each band, valence and conduction, has one hole added or an electron added. The density of states remains the same, but the Fermi level moves. Chemical or defect doping may also introduce states in the gap, but these can be taken into consideration easily, as with doped semiconductors. The total probability for removal of an electron (or adding a hole) changed with doping from being proportional to $2N$ to being proportional to $2N \pm 1$, depending on the type of doping. In the Mott–Hubbard model of an intrinsic semiconductor or insulator with one electron on each atom, each band has N states, as in the top row, center column of Fig. 2.23. In such an interacting system we cannot use the concept of density of states, for the energy of a one-electron "state" may now change when the occupancy of another "state" is changed. Only system states and energies have meaning. Instead, the spectral density function is introduced which is proportional to the probability of adding or removing an electron at a given energy. (See Chapter 3.) The spectral density function becomes the density of states in the limit of non-interacting electrons. If a hole is added to one of the atoms, i.e., an electron is removed from one atom, the integrated spectral density function of filled states is $N - 1$. The number of ways of adding an electron to the empty upper Hubbard band has also been reduced by one because now one of the atoms cannot accept an electron in addition to an

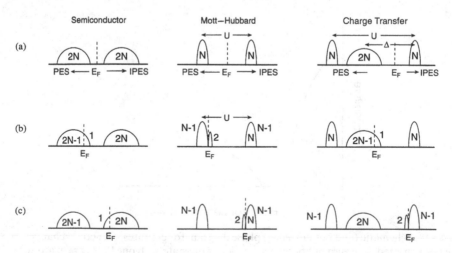

Fig. 2.23 Schematic picture of electron-removal and electron-addition spectra for three models of an insulator and how these spectra change with electron and hole doping. (a) is the intrinsic material, (b) is after a hole has been added, and (c) is after an electron has been added. (Meinders *et al.*, 1993)

existing electron to provide the proper system energy. The Fermi level then lies just below the top of the valence (lower Hubbard) band and there are two states just above it. Similarly if an electron is added, it goes at the bottom of the conduction band, the integrated spectral density function for adding an electron is decreased by one and there is one less way to add a hole to the valence band. The Fermi level then lies just above the bottom of the conduction (upper Hubbard) band, with two states just below it. For an intrinsic charge-transfer insulator (third column of Fig. 2.23) the gap between the two Hubbard bands has a band with $2N$ states in it based on the ligand atoms. These atoms are assumed not to have large correlation effects, so two electrons per atom fill the band. The Fermi level lies in the gap between the ligand band and the upper Hubbard band. Hole- and electron-doping proceed as in the Mott–Hubbard case, except that for hole doping the Fermi level moves to the top of the band based on the ligand atoms. In both the Mott–Hubbard and charge-transfer cases, doping causes a movement of the Fermi level across the gap, as in the independent-electron picture, but in addition, spectral weight is transferred across the gap, the equivalent of two electrons or holes per dopant.

Eskes *et al.* (1991) and Meinders *et al.* (1993) used a one-band Hubbard Hamiltonian on a small cluster, actually a 10-site ring. They varied the number of electron or holes about half filling for several fixed values of U and t. They integrated the spectral density function in the region of the band edges in Fig. 2.23 to determine the electron or hole spectral weight appearing in this region. At half filling, it was zero, of course, and for one added electron or hole per site it was 2, as shown in Fig. 2.23 (b) or (c). For intermediate doping, the spectral weight changes were not linear with doping, except in the limit of $U/t \to \infty$. For finite values of U/t, the spectral weight was larger than expected from a linear interpolation between 0 and 2. In this model there is electron–hole symmetry.

The dependence of the electronic structure of the cuprates upon hole concentration over the complete range achievable remains unexplained. The analogy with adding acceptors to a semiconductor is not valid. Photoemission studies of the evolution of the electronic structure with doping have not yet been carried out by starting with the insulating phase because surface charging of single-crystal insulators has precluded obtaining good spectra to date. Heating the crystal to increase its conductivity would reduce charging, but it may lead to surface instability, especially under the photon beam. Studies of superconducting samples with varying hole concentrations will be discussed later, but they are only beginning to lead to a clear picture of how the electronic structure of the insulating parent changes when the sample is doped to become metallic. What is known is that band structure calculations give some agreement with apparent energy bands obtained by angle-resolved photoemission for metallic cuprates. Agreement of these calculations is poorer with what is known about the experimental electronic structure of the insulating cuprates. This is a common problem, for the band gap of insulators is usually given incorrectly by LDA calculations. Corrected LDA calculations do much better, but they have not yet been applied to non-stoichiometric cuprates. The doping problem also occurs with 3d transition-metal oxides, and has been discussed clearly and extensively by Hüfner (1995).

2.7   Stripe phases

The structures described at the beginning of this chapter continue to be studied. Subtle differences have appeared upon further study. Over the years experiments have suggested that some of the cuprates are not homogeneous on a microscopic scale. These have been summarized by Egami and Billinge (1996).

   The cuprates may be viewed as antiferromagnetic insulators doped to produce holes. Sufficient doping destroys the static antiferromagnetic order. In the foregoing, the holes were assumed to be uniformly distributed, in the sense that in each crystallographic unit cell the hole wave functions were the same. Theoretical studies suggested that phase separation may be possible, with the holes distributed inhomogeneously (Emery et al., 1990; Kivelson et al., 1990; Emery and Kivelson, 1993). Evidence for such hole segregation in doped, but insulating, $La_{2-x}Sr_xCuO_4$ was deduced from nuclear magnetic resonance and nuclear quadrupole resonance studies, summarized by Johnston et al. (1996), who also described work on charge segregation in $La_2CuO_{4+\delta}$ involving oxygen diffusion. Recently direct evidence has been found that the hole distribution may be inhomogeneous on a scale of one or a few nanometers, the holes segregating into "stripe" domains, with antiferromagnetic domains between the stripes. Alternate antiferromagnetic domains are 180° out of phase, so the new unit cell is four stripes wide in one direction, containing two antiferromagnetic stripes and two metallic stripes. This segregation of holes need not be static. Instead it could be dynamic, i.e., occurring as fluctuations of this form. The new evidence for such charge separation comes from elastic neutron scattering, and is most secure for $La_2NiO_{4+\delta}$, $La_{2-x}Sr_xNiO$ and $La_{1-x}Nd_{0.4}Sr_xCuO_4$, with $x = 0.12$, 0.15, and 0.20 (Tranquada et al., 1996, 1997). (Neutrons do not scatter from the inhomogeneous hole density, but rather from the displaced nuclei the inhomogeneous hole density produces.) The charge ordering occurs at a temperature above the Néel temperature of the spins in the antiferromagnetic domains. Underdoped Y123 should be rather similar to $La_{1.48}Nd_{0.4}Sr_{0.12}CuO_4$. Superconductivity and the stripe phase apparently may coexist (Tranquada et al., 1997).

The regions with holes are presumably metallic or possibly semiconducting for smaller doping. Their sizes are limited by the buildup of Coulomb energy. The antiferromagnetic regions are presumably insulating. The stripe-phase unit cells are some 2 Cu–Cu spacings by 8, with the cell edges parallel to Cu–O bonds, as shown schematically in Fig. 2.24. The stripes may then be 8 Cu–Cu spacings wide and a number of unit cells long. For $La_{1-x}Nd_{0.4}Sr_xCuO_4$ the wave vectors of the magnetic neutron scattering peaks, incommensurate with the lattice wave vectors for $x > 0$, vary with x (Tranquada et al., 1997). These stripes need not be static, nor need they form a regular array over an extensive region. Their locations and widths may be influenced by charged impurities or defects. There may be domains with stripes running perpendular to those in other domains. The stripe directions in one CuO_2 plane may be perpendicular to those in the adjacent planes (Tranquada et al., 1996). It may be possible to view

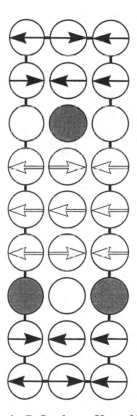

Fig. 2.24 Stripe phase unit cell in the CuO_2 plane of $La_{1.48}Nd_{0.4}Sr_{1.2}CuO_4$. Only the Cu atoms are shown. Arrows indicate magnetic moments, and the different shading denotes the antiphase domains. Shaded circles represent holes, although the holes are shared with oxygen atoms and they need not form the ordered array shown, nor be as localized as in the figure. (Tranquada et al., 1996)

a metallic stripe as nearly one dimensional. In one-dimensional Fermi liquids, spin and charge degrees of freedom separate (Lieb and Wu, 1968). A similar separation of spin and charge has been proposed for two dimensions in the case of the cuprates (Anderson, 1987). The consequences of this inhomogeneous charge distribution for various physical properties have not yet been fully worked out.

Chapter 3

Photoemission – Theory

3.1   Introduction

Photoemission studies may be said to have begun in 1887 with the observation by Heinrich Hertz that a spark between two electrodes was obtained more easily if the electrodes were illuminated (HeTrtz, 1887), although this occurred before the discovery of the electron. Improved experiments during the next several decades were important to the development of quantum mechanics. The modern era of photoemission spectroscopy arguably may be said to have started in 1964 with the papers of Berglund and Spicer (1964a,b). Although there had been prior experimental work and calculations in which Bloch electrons were assumed and dipole matrix elements calculated, it was these two papers, and papers by Gobeli *et al.* (1964) and Kane (1964), which stimulated a large amount of work, coming as they did nearly simultaneously with the widespread calculation of accurate electronic structures of many materials and the commercial availability of ultrahigh vacuum components. The first paper of Burglund and Spicer worked out what is called the three-step model for photoemission, and the second applied it to new data on Cu and Ag.

In the following we outline several approaches to the description of photoelectron spectra, starting with the most simple conceptually, then progressing to a more sophisticated picture. As we do this, we point out some assumptions and approximations often made, sometimes tacitly, and whether or not they may be important for photoelectron spectroscopic studies of high-temperature superconductors. Experimental considerations are presented in Chapter 4. When we present and interpret measured spectra in Chapters 5 through 10, we expand on some of the concepts presented below. In addition to the "theory of photoemission" we also present some of the large amount of photoemission lore useful for carrying out and interpreting experiments. These consist of special effects, e.g., resonance photoemission, and techniques, e.g., types of scans one can make. Our coverage will not be encyclopedic; we describe only effects and methods that have been applied to high-temperature superconductors, or which might prove useful in future studies. There exists a large number of books and review articles on photoelectron spectroscopy of atoms and molecules, solids, and surfaces with and without overlayers, covering both valence electrons and core electrons. Some of those dealing with solids are Smith (1971), Eastman (1972), Carlson (1975), Feuerbacher and Willis (1976), Brundle and Baker (1977, 1978), Wertheim (1978), Cardona and Ley (1978), Ley and Cardona (1979), Mahan (1978), Feuerbacher *et al.* (1978), Nemoshkalenko and Aleshin (1979), Williams *et al.* (1980), Inglesfield and Holland (1981), Wendin (1981), Leckey (1982), Plummer and Eberhardt (1982), Smith and Himpsel (1983), Margaritondo and Weaver (1983), Himpsel (1983), Courths and Hüfner (1984), Smith and Kevan (1991), Kevan (1992), Bachrach (1992), Hüfner (1995), and Braun (1996).

In most photoemission studies, the kinetic energies of emitted electrons are measured. For many years, the independent-electron model was adopted for the description of the electrons in the sample before and after the emission of the detected electron. In it, the measured kinetic energy, E, and the work function, Φ, give the energy E_f with respect to the Fermi level the emitted electron had inside the crystal after photoexcitation. Its energy before excitation, the initial energy E_i, was obtained from E_f by subtracting the known photon energy, $\hbar\omega$. (Fig. 3.1). The energies of these two states do not depend on their occupancy or on the detailed occupancy of other states in the independent-electron model. (In an interacting-electron picture, the energies E_i and E_f in Fig 3.1 refer to system energies, not single-electron energies, and the density of states must be replaced by a spectral density function. This will be discussed later.) The three-step model of Burglund and Spicer uses this picture in considerable detail to predict the spectrum of photoelectrons from a given band structure. This spectrum sometimes resembles the density of occupied electronic states weighted by a function that varies smoothly with energy. If the electron energy analyzer accepts only a small solid angle, the angle-dependence of measured spectra can be treated as well, leading to spectra of E_i vs. the wave vector of the initial state. This is what band structure calculations give, so a rather direct comparison between experiment and calculation can then be made. The use of the independent-electron model allows one to assign each measured photoelectron to an initial independent-electron state. The response of the other electrons to the presence of the photohole is not taken into account in this model. Despite this omission, we

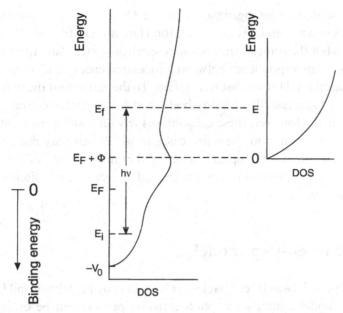

Fig. 3.1 Schematic density of states for electrons in a metal (*left-hand side*) and outside the metal (*right-hand side*). The origin of the energy scale for states inside the metal is not specified, but the origin of energy for states outside must align with the vacuum state inside, $E_F + \Phi$.

shall describe it in considerable detail for several reasons. It introduces many concepts useful in photoemission. In simple cases, e.g., nearly free-electron metals, spectra can be calculated analytically. It is still frequently used, perhaps implicitly, in qualitatively describing photoelectron spectra. A number of features of the angle-integrated photoelectron spectra of high-T_c superconductors have been interpreted successfully with this model, but certainly not all of them. It is clearly inadequate for the interpretation of the spectra of these materials and other materials in which the independent-electron model treats inadequately the effects of electron correlations. Finally, more sophisticated treatments reduce to the three-step model in certain limits. Moreover, these more sophisticated treatments rely on some of the concepts introduced in the three-step treatment, and sometimes on an experimental parameter or two obtained with the aid of the three-step treatment.

In reality, quantum mechanics does not allow separating the photoemission process into several steps separated in space and time. All one can say is that before photoabsorption, one has a system of N electrons and one incident photon. After the electron is emitted and headed for the analyzer, one has a system of $N - 1$ electrons in the solid and an external electron whose kinetic energy, direction of motion, and spin can be measured. The energy of the detected electron can be related to an eigenenergy of the $(N - 1)$-electron system. If the independent-electron model is not valid, the Hamiltonian for the $(N - 1)$-electron system is not the same as that of the N-electron system, and there need not be a one-to-one correspondence between a measured energy and a one-electron energy for the original (ground-state) system. To the extent that the two systems can be described successfully by many-body models, the photoelectron spectrum can be calculated. However, these calculations rely on another assumption, the sudden approximation, and there are cases in which this may not be a valid assumption. Fortunately it appears to be valid in most, or all, of the work to date on cuprates, but this was not demonstrated till recently. We discuss these in more detail below.

3.2   The three-step model

The three-step model was traced back to at least 1945 by Feibelman and Eastman (1974). This model assumes the photoemission process can be described by three sequential processes: (i) photoexcitation of an electron, (ii) transport of that electron to the surface without inelastic scattering, and (iii) escape of the electron into the vacuum. The primary photocurrent, $I(E, h\nu)$ is then the product of three factors, one from each of these processes: $I(E, h\nu) = P(E_f, h\nu)T(E_f)X(E_f)$, in the order just listed. (E is the kinetic energy outside the sample, E_f the energy of the excited electron inside the sample. Their relationship is given in Fig. 3.1.) We discuss each of these factors in turn, but in reverse order because P is the most complicated. In addition, some of the excited electrons will escape after one or more inelastic scattering events. These contribute to the secondary photocurrent $I'(E, h\nu)$ which forms a background, sometimes a large one, under the primary spectrum.

3.2.1   Escape

The escape of an electron at the surface is treated classically. The energy E_f of the excited electron inside the crystal is viewed as kinetic energy. One "component," $mv_z^2/2$, represents the motion parallel or antiparallel to the surface normal toward the the surface. If this exceeds $E_F + \Phi$, the potential barrier at the surface, and the motion is toward the surface, not away from it, the electron is assumed to escape with probability unity. Otherwise there is no escape. (A metallic sample is assumed here and E_F is the Fermi energy.) This ignores the possibility of quantum-mechanical reflection at the surface potential step. All electrons with $E_f > E_F + \Phi$ are assumed to escape if their velocities lie within a cone of half-angle θ with axis normal to the surface, and

$$\cos\theta = [(E_F + \Phi)/E_f]^{1/2}.$$

This satisfies the condition on the "z-component" of kinetic energy given above. If an isotropic distribution of electrons is incident on the surface from within, the fraction escaping is then

$$X(E_f) = (1/2)\{1 - [(E_F + \Phi)/E_f]^{1/2}\},$$

for $E_f > E_F + \Phi$,

$$X(E_f) = 0,$$

for $E_f < E_F + \Phi$.

  This is a structureless function plotted in Fig. 3.2. Thus in this model for an isotropic distribution of electrons, one uses an escape probability of 1 or 0, depending on energy and direction of travel.

3.2.2   Transport and inelastic scattering

An electron created with energy E_f at depth z below the surface has some probability of reaching the surface without inelastic scattering, i.e., of reaching the surface with its original energy E_f. "Inelastic" means the electron loses enough energy to emerge with an energy measurably different than if no scattering had occurred. Thus the definition depends on instrumental resolution. Scattering by phonon creation causes an energy loss of up to several tens of meV. As we shall see later, the instrumental resolution is determined by both the photon mono-chromator and the electron energy analyzer. The spectral bandpass can range from perhaps 5 meV (high resolution) to 300 meV (lower resolution). Scattering by phonons is then elastic for lower-resolution instruments. For high-resolution spectrometers, scattering by the more energetic phonons is inelastic. Elastic scat-tering is important, too, for it changes electron momenta, possibly enough to move electrons into or out of the escape cone described in the preceding section.

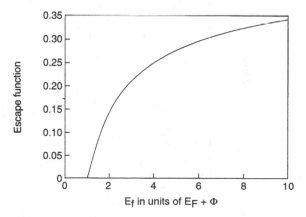

Fig. 3.2 Classical escape function, the probability of electron escape into the vacuum as a function of electron energy in units of $E_F + \Phi$.

It plays an even larger role in angle-resolved photoelectron spectroscopy. We defer our discussion of elastic scattering.

One universal inelastic scattering mechanism is electron-electron scattering, i.e., scattering by creating electron–hole pairs (Kane, 1967). A photoelectron of energy E_f scatters from an electron of initial energy E_s, with $E_s < E_F$. It loses energy, ending at energy $E_{f'}$ in exciting the other electron to a state with energy $E_{s'}$ with $E_{s'} > E_F$. The energy loss is $(E_{s'} - E_s) = (E_f - E_{f'})$ (Fig. 3.3). We ignore the (screened) Coulomb scattering matrix element for now. This neglect allows scattering processes that do not conserve crystal momentum. Then the rate for this scattering event is proportional to the product of three densities of states, those at energies E_s, $E_{s'}$, and $E_{f'}$. The same energy loss occurs for other pairs of levels with different values of E_s and $E_{s'}$. One must integrate over E_s or $E_{s'}$, subject to the restrictions of Fermi–Dirac statistics and energy conservation to get the rate at which the electron at E_f scatters to a state at $E_{f'}$. To get the total inelastic scattering rate this expression is then integrated over $E_{f'}$ in the range E_F to E_f. The result, proportional to the rate of inelastic scattering of an electron of energy E_f, is

$$\int_{E_F}^{E_f} N(E_{f'})\mathrm{d}E_{f'} \int_{E_F-E_f}^{E_F} N(E_{s'})N(E_{s'} - E_f + E_{f'})\mathrm{d}E_{s'},$$

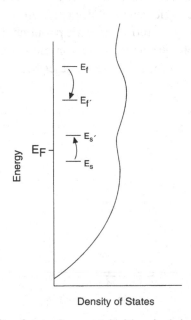

Fig. 3.3 Schematic density of states for a metal with a single inelastic scattering event marked.

where obtaining the magnitude would require a value for the average scattering matrix element, omitted from the expression above. The mean free path Λ for travel without inelastic scattering is taken to be the product of the electron (group) velocity and the mean scattering time, the reciprocal of the above, or an improved, scattering rate.

Since the density of states of a free-electron gas increases as $E^{1/2}$, the scattering rate in this model increases without limit as E_f increases, leading to a mean free path that always decreases as E_f increases. The inclusion of a scattering matrix element should produce a more accurate description of inelastic electron–electron scattering, but at lower values of E_f proper treatment of the screening of the Coulomb interaction is difficult; so difficult that calculations have not been useful in experimental work. At higher energies, screening is less of a problem and one finds (Heitler, 1954) that the scattering cross-section decreases rapidly enough with increasing E_f to overcome the increasing phase space for the scattering, so at higher energies the scattering rate falls with increasing E_f, i.e., the mean free path increases.

In addition to energy loss by electron–hole creation, the excited electron may also lose energy by exciting volume plasmons and surface plasmons and by exciting core electrons. These processes, of course, occur only when the electron energy is high enough. The contribution from core–electron excitation is usually neglibly small, unless one is studying photoemission from deep core levels in the presence of shallower core levels. Since plasmon energies are of the order of 10 eV, such losses must be included in most measurements of photoelectrons from valence bands when photon energies exceed 15 eV or so. If the resolution of the electron energy analyzer is high, energy loss by creating the more energetic phonons will also contribute. Electron–hole creation and plasmon excitation both contribute to the energy- and frequency-dependent longitudinal dielectric function, $\widetilde{\epsilon}_L(E, \mathbf{k})$. The probability that a fast electron loses energy E and momentum $\mathbf{k}$ in a material is proportional to $\text{Im}(-1/\widetilde{\epsilon}_L(E, \mathbf{k}))$ (Hubbard, 1955; Ritchie, 1957). This function can be measured in electron energy loss experiments (Raether, 1965; Daniels *et al.*, 1970; Schnatterly, 1970) and used to model energy loss by escaping photoelectrons. This method of obtaining the mean free path of escaping photelectrons has been used with some success, especially for core-level photoelectron spectroscopy, but it has not been used with much success in valence-band photoelectron spectroscopy.

Rather than attempt calculations of the magnitude and energy dependence of the mean free path for individual systems, the general shape of the curve, a minimum in the mean free path at some E_f, has been fitted to data obtained either by the transmission of electron beams through thin films or, more commonly, by the transmission of photoelectrons through overlayers. In these measurements, the word "attenuation length" is used instead of inelastic mean free path. It differs from the mean free path because elastic scattering in the overlayer may occur, while it is not considered in evaluations of the mean free path. Such elastic scattering increases the distance an escaping electron will travel before leaving the surface, thereby increasing the probability that an inelastic scattering event will occur. This makes the attenuation length shorter than the mean free path for inelastic scattering. These measurements, and others, fitting photoelectron spectra to a model involving attenuation, result in the so-called "universal curve" (Powell, 1974; Lindau and Spicer, 1974; Seah and Dench, 1979; Powell, 1985, 1988) shown in Fig. 3.4. Note that it is really a band rather than a curve.

There are a number of known deviations from the general shape of the average curve one might place through this band. One such occurs in insulators. If E_f is not more than E_g above the top of a band gap of width E_g, no scattering of the type envisioned can take place for lack of states into which the lower-energy electron can be excited. In fact, for insulators at low energies the mean free path rises rapidly above the curve of Fig. 3.4. It can reach very large values, e.g., over 100 nm for solid Ar (Schwentner, 1974; Saile *et al.*, 1974). Other deviations are the result of details in the scattering of electrons by the screened Coulomb interaction or of specific structure in the density of states. An example is Ba, where the minimum value of the mean free path is reported to occur at 15 eV, well below that of Fig. 3.4 (Jacobi and Astaldi, 1988). These material-specific deviations are not large, especially on a log–log plot like Fig. 3.4, but they give plots

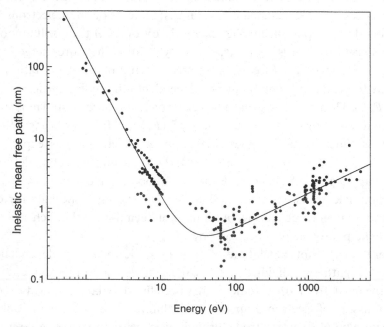

Fig. 3.4   The "universal" mean free path plot, electron mean free path for motion without inelastic scattering vs. electron energy. (Seah and Dench, 1979)

for individual materials that have locally different slopes and a different energy for the minimum than the universal curve. This can lead to incorrect conclusions about which of two spectra taken at different photon energies is more surface sensitive.

There is a large amount of literature on the measurement and calculation of attenuation lengths and mean free paths, for they are important in the stopping power of fast charged particles in matter as well as in photoelectron spectroscopy. They are especially important for analytical studies using x-ray-induced photoelectron spectroscopy of core levels. Many Monte-Carlo calculations have been carried out, including the important effects of elastic scattering. Most use calculated or measured electron energy loss spectra to describe the inelastic processes, and most consider electron energies of 100 eV or more. At lower energies a treatment by single-scattering processes is not reliable. Examples of such studies include Tougaard and Sigmund (1982), Tougaard (1990), Tougaard and Kraaer (1991), Puppin and Ragaini (1995) and Werner (1995). We do not pursue this further because at the moment no studies exist on the mean free path in high-temperature superconductors and because little is known about inelastic scattering effects in angle-resolved valence-band photoemission in any material. An experimental study by Liu *et al.* (1991) is one such, but it did not lead to a better way to deal with inelastic scattering of low-energy electrons in high-temperature superconductors. For the high-temperature superconductors we are limited to the use of the universal curve, which lacks two more concepts described in the next three paragraphs, both of possible importance for high-temperature superconductors.

The mean free path just described is isotropic. In non-cubic materials anisotropy is expected, in analogy with the anisotropy of the electrical resistivity due to scattering from impurity atoms. Since the calculation of the scattering rate by the screened Coulomb potential has not been carried out successfully, it is not surprising that no theoretical study of anisotropy has been made. In fact there appears to be no experimental work in which the anisotropy of the mean free path has been extracted from experiments on highly anisotropic materials, e.g., transition-metal dichalcogenides, even though the tensor components of $\tilde{\epsilon}_L(E, \mathbf{k})$ may have been measured. Such anisotropy could have consequences in the study of the electronic structure of cuprate superconductors. A discussion of the anisotropy of the scattering in cubic tungsten has been given by Tréglia *et al.* (1988).

The cuprates are highly anisotropic. For some purposes, they may be viewed as alternating nearly two-dimensional layers of metallic and insulating materials. Electron–electron and electron–phonon scattering in two-dimensional metals have been studied theoretically by Gurzhi *et al.* (1987, 1995) and qualitative differences found between two- and three-dimensional metals. Such differences have not yet been considered in thinking about mean free paths for inelastic scattering in electron spectroscopies of nearly two-dimensional materials.

The picture used for inelastic scattering is that of a homogeneous continuum. Since the minimum mean free path is about 10 Å, internal structure in the medium on this length scale could lead to departures from the description used above. The unit cell height in some cuprates is of the order of 30 Å. If we view these materials as stacks of conducting and insulating or semiconducting planes, one expects at low electron energies to have a longer mean free path in the insulating layers than in the conducting layers. An average mean free path for such a composite could be defined by averaging scattering rates (or reciprocal mean free paths). At the level of the use of the universal curve, this is probably of negligible importance, for the mean free paths of the individual layers and the averaged one probably all fall within the band on the universal curve. Pendry and Martín-Moreno (1994) have developed the first approach to this problem.

3.2.3   Photoexcitation

Consider an electromagnetic wave of known polarization incident at a known angle on a flat surface of the sample. The Fresnel coefficients can be used to obtain the amplitude of the wave(s) propagating into the sample. They can be found for isotropic media and anisotropic crystals exhibiting birefringence. In the simplest case for photoemission, an isotropic or cubic sample with optical absorption, the wave propagating inside is, in general, an inhomogeneous wave with planes of constant amplitude parallel to the surface and planes of constant phase not parallel to the surface. This is so also in insulators because for them photoemission spectra are taken with photon energies above the band gap and the dielectric function or complex refractive index at such frequencies has a finite imaginary part. At such energies, insulators may be as absorbing as metals.

For the case of a single wave, the electric field amplitude at a depth z below the surface is given by $\tilde{t}E_0 \exp(-\alpha z/2)$, where $\tilde{t}$ is a (complex) Fresnel coefficient for transmission through the surface, α is the absorption coefficient, and E_0 is the electric field amplitude of the incident beam. Both $\tilde{t}$ and α can be calculated from the complex refractive index or complex dielectric function, described below. In fact, many photoemission studies have been, and still are, carried out on materials whose optical properties at the appropriate frequencies are unknown. Photoemission is not a quantitative spectroscopy; one rarely measures the absolute photoyield as a function of electron energy. Lack of knowledge of the optical properties generally is not a problem; $\tilde{t}$ and α will not appear in the suitably normalized spectra. (Spectra are often normalized to the incident photon flux rather than the absorbed flux for lack of information on the optical properties of the materials being investigated.)

There are cases in which the foregoing picture is inadequate. The Fresnel rela-
tions for a vacuum–metal interface assumed a discontinuous decrease in the
longitudinal component of the electric field and a surface charge density. The latter implies surface charge, i.e., electrons, in a layer of zero thickness, which quantum mechanics does not permit. A more proper treatment repairs this defect,
and a non-local dielectric function is introduced, one depending on distance
from the surface, or, when Fourier analyzed, on the normal component of wave
vector. This can lead to surface-specific photoemission, even for a free-electron
gas model for a metal, for which there is no volume photoexcitation. (For an
introduction to the literature, see Kliewer (1978).) Such effects are not prominent
in high-temperature superconductors, and have not been studied in them.

The probability that the electromagnetic wave excites an electron in a differential layer dz thick at depth z is proportional to

$$\alpha dz |\tilde{t}E_0|^2 \exp(-\alpha z),$$

but this is not specific enough for photoemission, for it includes excitations from
a variety of initial states to many excited states, each pair being separated by the
same energy $h\nu$. We must separate α into contributions from individual electronic states.

The Hamiltonian for a system of electrons in an electromagnetic field is
obtained by replacing the momentum operator $\mathbf{p}$ by $[\mathbf{p} - (e/c)\mathbf{A}]$, where $\mathbf{A}$ is the
vector potential, e the electron charge, and c the velocity of light. The kinetic
energy operator then becomes

$$p^2/2m - (e/2mc)(\mathbf{p} \cdot \mathbf{A} + \mathbf{A} \cdot \mathbf{p}) + (e/2mc)^2 A^2.$$

In the linear optical regime, the last term is neglible and can be dropped. To
date, only pulsed laser sources can produce fields large enough to require retention of this last term to calculate photoexcitation. We discuss this briefly in
Chapter 12 under two-photon photoemission. The terms in $\mathbf{p} \cdot \mathbf{A}$ and $\mathbf{A} \cdot \mathbf{p}$ are
treated as a time-dependent perturbation in the Hamiltonian. In general, $\mathbf{A}$ and
$\mathbf{p}$ do not commute. The wavelengths of the electromagnetic wave are typically
100 to 2000 Å , while the amplitude of the wave falls off exponentially with a
decay length of a few hundred Å. The quantity $\nabla \cdot \mathbf{A}$, from $\mathbf{p} \cdot \mathbf{A}$, is largest at the
surface. In fact, it has been singled out for special study as the surface photoelec-
tric effect. This arose historically because a free-electron gas cannot be photoexcited except at a surface. The reason is that energy and momentum cannot both
be conserved during excitation within the volume of the gas, but the surface can
play a role in momentum conservation. The $\mathbf{p} \cdot \mathbf{A}$ matrix element is zero in the
interior, but the $\nabla \cdot \mathbf{A}$ term does not vanish near the surface, and it allows photoexcitation. The very small escape length means all primary photoelectrons come
from very near the surface. This surface sensitivity is not what is meant by the
surface photoelectric effect, nor is the role played by surface states. The surface

photoeffect arises from the effect of the $\nabla \cdot \mathbf{A}$ term at the surface in the photoexcitation matrix element. The calculation of $\mathbf{A}$ then should be carried out using a non-local dielectric function. The surface photoelectric effect was discussed extensively by Levinson and Plummer (1981) and Plummer and Eberhardt (1982). Except for a line shape issue in Section 3.5 we will not consider the surface photoelectric effect further because its role in work to date on the cuprates has not been identified.

The photoexcitation rate then involves the square of the matrix element $(\mathbf{p} \cdot \mathbf{A} + \mathbf{A} \cdot \mathbf{p})$ between the initial state i and the final state f. $\mathbf{A}$ will be of the form $\mathbf{A}_0 \exp(-\mathbf{k} \cdot \mathbf{r})$. Because the wavelengths are much longer than atomic and unit cell dimensions the exponential in $\mathbf{k} \cdot \mathbf{r}$ can be expanded in a Taylor series and only the first term, unity, kept. This is the electric dipole approximation and it is generally valid for all photoelectron spectroscopy to date on valence electrons. When higher-energy photons are used to study core levels, this approximation must be re-examined. The electric dipole approximation reduces $\mathbf{A}(\mathbf{r})$ to $\mathbf{A}_0$, $\mathbf{A}$ and $\mathbf{p}$ then commute, and the perturbation becomes $(e/mc)\mathbf{A}_0 \cdot \mathbf{p}$. ($\mathbf{A}_0$ will depend slowly on z, the depth below the surface.) The matrix element to evaluate is then just that of $\mathbf{p}$.

$$\mathbf{p}_{if} = \int \psi_f^* \mathbf{p} \psi_i \mathrm{d}^3 r.$$

If one has exact eigenstates, this can be put into several other forms involving other operators:

$$\mathbf{p}_{if} = \mathrm{i}m\omega\mathbf{r}_{if} = (\mathrm{i}/\omega)\nabla V_{if},$$

where $\omega = (E_f - E_i)/\hbar$, and V is the potential energy in the crystal (Manson, 1978). (For work on crystals, the form with $\mathbf{r}_{if}$ is not often used.)

With only approximate eigenfunctions, e.g, from a numerical calculation that may not be self-consistent, these will not all give the same value. $\mathbf{r}_{if}$ emphasizes the outer parts of the atomic wave functions more than the other two matrix elements and the "acceleration" matrix element, ∇V, emphasizes the regions closer to the nucleus. These integrals used to be difficult to evaluate accurately. Because of the oscillatory nature of the integrand, there is considerable cancellation. Small changes in either wave function changed the value of the integral disproportionately. More recent calculations with wave functions from self-consistent calculations agree much better with experiment. Several years ago Smith (1979) used an interpolation scheme to model the band structure, fitting an analytical model band structure to the results of a self-consistent band calculation. From this he obtained dipole matrix elements without integration by taking derivatives of Hamiltonian matrix elements with respect to wave vector. For high-temperature superconductors, there are significant effects due to changing magnitudes of the dipole matrix elements with photon energy and polarization.

These appear in the strong dependence of photoemission spectral peak amplitudes on photon energy and polarization. Calculations of the photoelectron spectra of Y123 show strong polarization effects (Bansil et al., 1992, 1993).

All the anisotropy of the optical properties of non-cubic crystals arises from the anisotropy of the dipole matrix elements, which also contributes to the anisotropy of photoexcitation. Group theory allows a quick identification of those transitions which are forbidden, i.e., it indicates which $\mathbf{p}_{if}$ are identically zero because of the symmetry of the initial- and final-state wave functions, without the necessity of carrying out integrations (Tinkham, 1964; Knox and Gold, 1964; Bassani and Pastori Parravicini, 1975). When the transition is allowed, $\mathbf{p}_{if}$ is non-zero, but group theory cannot give its magnitude; the integrals must be evaluated. For cubic and hexagonal crystals, electric-dipole-allowed transitions between Bloch states have been tabulated (Sachs, 1957; Cornwell, 1966; Eberhardt and Himpsel, 1980; Benbow, 1980; Niles et al., 1992), along with the polarization of the electromagnetic wave that allows them, in the case of non-cubic crystals. This has been done for both nonrelativistic electrons and for relativistic electrons. Borstel et al. (1981) have pointed out the importance of using relativistic selection rules even for bands without observable spin–orbit splitting. The selection rules are considerably less selective in the relativistic case; degeneracies often are smaller, and fewer transitions are forbidden. Relativistic selection rules are said to be required to explain part of the angle-resolved photoelectron spectrum of metallic Cu (Borstel et al., 1982), but this has been questioned (Goldmann, et al., 1982), then restated along with additional experimental data (Przybylski et al., 1983). This is a less important issue for photoemission from the Cu–O based bands of cuprates than for metallic Cu because of the lower symmetry of the cuprates, which tends to lead to reduced degeneracies and fewer forbidden transitions. We shall return to selection rules at several points further on.

Dipole matrix-element calculations have been carried out for most of the atoms in the periodic chart, and expressed as subshell photoionization cross-sections (Yeh and Lindau, 1985). The cross-section is proportional to the square of the dipole matrix element. For transitions between a bound atomic state with quantum numbers n, ℓ and continuum states with energy E and quantum numbers $\ell + 1$ or $\ell - 1$, it is

$$\sigma_{n,\ell}(\omega) = 4\pi\alpha a_0^3/(3\hbar\omega)[C_{\ell-1}R_{\ell-1}^2 + C_{\ell+1}R_{\ell+1}^2],$$

in which

$$R_{\ell\pm1} = \int\limits_0^\infty P_{n,\ell}\, r\, P_{E,\ell\pm1}\mathrm{d}r,$$

where the radial eigenfunctions are P/r, and the C coefficients are numbers resulting from degeneracies and integrations over angles. α is the fine-structure constant, and a_0 the Bohr radius. (See, e.g., Cooper 1962, 1975.) Knowledge of the shapes and energy dependence of these cross-sections is very useful, with an important exception noted below, and their magnitudes are usefully accurate. In the calculations of Yeh and Lindau (1985) the use of all three forms of the matrix element gave the same result within 1% for all core-level initial states. For most valence electrons differences of up to 10% appeared for photon energies above 100 eV. For core electrons these cross-sections are not accurate for atoms in molecules or solids for the first few eV above threshold, for in this region the final states of the free atom become bonding or antibonding states in the solid, and their wave functions change, thereby changing the cross-section. For valence electrons, even the initial states will change as the atoms are introduced into a molecule or solid. It is for those states that the atomic quantum numbers are the least appropriate, and one must look further. For example the five-fold degeneracy of the 3d atomic states is lifted by a cubic crystal field into separate two- and three-fold degenerate levels. (See Fig. 2.19.) The latter is split into two by a tetragonal field. The selection rules for these states in a solid will bear some resemblance to those of the atoms, but they will not be the same, for the angular distribution of the electrons has changed. The same is true for the radial integrals giving the absorption strength. Finally, since the calculations were based on an independent-electron model, there will be significant departures from experiment in some spectral regions for some atoms. When a core electron is excited to an incompletely filled shell, especially one with the same principal quantum number, the Coulomb and exchange interactions are poorly treated in the independent-particle model. This leads to resonances or "giant resonances" in both photoabsorption and photoemission (Wendin, 1982; Connerade *et al.*, 1987). These are especially important for the 3p → 3d excitations of the 3d transition metals and the 4d → 4f excitations of the rare earths. We discuss these later, for not only do rare earths appear in some of the cuprates, but there is a small resonance effect in the 3p → 3d excitations of Cu even though the 3d shell is nominally full in metallic elemental Cu.

Figure 3.5 shows the cross-sections for photoionization of the Cu 3d, and O 2p states, calculated for free atoms. These two states contribute most of the amplitude of the wave functions for the states near the Fermi level of all cuprate superconductors. Note that by using photon energies near 20 eV one maximizes the contribution of O 2p states to photoexcitation spectra, while at higher energies, the 3d states of Cu become increasingly more dominant.

For a collection of atoms with electrons treated in the independent-electron approximation, either a gas of atoms or the core electrons of the atoms in a crystal, the absorption coefficient α, or more correctly, the photoionization contribution to it, is simply $N\sigma$, the product of the atomic photoionization cross-section

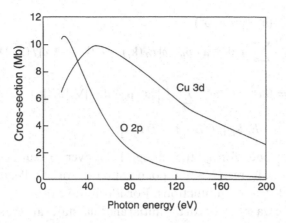

Fig. 3.5 Photoexcitation cross-section for the 3d levels of Cu atoms and 2p levels of O atoms as a function of photon energy. (Yeh and Lindau, 1985)

and the number density of atoms. The valence electron energies and wave functions change significantly when atoms form solids, and the dipole matrix elements change accordingly. Hence for the optical and photoelectron spectroscopy of valence bands, we cannot just add atomic cross-sections for each atom. The calculation of the probability that a photon of fixed energy excites an electron to a final energy E_f follows the calculation of the absorptive part of the optical properties of a solid, i.e., of the imaginary part ϵ_2 of the complex dielectric function $\tilde{\epsilon}$ or the real part σ_1 of the complex optical conductivity. The complex dielectric function is defined from one of Maxwell's equations,

$$\nabla \times \mathbf{H} = (4\pi/c)\mathbf{J} + (1/c)\partial \mathbf{D}/\partial t.$$

(Gaussian units are used.) Writing constitutive equations to describe the medium as $\mathbf{D} = \epsilon\mathbf{E}$ and $\mathbf{J} = \sigma\mathbf{E}$, with $\exp(-i\omega t)$ harmonic time dependence of the fields, the right-hand side of this becomes (Landau and Lifshitz, 1965; Stern, 1963; Wooten, 1972)

$$(\epsilon + i4\pi\sigma/\omega)(1/c)\partial\mathbf{E}/\partial t = (\tilde{\epsilon}/c)\partial\mathbf{E}/\partial t = (\epsilon_1 + i\epsilon_2)(1/c)\partial\mathbf{E}/\partial t.$$

For our purposes $\tilde{\epsilon}$ will depend on frequency or energy, but not position or wave vector. We write it as a scalar, but it is a Cartesian tensor for anisotropic crystals. $\tilde{\epsilon}$ can be replaced by a complex susceptibility $\tilde{\chi}$, $\tilde{\epsilon} = 1 + 4\tilde{\chi}$, or a complex conductivity, $\tilde{\sigma} = (i\omega/4\pi)(\tilde{\epsilon} - 1)$. ϵ_2 can be calculated by adding $(e/c)\mathbf{A}$ to the momentum operator in the Hamiltonian as above and treating the extra terms linear in $\mathbf{A}$ as a perturbation. The result (*overpage*) is (Bassani, 1966, Bassani and Pastori Parravicini, 1975, Bassani and Altarelli, 1983, Wooten, 1972)

$$\epsilon_2(\omega) = (\hbar^2 e^2/\pi m^2 \omega^2)$$

$$\sum_{i,f} \int d^3k |\mathbf{a} \cdot \mathbf{p}_{if}|^2 \delta[E_f(\mathbf{k}_i) - E_i(\mathbf{k}_i) - \hbar\omega] f(E_i)[1 - f(E_f)]$$

$$= (\hbar e^2/\pi m^2 \omega^2) \sum_{i,f} \int \left\{ |\mathbf{a} \cdot \mathbf{p}_{if}|^2 dS / |\nabla_k [E_f(\mathbf{k}_i) \right.$$

$$\left. - E_i(\mathbf{k}_i)]\} f(E_i)[1 - f(E_f)], \right.$$

where, in the second form, the integral is over a surface on which $E_f(\mathbf{k}_i) - E_i(\mathbf{k}_i) = \hbar\omega$. Here, $\mathbf{a}$ is a unit polarization vector parallel to $\mathbf{A}$ and the f are Fermi–Dirac distribution functions. For Bloch wave functions, the integral gives zero unless the wave vectors of the initial and final states are the same, *modulo* a reciprocal lattice vector. This was assumed when we wrote only $\mathbf{k}_i$ in the above expression, the result of wave-vector conservation and the small photon wave vector. Such transitions are called "direct" transitions. The wave-vector selection rule breaks down when the photon energy increases enough to make the photon wave vector a significant fraction of the Brillouin zone diameter or when the electron wave vector is no longer a good quantum number, e.g., at high temperatures when lattice vibrational amplitudes increase.

In the above integral, in each differential volume of the Brillouin zone there is a contribution from all pairs of states (bands) that satisfy the delta function in energy. Thus many final state energies may contribute to absorption at angular frequency ω. For photoemission we must count only those states with a particular final-state energy E_f. Thus the fraction of the transitions at ω resulting in the final state E_f is given by I_1/I_2, where I_2 is the integral above, and I_1 is the same integral but with an additional delta function, $\delta(E_f - E_{f'})$, multiplying the integrand. This gives an un-normalized probability which can be normalized by multiplying the integrand by dE_f and integrating, setting the result equal to unity. This integral gives us ϵ_2 back, so the normalized probability sought is

$$P(E_f, \omega) = (e^2/\pi m^2 \omega^2 \epsilon_2(\omega)) \sum_{i,f} \int d^3k |\mathbf{a} \cdot \mathbf{p}_{if}|^2 \delta[E_f(\mathbf{k}_i)$$

$$- E_i(\mathbf{k}_i) - \hbar\omega] f(E_i)[1 - f(E_f)].$$

At the time this analysis was developed, dipole matrix elements were difficult to calculate because of limited computer speed and memory. Moreover, without self-consistent calculations, the accuracy of the wave functions, hence of the matrix elements, was limited. A useful approximation was usually made, that of assuming the dipole matrix elements were constant, so they could be removed from the integral. It left the magnitudes of ϵ_2 and the photoexcitation probability indeterminate, but it was hoped that most of the features in the spectra would be described correctly. After factoring out the dipole matrix elements, the integral over the Brillouin zone could be replaced by an integration over surfaces of constant energy in the Brillouin zone to give

$$\omega^2 \epsilon_2(\omega) = \frac{8\pi^2}{m} |\mathbf{a} \cdot \mathbf{p}_{if}|^2_{\mathrm{avg}} \mathrm{JDOS},$$

where

$$\mathrm{JDOS} = \frac{1}{8\pi^3} \sum_{i,f} \int \mathrm{d}E_{if} \int \mathrm{d}S_{if} \delta(E_f - E_i - \hbar\omega)/|\nabla_k(E_f - E_i)|,$$

which is called the joint density of states. The normalized probability that the final state has energy E_f is

$$P(E_f, \omega) = [e^2/\pi m^2 \omega^2 \epsilon_2(\omega)] |\mathbf{a} \cdot \mathbf{p}_{if}|^2_{\mathrm{avg}} \sum_{i,f} \int \mathrm{d}^3 k \delta[E_f(\mathbf{k}_i) - E_i(\mathbf{k}_i)]$$

$$- \hbar\omega]\delta[E_f(\mathbf{k}_i) - E_i(\mathbf{k}_i)]f(E_i)[1 - f(E_f)]$$

$$= [e^2/\pi m^2 \omega^2 \epsilon_2(\omega)] |\mathbf{a} \cdot \mathbf{p}_{if}|^2_{\mathrm{avg}} \sum_{i,f} \int \mathrm{d}E_{if} \int \mathrm{d}S_{if} \delta(E_f - E_i$$

$$- \hbar\omega)/|\nabla_k E_f(\mathbf{k}_i) \times \nabla_k E_i(\mathbf{k}_i)|.$$

In the second equation, the sum and integral part is proportional to the energy distribution of the joint density of states. The surface integral in ϵ_2 is over a surface of constant energy difference in $\mathbf{k}$-space, and the line integral in $P(E_f, \omega)$ is over the curve of intersection of two surfaces in $\mathbf{k}$-space given by $E_{f'} = E_f$ and $E_f = E_i - \hbar\omega$, i.e., of two surfaces of constant energy difference. We illustrate this in an example below.

Structure in ϵ_2 is expected to come from the structure in the integrand, i.e., from the zeroes of the denominator, even when the dipole matrix element is not assumed to be constant. These zeroes occur at interband critical points in the band structure, either symmetry critical points at which $\nabla_k E_f$ and $\nabla_k E_i$ each vanish, or at general critical points, where neither vanishes, but the gradients are equal, causing the denominator to vanish. Structure in $P(E_f, \omega)$ arises when either of the gradients vanishes or when they are non-zero, but parallel or anti-parallel. Thus symmetry critical points cause structure in both ϵ_2 and the numerator of $P(E_f, \omega)$. The latter may be diminished by the structure in ϵ_2 in the denominator of $P(E_f, \omega)$.

3.2.4   Assembling the steps

The consequences of the three-step model can be worked out, following Berglund and Spicer (1964a), if we assume the photoelectrons are isotropically photoexcited and the mean free path is isotropic. The probability of exciting into differential angle $\mathrm{d}\theta$ at angle θ to the surface normal is $(1/2)\sin\theta\mathrm{d}\theta$. The probability that an electron isotropically excited to E_f at depth z and directed at angle θ will escape is

$$P_{esc}(E_f, z) = (1/2) \int\limits_0^{\theta_0} \exp(-z/\Lambda \cos\theta) \sin\theta d\theta,$$

with

$$\theta_0 = \cos^{-1}[(E_F + \Phi)/E_f]^{1/2}$$

if $E_f > E_F + \Phi$, and $P_{esc}(E_f, z) = 0$ otherwise.

The probability of exciting such an electron in dz at depth z is

$$P_{esc}(E_f, \omega, z) = \alpha \exp(-\alpha z) dz P(E_f, \omega),$$

where the $P(E_f, \omega)$ was discussed just above. The probability of excitation at any depth (by unit flux of photons entering the sample) is then

$$N(E_f, \omega) = \int\limits_0^\infty \alpha \exp(-\alpha z) dz P_{esc}(E_f, z) P(E_f, \omega) = T_{eff}(E_f) P(E_f, \omega).$$

The last factor does not depend on z. In T_{eff} the integral over z can be done first, then the integral over θ.

$$T_{eff}(E_f) = (1/2) \int\limits_0^\infty \int\limits_0^{\theta_0} \exp(-z/\Lambda \cos\theta) \exp(-\alpha z) \sin\theta d\theta dz$$

$$= X(E_f)[\alpha\Lambda/(1 + \alpha\Lambda)][\![1 + \{1/[2\alpha\Lambda X(E_f)]\}]\!] \ln [\![\{1 - [2\alpha\Lambda X(E_f)]\}/(1 + \alpha)]\!].$$

Here, α depends on the photon energy and Λ depends on the electron energy. $\alpha\Lambda$ may be $\gg 1$ when ultraviolet photons are used, for then α may be large and, with smaller values of E_f, Λ will be relatively large. For increasing photon energies α may be of the order of 1 or 2. However, $\alpha\Lambda \ll 1$ may occur if the absorption is rather weak.

Finally, the spectrum of photoelectrons escaping with energy E_f (measured inside the crystal) is

$$N(E_f) = P(E_f, \omega) T_{eff}(E_f).$$

The sharpest structure is expected to come from $P(E_f, \omega)$. (The actual number of photoelectrons escaping with energies in the range E_f to $E_f + dE_f$ is $N(E_f)$ multiplied by the incident photon flux and the fraction which enters the crystal.)

3.2.5   Other types of spectra

The spectrum $N(E_f)$ is called an energy distribution curve (EDC). The photon energy is fixed and the electron kinetic energy outside the sample is scanned. (Recall that this kinetic energy E is $E_f - \Phi - E_F$.) EDCs were the natural spectra to take with a light source that emitted a discrete spectral line, and even with newer continuum sources (synchrotron radiation sources) they continue to be the most frequently taken and frequently reported spectra. The EDCs have in them properties of both the initial and final states, and these are often difficult to unravel, for the dependence on the two types of states is not simple. There are spectra other than EDCs that one can take either in an angle-integrated or angle-resolved mode (Lapeyre *et al.*, 1974a,b). Instead of fixing the photon energy and scanning the electron energy analyzer, thereby scanning E or E_f, one can fix the analyzer energy and scan (i) the photon energy or (ii) both photon energy and analyzer in synchronism. Scan (i) yields a constant-final-state (CFS) spectrum for, by fixing the analyzer energy, E_f is held constant and the photon energy scan then becomes a scan of E_i only. Scan (ii) yields a constant-initial-state (CIS) spectrum, for in it, as the photon energy is increased, the final state energy is increased by an equal amount, thereby keeping E_i fixed. Both of these spectra are useful in separating the contributions of initial and final states. Both these scans are illustrated in Fig. 3.6, along with an EDC.

These three scans are special cases of a general "photoemission surface," shown in Fig. 3.7. The photocurrent detected depends on both the energy of the photoelectron and the photon energy, for a given polarization of the radiation. The three scans mentioned above are each the intersection of the surface $I(E, h\nu)$ with one of three planes, two of which are parallel to an energy axis, and the other at 45° to the axes if equal energy scales are used. This is also shown in Fig. 3.6. If one took enough of one kind of spectrum, e.g., EDCs at closely spaced photon energies, one would have enough data to generate the surface $I(E, h\nu)$ from which CIS and CFS plots could then be obtained without doing the special scans.

There is a special case of the CFS scan that is sometimes used for the study of core levels, not valence bands. If the photon energy exceeds about 20 eV and the electron energy analyzer is set to a low kinetic energy, just a few eV, the electrons collected are almost all secondary electrons, the result of multiple inelastic scattering and Auger electrons, some of which have also been scattered inelastically. A spectrum of this photocurrent as a function of photon energy is called a partial yield curve or spectrum. Sometimes all photoelectrons are collected by a positively biased electrode. The resultant total yield spectrum is nearly identical to the partial yield spectrum because the primary photoelectrons are only a few percent of the total photocurrent. In the photon energy region of core-level excitation these yield spectra resemble closely the absorption spectra, except, of course, the absolute magnitude of the absorption coefficient cannot be obtained from them. The positions, relative strengths, and shapes of structures in the absorption coefficient spectrum are given rather closely by the yield spectrum, although the continuum background may not have the same relative magnitude

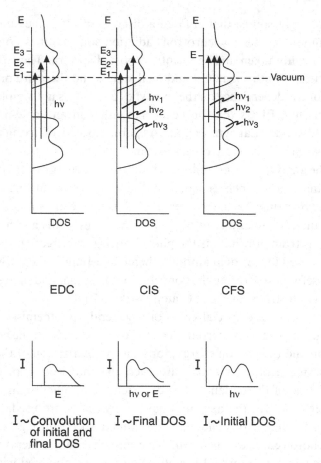

Fig. 3.6 Photoexcitation process in the three types of photoelectron spectra. In the EDC (*left*), monoenergetic photons excite a spectrum of emitted electrons. Its spectral shape is related to the density of states of both initial and final states. In the CIS (*center*) the photon energy and the energy analyzer are scanned in synchronism. The spectral shape is related to the density of final states. In the CFS (*right*), only the photon energy is scanned with a fixed analyzer energy. The spectral shape is related to the density of initial states.

and slope. There is no such resemblance at lower photon energies at which the most of the primary excitations are from the valence band. Yield spectra are easy to obtain during photoelectron spectroscopy of core levels. A different depth is sampled, however; yield spectra are not as surface sensitive as other photoelectron spectra. The photon penetration depth, α^{-1}, is typically a few hundred Å. Photoelectrons with a large spread of energies are produced in a depth roughly three times this large. Those headed for the surface produce many secondaries, possibly energetic ones, leading to many secondaries for each primary (hence to a large measured current). Some of the secondaries have too little energy to escape. Studies of several prototypical solids have shown that the

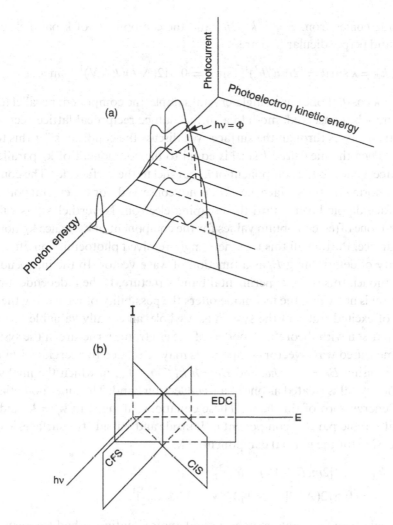

Fig. 3.7 (a) Schematic of the construction of count rate–photon energy–electron energy surface by taking a series of EDCs; (b) shows how the three common types of scans can be obtained by intersections of this surface with particular planes.

depth sampled is roughly 100 Å (Gudat and Kunz, 1972; Kunz, 1979). This should depend somewhat on the sample material.

3.3   Angle-resolved photoemission

If the electron energy analyzer accepts only a small solid angle and it is oriented at angle θ (now measured outside the crystal) to the surface normal, components of the electron's momentum or wave vector can be obtained. The wave-vector magnitude comes from $E = \hbar^2 k^2/2m$, and the components of **k** parallel to the surface and perpendicular to it are

$$k_\parallel = k \sin\theta = (2mE/\hbar^2)^{1/2} \sin\theta = 0.512(\text{Å}^{-1})(E/\text{eV})^{1/2} \sin\theta,$$

and $k_\perp = k \cos\theta$. If one has a single-crystal sample, the component parallel to the surface may be conserved, modulo a surface lattice reciprocal lattice vector, as the electron passes through the surface. We discuss the conditions for this to be so later. Then the measured $k \sin\theta$ is equal to the component of $\mathbf{k}_f$ parallel to the surface, hence to the component of $\mathbf{k}_i$ parallel to the surface, $\mathbf{k}_\parallel$. The component of **k** normal to the surface, i.e., $k_\perp$, is not conserved, for the crystal potential and surface dipole layer retard the escaping electron. Nevertheless, as will be described, one often can obtain values for the component of $\mathbf{k}_f$, hence $\mathbf{k}_i$, normal to the surface. Putting all this together, angle-resolved photoemission offers the possibility of determining E_i as a function of wave vector. In the independent-particle model, this is the experimental band structure. If the independent-electron model is not valid, the technique offers the possibility of measuring the dispersion of excited states of the system as a whole, an equally valuable quantity for comparison with theoretical models of the electronic structure of the system.

The measured wave-vector components may be related to energies within the crystal by using $E_f = E_i + \hbar\omega$ and $E_f = \hbar^2 k_f^2/2m - V_0$, in which the final state inside the crystal is treated as one in a free-electron band. The inner potential V_0 sets the energy zero of this band. These equations, along with $\mathbf{k}_f = \mathbf{k}_i$ and the continuity of the parallel component of **k** inside and outside the surface, lead to an expression for the normal component of **k**:

$$k_\perp = \hbar^{-1}[2m(E + V_0) - \hbar^2 k_\parallel^2]^{1/2}$$
$$= 0.512(\text{Å}^{-1})[(E + V_0)/\text{eV} - 3.84\text{Å}^2 k_\parallel^2]^{1/2}.$$

This free-electron final state may be a good approximation at high energies, but at lower energies the crystal potential will give a different dispersion for this band and introduce gaps. This will be discussed below.

Angle-resolved photoemission is handled in the three-step model by the insertion of an additional delta function, $\delta(\mathbf{k}_f/k_f - \widehat{\mathbf{k}})$, in $P(E_f, \omega)$, where $\widehat{\mathbf{k}}$ is a unit vector in the direction (θ, ϕ), specifying the detection angle. θ and ϕ are the polar and azimuthal angles of the outgoing electron measured inside the crystal. θ must be corrected for refraction at the surface ($k_\perp$ changes because of the effect of the work function) to get the external polar angle of the analyzer, unless $\theta = 0$. Figure 3.8 gives plots of several EDCs taken at either several angles from the surface normal, or for normal emission ($\theta = 0$), for several photon energies, in which case $\widehat{\mathbf{k}}$ is directed normal to the surface.

Angle-resolved photoemission also offers the possibility of determining wave-function symmetry as was explained by Hermanson (1977) and Jacobi *et al.*

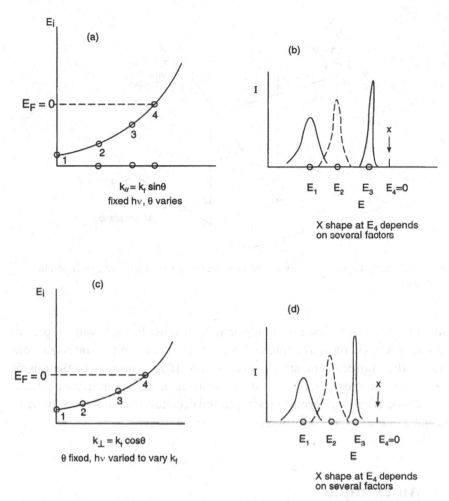

Fig. 3.8 Idealized angle-resolved photoelectron spectra from a valence band. In (a) four initial states are marked on a simple dispersing band, where the wave vector is parallel to the surface. By scanning θ, the angle at which electrons are collected, at fixed photon energy, photoelectrons from each of these states can be detected. (In practice the photon energy would also be varied to keep $k_\perp$ constant.) The four resultant spectra are shown superimposed in (b). The shape and area of spectrum 4 will depend on instrumental factors. If the electrons are collected only near the surface normal, spectra may be collected for various values of $k_\perp$ by varying the photon energy as shown in (c) and (d).

(1977). Consider the situation in which the angle-resolving detector is located in a mirror plane of the crystal, as sketched in Fig. 3.9. All initial and final states must be even or odd with respect to reflection in this plane. If an outgoing electron wave function were to have odd reflection symmetry, it would have a node in the plane, hence on the detector axis. To the extent that the angular acceptance of the analyzer is small, such photoelectrons will not be detected. Only even final states will be detected. Their contribution to the photocurrent is propor-

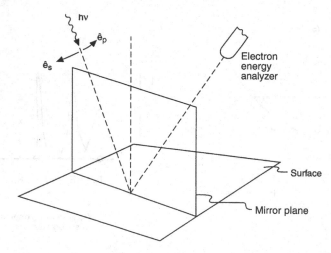

Fig. 3.9 Geometry for angle-resolved measurements with photoelectron detection in a mirror plane.

tional to $|\langle f|\mathbf{A} \cdot \mathbf{p}|i\rangle|^2$. The matrix element itself must be even with respect to reflection. If **A** lies in the mirror plane, it is even, so the only initial states that contribute to the photocurrent are the even states. If **A** is normal to the mirror plane, it is odd, and only odd initial states contribute. By changing the polarization, usually carried out by rotating sample and detector, the reflection symmetry of the initial states may be identified.

3.4   An example

Before discussing further the three-step model and examining extensions to it, and describing more sophisticated treatments of photoemission, we give an application of it to a real system which can be worked out analytically– photoemission from an alkali metal, first analyzed by Koyama and Smith (1970). A monovalent alkali metal has a simple band structure quite well described at low energies by applying a single Fourier coefficient of the pseudopotential, V_{110}, to the empty-lattice bands for the body-centered cubic structure. The energy eigenvalues and wave functions can be worked out by perturbation theory. The Fermi level is in the middle of the lowest band, far from the Brillouin zone surfaces, and the Fermi surface is rather accurately represented by a sphere, except for Li. The band gap at the nearest point on the zone boundary is $2|V_{110}|$, about 0.4 eV for Na. This point is along a [110] direction, at a wave vector of 1.414 in units of $2\pi/a$, while the Fermi wave vector is only 0.62 in the same units. The gap at the Brillouin zone boundary caused by the potential is small compared to the free-electron kinetic energy there, $\hbar^2 G^2/8m$, about 4.1 eV for Na. Thirteen bands are shown with solid lines in Fig. 3.10. These come from the empty-lattice bands (periodicity retained, but the potential set equal to zero) involving the twelve equivalent $\langle 110 \rangle$ reciprocal lattice vectors and $\mathbf{G} = 0$. This is all that is needed at lower energies, but bands produced by the six equivalent $\langle 200 \rangle$ reciprocal lattice vectors overlap those in the upper regions of Fig. 3.10. Some of these are shown as dashed curves. Vertical, $\mathbf{k}$-conserving transitions occur only for $\mathbf{k}_i < \mathbf{k}_F$, and in this region, the bands are accurately parabolic, and the wave functions are each a single plane wave with small admixtures of certain other plane waves. It is this admixture, proportional to V_{110}, that allows non-zero dipole matrix elements. Thus the energy bands, wave functions, and the geometry of $\mathbf{k}$-space are all easily expressed analytically, leading to straightforward evaluation of integrals.

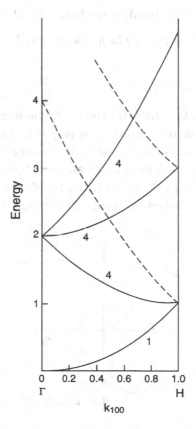

Fig. 3.10 Free-electron bands for a b.c.c. crystal. The energy is in units of $\hbar^2\pi^2/ma^2$ and momentum along the [100] direction in units of π/a. The twelve equivalent $\langle 110 \rangle$ reciprocal lattice vectors and $\mathbf{G} = 0$ give the solid bands. Larger values of $\mathbf{G}$ give the dashed bands. The numbers denote degeneracies.

The wave functions are

$$\varphi_k(\mathbf{r}) = V^{-1/2}\left[\exp(i\mathbf{k}\cdot\mathbf{r}) + (2m/\hbar^2)\sum_G \frac{V_G\exp(-i\mathbf{G}\cdot\mathbf{r})}{k^2 - (\mathbf{k}-\mathbf{G})^2}\right],$$

where

$$V_G = \int V(\mathbf{r})\exp(i\mathbf{G}\cdot\mathbf{r})d^3r,$$

and the $\mathbf{G}$ are the twelve $\langle 110\rangle$ reciprocal lattice vectors.

The dipole matrix elements become

$$\langle \mathbf{k}'|\mathbf{p}|\mathbf{k}\rangle = 0,$$

unless $\mathbf{k}' = \mathbf{k} - \mathbf{G}$, in which case

$$\langle \mathbf{k}'|\mathbf{p}|\mathbf{k}\rangle = \frac{-(2m/\hbar)V_G\mathbf{G}}{2\mathbf{G}\cdot\mathbf{k} - G^2}.$$

The surface of constant interband energy in reciprocal space is given by

$$\hbar\omega = E(\mathbf{k}-\mathbf{G}) - E(\mathbf{k}) = (\hbar^2/2m)[-2\mathbf{k}\cdot\mathbf{G} + G^2]$$

or

$$2\mathbf{k}\cdot\mathbf{G} = G^2 - (2m/\hbar^2)\hbar\omega.$$

All wave vectors with $k < k_F$ satisfy this and can be photoexcited at some ω. This equation describes a plane in k-space that is perpendicular to $\mathbf{G}$. If $\hbar\omega = 0$ the plane lies on the bisector of $\mathbf{G}$, i.e., at the Brillouin zone boundary. It does not intersect the Fermi sphere, so all points on it are unoccupied, and there is no contribution to the integral for ϵ_2. As $\hbar\omega$ increases the plane moves toward the origin, as shown in Fig. 3.11. When $\hbar\omega$ is such that it satifies the last equation for

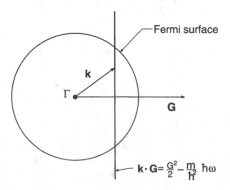

Fig. 3.11 Geometry for direct optical excitations between free-electron bands in the b.c.c. lattice. Transitions from any state inside the Fermi sphere occur with a phase space proportional to the area of the circle formed by the intersection of the Fermi sphere and the plane. As the photon energy increases the plane sweeps to the left.

ϵ_2 with $k = k_F$, absorption starts at $\hbar\omega_{min} = (\hbar^2/2m)G(G - 2k_F)$, 2.16 eV in Na. Larger $\hbar\omega$ sweeps the plane leftward and the area of integration increases, then decreases till it goes to zero at $\hbar\omega_{max} = (\hbar^2/2m)G(G + 2k_F)$, 31 eV in Na, when the plane passes through the further pole of the Fermi sphere. In the integral, the integration is over a disk of area $S = \pi\{k_F^2 - [G/2 - (m/\hbar^2)(\hbar\omega/G)]^2\}$, as in Fig. 3.11. On this disk, the dipole matrix element, given above, is constant, and $|\nabla_k(E_f - E_i)|$ is constant as well: $\hbar^2 G/m$. Putting all these together, for one **G** ϵ_2 can be written as

$$\epsilon_2(\omega) = (\hbar^2 e^2/\pi m^2)\omega^{-2}(m/\hbar^2 G)(4m^2/\hbar^2)(G^2|V_G|^2)\hbar^4/[4m^2(\hbar\omega)^2]$$
$$\times \pi[k_F^2 - (G/2 - m\hbar\omega/\hbar^2 G)^2]$$
$$= (4me^2|V_G|^2/G)(\hbar\omega - \hbar\omega_{min})(\hbar\omega_{max} - \hbar\omega)/(\hbar\omega)^4.$$

This must be multiplied by 12, for the twelve equivalent reciprocal lattice vectors and by $1/3$ which is the average of $\cos^2(\mathbf{a}, \mathbf{p}_{if}) = \cos^2(\mathbf{a}, \mathbf{G})$, appropriate if we have a polycrystal (but the result will be the same for a cubic single crystal). This was first worked out by Wilson (1936) and by Butcher (1951).

For photoemission we need $P(E_f, \omega)$, and this requires an integration over a curve which is the intersection of two surfaces of constant energy, one with $E_i < E_F$, the other with $E_f = E_i + \hbar\omega > E_F$. These are shown in Fig. 3.12. These surfaces will not intersect if drawn concentrically, but the lower energy sphere is associated with the point Γ, the origin in k-space, while the other derives from a parabolic energy band centered at a point **G** away from Γ, and must be drawn centered on this point. The intersection of the two is a circle. Also from the figure

$$|\nabla_k E_f - \nabla_k E_i| = 4(\hbar^2/2m)|\mathbf{k} \times \mathbf{G}|.$$

The integral for $P(E_f, \omega)$ can be done, again because the integrand is a constant. The result is

$$P(E_f, \omega) = [e^2/\pi m^2 \omega \epsilon_2(\omega)](|V_G|^2|\mathbf{a} \cdot \mathbf{G}|^2)$$
$$= \omega|\mathbf{a} \cdot \mathbf{G}|^2/[4\pi m(\hbar\omega - \hbar\omega_{min})(\hbar\omega_{max} - \hbar\omega)],$$

for $E_{min} + \hbar\omega < E_f < E_F + \hbar\omega$, the condition for intersection of the spheres, and

$$P(E_f, \omega) = 0$$

for E_f outside this range.

$$E_{min} = (\hbar\omega - \hbar^2 G^2/2m)^2/(4\hbar^2 G^2/2m).$$

This expression for $P(E_f, \omega)$, too, must be multiplied by $12/3 = 4$. This function sometimes is called the EDC, but its shape will be altered by energy-dependent transport and escape effects. For a given photon energy this function is a constant, whose magnitude and range in E_f for non-zero values depend on ω. Both $P(E_f, \omega)$ and $P(E_f, \omega)T_{eff}(E_f)$ are shown in Fig. 3.13. The latter is propor-

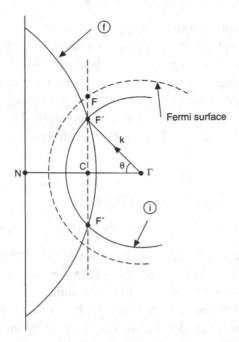

Fig. 3.12 Geometry for direct optical excitations between free-electron bands in the b.c.c. lattice but with a final state of fixed energy. The surfaces of constant initial-state energy and constant final-state energy are drawn centered at Γ and N, respectively, points separated by **G**. The straight dashed line is the intersection of the plane of constant interband energy. Direct optical transitions occur only for those states on a circular disk whose diameter is F′–C–F′. Those photoemitted with energy E_f lie on the circumference of this disk. (Koyama and Smith, 1970)

tional to the emitted angle-integrated photoelectron spectrum. Note that neither spectrum in Fig. 3.13 resembles the filled density of states for a simple metal like an alkali metal, proportional to $(E_f)^{1/2}$ and cut off at the Fermi level. In order to evaluate T_{eff}, an empirical energy-dependent inelastic mean free path of $(3\,\text{eV}/E_f)^{3/2} \times 100$ Å was used. This approximates the mean free path in Fig. 3.4. An energy-independent absorption coefficient of 500 Å^{-1} was used, although its spectrum could be calculated from the same model. These lead to values of $\alpha\Lambda \gg 1$ over the range of E_f values obtained with 5–25 eV photons. Finally, a work function of 2.7 eV was used.

The nearly-free electron model can be used to describe angle-resolved photoemission. This was first done by Mahan (1970) using a one-step formalism. In the extended-zone scheme, the electron is excited from a state of wave vector $\mathbf{k}_i$ to one of $\mathbf{k}_f = \mathbf{k}_i + \mathbf{G}$. Subtracting **G** from both sides gives $\mathbf{k}_i = \mathbf{k}_f - \mathbf{G}$. Conservation of energy gives

$$(\hbar^2/2m)k_f^2 - (\hbar^2/2m)(\mathbf{k}_f - \mathbf{G}) \cdot (\mathbf{k}_f - \mathbf{G}) = \hbar\omega.$$

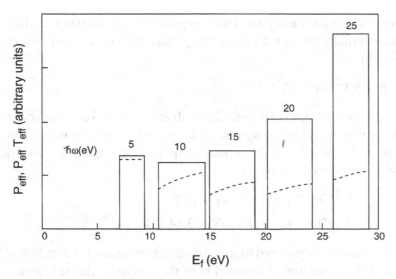

Fig. 3.13 Calculated angle-integrated photoelectron spectra for the nearly-free electron model with parameters typical of Na. The solid lines are the photoexcited spectra, $P_{eff}(E_f, \omega)$, for the photon energies indicated, with E_f the final-state electron energy inside the crystal measured with respect to the bottom of the conduction band. The dashed lines show the effect of T_{eff}, accounting for transport and escape through the surface.

Solving for k_f and evaluating $E_f = \hbar^2 k_f^2/(2m)$ gives

$$E_f = [\hbar\omega + \hbar^2 G^2/(2m)]^2/[(2\hbar^2 G^2/m)\cos^2\theta],$$

where θ is the angle between $\mathbf{k}_f$ and $\mathbf{G}$. This indicates that with a fixed photon energy and a fixed analyzer energy, θ is a constant and $\mathbf{k}_f$ lies on a cone with $\mathbf{G}$ as its axis. There are several $\mathbf{G}$ vectors equivalent by symmetry, although not all of them have a component directed toward the surface. Thus, just inside the surface, the photoelectrons of a given energy are directed along the surfaces of several cones. For a fixed photon energy, as the final state energy increases, the cone half-angle gets larger. The cones of electrons are distorted upon emergence from the surface by the refractive effect caused by the change in potential at the surface on the normal component of the wave vector.

We can sketch where these electrons are to be found in the nearly-free electron model we have been using. This should be quite accurate for alkali metals, where the Fermi surface is rather accurately a sphere and no states of concern for valence-band photoemission are near a Brillouin zone boundary. For more complicated electronic structures, it is sometimes assumed that the final states are free electron-like even if the initial states are not. We return to this complication shortly. For a free-electron final state as we have assumed in the preceding section we must know or assume the energy of the state at the bottom of the band of initial states. This is $E_F + \Phi$ below the vacuum level, the zero for energy measurements outside the crystal. This energy is the inner potential, V_0. (See Fig. 3.1.) The externally measured kinetic energy and the internal final state energy are related by

$$E = \hbar^2 k^2 / 2m = (\hbar^2 k_f^2 / 2m) - V_0.$$

Since E is measured, k is known. The analyzer angle, θ, from the surface normal is known, so $\mathbf{k}_\parallel$ is known. $\mathbf{k}_f = \mathbf{k}_\perp + \mathbf{k}_\parallel + \mathbf{g}$, where $\mathbf{g}$ is a surface reciprocal lattice vector. Solving for $k_\perp$, the normal component of $\mathbf{k}$ inside the crystal before and after photoexcitation, one gets

$$k_\perp = \hbar^{-1}[2m(E + V_0) - \hbar^2(\mathbf{k}_\parallel + \mathbf{g})^2]^{1/2}$$
$$= 0.512(\text{Å}^{-1})[(E + V_0)/\text{eV} - 3.84\,\text{Å}^2(\mathbf{k}_\parallel + \mathbf{g})^2]^{1/2}.$$

$\mathbf{k}_\parallel$ is $k \sin\theta$, where θ is measured from the surface normal and the direction of $\mathbf{k}_\parallel$ is determined by the azimuthal orientation of the analyzer. Allowed values of $k_\perp$ and $\mathbf{k}_\parallel$ are limited by energy conservation to fall on a circle in a plot of $k_\perp$ vs. $\mathbf{k}_\parallel$, as in Fig. 3.14, where only one quadrant is shown. Once θ is known, the values of $k_\perp$ and $\mathbf{k}_\parallel$ are limited to curves on this plot given by the above equation. If $\mathbf{g} = 0$, these curves all begin at $\mathbf{k}_\parallel = 0$, and $k_\perp = 0.512(\text{Å}^{-1})(V_0/\text{eV})^{1/2}$, as shown. At higher energies they asymptotically approach radii at angle θ in the plot. The intersection of one of these curves with an arc of the measured energy gives $k_\perp$ and $\mathbf{k}_\parallel$. These electrons are associated with a reciprocal lattice vector $\mathbf{G}$, and lie on one of the cones about it mentioned above. In the example, this is a $\mathbf{G}$ normal to the surface. There are similar cones about other equivalent vectors $\mathbf{G}$. It is assumed here that the cones do not overlap. If they do, several vectors $\mathbf{G}$ have to be treated simultaneously (Mahan, 1970).

A common experimental arrangement is to select photoelectrons emitted only along the surface normal. Then $\mathbf{k}_\parallel = 0$, and $k_\perp$ is simply $0.512(\text{Å}^{-1})[(E + V_0)/\text{eV}]^{1/2}$ if $\mathbf{g} = 0$. The surface is often chosen to be perpendicular to one of the shortest reciprocal lattice vectors $\mathbf{G}$. If $\mathbf{g}$ is not zero, a similar plot results but the origin of the curves is offset to $\mathbf{k}_\parallel = \mathbf{g}$ and $k_\perp = 0.512(\text{Å}^{-1})[V_0/\text{eV} - 3.84g^2\,\text{Å}^2]^{1/2}$, shown dashed in Fig. 3.14. For this example, $\mathbf{g}$ has been taken to be $(\pi/a)(\mathbf{i} + \sqrt{2}\mathbf{j})$, a reciprocal lattice vector for the (110) surface of b.c.c. Na. The direction of $\mathbf{k}_\parallel$ for the plot in Fig. 3.14 is then the direction of $\mathbf{g}$. This emission may be associated with a reciprocal lattice vector $\mathbf{G}$, and may be said to occur on a secondary cone about $\mathbf{G}$. (The three-step model has not been used to address the question of the relative strengths of secondary cone emission.) The surface reciprocal lattice vectors $\mathbf{g}$ need not be the same as the bulk reciprocal lattice vectors $\mathbf{G}$. For the b.c.c. lattice, the smallest non-zero reciprocal-lattice vectors are the twelve $\langle 110 \rangle$ vectors. The surface of the single-crystal sample must be specified before the $\mathbf{g}$s are known, and reconstruction of that surface can change the $\mathbf{g}$ vectors. For the (110) (closest-packed) surface of a

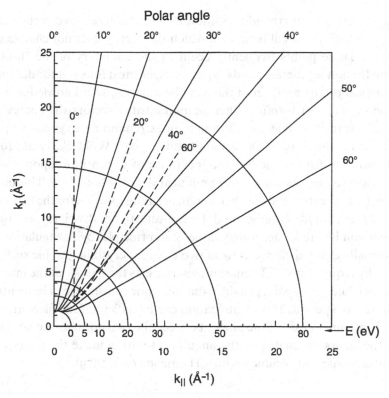

Fig. 3.14 Plot for determining **k** inside the crystal for photoemission from a (110) surface of Na, assuming free-electron final states. The circular arcs relate $k_\perp$ to $k_\parallel$. The other solid curves relate the measured kinetic energy and angle to $k_\perp$ and $k_\parallel$. The intersection of the circular arc for the measured energy with the curve for the known analyzer polar angle determines **k**. The dashed set of curves are for emission via a surface umklapp process

b.c.c. crystal, the simplest reciprocal lattice vectors for the primitive unit cell are of the form $(\pi/a)(\mathbf{i} \pm \sqrt{2}\mathbf{j})$. The vectors **G**, instead, have an effect on the electronic structure, the dielectric function and EDC in angle-integrated photoelectron spectroscopy, as well as other properties.

Angle-resolved photoemission becomes more complicated when applied to anything other than monovalent alkali metals or nearly two-dimensional materials. For the former, the assumption of free-electron final states is a good one, since no transitions take place near the zone boundary and $k_\perp$ can be obtained reliably from the free-electron bands. For the latter, $k_\perp$ is not important. In all other cases a way to determine $k_\perp$ must be found. Most methods involve some prior knowledge of the band structure. One method is to assume the calculated bands are reasonably close to reality. They may be off by a few eV, especially well above E_F, and the gaps at the zone boundary may be off by 20% or so. One can track a particular interband transition as a function of wave vector, especially $k_\perp$, and look for small regions in which the energy does not change with wave vector. These points frequently occur at the boundary of the Brillouin zone where for nondegenerate bands, $\nabla_k E = 0$. One must look out for degeneracies and critical points away from the zone boundaries, so a knowledge of the band structure is most helpful. When scanning wave vectors the energy and intensity of a transition may pass through an extremum at a symmetry point (Himpsel and Eastman, 1978, Dietz and Eastman, 1978; Wöhlecke *et al.*, 1984). One can also look for the sudden disappearance of a transition upon a small change in wave vector, one known to be not near the zone boundary. This occurs at the Fermi wave vector, and one then has an accurate location for the transition (Eastman *et al.*, 1978; Himpsel and Eberhardt, 1979; Rosei *et al.*, 1980). Finally, one can locate $k_\perp$ accurately, at least in principle, by triangulation. By taking normally emitted electrons from two different crystallographic surfaces, which usually requires two different samples, one gets the true $k_\perp$ at the intersection of the two lines in **k**-space, provided that the same transition can be identified in the two sets of spectra. This identification can be difficult if the faces are very different. It becomes easier if the angles between the two normals are not large, but then the errors on angle and the angular resolution make the intersection volume rather large and accuracy is lost (Heimann *et al.*, 1979).

3.5   Photoelectron and photohole lifetimes

The outgoing photoelectron may scatter inelastically. In the three-step model described above electron–electron scattering was considered the principal inelastic scattering process and we continue that view here. The finite lifetime τ_f for survival without inelastic scattering gives E_f a non-zero uncertainty or energy width, taken to be $\Delta E_f = \Gamma_f = \hbar/\tau_f$. In a many-body picture, Γ_f is equal to twice the imaginary part of the self-energy, Σ_2, for state f. If this were the only broadening effect and Σ_2 were approximately constant, the photoelectron "line" would be a Lorentzian of width Γ_f. Because of the large phase space for scattering when E_f is large, the photoelectron width can become large, several eV at 20 eV or so. The same processes scatter the photohole, so E_i also has a width, Γ_i. If the photohole is near E_F, Γ_i is small and its energy dependence would give information on the phase space available for scattering, hence on the possible quasiparticle nature of the hole. It is thus useful to try to extract Γ_i from measurements of the observed width Γ. In angle-integrated photoemission too many states contribute to the spectrum in the integral over the Brillouin zone and the effects of these widths are obscured. Valence-band lifetime effects are therefore studied only in angle-resolved photoelectron spectroscopy.

Expressions for the effect of the lifetime widths on the observed width have been stated or derived in the literature a number of times. Not all forms agree. We refer the reader to the paper of Smith et al. (1993) for references to previous work. Even that clear summary did not end controversy over how one should extract lifetimes from experimental values of Γ (Starnberg et al., 1993b; Hwu et al., 1993; McLean et al., 1994; Starnberg, 1995; McLean, 1995).

Refer to Fig. 3.15 which depicts states involved in angle-resolved photoemission with the electrons collected normal to the surface. The wave vector is normal to the surface and will be written as a scalar. E_f is scanned by scanning the photoelectron kinetic energy. (E_f is the final-state energy inside the crystal. Over the limited wave vector range of Fig. 3.15 it is assumed to vary linearly with wave vector.) The creation of an internal photoelectron with wave vector $\mathbf{k}_f$ and energy $E_f(\mathbf{k}_f)$ would give a delta-function peak in the spectrum to be measured. A finite energy width Γ_f causes a broader line, and it also leads to an apparent uncertainty in $\mathbf{k}_f$ of $\Delta\mathbf{k}_f = \Gamma_f/|\nabla_k E_f|$, where only the normal components of $\nabla_k E_f$ and $\mathbf{k}_f$ are used. ($\nabla_k E$ is $\hbar^{-1}$ times the group velocity of the outgoing electron.) The distribution of final-state energies around $E_f(\mathbf{k}_f)$ would then be approximately a Lorentzian:

$$I(E_f) = \int \{[E_f - E_f(\mathbf{k}_f) - \nabla_k E_f \cdot (\mathbf{k} - \mathbf{k}_f)]^2 + (\Gamma_f/2)^2\}^{-1} dk,$$

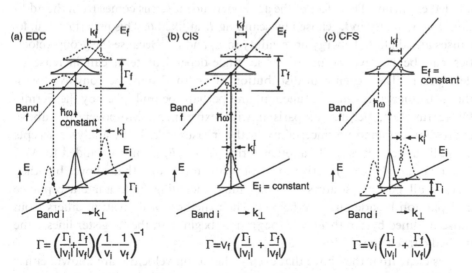

Fig. 3.15 k-space geometry for the derivation of lifetime-induced line widths in three modes of photoelectron spectroscopy: (a) energy distribution curve, (b) constant initial state, and (c) constant final state. The total width Γ (in energy units) is related to the lifetime widths Γ_i and Γ_f of the photohole and photoelectron, respectively, and the normal components of their group velocities as indicated. Expressions for Γ are for normal emission. (Smith et al., 1993)

where only the normal components of $\nabla_k E$, $\mathbf{k}$, and $\mathbf{k}_f$ are used. The finite lifetime allows a spread in values of $\mathbf{k}_f$, as shown in Fig. 3.15. The initial state also has a finite lifetime, hence a width Γ_i, uncorrelated with Γ_f. If the initial-state band is assumed to vary linearly with wave vector near $\mathbf{k}_i$, then

$$I(E_f) = \int \{[E_f - \hbar\omega - E_i(\mathbf{k}_f) - \nabla_k E_i \cdot (\mathbf{k} - \mathbf{k}_f)]^2 + (\Gamma_i/2)^2\}^{-1} dk,$$

where $\mathbf{k}_i = \mathbf{k}_f$. Each effect gives an independent Lorentzian distribution in energy. At a fixed E_f, the energy analyzer collects all values of $\mathbf{k}$, i.e., $k_\perp$, so the detected spectrum is the integral of the product of the above two Lorentzians with respect to $\mathbf{k}$. The final distribution in E_i is Lorentzian with width Γ:

$$\Gamma = \left[\Gamma_i + \Gamma_f \left| \frac{\nabla_k E_i}{\nabla_k E_f} \right| \right] = \frac{\Gamma_i/|v_i| + \Gamma_f/|v_f|}{1/|v_i| + 1/|v_f|},$$

where the group velocities v are only the components normal to the surface.

When the electrons are collected off-normal, the expression is somewhat more complicated. It has also been controversial (Grepstad *et al.*, 1982; Smith *et al.*, 1993)

$$\Gamma = \frac{\Gamma_i/|v_i| + \Gamma_f/|v_f|}{|(1/v_i)[1 - (mv_i/\hbar k)\sin^2\theta] + (1/v_f)[1 - mv_f/\hbar k)\sin^2\theta]|}.$$

In this, the parallel components of velocity and wave vector have been labeled explicitly. The effect of the angle-dependent terms comes from the additional uncertainty in $\mathbf{k}_\parallel$ caused by scanning E at fixed θ. This uncertainty in $\mathbf{k}_\parallel$ causes an additional energy uncertainty in E_i and E_f. Because the group velocities can be positive or negative the angle-dependent term can increase or decrease Γ. The Lorentzian distribution of width Γ must be convolved with the instrumental resolution function, and perhaps be multiplied by the Fermi–Dirac function, before comparison with experiment. Moreover, the above expression assumed no uncertainty in the measured $\mathbf{k}$. If the analyzer accepts a small range of $\mathbf{k}$, say $d\mathbf{k}$, another term, $|d\mathbf{k} \cdot \nabla_k E_i|$, must be added to ΔE_i. Fraxedas *et al.* (1990) derive an expression for Γ like that above, but they include all possible sources of broadening, including finite angle acceptance and phonon broadening (see below). Their method of derivation differs from those outlined by Smith *et al.* (1993); they begin with the "register lines," the dashed curved lines in Fig. 3.14.

It is clear from the above that even if the group velocities are known, either from calculated energy bands or from E vs.k measured by angle-resolved photoemission, a measurement of Γ cannot give either Γ_i or Γ_f individually. In special cases, only one of the two appears in the result. Obviously Γ_i is small near the Fermi level, and Γ_f dominates Γ, but it is Γ_i one may want to measure. If $|v_f|$ is sufficiently large, $\Gamma_i/|v_i|$ may dominate Γ and may be determined. In nearly two-dimensional materials, such as many cuprates, both inital- and final-state bands are fairly flat along the k_z-axis, and the normal component of both group velocities is small for a sample with the c-axis normal to the sample surface, the usual sample geometry. The final state is 15–20 eV above E_F, and more free electron-like than the initial state. It may then be that the photohole lifetime plays the major role in Γ, but one must be careful. Γ_i may be 50 meV or so, while Γ_f may be over 1 eV. The ratio of group velocities must then be 100 or so to make the Γ_f term negligible.

Smith *et al.* (1993) also derived expressions for the widths Γ_{CIS} and Γ_{CFS} to be found in the CIS and CFS modes, although angle-resolved measurements are infrequently carried out in these modes. For normal emission they find

$$\Gamma_{\mathrm{CIS}} = |v_f|(\Gamma_i/|v_i| + \Gamma_f/|v_f|),$$
$$\Gamma_{\mathrm{CFS}} = |v_i|(\Gamma_i/|v_i| + \Gamma_f/|v_f|).$$

They also generalized these to the off-normal emission case.

When the final-state energy is high, Γ_f can be large, 1 eV and more. At high energies, the bands in the reduced zone scheme become closely spaced in energy, and at some energy their spacing becomes of the order of Γ, so they become effectively a continuum when the width Γ_f is considered. At this point, the details of the final states become unimportant and the final states may be treated as a continuum. Even in angle-resolved photoemission, for any photon energy large enough, a direct transition from any initial state becomes allowed by the large smearing effect of the close spacing of final-state bands and their large lifetime broadening. This is the XPS (x-ray-induced photoelectron spectroscopy) region, and the EDC resembles more and more the density of states, still distorted somewhat by dipole matrix elements. EDCs taken at two photon energies differing by several, or tens of, eV, should be nearly the same. For a given material, however, it is not known *a priori* how large an energy is needed to reach this region. For example, in Au, the XPS region for valence electrons appeared to have been reached for photon energies above about 80 eV in a study extending up to 90 eV (Freeouf *et al.*, 1973). However, a later study extending to 275 eV showed the relative heights of the two features from the 5d bands varied in the 100–200 eV region (Lindau *et al.*, 1976).

The XPS region may be reached another way, by increasing the sample temperature. At increased temperatures, the amplitudes of lattice vibrations increase, making the crystal less periodic, thereby making the wave vector less precisely defined. This was discussed by Shevchik (1977) using the three-step model. The initial state was a tight-binding state and the final state an augmented plane wave. The square of the dipole matrix element was written as the square of an atomic matrix element multiplied by a structure factor. In the Debye approximation, the latter consists of two terms. One, representing direct, $\mathbf{k}$-conserving transitions, has a Debye–Waller factor, $\exp(-k_f^2 U_0^2)$ which decreases exponentially with increasing temperature because U_0 increases. The other, representing indirect transitions in which a phonon contributes wave vector conservation, increases with temperature as $[1 - \exp(-k_f^2 U_0^2)]$. U_0 is the atomic vibrational amplitude, proportional to $T^{1/2}$ when equipartition is valid. The XPS region is reached when the latter term dominates the former; the spectrum resembles a joint density of states weighted by matrix element effects. For low-energy (ultraviolet) photoemission, $\mathbf{k}_f$ is rather small and the XPS regime is reached only at high temperatures, while for x-ray-induced photoemission, $\mathbf{k}_f$ is larger, and the XPS region may be reached at lower temperature, including room temperature. A more extensive discussion of the effects of temperature on photoelectron spectra has been given by White *et al.* (1987) and by Vicente Alvarez *et al.* (1996). We shall return to this topic in Chapter 5, when discussing Cu to illustrate typical results from photoelectron spectroscopy.

White *et al.* pointed out that the motion of neighboring atoms was correlated with that of the photoexcited atom, so that phonon-induced indirect transitions should occur in a region of reciprocal space around the point of the direct transition, and the indirect spectrum should be nearly as structured as the direct one. Starnberg *et al.* (1993a) made a study of the temperature dependence of angle-resolved photoemission from VSe_2, a layered metallic compound. Spectra taken at 296 K were subtracted from spectra taken at 393 K to obtain the temperature-dependent contributions to the spectra. These were almost as structured as the 296 K spectra, although the peaks were broader, and they were strongly angle dependent. They contributed about half of the total spectra at room temperature. The authors concluded that the loss of the direct spectrum was described adequately by the Debye–Waller factor in this temperature range, and that the contribution of the indirect spectrum increased with temperature as described in the preceding paragraph.

As mentioned earlier, if the resolution of the spectrometer is high enough, scattering by the more energetic phonons must be viewed as inelastic, and electron–phonon scattering rates must be added to the others. Thus Γ_i and Γ_f will increase, and the increase due to scattering by phonons will be temperature dependent. Fraxedas *et al.* (1990) studied the phonon contribution to lifetime widths in the semiconductor InSb. They found Γ_f to increase with temperature by 0.4 ± 0.1 meV/K. Bulk initial-state (valence-band) thermal broadening due to phonons was negligible on this scale, as was broadening of the In 4d core level, but some surface states had comparable thermal broadening. Their paper contains an extensive analysis of broadening using a model different from that used earlier in this section. Several studies of lifetimes and of the phonon contribution to broadening in Cu will be described in Chapter 5.

Another issue is the change in lifetime width for states near the Fermi level due to the opening of a gap when a material goes superconducting. When a gap opens in a superconductor, even though it may not be a complete one, there is a rearrangement of the excitation spectrum for quasiparticles around the Fermi energy. A photohole created near E_F then has a different phase space for scattering. Γ_i should become smaller near E_F, since the quasiparticle excitation spectrum will be smaller near E_F in the superconducting phase. For slightly deeper initial states, Γ_i could become anomalously high. All this should occur on an energy scale of the order of the maximum gap energy Δ. Coffey and Coffey (1993a,b, 1996) calculated energy-dependent spectra for Γ_i for an s-wave and a d-wave superconductor. (The results depend not only on the symmetry of Δ but also on the electron–hole interaction assumed, the temperature, and the wave vector.) A fairly obvious anomaly in the superconductor–insulator–superconductor tunneling current between 3Δ and 4Δ, with a threshold at about 2Δ, was calculated for an assumed Δ of $d_{x^2-y^2}$ symmetry, and such a feature had been seen in tunneling studies of several different cuprates. The anomaly calculated for the photoelectron spectrum and the superconductor–insulator–normal tunneling current is much weaker, with small dips at about 3Δ. Actual spectra for $Bi_2Sr_2CaCu_2O_8$ often show a dip at about 3Δ. These spectra will be discussed in a later chapter.

The foregoing may give the impression that the line shapes in angle-resolved photoemission, although difficult to measure reliably, are reasonably well understood. This is not really the case. Miller et al. (1996) reported an asymmetric line shape for a photoelectron peak for a transition that should be simple. (This peak had been observed previously numerous times, but the line shape was either ignored or not explained.) The transition was between s–p bands in Ag, measured at normal emission with 7–9 eV photons on the (111) surface. The peak was asymmetric, with a tail toward lower binding energy, the opposite sense to the asymmetry of the peaks in spectra from cuprates. There was also a sharp peak at the Fermi energy due to a surface state, with a small step at a small binding energy, the energy of the onset of possible indirect transitions between the s–p bands. The region between them, up to 2 eV wide, contained the asymmetric tail of the main peak. They were able to model the line shape using the ideas described above, with the inclusion of the indirect transitions caused by the surface photoelectric effect. The term $\nabla \cdot \mathbf{A}$ does not conserve wave vector normal to the surface, leading to indirect transitions. The direct and indirect transitions contributed coherently. The consequent interference effect caused the observed asymmetry (Fano line shape; see Section 3.11.2). A similar asymmetric peak was later observed on the (100) surface of Ag and explained by the same effects (Hansen et al., 1997).

3.6   Core-level spectra

The foregoing has described photoemission from valence electrons. Some shallow core levels can provide photoelectrons when excited by a resonance lamp, and many core levels may contribute when synchrotron radiation is used for excitation, so that both core and valence electrons may appear in the same extended spectrum. Finally, photoemission from core levels can be excited by Al- or Mg-Kα characteristic radiation from electron beam-excited x-ray sources, the spectroscopy then being called x-ray-induced photoelectron spectroscopy (XPS) or electron spectroscopy for chemical analysis (ESCA) (Siegbahn *et al.*, 1967; Carlson, 1975; Brundle and Baker, 1977, 1978; Ertl and Küppers, 1985; Woodruff and Delchar, 1994). (Auger electrons also appear in these spectra, but we ignore them here.) Core-level spectra are useful for identifying surface species, either undesired or deliberately added adsorbates, and their reactions with the substrate. The apparent binding energies have chemical shifts which can be related to bonding. As a result, the apparent binding energies of the core states for atoms on the surface may differ from those in the bulk (Egelhoff, 1987; Jugnet *et al.*, 1987), and in ideal cases, even inequivalent surface sites may be distinguished. These binding energy shifts are labeled "apparent" because not all of the shift arises from a shift in an initial-state energy due to a change in local charge environment; some is due to relaxation of the entire system during photo-excitation (Shirley, 1978; Kotani, 1987), better described by the one-step model in the next section. An early application of core-level spectroscopy to cuprates was in a search for trivalent Cu ions in various cuprates. Core-level spectra are useful in determining the composition of the surface cleavage plane in the case of the high-temperature superconductors. Core-level photoexcitation and photoemission sometimes exhibit resonances as a result of interference with valence photoexcitation and photoemission Such resonances are helpful in determining the atomic wave function character of valence states. They have been useful in cuprates in identifying Cu 3d character in the states at the Fermi level.

In the three-step model, core-level spectra may be described along the same lines as valence spectra. In addition to electron–electron inelastic scattering, plasmon creation may contribute noticeably to the mean free path, also providing plasmon satellites in the spectrum. The plasmons may be intrinsic (created in the excitation process) or extrinsic (created as the fast electron travels to the surface) (Steiner *et al.*, 1979). The most important change is in the dipole matrix element, for the core levels are localized, having average radii well below interatomic spacings. If the final state is a Bloch function or a plane wave, transitions to any possible wave vector are possible, for a Fourier decomposition of the initial-state wave function, the core-level wave function, provides components with all wave vectors in the Brillouin zone. In other words, wave-vector conservation does not occur. The core states usually have angular momentum as a good quantum number. A final state may be expanded in spherical harmonics, and transitions with a change of $+1$ or -1 in the orbital angular momentum quantum number ℓ may occur from the core level, provided the final state overlaps physically with the atom under consideration.

If the energy of the emitted photoelectron is high, its wave function may be approximated by a plane wave of wave vector $\mathbf{k}$. The core level is a localized atomic orbital $\phi(r)$. The dipole matrix element for photoexcitation may be written as

$$\mathbf{p}_{if} = \mathrm{i}\mathbf{k} \int e^{\mathrm{i}\mathbf{k}\cdot\mathbf{r}} \phi(r) \mathrm{d}^3 r,$$

proportional to the Fourier transform of the atomic orbital (Gadzuk, 1974). Wave-vector conservation has been lost. Shevchik (1977) showed that an augmented plane-wave final state was a significant improvement over the plane wave.

Core-level photoelectron spectroscopy usually is carried out in the angle-integrated mode. Dispersion effects in core-level spectroscopy are not expected, except perhaps at resonance. However, angle-resolved studies of core spectra are used to vary the sampling depth by varying the collection angle. Also, there are interference effects in the outgoing electron's wave function involving scattering by the neighbors of the ionized site. This is called photoelectron diffraction (PHD) or X-ray photoelectron diffraction (XPD). The analysis of such spectra, scanned in energy or angle, both with angle resolution, gives geometrical information about the surroundings of the ionized atom (Smith, 1978; Woodruff and Bradshaw, 1994). Studies by this technique have not yet been made on high-temperature superconductors.

The spectrum for a given core level may consist of several peaks. First, spin–orbit splitting may be partially or completely resolved. The splitting is usually known from atomic data, but the relative intensities of the two components may differ from the expected statistical ratio of the values of $(2j + 1)$ (Sieger et al., 1995, and references therein). Even if we have only one site per unit cell for the atom under study, the binding energy and screening of the photohole may be measurably different for the surface layer and perhaps for the layer just below. Thus there may be two or more spin–orbit-split pairs of lines, one from the bulk, one from the surface, and perhaps more, if one can resolve peaks from the next layer under the surface or if there is a surface reconstruction creating inequivalent surface sites. Each pair will have the same spin–orbit splitting and relative intensities of the two components, but different apparent binding energies. The shift in binding energy for the atoms on the surface may have either sign. Not all of the observed surface shift arises from the different binding of the initial state. The final state, a photohole, may be partially screened by other electrons, causing a shift to lower binding energy. The screening can be different at the surface from that in the bulk due to differences in electron density, and to differences in wave function character, e.g., more, or less, d-character at the surface than in the bulk. Such shifts tend to be smaller than the shifts in binding energy, but they are not negligible. They may occur in insulators as well as metals. In the former the screening is by electrons on neighboring atoms or ions, or in covalent bonds, both of which are modified at the surface.

The shapes of each component, aside from instrumental effects, is likely to be a Lorentzian, but the initial-state widths Γ_i need not be the same for each pair. The temperature dependence of the widths may not be the same either, for surface states may interact with surface phonons of lower frequency. In some cases there may be resolved or unresolved multiplet structure, especially if the final state is localized. There may be additional core-level peaks, called satellites, due to neglected many-body effects in which the solid is left in an excited state, e.g., a photohole and an electron–hole pair (Kotani and Toyozawa, 1979; Kotani, 1987). Plasmon satellites may be identified by their location in energy. Finally, for a metal, the intrinsic line shape may not be Lorentzian. In responding to the sudden creation of a core hole, the electrons rush in to screen the hole, creating many electron–hole pairs. The phase space for such processes diverges as the electron–hole energy approaches zero. This leads to a possible singularity at the threshold for core-level photoexcitation, the "x-ray edge singularity." The absorption coefficient at and above the edge threshold E_0 varies as $(E - E_0)^{-\alpha_\ell}$, where α_ℓ, the singularity index, depends on the angular momentum ℓ of the core hole and on the screened Coulomb interaction between the core hole and the conduction electrons. If α_ℓ is positive, the edge is singular. The x-ray edge problem has been reviewed extensively in Ohtaka and Tanabe (1990). In core-level photoemission, the line shape is a convolution of this singular power-law function with a Lorentzian of width Γ_i . The convolutions can be completed in closed form if the density of states near E_F may be assumed constant. The result is the Doniach–Šunjić line shape (Doniach and Šunjić, 1970), normalized to unit area:

$$I(E) = \frac{\Gamma_m(1 - \alpha) \cos[(\pi\alpha/2) + (1 - \alpha) \arctan((E_0 - E)/\Gamma_i)]}{[(E_0 - E)^2 + \Gamma_i^2]^{(1-\alpha)/2}},$$

where Γ_m represents the gamma function, Γ_i the photohole lifetime broadening, and α the asymmetry parameter, related to the α_ℓ. E_0 is the core-level binding energy. When α is zero, the Lorentzian shape is retrieved. Wertheim and Walker (1976) (see also Wertheim and Citrin, 1978) have carried out the convolution for several other assumed energy dependences of the density of states near E_F. The spectral shape is skewed to higher binding energies for non-zero values of α. The parameter α can be obtained from calculations of the scattering of the conduction electrons by the screened Coulomb potential of the hole, a formidable many-body problem. It is proportional to a weighted sum of the squares of partial screening charges, "partial" referring to a decomposition of the scattering into spherical harmonics. In non-metallic systems, even poorly conducting metals, these charges are small and α is very small, effectively zero. For all but a few nearly-free electron metals, α is not calculated, but obtained by fitting core-level spectra.

The sudden appearance of the charged core hole can excite phonons as well as electrons. Phonon sidebands are not resolved in the spectra, but they contribute a Gaussian broadening. Thus the observed line shape is the Doniach–Šunjić line shape (or something similar), convolved with a Gaussian, then convolved with the instrument function. For a single peak, there are shape-fitting parameters α, Γ_i, and the width of the Gaussian. In an insulator α vanishes. In a metal, if the density of states at E_F is low, or even if it is not but the states near E_F have very litle wave function component on the site of the core level under study, α will be very small.

Additional reviews of core-level spectroscopy may be found in Campagna *et al.* (1979), Flodström *et al.* (1992), Hüfner (1979), Steiner *et al.* (1979), and Spanjaard *et al.* (1985). Angle-resolved core-level spectroscopy is also useful, for it can give structural information about surfaces (Fadley, 1984, 1992; and Woodruff and Bradshaw, 1994).

3.7   The one-step model

The three-step model has a number of defects. Nonetheless it has been used with considerable success. As mentioned previously, it does not include the surface photoeffect. More importantly, it does not include any surface-specific effects, due, for example, to surface states which may be present even on ideally terminated surfaces as well as on relaxed or reconstructed surfaces. Most of the photoelectrons that escape without inelastic scattering come from a depth of about three mean free paths, and for many studies of valence bands, this means much of the depth sampled may be viewed as surface rather than bulk. Nonetheless, EDCs for many materials have been taken and analyzed with the three-step model, often with good agreement with predictions from energy bands calculated for the bulk. This agreement requires more explanation. Some extra spectral features have appeared, and these were associated with surface effects, especially if they disappeared when foreign atoms were adsorbed on the surface, the so-called "crud test."

There are more fundamental problems with the three-step model. The very small depth sampled causes a conceptual difficulty, for transport to the surface and escape from the surface are not separable. Indeed, excitation of an electron whose final state is a wave packet centered within a typical escape depth Λ

below the surface already has significant final-state amplitude at the surface when it is created. Energy-dependent probabilities between 0 and 1 of reflection and transmission, at the surface were not considered. These difficulties and the growing popularity of photoelectron spectroscopy led to one-step calculations, (Mahan, 1970, 1978; Schaich and Ashcroft, 1971; Caroli *et al.*, 1973; Feibelman and Eastman, 1974; Pendry, 1976; Liebsch, 1976, 1978, 1979a,b; Schaich, 1978) which were analyzed to determine under what conditions they gave the same result as the three-step model (Feibelman and Eastman, 1974). The improved calculations take into account the possibility that in angle-resolved photoemission, the component of wave vector parallel to the surface may be augmented by a reciprocal lattice vector, so the photoelectron appears at a different angle, as described above. They also include the possibility of interference effects between "surface" and "volume" photoemission, making their separation difficult. Such calculations also can be applied to a many-body system (Caroli *et al.*, 1973), not just to one described by the independent-electron approximation. In recent years the one-step formalism has been used to treat electron correlations better. The limitations in so doing are then imposed by how realistic the many-body model may be for describing the actual system and by the sudden approximation (see below) for the photoemission process, which may not always be valid for valence-band photoelectron spectroscopy.

One-step calculations can be applied to the independent-electron model for the system or to as elaborate a many-body model for the system that one cares to work out. The initial state is an eigenstate of the N-electron system. Its wave function is localized within the crystal, perhaps with an evanescent tail extending into the vacuum. The final-state wave function is also an eigenstate of the N-electron system in the independent-electron picture, but more generally, it is an eigenstate of the ionized $(N - 1)$-electron system, which usually is not an eigenstate of the un-ionized N-electron system. In either case, this final state must have an amplitude inside the crystal to provide some spatial overlap with the initial state, and a propagating component in the vacuum to provide an electron at the detector. The effects of the surface appear in the matching of these, and other, components at the surface. The inelastic scattering losses must also appear in this final state, either by using complex wave vectors or complex crystal potentials (Mahan, 1970, Schaich and Ashcroft, 1971), or by including explicitly the inelastic scattering processes (Caroli *et al.*, 1973). A steady-state scattering formalism (Adawi, 1964) is used in which the initial state is changed to the final state by the absorption of a photon via the interaction Hamiltonian H', introduced in Section 3.2.3. The final state evolves by interaction with the crystal potential, including the termination at the surface, and by interaction with whatever inelastic scattering centers one wants to include. One component of the final-state wave function represents the electron propagating to the detector, and the current from this component is calculated. In this picture, the final-state wave function is given by

$$\psi_f(\mathbf{r}) = C \int G(\mathbf{r}, \mathbf{r}', E_f) H' \psi_i(\mathbf{r}) d^3 r,$$

or

$$\psi_f = G H' \psi_i.$$

H' is the electron–photon interaction Hamiltonian used in the three-step model, G is the (complex) Green's function for an electron in the interacting system, and ψ_i is the initial-state wave function, an eigenstate of the initial system, unperturbed by the photon and inelastic scattering. G includes the effect of the crystal potential, including its termination at the surface. G may include other interactions of the photoelectron, e.g., electron–electron scattering which not only reduces the amplitude of the wave function near the surface, but also broadens the contribution of a single photoemitted electron to the spectrum by giving the hole a finite lifetime from electron–electron scattering. The form of the final-state wave function must be chosen so that at large distances from the crystal there is a component of the wave function that gives a current directed at the detector. Thus one begins with an angle-resolved analysis.

In the independent-electron model, G is known. Outside the crystal at large distances it is

$$G_0(\mathbf{r} - \mathbf{r}') = C \exp[i\mathbf{k} \cdot (\mathbf{r} - \mathbf{r}')]/|\mathbf{r} - \mathbf{r}'|,$$

with C a constant. If $\mathbf{r}'$ is taken to be the vector to the detector, large compared with $\mathbf{r}$, which is near the surface of the crystal, and $\mathbf{k}$ is parallel to $\mathbf{r}'$, G_0 simplifies to

$$G_0(\mathbf{r} - \mathbf{r}') = (C/r') \exp(ikr') \exp(-i\mathbf{k} \cdot \mathbf{r}).$$

For interacting electrons, the asymptotic form of G at the detector will be the same, except for a reduced amplitude and a phase shift, both due to the interactions within the crystal.

There must be a final-state amplitude inside the crystal in order to provide some overlap with the initial-state wave function. The types of initial- and final-state wave functions are shown schematically in one dimension in Fig. 3.16. The initial states include surface states (gap states) and surface resonances, as well as bulk Bloch states. The former two will be discussed later. The proper asymptotic form is often described as a time-reversed LEED wave function. In LEED (low-energy electron diffraction) the asymptotic form of the wave function is an incoming plane wave in the vacuum, outgoing spherical waves in the vacuum, and Bloch waves inside the crystal with group velocities directed into the crystal (Liebsch, 1976, 1978, 1979a; Schaich, 1978). The time-reversed LEED wave function has the Bloch functions inside the crystal with group velocities directed

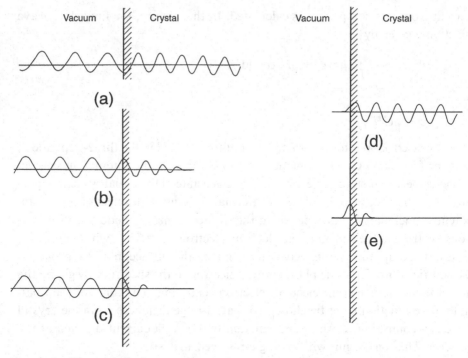

Fig. 3.16 Schematic final-state un-normalized wave functions, ψ_f, (a–c) and initial-state wave functions, ψ_i, (d,e) in photoemission. (a) Bloch-function final state, (b) Bloch-function final state, but with short mean-free path, or surface-resonance final state, (c) surface-state final state, (d) Bloch-function initial state, (e) surface-state initial state.

toward the surface, a plane wave in the vacuum headed for the detector, and incoming spherical waves approaching the surface from the vacuum. This is also the correct asymptotic form of the final-state wave function in angle-resolved photoemission.

For a given model, e.g., the independent-electron model, several techniques, e.g., Fermi's golden rule and calculating a current from the wave function, have been used to calculate the photocurrent into the detector. These techniques have been shown to be equivalent (Pendry, 1976). The one-step calculation generally does not give the same result as the three-step model. Feibelman and Eastman (1974) have shown under what circumstances the one-step calculation can be replaced by the three-step calculation. The mean free path for inelastic scattering must be long, longer than is generally found when using 10–50 eV photons. Caroli *et al.* (1973) have shown that since the three-step picture is not valid in general, one cannot have ϵ_2 playing a role in photocurrent calculations such as presented above in the three-step model.

Many of the calculational schemes just mentioned treat inelastic scattering by making a wave vector or a potential complex instead of real. Caroli *et al.* (1973)

avoided this phenomenological treatment of inelastic scattering in their formalism by incorporating inelastic scattering processes directly. This allowed them not only to obtain the effect of inelastic scattering on the primary spectrum, but to calculate the spectrum of inelastically scattered electrons. They worked out the formalism for inelastic scattering by phonons, but found the direct incorporation of electron-electron scattering rather difficult and did not complete its evaluation. The treatment of Caroli *et al.* gave an angle-integrated photocurrent. However, the angle-resolved version has been worked out. In the calculation of Caroli *et al.*, a response-function formalism was used. This leads to an expression for the photocurrent proportional to

$$-(1/\pi)\mathrm{Im}\langle \mathbf{k}|G_f^+ H' G_i^+ H' G_f^-|\mathbf{k}\rangle,$$

where $|\mathbf{k}\rangle$ is a plane wave directed at the detector. The superscripts $+$ and $-$ indicate that the Green's function is built from outgoing or ingoing waves, respectively, and the i and f refer to the hole and electron. This form of the photocurrent can be derived from the other one-step expressions. This expression is also the spectral density function, to be introduced below. The spectral density function is the density of excitations of two types, an electron removed from below the Fermi energy, as in photoemission, and an electron added to the system, as in inverse photoemission.

For one-step calculations of the angle-resolved photocurrent in the independent-electron model, the analogy to LEED calculations was exploited by Pendry and colleagues, resulting in a computer code that is used frequently (Hopkinson *et al.*, 1980). The crystal potential in this code is a "muffin-tin" potential, which is zero outside the muffin-tin spheres. Grass *et al.* (1993, 1994) have generalized the Pendry formalism to deal with non-muffin-tin potentials, better suited for many complex materials. Das and Kar (1995) have augmented Pendry's treatment to include the depth dependence of the vector potential near the surface, using a parameterized form for the latter.

3.8 Many-body treatment

The one-step model can be used to include many-body effects. When electron correlation is greater than is included in an average way in the independent-electron model, features may appear in a photoelectron spectrum that can have no origin in the one-electron band structure. An example is a peak where none was expected, a "satellite." Such satellites usually occur on the low kinetic energy (high binding energy) side of the main spectrum, and often are called "shakeup" features in the spectrum because when these photoelectrons leave the system, the $(N-1)$-electron system is left in an excited state. Not only does such a feature require explanation, but the parts of the spectra that do seem to be explained by band theory need to be reexamined for possible reinterpretation. These satellites and other features require a better treatment of the many-body aspects of the system than the independent-electron model provides. Although some of the calculations we have just described can be applied to a many-body system, often another formalism is used, one that is more readily and accurately applied to core levels than to valence electrons, but which still can be used on the latter.

Hedin and Lundquist (1969), Wendin (1981), and Almbladh and Hedin (1983) review a formalism for treating the interacting-electron system. If we start with the N-electron initial system, photoemission removes an electron, while inverse photoemission (see below) adds an electron. Both processes involve the electron–photon interaction. These two processes can be treated together in the spectral density function or spectral function, $A(\mathbf{k}, E)$, where

$$A(\mathbf{k}, E) = -(1/\pi)\mathrm{Im}G(\mathbf{k}, E),$$

and G is the single-particle Green's function for the interacting system. Written as $G(\mathbf{k}, E)$, it is the Fourier transform of $G(\mathbf{r}, \mathbf{r}', t)$ for a periodic system in which only $\mathbf{r} - \mathbf{r}'$ is meaningful. The photoelectron spectrum can be obtained from $A(\mathbf{k}, E)$ for $E < \mu$, the chemical potential, and the inverse photoemission spectrum from $A(\mathbf{k}, E)$ for $E > \mu$. For $E < \mu$, $A(\mathbf{k}, E)$ is proportional to the probability that one can remove an electron from the state $|\mathbf{k}, E\rangle$, while for $E > \mu$, it is proportional to the probability that an electron can be added to the state $|\mathbf{k}, E\rangle$. The spectral density function plays a role similar to that of a density of states. This probability is modified by the details of how the removal is to be done. All of the processes just described have a probability proportional not just to $A(\mathbf{k}, \omega)$ but to the product $A(\mathbf{k}, \omega)|\mathbf{p}_{if}|^2$, assuming the sudden approximation is valid. For photoemission in the independent-electron model, $A(\mathbf{k}, E)$ is proportional to $\sum_i \delta[E - E_i(\mathbf{k})]$, a series of infinitely sharp peaks at the one-electron energies of the filled states. For $E > \mu = E_\mathrm{F}$, it is proportional to $\sum_f \delta[E - E_f(\mathbf{k})]$, and the peaks are determined by the empty-state eigenvalues. More generally, $A(\mathbf{k}, E)$ will have peaks of finite width, representing the excitation of quasiparticles, electrons or holes with accompanying clouds of screening electrons. The quasiparticles have finite lifetimes due to the interactions which are not treated in the independent-electron model. The background between these peaks may be finite, and large enough to measure. The interactions may be so large that the peaks become so broad that they are hardly recognizable, in which case quasiparticles become ill defined. There need not be a one-to-one correspondence between the peaks in $A(\mathbf{k}, E)$ and the peaks expected from the independent-electron model, for there may be new satellite peaks in A, and some of the peaks from the independent-electron picture may be broadened beyond recognition.

If G_0 is the single-particle Green's function for the non-interacting system, G can be obtained by expanding

$$G = G_0 + G_0 \widetilde{\Sigma} G,$$

where $\widetilde{\Sigma} = \Sigma_1 + i\Sigma_2$ is the self-energy, which contains all of the electronic interactions. If this is solved iteratively

$$G(\mathbf{k}, E) = 1/[G_0^{-1} - \widetilde{\Sigma}(\mathbf{k}, E)] = 1/[(E - E_0(\mathbf{k}) - \widetilde{\Sigma}(\mathbf{k}, E)],$$

where E_0 is an eigenstate of the independent-electron system.

$A(\mathbf{k}, E)$ is proportional to

$$\frac{\Sigma_2}{[E - E_0(k) - \Sigma_1]^2 + \Sigma_2^2},$$

in which Σ_1 and Σ_2 may or may not vary slowly enough with energy to be treated as constant for each peak in $A(\mathbf{k}, E)$. If they can be considered constant, such a peak has a Lorentzian line shape. A peak near one of the single-particle energies, E_0 is shifted by Σ_1 and broadened by Σ_2. In the full $A(\mathbf{k}, E)$ there also may be peaks not associated with any particular E_0. The photoelectron spectrum can be obtained from the spectral function by invoking the sudden approximation, to be discussed below. The spectrum then is proportional to

$$A(\mathbf{k}, E)|\langle f|H'|i\rangle|^2,$$

where $\mathbf{k}$ is the wave vector of the photohole. The second factor leads to the square of the dipole matrix element.

The most difficult effort is in obtaining a realistic self-energy for a given material. The electron–electron interaction is screened by the other electrons, and a screening function (dielectric function) must be used that depends on the electron–electron separation and on time. Its Fourier transform is $\widetilde{\epsilon}(\mathbf{k}, \omega)$. It is not known a priori, and must, in principle, be calculated for any new system to be studied. The calculation (Hedin and Lundqvist, 1969; Godby, 1992) involves solving four coupled equations self-consistently. Godby (1992) and Zaanen (1992) have summarized results on materials with delocalized and localized electrons, respectively. The former author used the LDA results as input to the calculation. These already incorporate an average of some of the correlation effects. Wendin (1981) describes calculations of the self-energy in atomic systems. Partially localized states are much harder to treat. The self-energies of crystalline elemental transition metals have not been calculated to the same accuracy as have those of, e.g., Na or Si.

Figure 3.17(a) shows schematically part of the independent-electron band structure for a nearly free-electron metal with a narrow d band, represented as flat. The states in such a flat band are well localized on the atomic sites. Also shown, Fig. 3.17(b), is the corresponding density of states $n(E)$, which would

INDEPENDENT ELECTRONS

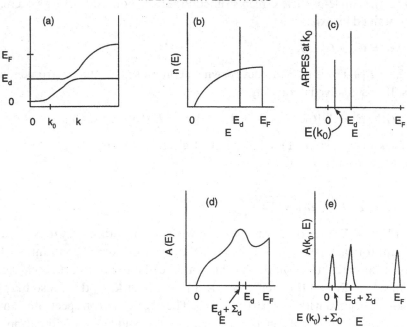

WITH ELECTRON-ELECTRON INTERACTIONS

Fig. 3.17 (a) Schematic independent-electron band structure for a simple metal with a localized-electron state at energy E_d. (b) The resultant density of states and (c) angle-resolved photoemission spectrum taken at wave vector $\mathbf{k}_0$. Panels (d) and (e) show the spectral density functions $A(E)$ and $A(\mathbf{k}_0, E)$ for the same hypothetical system when electron–electron correlations are turned on.

give, apart from weighting factors due to dipole matrix elements, the angle-integrated photoelectron spectrum. Figure 3.17(c) shows the angle-resolved photoelectron spectrum for a fixed electron energy analyzer angle and photon energy. (This is a common plot from experiment, but as E varies, $\mathbf{k}$ is not kept constant if θ does not change.) If the interelectron Coulomb interaction is turned on, the spectral density function must be calculated. $A(E)$ typically will have a peak at an energy generally near that of the localized level, but shifted and broadened, as shown in Fig. 3.17(d). This peak sometimes is called the coherent peak. In some metallic systems there may be a second peak apparently at the Fermi energy. This peak represents a final state with a delocalized hole. It may be viewed as the result of the near-perfect screening of a localized photohole by the electron gas, leaving the hole at the Fermi level. In addition to the above peaks, and the region between them, there is also a free electron-like contribution to the spectral density function similar to that of the free-electron gas alone. It arises from that part of the free-electron gas that cannot, for reasons of symmetry, interact with the localized hole. The angle-resolved photoelectron spectrum is also shown as the spectral density function $A(\mathbf{k}, E)$. In a system with localized states or a flat band but with electron-electron repulsion, the coherent part of the photoemission spectrum is that corresponding to removal of a quasiparticle, a peak, possibly quite broad, somewhat near the expected independent-electron peak. The incoherent part is that part of the spectrum, usually structureless, corresponding to removal of a quasiparticle with simultaneous excitation of other quasiparticles, the equivalent of the shakeup part of the spectrum, and many-electron excitations not coupled to the quasiparticle. As mentioned in Chapter 2, the area of the coherent peak is reduced by a factor of $Z(< 1)$, the renormalization coefficient, from that of the corresponding peak in the hypothetical independent-electron system.

Integrating the product of the spectral density function and the Fermi–Dirac distribution function over energy gives the momentum distribution function, $n(\mathbf{k})$. The sudden approximation is still required. In mapping the Fermi surface, one sets the angle-resolving electron energy analyzer to pass a small range of energies around E_F, or, alternatively, chooses this window in EDCs. The step in $n(\mathbf{k})$ at $\mathbf{k}_F$ is smaller by a factor of Z than in the independent-particle description. This makes the determination of $\mathbf{k}_F$ difficult when Z is small, for the step then is difficult to detect. The problem becomes even worse when dealing with narrow bands, for then the step at $\mathbf{k}_F$ remains in the energy window for a wider range of angles. Straub et al. (1997) addressed the problem of locating $\mathbf{k}_F$ in such cases. They took a very extensive data set of photoelectron current as a function of angle, using an energy window of 100 meV centered on E_F and an angle resolution of $\pm1.75°$. The samples were Cu and $TiTe_2$, the latter having a flat band crossing the Fermi level. The photocurrent distribution, $w(\mathbf{k})$, was proportional to $n(\mathbf{k})$, and the Fermi surfaces could be seen easily in some parts of the Brillouin zone, and with difficulty in other parts where the dipole matrix element was small. The step at E_F was then sought by plotting $|\nabla_{\mathbf{k}} w(\mathbf{k})|$. This showed a peak at $\mathbf{k}_F$ where a narrow band crossed the Fermi surface, and two well-separated peaks where a wide band crossed, one at the $\mathbf{k}$ where the band entered the energy window and one at $\mathbf{k}_F$. The latter can be identified as the correct measure of $\mathbf{k}_F$ even in the absence of a calculated Fermi surface because it is narrower and does not shift location as the energy window is increased.

A large number of calculations of photoelectron spectra, or at least spectral density functions, have been carried out for clusters of Cu and O atoms in which some many body effects have been treated, often at the expense of a simplified picture of the one-electron electronic structure (Dagotto, 1994). These clusters have been rather small, but if correlation effects are dominant, perhaps not much of importance is lost by the small cluster size, limited by the enormous number of states possible for just a few atoms with several atomic states on each, even after maximal use of symmetry has been made. Moreover, many features of such a calculation, but not all, do not depend much on cluster size. Chen and his collaborators have made calculations on clusters in which they treat the single-particle and some many-body effects essentially on an equal footing (Chen, 1990, 1992a,b, 1993; Zheng and Chen, 1995). The extended Hubbard Hamiltonian used by Zheng and Chen (1995) included hopping between first- and second-nearest neighbors, with coefficients t and s, respectively, on-site Coulomb repulsion U, and intersite Coulomb repulsion K between nearest neighbors. It was applied to an eight-site cluster with periodic boundary conditions, which approximates a Cu_4O_8 planar cluster after a suitable choice of atomic orbitals. With only two single-particle orbitals per site, there are 11440 multi-electron orbitals for the seven-electron initial state, 8008 final state orbitals in photoemission, and 12870 final states in inverse photoemission. The use of symmetry gave a block-diagonal Hamiltonian matrix with the largest block just smaller than 300×300. The Hamiltonian was then diagonalized and angle-integrated photoelectron spectra calculated for various combinations of the parameters. As U increases with the other parameters held constant, the photoelectron spectrum broadens and its centroid shifts to larger binding energies. A satellite at large binding energy begins to appear for larger U. For a fixed U, increasing K had the opposite effect to that of increasing U. There are also shape changes in the spin-resolved components of the spectrum as U and K vary. The effect of doping was also studied by starting with six electrons instead of seven, and the effects of charge transfer studied by placing "d" orbitals on half the sites and "p" orbitals on the others. This is an informative model study, but the effects of correlation may not be as large in spectra measured with low-energy photons as were calculated here because of the possible weakening of the sudden approximation.

3.9   Sudden vs. adiabatic limits

If the photoelectron has barely enough energy to escape through the surface, it may move slowly enough to allow the remaining electrons to relax to the lowest energy state, the ground state, of the final $(N - 1)$-electron system. Conservation of energy requires that if the system is left in its lowest energy state, the escaping photoelectron has its maximum kinetic energy. If the energy of the photoelectron is reported as a binding energy then, in this picture, the binding energy is a minimum for this photoexcitation process. The hole appears to be as near the Fermi level as possible. This is the adiabatic limit.

   The other extreme occurs for high-energy photoelectrons. In this limit the sudden approximation is invoked and the $(N - 1)$-electron system is left in its ground state or in one of its excited states. The excited states consist of the hole left by the photoelectron, and one or more electron–hole pairs. Core-level spectra thus may show several satellites. (It may not be possible to tell which peak is the "main peak" with just one hole, and which are satellites. The "main peak" may not be the most intense. This is most likely to occur in transition metals and rare earths.) The peaks with least binding energy result from photoemission with maximal screening of the core hole, possibly representing the adiabatic process. The other peaks, often called "shakeup" satellites, have more poorly screened holes in the final state. The point of mentioning the transition from adiabatic to sudden here is that the sudden approximation is used in many-body calculations of the photoelectron spectra of valence electrons in systems of interacting electrons, and it may not be valid for low-energy photoelectrons. Photoemission is much more difficult to treat in the adiabatic limit, for one must account for the detailed screening of the photohole by the electrons. An example of the change in line shape as the photoexcitation changes smoothly between the adiabatic and sudden limits for core level excitations in a metal has been given by Gadzuk and Šunjić (1975). The sudden limit is the Doniach–Šunjić line shape discussed in Section 3.6. The line shape in the adiabatic limit is Lorentzian with its peak is shifted toward the Fermi level from that of the Doniach–Šunjić shape.

Little is known about how large the energy of a photoelectron from the valence band must be in order to reach the region of validity of the sudden approximation. Core-level studies have shown that the answer is dependent on the system being studied, not only the atom, but also its chemical environment. Stöhr et al. (1983) studied the photon energy dependence of the intensity of a shakeup satellite in the N 1s spectrum for N_2 adsorbed on a Ni surface. They found that the sudden approximation was valid at 15 eV above threshold, much lower than the 200 eV Carlson and Krause (1965) had found for Ne core spectra. Studies on CO have shown the C 1s main line in CO molecules reaches a limiting intensity about 30 eV above threshold, but the satellite intensities require 60–70 eV before reaching a constant relative intensity (Reimer et al., 1986). At near-threshold excitation, the excitation of these satellites can be resonantly enhanced by a factor of ten (Ungier and Thomas, 1984), indicating the complexity of processes near the adiabatic limit. These C 1s spectra for CO have been calculated near threshold by Schirmer et al. (1991) and Bandarage and Lucchese (1993).

More recently there has been evidence that in valence-band photoemission studies of high-temperature superconductors the sudden approximation is indeed valid. Randeria et al. (1995) showed that angle-resolved photoemission spectra of $Bi_2Sr_2CaCu_2O_8$ using 19-eV photons were consistent with their description of them as a spectral density function $A(\mathbf{k}, E)$ weighted by dipole transition matrix elements. This implies that the sudden approximation was valid for the photon energies used. Since the valence bands involved are derived almost completely from O 2p and Cu 3d states on the atoms of the CuO_2 planes in $Bi_2Sr_2CaCu_2O_8$ and in the other high-temperature superconductors, it is likely the sudden approximation may be made safely for all cuprates for excitation with photon energies of 19 eV and above, and possibly at somewhat lower energy.

3.10   Inverse photoemission

If an electron is added to the crystal from outside with an energy above the Fermi energy, it may decay to an empty state above the Fermi energy by emission of a photon. The dipole matrix elements are those of photoemission, but the states involved are both excited states. As with many techniques, it is rather old (Ohlin, 1942), but it underwent a renaissance with modern vacuum and electron spectroscopic instrumentation. Pendry (1980, 1981) called it to the attention of the solid-state and surface-science experimental communities, which began a new round of studies. It was called brehmstrahlung isochromat spectroscopy (BIS), a name still used when photon energies are in the keV range, while inverse photoemission tends to be used to describe the detection of lower-energy photons, usually under 30 eV. It may be carried out in the angle-integrated mode or angle-resolved mode. In the angle-resolved mode, it is sometimes called **k**-resolved inverse photoelectron spectroscopy (KRIPES). In the independent-electron model, both states are one-electron states, and the lower one may lie between E_F and $E_F + \Phi$, a region not accessible in photoelectron spectroscopy. This alone makes the technique important for it can access states that photoemission cannot. In the many-body picture, the final state is an $(N + 1)$-electron state, while that in photoemission is an $(N − 1)$-electron state. This difference affords further discrimination between many-body models for materials when calculations are compared with experiment. Often the Coulomb repulsive energy for two d or f electrons on one site can be obtained from the combination of a photoemission spectrum and and inverse photoemission spectrum. As we shall see, inverse photoemission is a low-probability process, leading to very low count rates, hence to the use of a large spectral bandpass. The low-resolution spectra are useful in studying many-body effects and the electronic structure on a rather coarse scale, but they are not useful for elucidating details of the electronic structure in a region several $k_B T_c$ wide just above E_F in high-temperature superconductors. Several reviews of inverse photoemission and BIS exist (among them Dose 1983; Nilsson and Kovacs 1983; Smith and Woodruff 1986; Borstel and Thörner 1987; Smith, 1988; Himpsel 1990; Andrews *et al.* 1992; Fuggle 1992; Schneider and Dose 1992; Johnson 1992; and Skibowski and Kipp, 1994).

Figure 3.18 shows schematically the angle-resolved inverse photoemission process for the emission of ultraviolet photons. The incident electron must have the wave vector and energy of a suitable Bloch function in order to enter the crystal. In fact, in order to match boundary conditions, the electron enters the crystal with a wave function that is a linear combination of Bloch functions with wave vectors differing by a reciprocal lattice vector. If there is not a matching internal wave function, the probability for reflection of the electron will be high, as in a LEED experiment. Once in the crystal it is subject to inelastic electron–electron scattering, but it may also emit a photon. Photon emission follows from the same Hamiltonian H' as used for excitation, except that the electromagnetic field must now be quantized. In an elemental photoemission process, there are many incident photons and only one is destroyed. Because of the large number, the field may be treated classically. In inverse photoemission, one starts with no photons and ends up with just one, too few for a classical treatment. The same H' leads to identical selection rules for the two processes, although non-zero matrix elements won't be the same, for different states may be involved (Johnson and Davenport, 1985). The Pauli principle dictates that the final state be empty initially, i.e., be above E_F. The Fermi golden rule result for the transition rate gives an emission probability proportional to the density of final states. In comparing photoemission and inverse photoemission the dipole matrix elements are nearly the same, so the major difference will arise from the densities of final states. For inverse photoemission one measures the number of photons of energy $\hbar\omega$ emitted per unit energy interval into a differential solid angle per incident electron. For photoemission it is the number of electrons of energy E emitted per energy interval into a differential solid angle per incident photon.

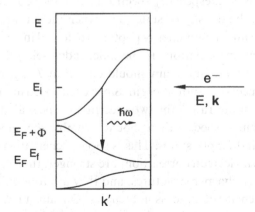

Fig. 3.18 Schematic picture of angle-resolved inverse photoemission. The electron may enter the crystal in state i and decay radiatively to state f. Note that for the E_i chosen the final state is one that cannot be reached in photoemission.

For equal solid angles the ratio of the two is, ignoring factors in emission angles,

$$J_{ph}/J_{el} = (k_{ph}/k_{el})^2 = (\lambda_{el}/\lambda_{ph})^2,$$

where the λ are wavelengths (Pendry, 1980). For 10 eV photoelectrons and 12 eV photons emitted, this ratio is 1.5×10^{-5} for comparable energy and angle resolutions. For keV photoelectrons, as in XPS, and photons of the order of 1 keV energy emitted in the inverse process, this ratio is 10^{-3}. If we use realistic incident electron and photon currents to produce the measured current densities with comparable energy spreads, the ratio of the count rates in the two angle-resolved spectroscopies is still unfavorable for inverse photoemission, but not prohibitively so. An additional complication is that in inverse photoemission a new type of electron state may play a role, one that is not included in band calculations, even when surface states may be included. These are the "image potential" states, caused by the attractive potential of the induced surface charge as the slow electron approaches the surface (Chen and Smith, 1987; Smith, 1988).

As we shall see in the next chapter, experimental systems used in inverse photoemission are of rather low resolution, below that possible in photoemission, and too low to allow accurate studies of the Fermi surface and states with energies within a few $k_B T_c$ of it. The use of a monochromator offers several advantages, but the very low count rates in inverse photoemission usually mean that the resolution still does not approach that of photoemission. To work with comparable resolution to that of photoemission requires inordinately long times for data collection. Nonetheless, inverse photoemission is an important spectroscopy which leads to a more complete understanding of the electronic structure of any material.

From the combination of angle-integrated valence-band photoemission and inverse photoemission, especially spectra taken under conditions for which the spectra resemble the density of states, i.e., when the XPS region is approached in photoemission and a high-energy photon is detected in inverse photoemission, one can gain new information. If the independent-electron approximation is valid, the valence band spectrum should cut off at E_F, just where the inverse photoemission spectrum should begin. Since each spectrum is measured in arbitrary units, i.e., "count rate," the two spectra must be scaled to match ordinates at E_F. The combined spectra then should resemble the one-electron density of states for filled and empty states. This is shown schematically in Fig. 3.19. For materials in which electron correlations are stronger, those with d or f electrons, one expects rather sharper structures, and they are found. Their interpretation as peaks in the density of states is not tenable. One must use the spectral density function. A common result in such systems is that there is a peak at an energy Δ_- below E_F in the photoemission spectrum, representing the removal of an f or d electron with concomitant response of the rest of the electronic system, but

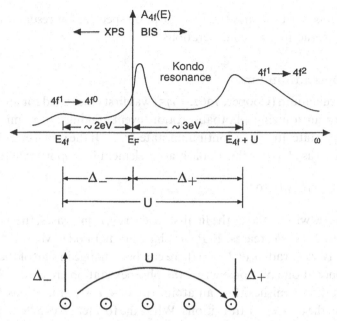

Fig. 3.19 Spectral density function for a single 4f electron hybridized with a continuum, as in Ce. $E_F - E_{4f}$ is the energy to excite the 4f electron to the Fermi level before hybridization and $E_{4f} + U - E_F$ is the energy to take an electron from the Fermi level to the 4f site, giving it a $4f^2$ configuration. The solid curve is the spectral density function after hybridization. It gives the XPS and BIS spectra if dipole matrix elements can be ignored.

not complete screening. The inverse photoemission spectrum has a peak at energy Δ_+ above E_F, representing the addition of an f or d electron to the system and the system's response. The energy $\Delta_- + \Delta_+$ is equal to U, the energy to remove an f- or d-electron from one site and place it on a distant identical site. This is shown at the bottom in Fig. 3.19. There may be multiplet structure giving rise to several peaks of each type because there is not necessarily a single value of U independent of multiplet structure. Also, each individual spectrum may have a peak "at" E_F. This is labeled "Kondo resonance" in the figure. When the two spectra, electron removal and electron addition, are combined after removal of the effects of finite resolution the Kondo peak is one whose location is just above or below E_F. It is the result of many-body effects as the system screens the added electron or hole.

3.11   Additional aspects of photoemission

We turn now to two more effects that are often used in photoelectron spectroscopy, the Cooper minimum and "resonance" photoemission. Both are applicable only to certain elements, and both may be discussed in terms of the dipole

matrix elements, for they show up in absorption spectra, but resonance photoemission requires additional considerations.

3.11.1 Cooper minimum

The Cooper minimum (Cooper, 1962, 1975) was first elucidated for atomic spectra. Consider an ionizing photoabsorption event, taking an atomic electron
from a discrete state, $|n,\ell\rangle$ to a continuum state, $|\epsilon,\ell'\rangle$. If the radial eigenfunctions
$R(r)$ are written as $P_{n,\ell}(r)/r$ the dipole matrix element is proportional to

$$\int P_{n,\ell}(r) P_{\epsilon,\ell'}(r) r \, dr.$$

$P_{\epsilon,\ell'}(r)$ oscillates with r, and as the final-state energy, ϵ, increases, the wavelength
of oscillation, $2\pi/k$, decreases. If $P_{\epsilon,\ell'}(r)$ has a radial node, when k^{-1} is of the
order of the average radius of $P_{n,l}(r)$ there can be a change in sign of the integral.
At this ϵ, hence at one photon energy, the photoexcitation cross-section for this
type of excitation vanishes. For an atom, the $\ell \rightarrow \ell + 1$ transitions are much
stronger than the $\ell \rightarrow \ell - 1$ transitions. When the former gives a "Cooper zero"
in the cross-section, the latter will not be zero. The result is a minimum, instead
of a zero, in the photoexcitation cross-section from the state n,ℓ. As shown by
Cooper, there are conditions for obtaining such a minimum. The initial-state
radial wave function must have at least one node. There will be no effect from
1s, 2p, 3d, 4f,... initial states. An example is Ag. The 4d $\rightarrow$ ϵf transitions give a
zero in the calculated dipole matrix element at about 102 eV. The minimum in
the total photoabsorption cross-section occurs about 6.8 eV higher due to the
energy dependence of the photoexcitation from other valence electrons and
from the 4d $\rightarrow$ ϵp excitations. These atomic effects persist in solids, although
altered somewhat (Abbati *et al.*, 1983, Rossi *et al.*, 1983, Cole *et al.*, 1994). The
Cooper minimum may be used to reduce the contribution of a subshell to the
photoelectron spectrum thereby allowing the contributions from other electrons
on the same atom, or a different kind of atom, to be seen.

3.11.2 Resonance photoemission

Resonance usually means that a response to a frequency-dependent stimulus
goes through a maximum at some frequency. Resonance photoemission in a
sense is doubly resonant. The photoabsorption coefficient goes through a very
pronounced maximum, often preceded by a minimum, the signature of an interference effect. In addition, there are preferred de-excitation modes leading to
peaks in the photoelectron current per absorbed photon. The largest and most
important such effects occur in 3p $\rightarrow$ ϵd transitions in 3d transition metals, 4d
$\rightarrow$ ϵf transitions in rare earths, and 5d $\rightarrow$ ϵf transitions in actinides. In all cases
the initial state is a core level. The phenomenon occurs for atoms, molecules,

and solids, differing only in details of the spectra (Wendin 1981, 1982; Connerade et al., 1987).

Consider the 4d → 4f excitations in the rare earth atoms, and the elements immediately preceding them. These take place at about 100 eV photon energy. The lowest unoccupied 4f states are localized on the atom and might be expected to give discrete absorption lines. The single 4d hole and the 4f electrons, both those initially present and the newly promoted one, have similar average radii. This leads to large Coulomb and exchange interactions among these electrons and the core hole, interactions of the order of 10 eV. These interactions spread the final states over a large range in energy, and those pushed up may become degenerate with the continuum into which they autoionize. An example of this, the absorption spectrum of CeF_3, is shown in Fig. 3.20. Several sharp Ce 4d → 4f lines appear below the large peak, which is due to other, stronger 4d → 4f transitions which autoionize to give the continuum. This large peak, well above threshold, is associated with the centrifugal potential, the $\ell(\ell + 1)/r^2$ term in the Hamiltonian expressed in spherical coordinates. This term tends to push the radial wave functions for states of large ℓ away from the nucleus, causing poor radial overlap with the 4d states. Consequently the 4d → 4f dipole matrix elements tend to be small until the ϵf state energy is large enough that the final-state wave function can penetrate the weak potential barrier produced by the sum of the radial potential energy and the centrifugal potential term. Then overlap is better and the matrix element increases. The delayed onset of absorp-

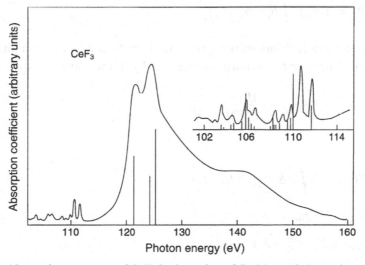

Fig. 3.20 Absorption spectrum of CeF_3 in the region of Ce 4d → 4f absorption. The "fine structure" is due to excitation to $4d^9 4f^2$ multiplet states below the ionization threshold, and the broad "resonant" continuum to the excitation of $4d^9 4f^2$ multiplets above this threshold. The lines are the spectrum calculated for the free Ce^{3+} ion. (Olson and Lynch, 1982)

tion is observed for the 4d excitations of not only the rare earths, but also several elements preceding them in the periodic chart, I, Xe, Cs, Ba, and La. The peak in absorption is sometimes called the giant resonance. It is made even larger by an interference effect (Fano and Cooper, 1968).

The 4d → 4f transitions leave an excited state which decays in several ways, one of which is the following:

$$4d^9 4f^{n+1} \rightarrow 4d^{10} 4f^{n-1} + \epsilon.$$

The decay is by an Auger process brought about by the screened electron–hole Coulomb interaction. The final state is the same as that reached by direct photo-ionization of a 4f electron:

$$4d^{10} 4f^n \rightarrow 4d^{10} 4f^{n-1} + \epsilon.$$

Although the latter transitions begin at much lower energies, they provide a continuum extending through and beyond the region of photoabsorption due to the start of 4d photoexcitation. Both processes lead to the same final state. The amplitudes for each process must be squared in the absorption cross-section, and this can lead to interference effects. Such interference between a discrete excitation and a continuum in atomic absorption was studied by Fano (1961), Fano and Cooper (1965) and reviewed by Fano and Cooper (1968).

If the discrete excited state has wave function ϕ and energy E_f, and the continuum wave function is ψ_E, the excited-state wave function becomes

$$\Psi_E = a_E \phi + \int b_{EE'} \psi_E dE,$$

when the interaction (Coulomb) between the discrete and continuum final states is turned on. The interaction matrix element is V_E. Fano solved for a_E and $b_{EE'}$:

$$|a_E|^2 = \frac{|V_E|^2}{[E - E_\phi - F(E)]^2 + \pi^2 |V_E|^4},$$

where

$$F(E) = P \int dE |V_E|^2 / (E - E')$$

and P denotes "principal part";

$$b_{EE'} = V_E / (E - E').$$

To the extent that $F(E)$ and V_E can be considered constant, a_E is a Lorentzian whose peak is shifted from E_f by the interaction with the continuum, and whose width is determined by this interaction (autoionization). The discrete final state is thus smeared out over a range of energies of the order of $|V_E|$.

Fano then calculated the matrix element of a transition operator T (which can be the momentum operator $\mathbf{p}$) between the initial state i and a final state:

$$\langle \Psi_E | T | \phi_i \rangle = a_E^* \langle \phi | T | \phi_i \rangle + \int b_{EE'}^* \langle \psi_E | T | \phi_i \rangle \mathrm{d}E.$$

Taking the absolute square of this to get the cross-section gives an interference effect. The ratio of this matrix element squared to the square of that of the original continuum, $\langle \psi_E | T | \phi_i \rangle$ gives the ratio of the cross-sections or absorption coefficients with and without the interaction:

$$\sigma(E)/\sigma_0(E) = (q + \epsilon)^2/(1 + \epsilon^2) = 1 + (q^2 - 1 - 2q\epsilon)/(1 + \epsilon^2),$$

where ϵ is a normalized unit of energy:

$$\epsilon = [E - E_\phi - F(E)]/\pi |V_E|^2.$$

F and V_E depend on energy, but are usually taken as constants over the width of a spectral feature. The lineshape parameter q is the scaled ratio of transition operator matrix elements for transitions to the modified discrete state to those to a band of original continuum states, both from the initial discrete state:

$$q = \frac{\langle \Psi_E | T | \phi_i \rangle}{\pi V_E^* \langle \psi_E | T | \phi_i \rangle}.$$

The cross-section ratio, often called a "Fano line shape," is shown in Fig. 3.21 for positive q. The zero values of the absorption coefficient are the result of perfect destructive interference. Note that as q gets below about 0.5 there is no peak in the photoexcitation spectrum, but the zero remains. When q is negative, the shape is a reflection about $\epsilon = 0$ of the plots in Fig. 3.21. Fano extended his analysis to several discrete states interacting with several continua (Fano, 1961; Davis and Feldkamp, 1981).

The decay of the $4d^9 4f^{n+1}$ excited state listed above is not the only possible decay; other core electrons or valence electrons may be involved. Several possible decays are:

$$4d^9 5p^6 4f^{n+1} 6s^2 6p \rightarrow 4d^{10} 5p^6 4f^{n-1} 6s^2 6p + \epsilon$$
$$\rightarrow 4d^{10} 5p^6 4f^{\,n} 6s6p + \epsilon$$
$$\rightarrow 4d^{10} 5p^6 4f^{\,n} 6s^2 + \epsilon$$
$$\rightarrow 4d^{10} 5p^5 4f^{\,n} 6s^2 6p + \epsilon.$$

These are shown in Fig. 3.22. The continuum electrons can escape; they are those collected in resonance photoemission.

Returning now to the Fano line shape, we can see that by using a photon energy at the minimum of the cross-section, in an ideal situation we do not excite the 4d or 4f electron at all, for the cross-section is zero. There will be absorption from other states that do not interact with the 4d or 4f. Then by using a photon

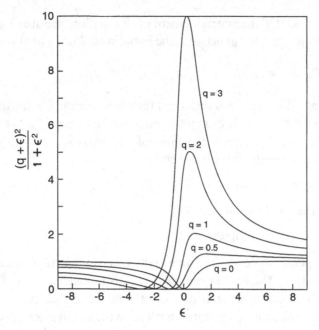

Fig. 3.21 Fano line shapes for various values of the parameter q. Reverse the direction of the ϵ scale for negative q. (Fano, 1961)

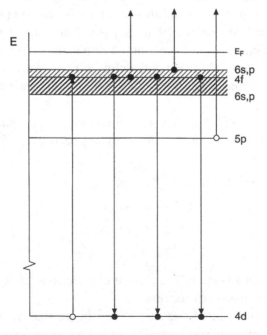

Fig. 3.22 Schematic picture of possible 4d → 4f resonance photoemission processes. The 4f level is shown below the Fermi level. In fact, its location will change with occupancy. Moreover, with a 4d hole, the spread in 4d–4f multiplet structure can be over 10 eV.

energy at the maximum of the Fano line shape, the excitation of 4d or 4f electrons is maximized. These two energies are rather close, so the background cross-section from other electrons does not change much. Then by taking the difference, "on resonance" minus "off resonance," the other processes are removed from the spectrum. The difference spectrum contains only photoelectrons from the decays of the excited states listed above or from the fairly weak 4f direct ionization continuum.

Actual situations are more complicated. First there is the multiplet structure of the $4d^9 4f^{n+1}$ states to consider. This was treated by Fano and students (Dehmer et al., 1971; Dehmer and Starace, 1972; Starace, 1972; Sugar, 1972) in absorption. Then there is the untangling of the individual decay processes listed above. This was tackled by Zangwill and Soven (1980) for Ce (see also Zangwill, 1982), with subsequent verification by photoemission on Ce vapor by Meyer et al. (1986) and Richter et al. (1988). The photoemission of an electron from one subshell may not have the same spectral dependence within the resonance absorption envelope as that from another subshell. This appears in both the calculation of Zangwill and Soven and in the measurements of Richter et al., shown in Fig. 3.23.

Figure 3.24 shows an example of the use of resonance photoemission. The angle-integrated valence-band photoelectron spectrum of NiO, Fig. 3.24(a), has three structures, labeled A, B, and C. Peak A arises from states composed primarily of Ni 3d atomic states, and C is a satellite associated with these states, a two-hole final state, to be discussed in Chapter 5. Region B is from bands based mainly on O 2p states. As the photon energy is scanned through the Ni 3p core excitation threshold the relative intensities of peaks A and C change with respect to that of B, as can be seen in Fig. 3.24(a). This resonance effect marks those parts of the valence band which have significant Ni 3d character. CIS spectra, shown in Fig. 3.24(b) for initial states in the regions of structures A, B, and C, show Fano line shapes. These shapes are different for peaks A and C. This difference can be understood theoretically (Fujimori and Minami, 1984). Peak C also shows slight structure at the Ni 2p threshold, accounted for by a small amount of Ni 3d character in this region of the valence band.

3.12   Core-level satellites

For a relatively simple core level, e.g., a 2p level, in the independent-electron picture one expects two peaks separated by the spin–orbit splitting. The apparent binding energies are those of the two one-electron levels calculated with the other electrons frozen in their ground-state orbitals. This is Koopmans' theorem. We have mentioned above that the spectra rarely are this simple due to the

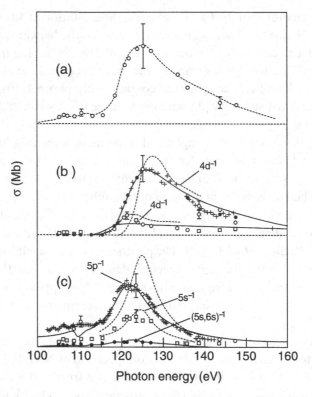

Fig. 3.23 Photoabsorption spectrum of atomic Ce in the region of 4d → 4f excitation. The curve in (a) is the total measured cross-section and the circles represent the sums of the measured partial cross-sections. (b) and (c) show measured (points) and calculated (curves) partial cross-sections for various final states. (Richter *et al.*, 1988)

screening by conduction electrons or electrons on ligands, which may also be viewed as creating a second hole in the valence band or on a neighbor and filling, or partially filling, the original photohole, all during the photoemission process. There are additional many-body effects which lead to additional discrete peaks in the core hole spectrum. These peaks, called satellites, often were called shake-up peaks in the early literature.

The independent-electron approximation is inadequate for giving an accurate explanation of observed core-level spectra because it neglects the response of the system to the core hole. The $N - 1$ electrons in the final state have a different Hamiltonian than the N electrons had in the initial state. In the sudden approximation any of the final states may be reached, subject to dipole selection rules and spatial overlap with the core hole. One of the final states, that with lowest energy, may be the ground state of the $(N - 1)$-electron system. The others are all excited states. The spectrum may then consist of several lines, and perhaps a continuum at higher apparent binding energy. In the one-electron picture, the

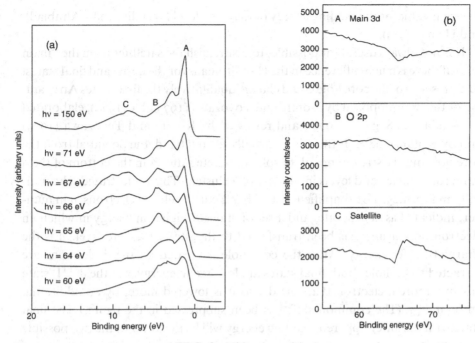

Fig. 3.24 (a) Angle-integrated EDCs for NiO taken with photon energies spanning the Ni 3p photoexcitation threshold. (b) CIS spectra for initial states in peaks A, B, and C. (Oh *et al.*, 1982)

system wave functions can be written as products on one-electron functions. The photon then excites one electron from state ϕ_i to state ϕ_f. The other electrons have wave functions Φ_i and Φ_f, where each Φ is a product of $N - 1$ one-electron functions, but the Hamiltonians for the the states i and f are for N and $N - 1$ electrons, respectively. In the sudden approximation, the photocurrent will be proportional to the square of the matrix element

$$\langle \phi_f | \mathbf{p} | \phi_i \rangle \langle \Phi_f | \Phi_i \rangle.$$

The second factor, a multi-dimensional overlap integral, would be zero if the states i and f were not identical, but both were eigenstates of the same Hamiltonian. Because the Hamiltonians are different, a number of final states Φ_f can give non-vanishing integrals, and these states will have different energies, giving rise to final states with one or more electrons excited in addition to the one that is emitted. Each will have a strength proportional to the square of the above overlap integral. These states need not all be discrete. If one could separate accurately the continuum of such states from the background of inelastic electrons so that the average energy of all the contributions to the spectrum from one core level can be obtained, the result should be equal to the Koopmans'

theorem value for the binding energy of that core level (Wendin, 1981; Almbladh
and Hedin, 1983).

It is not always useful (or possible) to distinguish the satellites from the "main
line" if there is much difference in the Hamiltonians for the initial and final states.
The presence of the core hole could change qualitatively the final states An example of this was proposed by Kotani and Toyozawa (1973a,b,c, 1974), elaborated
by Asada and Sugano (1976), and reviewed by Kotani and Toyozawa (1979)
and by Kotani (1987). If there is a partially filled d band, the potential from the
core hole may be strong enough to split a discrete state from the bottom of the d
band (or complex of d levels in the case of a cluster). This is illustrated schematically in Fig. 3.25. The simplified system has a core level, c, d electrons, n in number, indicated as degenerate, and a set of states Δ higher in energy in which an
electron on a ligand has been transferred to the site of the core level. Δ is the
charge-transfer energy. After the core hole is created all the d electrons are
attracted to the hole. Both final states are lowered in energy, but the $d^{n+1}L$ state
has one more d electron than the d^n, so it is lowered more, by an additional
amount U_{cd}. (The common shift has been suppressed in the figure.) The final
state with the lower apparent binding energy will be the lower of the two possible

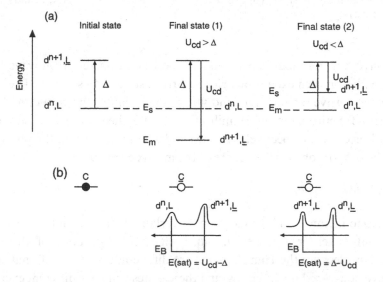

Fig. 3.25 Energy-level diagram for a transition-metal compound before and after
removal of a core hole. The two final states (*right*) are the ground (d^n) state and the state
(d^{n+1},$\underline{L}$) produced by transferring an electron from a ligand orbital to the transition
metal, a process requiring energy Δ. (Hybridization has not been included.) The final states
have a core hole which lowers the energy of the state $|d^{n+1}\underline{L}\rangle$ an amount U_{cd} more than it
lowers that of the state $|d^n\rangle$ because the former state has more electrons localized near the
core hole. Which final state is lower in energy depends on whether U_{cd} is larger (*center*)
or smaller (*right*) than Δ.

final states, but this depends on the magnitudes of Δ and U_{cd}. If U_{cd} is less than Δ, the lower final state is d^n, giving what might be called the main peak, and the peak due to $d^{n+1}\underline{L}$ will appear $\Delta - U_{cd}$ higher in binding energy, possibly called a satellite or shake-up satelllite. If U_{cd} is larger than Δ, the lower final state is $d^{n+1}\underline{L}$, which has an extra d electron. It corresponds to the state split off below the d complex by the core hole potential. Then the d^n state appears at $U_{cd} - \Delta$ higher binding energy in the spectrum. The spectra, which look alike, are shown in the bottom of Fig. 3.25.

3.13 Gap states

Frequently, the angle-resolved EDC of a metal may show some intensity at and below the Fermi level, in cases where no final states are possible for initial states near E_F when one considers only the bulk band states and surface states. After eliminating the possibility of the involvement of a non-zero reciprocal lattice vector, and the indirect processes from the $\nabla \cdot \mathbf{A}$ part of H', described above, one concludes that the final states must be gap states, or surface states. These states have energies in the band gap. Such states are well known. They are described by a complex wave vector, the imaginary part of which is directed normal to the surface, causing the wave-function amplitude to decay with distance into the solid, usually within a few lattice spacings. They have been seen and identified by a number of authors, for example on Cu by Dietz and Himpsel (1979), and discussed more extensively by Courths *et al.* (1989).

3.14 Other spectroscopies

We have described photoemission and inverse photoemission spectroscopies. They involve the absorption of a photon and the removal of an electron, and the addition of an electron and the emission of a photon, respectively. The electric dipole matrix element plays a role in both spectroscopies in both the independent-electron picture and any treatment including many-body effects. There are other spectroscopies which give similar, but never the same, information. These spectroscopies also have descriptions in the independent-electron and many-body frameworks. Some do not involve the dipole matrix element. Photoelectron spectroscopy of a material should not be studied in isolation. Attention should be paid to all other available information, especially the other spectroscopies reported below, for they involve very similar, but not identical, excitations of the system. As we shall see, however, inter-relating data from different spectroscopies is not always easy.

The oldest, and experimentally simplest, experiment is optical absorption. Infrared, visible, and near-ultraviolet absorption spectroscopy are widely practiced, but any extensive discussion of such spectra will take us too far afield. Interband optical absorption occurs at a rate proportional to the square of the dipole matrix element between initial and final states and the densities of initial and final states, in the independent-electron approximation. For Bloch states, wave vector is conserved, as discussed for photoexcitation. For direct band-gap insulators, the conduction-band minima and the valence-band maxima are at the same point(s) in the Brillouin zone and the absorption coefficient rises rapidly with increasing photon energy at the band-gap energy, reaching values of 10^4 cm^{-1} and more. For indirect-gap insulators, the wave vectors of the conduction-band minima and valence-band maxima are at different points in the Brillouin zone. The dipole matrix element between them is zero. The band-edge absorption occurs in second order, with the participation of a phonon to conserve wave vector, and the absorption edge is consequently weaker. In either case at the band edge, the electron–hole Coulomb interaction leads to exciton creation below the band-gap energy, and to distortions in the spectra above the band edge. If these are weak, the Mott exciton picture is valid and the exciton binding energies are small. If they are strong, one is in the Frenkel exciton regime, and the exciton binding energy can be rather large, up to an eV or so, with appreciable distortion of the absorption edge for an eV or so above the band-gap energy. The Mott exciton is treated in the effective mass approximation and the Coulomb interaction is screened by a presumably large dielectric function. A Frenkel exciton resembles an excited atom or molecule, and the rate of transfer of the excitation to equivalent neighboring atom sites is slow. The exciton band width is small. These two limiting cases are analogous to the band and localized starting points in the treatment of the electronic structure of the cuprates.

At higher photon energies, excitation of core electrons can occur. Soft x-ray or x-ray absorption spectroscopy is widely practiced by several experimental techniques and the results often compared with those of other spectroscopies. The probability of absorbing a soft x-ray photon by exciting a core level is proportional to the square of the electric dipole matrix element and the density of final states at and above the Fermi level in the independent-electron picture. The initial state is the core level state with a wave function well localized on a particular type of atom and with a definite angular momentum, for the potential from the nearest ion core far exceeds that from the rest of the crystal. The final states are only those with a large amplitude on the atom of interest and with the proper angular momentum quantum number, larger or smaller by unity from that of the core level under study. Thus the threshold absorption for the O 1s level is to final states with large amplitude on the O sites and with p-like symmetry about those sites. For the Cu 2p level, the final states are of s and d symmetry about the Cu site. At the absorption edge, barring exciton effects, states just above E_F with a particular symmetry about a particular ion are the final states in the dipole matrix element. The use of polarized radiation and single crystals allows selection rules on Bloch states or localized orbitals to further help identify the final states.

Measurements have been made three general ways. The most straightforward is by attenuation of a photon beam by a thin-film sample. Other optical techniques suitable for the visible and near-ultraviolet regions do not work well at high energies because the dielectric function has a real part very near to unity and a small imaginary part. This, however, makes corrections for reflections in the transmission measurement negligible and one can get ϵ_2 directly. Total (or partial) photoelectron yield spectra resemble ϵ_2 spectra in the region of core-level excitation, except very shallow core levels. Although the magnitude of ϵ_2 cannot be obtained this way, the energies of thresholds and peaks may be obtained rather easily. Finally, electron energy-loss spectra (EELS) can give core-level ϵ_2 spectra rather directly. The measured quantity is $Im(-1/\tilde{\epsilon}) = \epsilon_2/(\epsilon_1^2 + \epsilon_2^2)$, but at energies above 20 eV or so, $\epsilon_1 \simeq 1$ and $\epsilon_2 \ll 1$, so this function is very close to just ϵ_2. The loss function, $Im(-1/\tilde{\epsilon})$, actually contains the longitudinal dielectric function, not the transverse one, and it depends on wave vector as well as energy. However, if the measurements are made for momentum transfers of nearly zero, i.e., in transmission with thin samples, the longitudinal dielectric function and the transverse are essentially equal. If the energy of the electron beam used for the measurement is well above threshold for the excitation of a particular core level, electric dipole selection rules are valid.

The "inverse" of optical absorption is fluorescence. The same electric dipole selection rules apply, but now to initial states that are in the filled valence bands below E_F to holes localized on one type of atom or ion. The hole must be created. Both photons and electrons have been used, although as the resolution of the experiment is increased, monochromatic photons are preferred, for they can be tuned through the threshold for core production. Both absorption and emission spectroscopy can be viewed as conserving the number of electrons, at least at threshold, where they are also viewed as adiabatic transitions as far as the electronic system is concerned. They may not be adiabatic with respect to phonons; the emission and absorption thresholds may differ by an energy representing the emission of phonons in each of the processes. (Phonon absorption is possible at high temperatures.) Well above threshold, the transitions in absorption may be viewed as sudden, but the spectrum there is a continuum with little structure.

Auger spectroscopy measures the spectrum of Auger electrons produced when a core hole, usually either electron- or photo-induced, is filled by an electron from a subshell of lesser binding energy, with energy transferred to the emitted electron. (Carlson, 1975; Fuggle, 1981; Weissmann and Müller, 1981; Weightman, 1982; Ertl and Küppers, 1985; Woodruff and Delchar, 1994.) If all the electrons involved are core electrons, the process is denoted $CC'C''$, where C identifies the original core hole. An example is the KL_1L_2 process in which a hole is produced in the K shell, and filled by an electron from the L_1 shell with emission of an electron from the L_2 shell. The atom is left with two holes. The spectrum from such a process involving three core states does not change much when the atom is placed in different environments, i.e., different compounds, even with valence differences, because relative core-level energy shifts are very small. $CC'V$ and CVV' processes may be more useful. V stands for a valence electron. Since the valence-electron distribution depends on environment these spectra may be different for different compounds containing the element whose Auger spectrum is studied. Such spectra may be rather broad, up to twice the filled valence-band width, but some structure, usually rather broad, may appear.

The dipole matrix element does not play a role, except in photoexcitation, and except at threshold excitation, it is not considered. (Note that in any photoemission spectrum, photoexcited Auger electrons may appear. They may be distinguished from photoelectrons for their energies do not depend on photon energy.) Except at threshold, the spectra are not influenced by the energy spread of the exciting particles. The interaction leading to Auger emission is the Coulomb interaction between the core hole and an electron and between two electrons. For the CVV' Auger emission the transition emission rate is proportional to the square of the matrix element

$$\int \psi^*(\mathbf{r}_1)\psi_V^*(\mathbf{r}_2)(e^2/r_{12})\psi_C(\mathbf{r}_2)\psi_V(\mathbf{r}_1)d^3r_1d^3r_2 + \text{exchange term},$$

where the unsubscripted wave function is that of the electron that is detected. The wave functions subscripted C and V are those of the singly core-ionized atom, while the other two are for the doubly ionized atom. One can see that radial overlap of the wave functions is important in obtaining a large Auger transition rate. For $CC'C'$ transitions, when C and C' are in the same shell the overlap is good, and the intensity of these Coster–Krönig transitions is high. The even more intense super-Coster–Krönig transitions occur when C, C', and C' are all in the same shell. There are some selection rules based on angular momentum that simplify the calculation of the spectra.

With the neglect of the Coulomb matrix element, the $CC'V$ spectra should resemble the density of states in the valence band below E_F, and the CVV spectra should resemble the self-convolution of the valence-band density of states below E_F. The former case is often observed, but it is the valence band in the presence of the hole or holes. For CVV spectra, comparison with experiment shows that the self convolution is close to experimental spectra for a number of materials, but not at all close for others. When this picture fails completely, the Auger spectrum is narrower than twice the filled valence-band width expected from the self-convolution of the filled part of the valence band. Examples of such a case are the L_2VV and L_3VV Auger spectra of Cu. Besides the fact that the main peak is narrower than twice the filled valence-band width, even with the inclusion of matrix elements in the calculated spectrum, there are some satellite structures reminiscent of XPS satellites. These suggest that electron–electron Coulomb repulsion effects are stronger than in those cases for which the self-convolution of the filled valence band is reasonably close to the experimental Auger spectrum. In such cases the Coulomb repulsion, U, of two holes on the same site is large compared to W, the width of the valence band, leading to the splitting off of a two-hole bound state from the valence band. This final state can have much more intensity in the spectrum than the final state with two "non-interacting" valence-band holes in the independent-electron approximation, thus dominating the spectrum. This intensity arises because the two valence-band wave functions in the matrix element are no longer Bloch functions, but functions localized on the site of the original hole, and overlap is thereby enhanced. The spectrum then resembles the Auger spectrum of the free atom, for all three states, C, V, and V', are more and more localized on one atom in the solid as U/W increases (Cini, 1976, 1977, 1978; Sawatzky, 1977; Sawatzky and Lenselink, 1980). When this happens it is difficult to relate the energies of features in the Auger spectra to the Fermi energy, making comparison with other spectroscopies more difficult.

Auger emission competes with fluorescence for the de-excitation of a system with one core hole. Auger emission is dominant for elements with an atomic number less than about 33 (As); fluorescence dominates for elements of higher atomic number. However, fluorescence, though weak, can be measured from lighter elements, and Auger spectra for heavier elements can be obtained.

The spectroscopies mentioned above have similarities, in that some involve the dipole matrix element, and differences, in that Auger spectroscopy does not. They also differ in the number of electrons in the final state, a fact not important in the independent-electron picture, but important in most actual cases. If we start with an N-electron system, the final state in photoemission from either valence or core levels has $N - 1$ electrons, while for inverse photoemission the final state has $N + 1$ electrons. In soft x-ray absorption, including by EELS, the final state has N electrons at and near threshold, and $N - 1$ at higher energies. In soft-x-ray fluorescence, the final state has $N - 1$ electrons, since hole creation usually removes an electron from the sample. Finally, Auger spectroscopy leaves the sample with $N - 2$ electrons.

Baer (Baer, 1984; Baer and Schneider, 1987) developed a pictorial representation for these spectroscopies, shown in Fig. 3.26(a) and (b). In each figure, the regions below the horizontal bar show the energies of the incident and outgoing particle(s), measured downward from the Fermi level of the N-particle system.

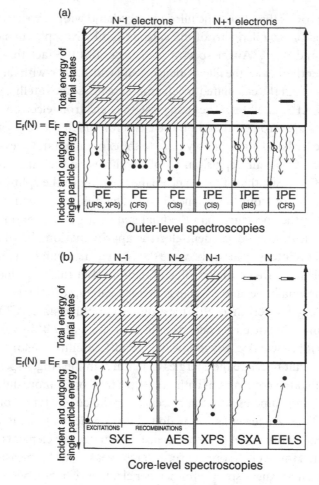

Fig. 3.26 Energy-level diagram for several electron and photon spectroscopies: (a) is for valence-band spectroscopies, and (b) for core-level spectroscopies. In each figure the regions below the horizontal bar show the nature of the incident particles (straight line for electrons, wiggly line for photons) and the energies of the outgoing particles, measured downward. The regions above the bar show the final-state energy of the system left behind, with an open rectangle denoting a hole and a filled rectangle representing an additional electron. PE = photoelectron emission, IPE = inverse photoemission, SXE = soft x-ray emission, AES = Auger electron spectroscopy, SXA = soft x-ray absorption, EELS = electron energy loss spectroscopy. (Baer and Schneider, 1987)

Straight lines represent electrons and wiggly lines, photons. An arrow crossing a line indicates a variable energy. The regions above the bar indicate the corresponding energies of the final states for the processes shown, with respect to the energy of the initial, N-electron ground state. Open rectangles represent holes, filled rectangles represent electrons, and the half-filled rectangles represent excited N-electron states with no change in electron count.

Another spectroscopy, not yet applied to cuprates, is target current spectroscopy (Schäfer *et al.*, 1987; Otto and Reihl, 1990; Drube *et al.* 1992; Komolov, 1992). It is a relatively little-used technique, but it can yield data similar to angle-resolved inverse photoemission. A collimated beam of low-energy electrons is incident on a single-crystal sample. The electron gun in a commercial low-energy electron diffraction apparatus may be used. The energies may be reduced to nearly zero by a retarding field. The incident electrons have a definite energy and wave vector. They will enter the crystal if there are suitable states available. If not, they are reflected, forming a LEED pattern and an elastic and inelastic background. The spectrum of current into the sample vs. beam energy shows structures due to bands above E_F. Varying the angle of incidence can yield a band map similar to that of inverse photoemission. To date, little data exist and the limitations of the technique have not been explored extensively.

Chapter 4

Photoemission – Experimental

The key instrumentation necessary for photoemission consists of a radiation source, an electron energy analyzer, and a means of preparing and maintaining a clean sample surface in ultrahigh vacuum. Ancillary sample characterization capability, both *in situ* and *ex situ*, is also important, especially for non-stoichiometric samples like high-temperature superconductors. We discuss each of these in turn, emphasizing some of the important attributes of each. We do not go into enough depth to allow a reader to construct the instrumentation. We emphasize the crucial characteristics of resolution and signal-to-noise ratio, and how they are limited by instrumentation.

4.1   Radiation sources

Work functions may be as high as 5 eV, so light sources for photoemission must emit photons of energies higher than this. Since O_2 and N_2 absorb photons starting just above 6 eV, the entire optical path must be evacuated to a pressure below about 10^{-5} Torr in order to ensure significant transmission over light paths of a meter or more. For this reason, the spectral range above about 6 eV is called the vacuum ultraviolet region, Its name changes at around 50–100 eV to the soft x-ray region, for which the optical path must also be in vacuum.

Radiation sources used in the earlier valence-band studies (Berglund and Spicer, 1964b) often were hydrogen discharges. These gave a weak continuum from the H_2 molecule and a strong emission line, the Lyman-α line at 10.2 eV, from the H atom (Samson 1967). A grating monochromator was used to disperse the continuum and a window of LiF was used to separate the gas discharge from the vacuum of the monochromator, with perhaps a second such window to separate the relatively poor vacuum of the latter from the ultrahigh vacuum (UHV) of the sample chamber. Bulk LiF transmits further into the vacuum ultraviolet than any other bulk material. A plate 2 mm thick will transmit at least 10% up to 11.9 eV (Samson, 1967). (Thin-film windows cannot support a large pressure differential.) Thus the photon energy range was rather limited. This sometimes was increased in effect by decreasing the work function. An overlayer of about 1/3 of a monolayer of Cs can lower the work function of many surfaces by several eV. In the 1960s no other effects from the Cs were discerned.

Two improved sources soon came into use. Eastman and Cashion (1970) demonstrated that the He resonance line could be used without a monochromator and with no windows, without contaminating the sample chamber vacuum with He gas, by installing several small apertures and differential pumping between the lamp and the sample chamber. In fact, there are two He resonance lines, one (He I) from the atom at 21.2 eV and one (He II) from the He^+ ion at 40.1 eV. These lines have widths of 10 meV or less. The relative intensities of these lines could be varied by changing the He pressure in the lamp. By using the two He lines, one could discriminate partially between emission from different atomic subshells, making use of the different dependences of the photoexcitation cross-sections on energy. Helium discharge lamps have been commercially available for many years. Resonance lines of other rare gases can be used to excite at different photon energies, but this is no longer done frequently. The He resonance lines are still widely used for angle-resolved photoemission, including work on high-temperature superconductors.

Very few core levels can be excited by He radiation. The standard sources for XPS have been electron-beam excited Mg-$K\alpha$ and Al-$K\alpha$ radiation at 1245 and 1487 eV, respectively (Carlson, 1975; Rivière, 1983). The radiation may be used as emitted, after much of the continuum has been filtered by a thin Al film. The line widths are then 0.8 and 0.9 eV, respectively, but there are also weak satellite lines ($K\alpha_3$, $K\alpha_4$, ..., $K\beta$) which cause additional spectral features. These can be eliminated by using a monochromator, which also has the virtue of producing a narrower $K\alpha$ line. The dispersing element is usually a quartz crystal. However, the flux on the sample is reduced considerably. The flux can be increased at little loss in resolution by the use of many quartz crystals, as many as 30, to increase the solid angle collected from the source by the monochromator. Such sources are now available commercially. The best spectral bandpasses are about 0.2 eV. This makes them suitable for core-level spectroscopy, but not for high-resolution valence-band studies.

The second improved source was synchrotron radiation (Winick, 1980; Koch et al., 1983; Margaritondo, 1988; Winick, 1995). In the late 1960s synchrotron radiation was used for photoemission studies at several sites around the world, using beam lines built on electron synchrotrons operated for studies in particle physics or on electron or positron storage rings either operated as facilities for particle physics or operated primarily as sources of vacuum ultraviolet, soft x-ray, and even hard x-ray radiation. (A synchrotron accelerates charged particles to some maximum energy, then extracts them, after which more particles are injected and accelerated. A storage ring is a synchrotron that does not extract the particles, but accelerates them and holds them at fixed energy for long periods of time, with half-lives that can exceed 24 hours.) These facilities are called first-generation sources. Second-generation sources are storage rings usually designed, built, and operated exclusively to serve as sources of synchrotron radiation, which is emitted from the bending magnets, where the electron beam characteristics were optimized, within existing technology. These rings also had short straight sections of vacuum tank between bending magnets, with quadrupole and sextupole magnets for focusing. It was soon realized that one could put certain devices, "insertion devices," on the straight sections to produce more intense radiation than that produced in the bending magnets. The radiation spectrum of these devices can be tailored to be quite different from that of the bending magnets. Third-generation sources, which began operating in the 1990s, are designed to optimize the characteristics of the electron or positron beam in the straight sections, which were made more numerous and longer in order to accommodate more and longer insertion devices. We describe insertion devices in more detail later.

Synchrotron radiation is electromagnetic radiation emitted by charged particles with relativistic velocities as they are accelerated radially by a static magnetic field, originally the field of a bending magnet in the accelerator. (There is also synchrotron radiation from astrophysical sources.) The radiated power, its spectrum, and its angular dependence can be calculated in closed form (Jackson, 1975; Kunz, 1979a,b; Winick, 1980; Koch et $al.$, 1983; Krinsky et $al.$, 1983; Brown and Lavender, 1991; Raoux, 1993). As with all electromagnetic radiation, the radiated power is proportional to the square of the charge-to-mass ratio of the emitting particles, largest for electrons and positrons. (Positrons, though more difficult to produce, offer advantages in beam stability and lifetime, and are sometimes used, despite the requisite increase in capital cost of a facility.) Relativistic electrons are described by a parameter $\gamma = E/m_0 c^2$, the ratio of the total energy of the electron to its rest energy. For an electron energy E of 1 GeV, γ is about 2000. The synchrotron radiation spectrum is described by a characteristic wavelength $\lambda_c = (4\pi R/3)\gamma^{-3}$, where R is the local bending radius of the electron orbit, determined by E and the local magnetic field. For 1 GeV electrons and $R = 3.5\,\mathrm{m}$, λ_c is 18.3 Å.

For a single electron the radiation is confined to a narrow cone centered on the forward tangent to the electron's path. For wavelengths well below λ_c, the power distribution with angle falls off from this tangent approximately as a Gaussian with a full-width at half maximum (FWHM) of $2\gamma^{-1}(\lambda/\lambda_c)^{1/3}$. For wavelengths much longer than λ_c, the width increases as $2\gamma^{-1}(\lambda/3\lambda_c)^{1/2}$. If the electron energy E is 1 GeV, γ^{-1} is only 5×10^{-4} radians. In the plane of the electron orbit ($\psi = 0$), the radiation is linearly polarized, as shown in Fig. 4.1. Above and below this plane, the radiation is elliptically polarized, becoming closer to circular polarization as the angle increases (and the power falls). In an ideal case, N electrons all following exactly the same path, the radiated power would be N times that of a single electron; coherence effects usually not being present. The synchrotron radiation spectrum can be calculated. It is shown in Fig. 4.2. The abscissa is directly proportional to the current in the storage ring. The peak in the spectrum shifts to higher photon energy as the electron energy E increases, and the total radiated power increases.

Descriptions of storage rings and electron (or positron) motion in them may be found in Rowe (1979), Krinsky *et al.* (1983), Wiedemann (1991), and Winick (1995). The electrons cannot all follow the same path. This increases the apparent source area and can broaden the angular distribution. Any energy spread will cause a spread in the bending radii of the equilibrium orbit, and at the least, electrons lose some energy by radiation emission which causes such a spread. The radio-frequency cavity restores some of this, but until it is reached in each period, some electrons will be slower than others, hence on different orbits. Gas scattering and the focusing effect of magnets cause electrons to leave their original orbits. As a result of these and other effects the electrons execute transverse oscil-

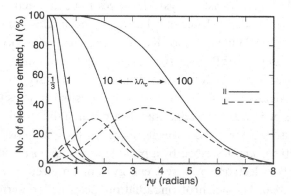

Fig. 4.1 Universal plot of the angular distribution of synchrotron radiation above (and below) the orbital plane. The width of the distribution varies with wavelength as shown, with wavelengths in units of the critical wavelength λ_c. ψ is the angle from the orbital plane, and typically is a few milliradians. The electric vector is indicated as parallel or perpendicular to the orbital plane.

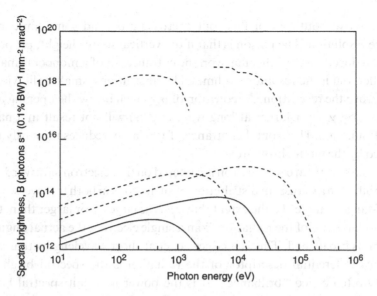

Fig. 4.2 Spectral brightness of synchrotron radiation from the bending magnets of several storage rings. Lower solid curve: 800 MeV uv ring at the National Synchrotron Radiation Source (NSLS), Brookhaven National Laboratory. Upper solid curve: 2.5 eV x-ray ring at NSLS. Dashed curves: 1.5 GeV Advanced Light Source (ALS) at Lawrence Berkeley National Laboratory; and 7 GeV Advanced Photon Source (APS) at Argonne National Laboratory. The upper dashed curves are for undulators mounted on the ALS (*left*) and APS (*right*), whose meaning is explained in the text accompanying Fig. 4.5.

lations about their equilibrium orbits in the ring. At each point around the circumference a spread in electron displacements from the equilibrium orbit and in transverse velocities is found. The distributions in transverse position and velocity or momentum are taken to be Gaussian distributions of widths (standard deviations) σ_x, σ_y, $\sigma_{x'}$, and $\sigma_{y'}$. The products $\sigma_x\sigma_{x'}$ and $\sigma_y\sigma_{y'}$ are called the horizontal and vertical emittances, ϵ_h and ϵ_v. Each emittance is a conserved quantity around the ring (from Liouville's theorem). If focusing reduces the vertical height of the electron distribution at a point, it is at the expense of a larger distribution in angles of the orbits above and below the direction of the ideal orbit. The maxima in distance from the equilibrium orbit occur at minima in angular spread, i.e., σ_i is maximum where $\sigma_{i'}$ is minimum and vice-versa. Moreover, because the focusing elements in a storage ring focus only in one plane and defocus in the perpendicular plane, the regions of minimum σ_y and maximum $\sigma_{y'}$ are the regions of maximum σ_x and minimum $\sigma_{x'}$. The horizontal emittance is determined by the design of the storage ring. The vertical emittance is usually a fraction, e.g., 10% or 1%, of the horizontal emittance, determined primarily by the quality of the alignment of the ring, e.g., how accurately parallel are the fields of different magnets. Vertical emittance should be small for use with high-resolution monochromators without loss of flux, but decreasing beyond some limit may not improve resolution. The reason is that if the vertical source height, proportional to σ_y, is reduced further, the image on the entrance slit of a monochromator will be smaller, but if the resolution is limited by aberrations, smaller slit widths will not improve the resolution. A reduction of $\sigma_{y'}$ much below the opening angle of the radiation γ^{-1} (or larger at long wavelengths) will not result in a narrower cone of radiation. Horizontal emittance, if too large, reduces primarily the flux collected by the monochromator.

The emission from each electron is centered on that electron's forward tangent to its orbit, with a Gaussian distribution of such orbits in both transverse position and off-tangent angle. To the extent that $\sigma_{y'}$ is as large as, or larger than, the natural angular width of the radiation from a single electron, the actual angular distribution is broadened. This is a larger effect at short wavelengths than at long. The most differential descriptor of the radiation is the spectral brightness, a quantity often called "brilliance." It is the power in a unit spectral bandpass radiated per unit area into a unit solid angle. Thus area $dx dy$ radiates $B(x, y, \theta, \phi) dx dy d\Omega d\lambda$ of power between wavelengths λ and $\lambda + d\lambda$ into solid angle $d\Omega$ oriented at angles θ and ϕ to the forward tangent. This then may be integrated over area, over solid angle, or both, to obtain other quantities. The units are $W/(m^2 \text{ sr-m})$ in SI units. Replacement of watts by photons per second, wavelength by photon energy, etc., is often done (see, e.g., Fig. 4.2). Spectral brightness and brightness, its integral over wavelengths, are conserved quantities in ideal optical systems. Optical systems can reduce B through absorption, scattering, and aberrations, but not increase it. At a bending magnet, a typical value of σ_y is 100 μm, while the opening of the entrance slit of a monochromator might be 10 μm. (The plane of dispersion of the monochromator is taken to be vertical just for the reason that σ_y is usually smaller or much smaller than σ_x.) The mirror imaging the source on the entrance slit then should produce an image demagnified by a factor of 0.1. Less reduction causes loss of intensity due to radiation blockage by the slit blades. The angular spread, initially $[\sigma_{y'}^2 + (\gamma^{-1})^2]^{1/2}$, is increased by a factor of 10. The diffraction grating must then have a ruling width adequate to capture this increased angular distribution.

The radiation from one electron at a bending magnet, confined to vertical and horizontal angles of the order of γ^{-1} about the forward tangent to the orbit, is spread horizontally into a strip as the electron moves on its arc in the magnet. To send more radiation into a monochromator, a collection mirror is needed. Such mirrors must focus in the horizontal plane, and may also focus in the vertical plane. They typically collect from a few horizontal milliradians of radiation to as much as 50–100 mrad. Collecting more might be possible, but it is worth doing only if a large fraction of the additional radiation can be passed through the succeeding optical system.

The magnetic fields of the numerous bending magnets are limited by the magnetic saturation of iron. Superconducting bending magnets rarely have been used. The bending radius, hence the size of the ring, then is determined by the energy of the electrons. In an insertion device, magnetic fields of almost any achievable magnitude may be used, including those much smaller or much larger than those used in the storage ring itself. An insertion device is a series of magnets of alternating polarity placed with fields usually vertical in a straight section of a storage ring (Kunz, 1979a,b; Spencer and Winick, 1980; Krinsky *et al.*, 1983; Howells and Kincaid, 1994). The magnets may be permanent magnets, electromagnets or superconducting magnets. The formerly straight-line path now becomes oscillatory on one side of the original path. The amplitudes of the oscillations may be small, only a few μm, smaller than σ_x. The maximum angle δ the new path makes with the original path may be smaller or larger than γ^{-1} (Fig. 4.3). In an actual insertion device, half-field magnets are placed at each end so the electrons return to the original path upon exiting. If we look along the forward tangent to the oscillatory path in an insertion device with N periods, we see N sources, or $2N$, each emitting the same synchrotron radiation spectrum, determined by the electron energy E and the maximum magnetic field of the device. The latter is tunable by varying the current in devices built with electromagnets or by altering the gap in a permanent-magnet device. This allows tuning the spectrum, perhaps to create shorter wavelengths than are produced in the bending magnets, done by increasing the insertion device field. If the radiation from each maximum orbit excursion adds incoherently to that of the others, the output power is N times that of a single magnet, and the device is called a "wiggler." N can be as large as 100. The more interesting case for us is if the fields can be added, for in such a case of high coherence, the output in part of the spectrum can be N^2 times as great as for one magnet alone, while in other parts of the spectrum, it is far less than from one magnet. Such a device is called an "undulator," and the interference peak or peaks can be tuned in wavelength by

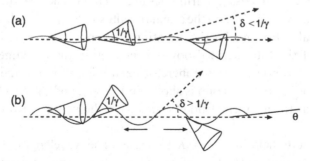

Fig. 4.3   Path of an electron in an insertion device of period L. The maximum slope of the orbit is δ and the opening cone half-angle is $1/\gamma$. (a) is for an undulator, $\delta < 1/\gamma$; (b) is for a wiggler, $\delta > 1/\gamma$.

scanning the magnetic field. For a variety of reasons the coherence is not as perfect as in the ideal case, and the brightness from an undulator depends on a power of N less than 2, but larger than 1.

Refer to Fig. 4.3. Between successive maximum excursions of the electron path, forward-emitted radiation travels in a straight line at velocity c. The electrons travel along a path that is slightly longer, and they travel at a velocity just below c. They arrive at the next point of maximum excursion several optical periods later than the photon. The radiation emitted in the forward direction at one point of maximum excursion can interfere constructively or destructively with that from the previous equivalent point if the difference in the arrival times of electron and photon is small enough that the photons overlap within a longitudinal coherence length of the radiation. The condition for this is given by the ratio of the maximum orbit deflection angle δ to the characteristic opening angle of the radiation, $K = \delta\gamma$. From the solution to the equation of motion for the electron, this ratio also turns out to be $K = eB_0L/2\pi m_0c$, where B_0 is the amplitude of the fundamental Fourier component of the magnetic field of the insertion device and L is the period of the device. When K is very small, coherence effects are large. The on-axis spectrum consists of a peak at wavelength $\lambda = L/(2\gamma^2)$. The factor γ^{-2} arises from the Lorentz-contracted period of the device when viewed in the electron frame and from the Doppler shift of the electron's radiation when viewed in the laboratory. At higher values of K, harmonics begin to appear, but they are harmonics of a fundamental that shifts to longer wavelengths with increasing K. Since the interference conditions depend on path length differences, there is also an angular dependence of the wavelength in the interference peaks. If θ is the angle in the horizontal plane of viewing with respect to the original path, the undulator spectral peaks are at wavelengths

$$\lambda_n = (L/2n\gamma^2)(1 + K^2/2 + \gamma^2\theta^2),$$

where n is an integer. In an ideal device the on-axis radiation arises only from odd harmonic indices n, but in practice, imperfections permit some on-axis radiation of even harmonics. Off axis, all harmonics are radiated. The peak heights in the emitted spectrum vary as N^2 and the angular widths vary as γ^{-1}/N both horizontally and vertically. For small K the radiation in harmonics is weak. The fraction $(1 + K^2/2)^{-2}$ of the total emitted power is emitted in the fundamental. As K increases, the total power emitted increases and the fraction in the harmonics increases. An undulator spectrum is shown in Fig. 4.4, and the envelopes of the spectral peaks as the magnetic field is scanned are shown for several undulators in Fig. 4.5

As the magnetic field increases, K increases. The wavelength of the fundamental increases and the fraction of radiation in the harmonics increases. The total radiated power also increases. As K increases through 1 and becomes larger, the power in the nth harmonic grows, the harmonics become more closely spaced in frequency as the fundamental frequency decreases, and finally the harmonics merge into a continuum. Also, coherence effects are lost and the radiation becomes N times that of one arc of the device, now called a wiggler. The radiation is confined to a half-angle of order γ^{-1} vertically, and δ horizontally. Wiggler spectra can be seen in Figs. 4.4 and 4.5.

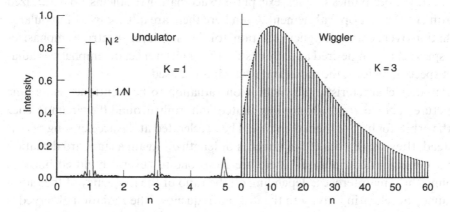

Fig. 4.4 Calculated on-axis spectrum emitted from an N-period undulator (*left*) and a wiggler (*right*), $K = 1$ and 3, respectively. The abcissa is the harmonic number n and the energy of the fundamental scales with K as $1/K^2$.

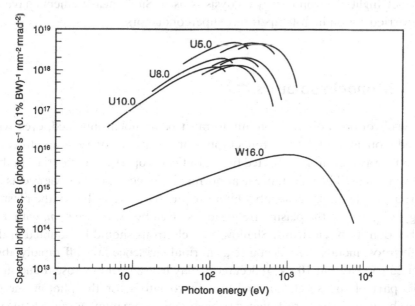

Fig. 4.5 Track of the spectral brightness peak of the fundamental emission of undulators, U, and a wiggler, W, at the Advanced Light Source as K is scanned. Compare the ordinates with those of Fig. 4.2. The numbers after U and W are the periods in cm.

Despite the relatively narrow peak in the undulator spectrum, a peak which can be scanned, a monochromator is still needed to provide a spectral bandpass adequately small for photoelectron spectroscopy. However, the peak in the spectrum provides a greater flux of photons than one can get from a bending magnet, even when one uses a mirror to collect radiation from a large arc of the path in a bending magnet. Moreover the high collimation of undulator radiation makes the optics of beam lines simpler, except for exacerbating problems from localized heating of the first optical element. Even here there are advantages of undulator radiation over bending magnet radiation, for the undulator spectrum emphasizes the spectral region desired and, by destructive interference, de-emphasizes adjacent spectral regions, the radiation in which is not used.

Another characteristic of synchrotron radiation to be mentioned is its time structure. The electrons are not distributed uniformly around the circumference of the orbit, for half of them then would be decelerated at the accelerating cavity. Instead, they are distributed in bunches of length σ_z, again a standard deviation for a Gaussian distribution. There are from one to several hundred bunches around the circumference, depending on the ratio of the frequency of the radio-frequncy accelerating cavity to the orbital frequency. The radiation observed is then a series of pulses of duration σ_z/c, which is 3 ns for 1-cm long bunches. The pulse separation in time is determined by the ring circumference and the number of bunches. The time structure is of use in several measurements other than photoelectron spectroscopy. In photoelectron spectroscopy, it plays a role only if time-of flight electron energy analysis is used. Such measurements have not been carried out on high-temperature superconductors.

4.2   Monochromators

The synchrotron radiation continuum must be monochromatized, even when emitted from an undulator. Monochromator design for the vacuum ultraviolet and soft x-ray regions is restricted by the optical properties of materials in these spectral regions. The reflectance near normal incidence of metals, viewed as free electron-like, remains reasonably high for photon energies below the plasmon energy, $\hbar\omega_p$, where the plasma frequency is given by $\omega_p^2 = 4\pi n e^2/m$, with n the number density of electrons. Shallow core electrons should be counted in this, and for most metals the reflectance at normal incidence falls off rapidly above about 25 eV. Even at 10 eV, reflectances may be well below those found in the visible part of the spectrum, and monochromators for the photon energies above about 6 eV often are designed to have a minimum number of reflecting surfaces. Reflectances can be enhanced in limited spectral regions by interference effects, using a coating on the grating, but the cost is reduced reflectance in other parts of the spectral region. The use of multiple gratings on a turret is a good solution and allows enhanced reflectances to be used on each grating. Discussions of monochromators more detailed than that below can be found in Samson (1967), Gudat and Kunz (1979), Johnson (1983), and West and Padmore (1987).

A common monochromator for the region below 30 eV is the near-normal incidence monochromator using a spherical reflection diffraction grating with the center of the grating surface and the entrance and exit slits all kept on the Rowland circle, a circle whose diameter equals the radius of curvature of the grating. The grating is rotated to scan wavelengths (Fig. 4.6). The wavelengths in focus at the exit slit are given by the grating equation, $n\lambda = d(\sin\alpha + \sin\beta)$.

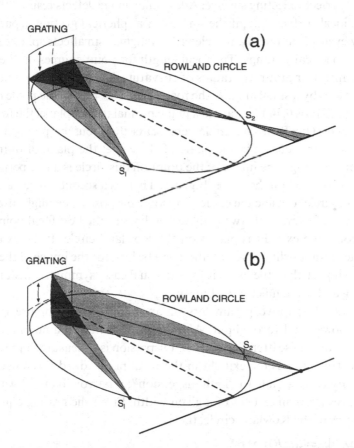

Fig. 4.6 Focusing by a spherical grating. The grating grooves, parallel to the arrows, are perpendicular to the Rowland circle. A point source at S_1 is imaged astigmatically by the grating. (a) Rays in the plane of the Rowland circle are imaged at S_2 on the circle. (b) Rays normal to the plane of the Rowland circle are imaged on the tangent to the Rowland circle where the grating normal (dashed) intersects it. A vertical exit slit would be located at S_2.

The integer n is the order of diffraction, and for photoelectron spectroscopy it is almost always ± 1. α and β are the angles of incidence and diffraction measured from the grating normal. With fixed entrance and exit slits, rotating the grating keeps $(\alpha - \beta)$ constant (in the geometry usually used, β is a negative angle), typically 15°. Maintaining focus necessitates translating the grating, as well as rotating it if the entrance and exit slits are kept fixed in position. The advantages of the Rowland circle monochromator are the use of only one reflecting element, the grating, and excellent focusing in the plane of the instrument dispersion, the plane of Fig. 4.6. With the small angle between the entrance and exit slits, imaging of the entrance slit on the exit slit is good, the main aberration being astigmatism which affects the flux out the entrance slit but not the resolution. Aberrations refer to defects in the image of a point due to the nature of the geometrically perfect reflecting surface. Additional image defects result if the surface is not the ideal surface, e.g., if the surface of a spherical grating departs from a sphere by even a fraction of a wavelength. Astigmatism affects spherical mirrors as well as spherical gratings. The focal length for point sources on the axis of a spherical mirror or grating of radius of curvature R is $R/2$. If the point source is located off axis by a small angle θ, the focal length for rays in the plane of the dispersion is $(R/2)\cos\theta$, but for rays in a plane normal to that plane, the focal length is $R/(2\cos\theta)$. The Rowland circle represents the locus of point-objects and point-images for focusing in the plane of the circle, the plane of diffraction in the case of a grating. The image of the point on this circle is a line perpendicular to the circle, located on the circle, Fig. 4.6. There is a second image, a line in the plane of the circle, outside the circle. For some purposes, one might use the "circle of least confusion" in between these two lines as the best focal point but, for spectroscopy, the exit slit is placed on the Rowland circle. If the exit slit is as high as the entrance slit, some radiation will be lost, for the image of the entrance slit will be higher than the exit slit for the usual case, symmetric distances from the grating and a magnification of unity.

Three key descriptive parameters of any monochromator are dispersion, resolving power and throughput. Also important are the amounts of higher-order radiation and scattered radiation. Dispersion is a measure of the spreading of the spectrum across the exit slit in the focal plane, $d\lambda/dx$. A small value of this quantity corresponds to "high dispersion"; a narrow band of wavelengths then emerges from an exit slit of a given width. From the grating equation and the geometry of the Rowland circle, this is

$$d\lambda/dx = (d/R)\cos\beta,$$

where R is the radius of curvature (Rowland circle diameter) of the grating and d is the spacing between adjacent rulings. Thus high dispersion is achieved by a high ruling density and long focal length. A small value of $\cos\beta$ is not used because at large values of β aberrations cause poor imaging. A common value for d is 0.833 µm, while R is 1–5 m. The quality of gratings with smaller d values can be lower than that for the ruling just quoted, while larger values of R lead to very large instruments. To preserve polarization with synchrotron radiation and to better match to the source shape, monochromators usually are mounted so the slits and grating rulings are horizontal, and the dispersion plane vertical.

The resolving power often quoted is that given by the Raleigh criterion for "just resolving" two spectral lines of equal intensity separated by dλ when their average wavelength is λ. For a grating of N illuminated lines of spacing d used in order n

$$R = d\lambda/\lambda = nN.$$

This is rather misleading for photoelectron spectroscopy, having been derived for resolving spectral lines with a plane grating, and is not often used or achieved (Samson, 1967; Gudat and Kunz, 1979). We replace it below. The throughput, acceptance, or entendu of the grating is defined as the product of the slit area and solid angle subtended by the grating at the center of the entrance slit. It is a measure of the radiation that can be collected usefully from the source.

One may use these expressions in describing an instrument, but a better description is afforded by ray tracing. Ray-tracing computer programs account for the progress of thousands of rays through an optical system, and give one a much better picture of the emerging beam of radiation. They also can be used to treat the effects of resolution-enhancing apertures on the throughput of the monochromator. Modern ray-tracing programs such as SHADOW (Cerrina and Lai, 1986) can work with nearly any optical component in existence. Finally, measurements of these quantities are desirable, especially the shape of the instrument response function, the spectrum sent out for an input as close to a delta function as possible. If the output for a delta function input, $\delta(\lambda - \lambda_0)$ is $S(\lambda - \lambda_0)$, then the output for an input spectrum $I(\lambda)$ is

$$I(\lambda) = \int S(\lambda - \lambda')I(\lambda')d\lambda'.$$

Atomic emission lines are usually adequate for measurements of S, although scanning the grating through the zero-order "white" spectrum is widely used. S depends on polarization. The function S used above represents an integration over solid angle of a more differential quantity upon which the output depends where on the grating a ray is incident. (Such a function could be measured by moving a small mask over the surface of the grating, but this is not easy under vacuum and it is rarely done. Ray tracing could approximate it if needed.) The full width of $S(\lambda)$ at half maximum is what is most frequently quoted as the spectral bandpass or "resolution" of the monochromator. It is usually much larger than that from the Raleigh criterion.

From the above, high resolution should be achieved by using a high-dispersion instrument (small d, large R) and opening the entrance and exit slits only slightly, e.g., to 5 μm. However, the flux out of the exit slit, assuming a continuous spectrum at the input, varies as the square of the slit widths, assuming equal widths for both slits. One then may have high resolution, but inadequate photon flux to carry out the measurement with the requisite signal-to-noise ratio in the desired amount of time. (The time could be limited by sample deterioration or radiation source stability.) Moreover, aberrations in the image will increase the spectral bandpass, especially at small slit widths, well beyond that given by the product of slit width and linear dispersion.

Second- and higher-order radiation contributes undesirably to the output of the monochromator. They may be present in the input spectrum with more intensity than that of the first-order radiation. The relative intensities at the exit slit of different orders are determined by the grating groove shape and the reflectance spectrum of the coating. A sinusoidal transmission grating diffracts only in the zeroth and ± first orders, but reflection gratings always diffract higher orders. Gratings may be ruled gratings (or made as replicas from ruled gratings) or "holographic" gratings, made by the interference of two beams from a laser on a photoresist coating . The former are faceted ("blazed"). The faceting may be angled to give specular reflection from the broad face of the facet of the radiation diffracted in first order at a certain wavelength; the grating is blazed for this wavelength. This may not reduce higher-order radiation, but it enhances the first-order radiation. Holographic gratings may be blazed during processing. The reflectance of the grating surface at near-normal incidence is usually much lower for higher-order radiation than for the first order, especially when second-order radiation falls at energies above $\hbar\omega_p$. If it does not, then transmission filters are often inserted in the beam to absorb preferentially higher-order radiation. Otherwise, rather uncertain corrections may have to be made to the data, uncertain because the spectrum of higher-order radiation is difficult to measure.

Scattered radiation has a spectrum similar to the input spectrum, modified by the wavelength dependence of the scattering cross-section of the scattering centers: dust particles on slit or grating, tiny scratches, and most importantly, pileup of material on the edges of the rulings in a ruled or replica grating. Holographic gratings tend to produce far less scattered radiation. Suitable baffles in the instrument, and sometimes filters, can be used to reduce scattered radiation.

At higher photon energies, grating reflectances are intolerably low at normal incidence, but remain high at larger angles of incidence, roughly to energies of $\hbar\omega_p / \cos\alpha$. By going to angles of incidence near 90°, i.e., a few degrees from grazing, one can have acceptable reflectances up to $20\hbar\omega_p$ or so. Rowland-circle grazing-incidence instruments have been constructed, but the scanning is more complex than at near-normal incidence if one wants to keep both entrance and exit slits fixed. The former is desirable because the storage ring is immobile and the latter desirable to keep a fixed sample and electron energy analyzer. Monochromators that allow one or the other to move have been built successfully. The problem with using a spherical grating at grazing incidence is that imaging is terrible due to large aberrations. Astigmatism is large, but resolution can be maintained by keeping the grating and slits on the Rowland circle. This is done with a high cost in intensity, as the image of the entrance slit at the exit slit is now very long, much longer than the exit slit. Samson (1967) quotes an example: at 2° grazing incidence on a 600 line/mm grating with rulings 2 cm long, the astigmatic image is 16 cm long. The equations for dispersion, resolution and throughput given above remain valid for grazing-incidence monochromators, but the resolving power will be far worse than that given by the simple formula. Throughputs tend to be rather small, as larger acceptance angles lead to large aberrations and lower resolution. Grazing-incidence monochromators are more subject to scattered radiation problems because the near-forward scattering cross-section is larger than that for the large angles of scattering required to have scattered radiation reach the exit slit in a normal-incidence monochromator. Higher-order radiation may be enhanced or diminished in grazing-incidence monochromators, depending on the design of the instrument.

Non-Rowland-circle instruments have also been used. These do not have exact focusing in the plane of dispersion for all wavelengths. The Seya–Namioka mount uses a spherical grating with an angle of incidence of about 35° $(\alpha - \beta = 70°32')$, so it is intermediate between normal and grazing incidence. As a result, it can be used to higher photon energies than a normal-incidence instrument, going to about 40 eV instead of 30 eV. Typically, several gratings would be used to cover the 5–40 eV range. There is not exact focusing at any wavelength, but imaging is fairly good, except for astigmatism. To increase throughput, expected to be low because of astigmatism, long exit slits have been used, and cylindrical focusing mirrors used outside each slit (Rehfeld et al., 1973), mirrors which also serve to make the exiting radiation parallel to the incident. As with grazing-incidence monochromators, the simple expression for resolving power is overly optimistic. Improved resolution has been achieved by masking off the peripheral regions of the grating, reducing throughput, however.

Toroidal grating monochromators use gratings ruled on toroidal surfaces. They focus in the dispersion plane at two wavelengths and are close to focusing for wavelengths in between and in a region beyond each. They are used near grazing incidence in the soft x-ray region. The astigmatism is lower than that of a spherical grating, so with larger gratings, a larger solid angle can be effectively passed through the exit slit compared with a spherical grating instrument. Hence they have been popular for photoemission work, offering a high flux at acceptable resolution. Finally to be mentioned here, but not exhausting a description of all the monochromators that have been used, is the plane-grating monochromator. Such an instrument uses a plane grating as a dispersion element, but requires a focusing mirror between the grating and the exit slit. This mirror must move and rotate as the grating is rotated to scan the wavelength. The focusing mirror may be spherical, paraboloidal, or ellipsoidal.

As monochromator and component technology have advanced, there has been a tendency for all optical surfaces to be plane or spherical, other surfaces being too difficult to make and test adequately for departures from the desired figure. Improved resolution and throughput have come about by more precise tolerances on the optical elements and their mounts, and smoother coatings. This has not been an easy task, for with third-generation storage rings, the heat load on a mirror can be very large, and thermal expansion can easily spoil the figure if the mirror does not have adequate heat transfer and cooling. Spectral bandpass scales with slit opening and the focal length of the grating (or focusing mirror for a plane-grating monochromator), as long as manufacturing tolerances, sag due to external forces, e.g., gravity, and distortion due to thermal expansion caused by absorbed radiation do not degrade the figure of the focusing element. As optical surfaces approach perfection, the performance of a monochromator becomes limited by the inherent aberrations. In designing instruments, sometimes one aberration can be reduced at the expense of another or at the expense of a more complicated optical element. One example of this is that coma can be reduced in a grazing-incidence monochromator by ruling the grating with a spacing that varies across the face of the grating (Harada and Kita, 1980; Aspnes, 1982). Such new instruments, along with new synchrotron radiation sources, may permit future photoelectron spectroscopy at yet higher resolution with acceptable count rates.

4.3   Other optical elements

Synchrotron radiation beam lines require other optical elements in addition to monochromators (West and Padmore, 1987; Howells, 1994). Mirrors are needed to focus the radiation on the entrance slit, and to focus diverging monochromatic radiation on the sample. Generally these are used at grazing incidence, and they may be up to a meter in length. The first such mirror receives the full synchrotron radiation spectrum. The highest-energy part of the spectrum, unless it is to be used, is filtered out by this mirror through a choice of angles and materials. However, this means that the power not reflected remains in the mirror as heat. Unless this heat is removed, the figure of the mirror may be degraded, and the photon flux through the rest of the beamline seriously reduced. A large effort has gone into the technology of mirrors, and the achievable tolerances on figure error and smoothness have improved markedly. In first- and second-generation synchrotron radiation sources, the first mirror(s) may be figured either by grinding and polishing or by bending a plate of float glass whose shape has been altered to give a cylindrical surface upon bending. (Ground and polished mirrors may be rougher than bent float glass, leading to radiation loss by scattering from the "rough" surface.) Sometimes a combination of the two techniques has been used. Third-generation sources require cooling of the first mirror, which may be made of one of a number of exotic materials. All mirrors before the monochromator tend to have rather large radii of curvature. The mirror after the exit slit, used to image the slit on the sample, often has a small radius to produce an image of reduced size. In specialized applications, e.g., photoelectron microscopy, to be discussed in Chapter 11, a mirror or several mirrors of short focal length and high normal-incidence reflectance is required. These have been made of multilayers of two materials, usually one of high atomic number and one of low (Spiller, 1983). These can have quite respectable normal-incidence reflectances in a limited spectral region, some tens of percent, produced by interference.

4.4   Electron energy analyzers

The continuous spectrum of photoelectron kinetic energies must be measured. This is done by an electron energy analyzer. Many types of electron energy analyzers exist, and variations on them are still being invented. For photoelectron spectroscopy only a few types are widely used. All of these are electrostatic analyzers. Three types are deflection analyzers, and the other is based on one or more retarding grids. Analyzers using a magnetic field to deflect and disperse the electrons, commonly used in other fields of physics for electrons in the keV energy range, have been used for photoelectron spectroscopy, but we will not describe them here. More detailed descriptions than those below may be found in a number of references: Sevier (1972), Roy and Carette (1977), Smith and Kevan (1982), Ghosh (1983), Granneman and van der Wiel (1983), Rivière (1983), and Erskine (1995). They are described by characteristics similar to those used for monochromators for photons: resolution, dispersion, and throughput, all of which can be found analytically from the equations of motion for an electron passing through regions with electric fields having certain symmetry. And, as for monochromators for photons, all these can be replaced by an instrument function, obtained by detailed ray tracing or, in part, by experiment. Differences between the measured instrument function and the one calculated can be assigned to differences in geometry, i.e., to imperfect manufacture, even though it may be on the scale of a micrometer or so, and to stray electric and/or magnetic fields not properly considered in the ray tracing.

A retarding electrode was the energy analyzer used when the cutoff on the photoelectron spectrum was first discovered (Lenard, 1900). A retarding-grid analyzer was used in the measurements of Berglund and Spicer (1964b), so influential in making photoelectron spectroscopy so popular. Most measurements in the last two decades were made with deflection analyzers. However, the retarding-grid analyzer plays a role in several special analyzers currently in use, so we describe it here, see Fig. 4.7.

If the sample, assumed to be electrically conducting and grounded, were a point located at the center of a spherical mesh at ground potential, the photoelectrons would travel radially outward in a field-free region, passing through the holes in the mesh, which may transmit 90% of them. A concentric mesh of larger radius is held at a scannable retarding potential, $-V_R$. Only the more energetic photoelectrons pass through this grid to be accelerated to the collector sphere, where the current is measured as a function of the retarding potential. This current represents an integral of the desired spectrum, extending from kinetic energies between V_R and the cutoff energy. The desired spectrum is the derivative of this spectrum with respect to energy, which can be taken in an analog mode by modulating the retarding potential and using a synchronous detector (lock-in amplifier), or taken digitally. A deficiency of all retarding-grid analyzers is that their signal-to-noise ratio is inherently lower than that of deflection analyzers. A deflection analyzer measures a current, $I(V_R)$, due to electrons with energies in the range V_R to V_R+dV_R. The statistical noise is proportional to the square-root of $I(V_R)$, so the signal-to-noise ratio varies as the square-root of $I(V_R)$. The signal measured from a retarding-grid analyzer is much larger than $I(V_R)$; it is the integral of $I(V)$ from V_R to the cutoff. The noise is thus the proportional to the square-root of a much larger number, while the signal is only that part of the measured signal in the interval dV_R, taken by differentiating the current spectrum, leading to a smaller signal-to-noise ratio.

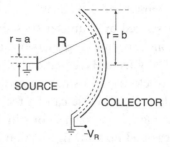

Fig. 4.7 Retarding-grid electron energy analyzer.

The retarding-grid analyzer obviously is an angle-integrating device. However, Gerhardt and collaborators (Becker *et al.*, 1975) have converted it into an angle-resolving spectrometer by placing six individual detectors on the collecting hemisphere and measuring the output of each. Each detector has a small angle acceptance. The chief limitations of this arrangement for angle-resolved spectra are that only a few angles are sampled, and that the energy resolution is generally poorer than that of other types of analyzer.

The illuminated area of the sample is finite, so some photoelectrons do not leave from the center of curvature of the spherical grids. They do not travel along a radius, but have a small component of velocity perpendicular to the radius. This component is not changed by the retarding field, so it contributes to a degradation of energy resolution. Photoelectrons of energy E emitted from the center of the sample cannot be distinguished from electrons of energy $E + dE$ emitted from off-center spots, with dE increasing with distance from the center. (This effect degrades the resolution of cylindrical retarding-grid analyzers so much they are not often used, despite their ease of manufacture.) Spherical retarding grids are commonly used for LEED, for which three grids are usually employed. Many LEED instruments have a fourth grid, allowing the same electron optics to be used for Auger spectra as well, subject to the same signal-to-noise problem mentioned above. In fact, the analyzer shown schematically in Fig. 4.7 is one such.

If the retarding-grid potential is modulated, the signal at the modulation frequency will be proportional to the modulation amplitude. However, the range of electron energies modulated, hence the resolution or bandpass, is also proportional to this amplitude. As with all spectrometers, there is a trade-off between resolution and output flux or signal.

The cylindrical deflection analyzer or spherical (actually hemispherical) deflection analyzer, the latter being shown in Fig. 4.8, are widely used. The

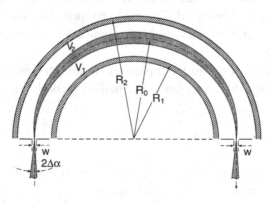

Fig. 4.8 Spherical capacitor electron energy analyzer.

former is more often used for low-energy EELS, and the latter more for angle-resolved photoelectron spectroscopy, as well as XPS. Other names are spherical sector analyzer or spherical capacitor analyzer. In use, there are electron lenses at the entrance and exit planes, but we defer discussion of them until later. Again, the analysis begins with a point source of electrons of a single energy E_p, the pass energy. The equations of motion for electrons in the radial electric field varying as $1/r^2$ can be solved. Consider inner and outer concentric hemispheres of radii R_1 and R_2 with a potential difference ΔV between them, the outer sphere the more negative. A point source of electrons is located at a radius $R_0 = (R_1 + R_2)/2$. For an electron emitted with E_p and directed normal to the electric field, a proper setting of ΔV can result in a circular orbit in which the electron remains equidistant from each of the two electrodes: $e\Delta V = E_p(R_1/R_2 - R_2/R_1)$. It has been shown that for a deflection of 180° the spherical analyzer is doubly focusing. This means that electrons from a point in the gap center at the entrance plane, emitted at an angle to the circular equilibrium orbit, either in the plane of Fig. 4.8 or normal to it, will converge to a point 180° away on the equilibrium orbit. The angular acceptance can be rather large, an advantage in many techniques, but not in angle-resolved photoelectron spectroscopy. With the same potentials, electrons emitted from the same point with different energies focus to different points in the focal plane, higher-energy electrons nearer the outer electrode and lower-energy electrons at smaller radii. Thus an aperture placed in the focal plane passes electrons within a small range of energies around E_p, the size of the range determined by the aperture size and the dispersion in the focal plane.

One can scan this analyzer in two modes. Increasing ΔV allows the pass energy E_p to increase, so one can scan a spectrum. However, as we shall show, the resolution then will decrease as one scans. The other type of scan keeps ΔV and E_p constant, scanning by retarding or accelerating the incident electrons, thereby making a different part of the spectrum have energy E_p. This keeps the resolution constant and is the scan normally employed in photoelectron spectroscopy of valence bands. The retardation could be applied by adding the scan voltage to both hemispheres (or the sample), keeping ΔV constant, but it is usually done by an input lens.

If an entrance aperture of width w is placed at the object point, and an identical one placed at its image in the exit plane, the half-width of the energy distribution emerging is

$$\Delta E = E_p(w/R_0 + \alpha^2/4),$$

where α is the largest off-axis angle the analyzer accepts. From this, it is clear that for high resolution, small slit widths and smaller acceptance angles are needed. Both of these result in a small éntendu, i.e., low acceptance. Moreover, at the same pass energy a large analyzer will have a smaller bandpass than a small one. However, the size is limited by the need to move the analyzer over a hemisphere on an in-vacuum goniometer, if one wants a flexible angle-resolved photoemission system, and to provide for good electrostatic and magnetic shielding. Finally, the bandpass is smaller for smaller pass energies. However, low-energy electrons are more susceptible to orbit perturbation by stray magnetic fields than high-energy electrons. Typical electron analyzers of the spherical capacitor type have radii R_0 of 50–100 mm, angular acceptances of 1–2° (in two dimensions), determined by apertures in an input lens, and entrance and exit apertures of 1 mm. They are electrostatically and magnetically shielded and they are often placed in vacuum chambers with additional shielding. Sometimes three pairs of Helmholtz coils are used to cancel the earth's magnetic field, which may be modified by the vacuum chamber and its surroundings. Pass energies down to about 1 eV can produce ΔE of 5 meV. The importance of magnetic shielding can be seen from an analysis of Roy and Carette (1977). The maximum magnetic field in G averaged over the path of length ℓ (cm) of an electron of energy E (eV) is $B = 6.74E^{1/2}d/\ell^2$, where d is the largest acceptable deviation of the beam at the exit aperture, presumably some fraction of w. For a 2 eV pass energy, a path length of 180° in a spherical capacitor of 50 mm radius plus 5 cm of travel through a lens, a displacement of 0.1 mm requires that B not exceed 5 mG.

Mann and Linde (1988) showed that by using two hemispherical analyzers in series, each of radius R_0, one has a system with the dispersion of a single analyzer of radius $2R_0$. Baraldi and Dhanak (1994) analyzed such a system further with an eye to multichannel detection for the rapid accumulation of entire core-level spectra.

The third analyzer we discuss is the cylindrical mirror analyzer (CMA), shown in Fig. 4.9. Such analyzers were widely used for photoelectron and Auger spectroscopy, and are still in use, but commercial spectrometers now tend to use

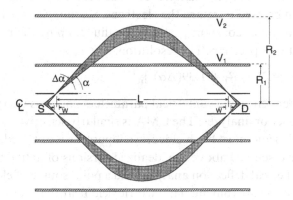

Fig. 4.9 Cylindrical mirror analyzer.

spherical analyzers. They are not suitable for angle-resolved spectroscopy as shown in Fig. 4.9 but, by adding an internal drum with an aperture, electrons in only a small range of azimuthal angles are passed to the detector, allowing some angle resolution (Knapp *et al.*, 1982). One cannot cover much of the 2π solid angle above the sample with such an arrangement. The azimuthal angle can be varied by rotating the drum, but changing the polar angle requires tipping the sample because CMAs are fixed in position. The double-pass CMA is rather long and not suitable for mounting on a goniometer, so usually the sample has been rotated in angle-resolved studies with CMAs.

The CMA consists of two concentric hollow cylinders of radii R_1 and R_2 with a potential difference ΔV between them, the outer cylinder being the more negative. In the single-pass analyzer shown there is an entrance aperture in the inner cylinder. Ray tracing has shown that for a point source S located on the axis, electrons emitted with kinetic energy E_p at an angle of $42°18'$ to the axis will be focused to second order at a point D on the axis, as shown, if

$$\Delta V = (E_p/1.31e)\ln(R_2/R_1).$$

The source and image (detector) points are then a distance $L = 6.13R_1$ apart along the axis. For the single-pass CMA shown in Fig. 4.9, there would be an exit aperture of longitudinal length w in the inner cylinder. Such analyzers are used for Auger spectroscopy but not for photoelectron spectroscopy. Double-pass CMAs, used for photoelectron spectroscopy, have a second stage after an intermediate aperture. In practice the CMA accepts a small range of angles, about 6°, around $42°18'$, collecting thus a differential cone of electrons from the point source. This angle range, $\Delta\alpha$, is limited by the aperture diameter as $2w = 10.56R_1(\Delta\alpha)^2$.

The single-pass CMA usually has the inner cylinder grounded to provide a field-free path for the photoelectrons from the sample. It is scanned by varying the potential on the outer cylinder. Double-pass CMAs usually have a pair of spherical grids surrounding the sample. The one nearer the sample is grounded and the other can be used to retard the electrons. Scanning the potential on this grid allows operation at constant pass energy, thus keeping the resolution constant throughout the spectrum. The resolution, given by

$$\Delta E = E_p[0.36w/R_1 + 1.39(\Delta\alpha)^3],$$

has a similar dependence on geometrical parameters and pass energy as that for the spherical capacitor analyzer. The CMA is similarly sensitive to stray electric and magnetic fields and must be shielded. An input lens is not used.

The analyzers discussed above are idealized versions of actual analyzers. As described, the spherical deflection analyzer had a point source of electrons in the gap at the edge of the instrument. The electric field near that point is not purely radial, for there is a fringing field. Actually placing a sample at this point would distort the fields and it would be difficult to get the photon beam to hit the side of the sample facing the analyzer. Similarly, the point source for the CMA was inside the CMA, and cutting away parts of the cylinders to allow radiation to reach a sample at that point will cause the fields to depart from purely radial. (The electron paths can be accommodated by placing holes on the inner cylinder, covered with grids to provide an equipotential surface where the material was removed.) Terminations at the entrance and exit ends have been developed to reduce fringing field effects. The spheres and cylinders must be sufficiently concentric or coaxial that no resolution degradation is introduced (Wannberg, 1985; Baltzer et al., 1991). The materials must have work functions uniform over their inner surfaces to avoid microfields, and they must not produce many secondary electrons. The latter is accomplished by choice of materials or a coating of colloidal carbon. Finally we have not yet discussed electrons which leave the sample from a point not coincident with the source point. We have discussed only electrons leaving the source point at different angles and with different energies.

Off-axis source points can be modeled analytically, if they are not far off axis, or by numerical ray tracing. These points will focus to points near the images of the on-axis source points, but the acceptance cone will be smaller, so the transmission of the analyzer falls off with distance of the source point from the axis. In actual photoelectron spectroscopy there are two limiting cases. One is when the area of the sample from which the analyzer accepts electrons is much smaller than the area that the photon beam illuminates. This is common in XPS without a monochromator. The other is the reverse, the illuminated portion of the sample is much smaller than the area from which the analyzer collects electrons. This may occur in some forms of photoelectron microscopy, but it is normally not a common situation. More common is the case in which a sample is inhomogeneously illuminated with a flux $I(x, y)$, whose distribution and outer boundary may depend on photon energy, coupled with an analyzer which has a transmission $T(x, y)$ from different points on the sample, the boundaries and form of which may depend on the electron kinetic energy. Moreover, $T(x, y)$ may depend on the angles of emission of the electrons with respect to the analyzer axis and on the electron energy. Neither I nor T is well known and it is for this reason that photoelectron spectroscopy is not quantitative. I can be measured by scanning the photon beam with a knife-edge or a pinhole. The energy dependence of T can be found from spectra taken on a Fermi edge or on a sharp line from an atomic vapor. Not only do we not know I and T well, but usually we cannot reproduce them from one sample to the next or even for the same sample if we move the sample between scans, then replace it as best we can in its original position. Seah (1995) describes what is necessary to make an accurate calibration of the intensity from an electron spectrometer.

Spherical capacitor analyzers require an electron lens to image the source point on the sample to the source point shown in Fig. 4.9, which is in the gap of the analyzer. (This lens may be used in other modes.) The lens serves several other purposes. The end of the lens nearer the sample is at ground potential to help produce a field-free region for the electrons leaving the sample. If this end is not too close to the sample, a reasonable working volume around the sample is available. The end nearer the analyzer serves to accelerate or retard the electrons to the desired pass energy. The apertures in the lens serve to define the area on the sample accepted by the analyzer and the angular acceptance. This dual function requires that the electron lens have two separate apertures, see Fig. 4.10. The electron lenses used in electron spectroscopies are almost always electrostatic lenses. The geometric optics of such lenses is similar to that of optical lenses except for several complicating features (Harting and Read, 1976, Moore *et al*., 1983). The role of the refractive index is taken by the (nonrelativistic) electron velocity. Most important is that the focal lengths depend on potential ratios between elements, so they may change when any scanning is done. One usually must scan the potentials on several lens elements to make the overall system focus as desired if the potentials at each end of the lens are to be held constant, as is the case when using the lens to retard electrons in the emitted spectrum to a fixed analyzer pass energy. The electron paths are curved over extensive lengths and the lenses must be viewed as thick lenses. Aberrations may be rather large. Nonetheless there have been a number of electron zoom lenses reported in the literature and sold as part of commercial electron energy analyzers.

The lens must image the source on the entrance aperture in such a way that electrons of the desired kinetic energy reach the entrance slit at the pass energy. Ideally it will do this over a wide range of energies, keeping the transmission constant, and at constant angle acceptance, without unreasonable values of applied potential. The resolution-defining entrance slit may, in fact, be a virtual one. As with the matching of a monochromator and other elements of a beam line to the emittance of the beam in a storage ring, the electron energy analyzer and its input (and output) lenses should be considered as a unit (Plummer, 1980; Smith and Kevan, 1982; Ovrebo and Erskine, 1981; Stevens *et al*., 1983; Mårtensson

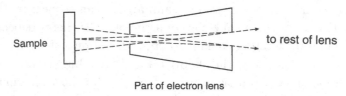

Fig. 4.10   Aperture pair defining the angular and area acceptance from the photoelectron source. The source region is at the left, and the entire region is field free. The first aperture defines the area of the source accepted, and the second, the angular acceptance.

et al., 1994). Elements in phase space for the electrons are conserved in the entire ideal electron optics–analyzer system. This is expressed by the Helmholtz–Lagrange law: $x\alpha E^{1/2} = $ constant, where x is the displacement from the axis of an electron of energy E whose velocity makes angle α with the axis. Losses occur due to grids. Plummer (1980) has given a general description of a lens–analyzer combination. The two apertures defining the sample area and angular acceptance of the system (Fig. 4.10) should each be in a field-free region in front of the analyzer in order to make their function independent of the potentials on lens elements or analyzer electrodes. The Helmholtz–Lagrange relation "squared" to describe the two transverse coordinates gives $A\Omega E$ at the sample equal to $A_a\Omega_a E_p$ at the analyzer entrance. For a matched system the detected signal is proportional to $A\Omega$, if no electrons are lost. For a hemispherical analyzer with its angular acceptance limited to $\alpha_a^2 = w/2R_0$,

$$A\Omega \approx R_0^2 (E/E_p)^2 (\Delta E/E_p)^3,$$

where the expression for ΔE from above has been used (Plummer, 1980). The factor E/E_p is the retarding (or accelerating) ratio of the lens. (For high-resolution valence-band spectroscopy the lens is always retarding.) With a fixed analyzer pass energy, the scanning is done by varying the retarding ratio, and the signal increases as its square. A big challenge in the design of lenses, especially for XPS, is to increase the range of the retardation ratio without reducing the transmission of the lens or mismatching to the analyzer. Kevan (1983a) describes the design and performance of a small ($R_0 = 50$ mm) hemispherical electron energy analyzer with a matched zoom lens that allows variable energy and angle resolution. The design of another lens–analyzer system ($R_0 = 30$ mm) is described by Ovrebo and Erskine (1981) and Stevens *et al.* (1983). If motion of the analyzer on a goniometer is not needed a large fixed analyzer may be used. One modern commercially available hemispherical analyzer has $R_0 = 200$ mm (Mårtensson *et al.*, 1994), used for high-resolution XPS.

There are several other types of analyzers in use. Because synchrotron radiation is pulsed, time-of-flight electron energy spectrometers can be used. One such spectrometer was reported by Bachrach *et al.* (1975), but more recent use of time-of-flight spectrometers has been in photoemission from gases rather than solids. Eastman *et al.* (1980) developed a display analyzer, shown in Fig 4.11. Versions of this analyzer (Schnell *et al.*, 1984; Böttner, *et al.*, 1990; Santoni *et al.*, 1991) are currently in use. The sample is located at the focus of an ellipsoidal mirror which is biased negatively at V_1. Only electrons with energies lower than eV_1 are reflected from this mirror, passing through the aperture at the other focus of the ellipsoid. They then pass through a spherical-grid retarding analyzer set to pass electrons of energy greater than eV_2. Thus only electrons in the energy window $e(V_1 - V_2)$ are accelerated to the phosphor screen. The

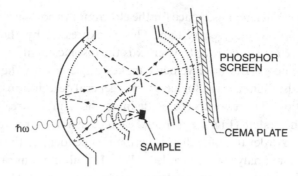

Fig. 4.11 Display analyzer. The left-hand element is an ellipsoidal mirror with a grid in front of it to suppress secondary electrons. The right-hand system is a retarding-grid analyzer with a channel-plate multiplier (CEMA plate) and phosphor-screen display.

angles of emission are conserved, accurately for a point sample, and the pattern on the screen is the angle distribution, polar (θ) and azimuthal (ϕ), of the electrons in the energy window, which may be scanned. The screen display may be recorded by a video camera.

A different type of display system has been developed by Osterwalder *et al.* (1991). It has been used with spectral line sources to date. The analyzer is a conventional hemispherical analyzer (1° angle resolution) on a goniometer, and the sample is on a goniometer with both axes driven by stepping motors. An analyzer energy is set and the sample is maneuvered so that about 3500 differential solid angles are sampled in the 2π solid angle above the sample. The data-handling system displays the results with a gray scale on a stereogram. The results resemble the output of the display analyzers discussed above, but the solid angle covered is larger. In operation, rather than make use of symmetry to reduce the range of solid angle to be sampled, the authors prefer to sample the full range as a check on the data. Note that in this scan mode the angle of incidence of the radiation on the sample, and perhaps its polarization, changes during the angular scan.

Toroidal capacitor deflection analyzers have been used with multichannel detection to provide spectra in an energy window over a wide range of polar angles (Engelhardt *et al.*, 1981; Leckey and Riley, 1985), but not azimuthal angles simultaneously, as in the system of Osterwalder *et al.*

Going in the direction of yet higher resolution, a commercial hemispherical analyzer has achieved 2.7 meV resolution and an angle-resolution of 0.2° (Mårtensson *et al.*, 1994). This is a large, heavy instrument, so angle-dependent studies usually will require angular motion of the sample. Ibach (1993) has achieved a resolution of 0.5 meV with a cylindrical capacitor electron energy analyzer for use in ultrahigh-resolution surface vibrational spectroscopy by electron energy loss. Shaping electric fields to reduce aberrations was the key to this achievement. Such an instrument has not yet been applied in photoelectron spectroscopy, but its use is anticipated. It, too, is available commercially.

There are several other considerations for high-resolution photoelectron spectroscopy. One is the stability of the electronics (Mårtensson *et al.*, 1994). Drift and ripple in some of the power supplies for the analyzer and input lens have the effect of reducing the energy resolution. Drift often can be detected by frequent scanning of the Fermi edge of a second sample, e.g., Pt, mounted in electrical contact with the sample under study. A few meV drift per day is possible. The charging of insulating samples is also a problem. Photoemission depletes the surface of electrons under the illuminated region of the surface. These are readily replaced from the bulk in metallic samples. For insulators, such replacement may not occur until large electric fields build up from the surface charge. This shifts the measured energies of all spectral features, and to the extent that the surface charge is inhomogeneous or time-dependent, broadens them. Such effects often are reduced by irradiating the surface with low-energy electrons from a "flood gun". Optimizing the use of such a flood gun has been discussed by Bart *et al.* (1994). Even so, the surface charge-density and potential are not homogeneous (Coluzza *et al.*, 1994). Very thin insulating films, tens of Å thick, on conducting substrates probably do not charge enough to alter photoelectron spectra at the resolution currently in use. Higher resolution may reveal smaller effects from charging.

Of the analyzers we have described, the retarding-grids clearly give a spectrum integrated over angles. Placing discrete detectors in one allows angle resolution, but usually at a few fixed angles. The spherical and cylindrical capacitors with input lenses are obviously angle-resolving analyzers. They may be used for angle-integrated spectra three ways. The first, not used, is to take spectra at many different orientations and average. The second, more commonly used, is to use a thin-film sample. This assumes randomly oriented crystallites, but many films are textured, with close-packed crystallographic planes in the plane of the film, leading to incomplete angle averaging. The third, the use of a high photon energy, puts the measurement in the XPS regime where wave-vector selection rules are inoperative. The CMA is neither angle-resolved nor truly angle-integrating. Angle resolution may be achieved via the apertured drum method mentioned above. Angle-integrated spectra may be obtained with polycrystalline films or high photon energies, as just described. Partial angle integration may be achieved by suitable orientation of the axis of the CMA with respect to the radiation beam axis and the sample normal. If these angles are 78° and 42.3°, respectively, the angle-integration effect of sampling in the differential cone the CMA accepts is maximized (Margaritondo *et al.*, 1979).

The instrument function for the electron energy analyzer alone, if it is measured, is measured by recording a photoelectron spectrum that is as close to a delta function as possible, i.e., from free atoms. A common measurement is of photoelectrons from a rare gas excited by the resonance line of He. The kinetic energy spectrum is then a peak with a width of the order of several meV, up to 10 or so, coming from the width of the resonance line, the lifetime of the photo-hole, and Doppler broadening. This is no problem for determining analyzer reso-lutions of 30 meV or so, but for instruments of higher resolution, narrower photoelectron peaks have been found (Mårtensson *et al.*, 1994). The angle dependence of the instrument function for the analyzer is rarely measured, although Mårtensson *et al.* (1994) report such a measurement for a particular analyzer. More commonly, the overall instrument function for the monochro-mator–analyzer combination is measured by recording the photoelectron spec-trum of the Fermi-edge region of a metal, perhaps at low temperature to sharpen the edge. The Fermi function is convolved with an assumed instrument function till the measured spectrum is matched.

Detectors for the energy-analyzed electrons are nearly always continuous-electrode electron multipliers. The simplest is a channel electron multiplier (CEM). This is a hollow curved glass tube about 1 mm in diameter whose inside wall is coated with a high-resistance semiconducting film. It is placed behind the exit aperture of the analyzer, possibly with an output lens interposed. The entrance opening may have a collection funnel a few mm in diameter. This end is biased at 100 V or so above the potential of the exit aperture to accelerate the electrons so that they produce secondary electrons upon collision with the coat-ing on the inner wall. The other end of the tube is biased about 3 kV more posi-tively. The electric field and the curved shape ensure a number of collisions of an electron with the wall, each generating more electrons. The gain can be as high as 10^6. The output pulses are collected on a separate anode and counted. Saturation and dead-time corrections are needed at high count rates, above 10^5/s, or so, but this rate is not reached in high-resolution valence-band work.

Single-channel counting can be slow. Each spectral element is counted for a fixed time or until a particular number of counts is reached, usually the former. The signal is the number of counts, N. Under the usual conditions for photoelec-tron spectroscopy the statistical noise usually averages to $N^{1/2}$, so the signal-to-noise ratio scales as $N^{1/2}$. It is thus improved only by counting for a longer time or improving the count rate, which presumably has already been maximized for the beam line and sample. The improvement with longer counting time is slow; doubling the signal-to-noise ratio requires a fourfold increase in counting time. It can also be improved by counting n spectral elements simultaneously, keeping the count in each one separate. This is multichannel or multiplex detection and it improves the signal-to-noise ratio by a factor of $n^{1/2}$ if equal amounts of time are spent in scanning in each of the two modes. n may be 20–30. Multichannel scanning is most easily accomplished with a hemispherical capacitor analyzer

(Hadjarab and Erskine, 1985). One removes the output aperture plate and replaces it with a position-sensitive detector. Such a detector is commonly a channel array or channel plate, a honeycomb of CEM tubes (without input funnels); see Smith (1995). The output goes to individual collector anodes, each of which must have its own amplifier and discriminator circuit, or to a resistive anode. The latter is a resistive film. Upon being struck at a point by an electron bunch, a signal pulse propagates toward each end of the strip. Both are detected and the ratio of their charges gives the position where the anode was struck. This gives the energy with respect to the pass energy of the electron that originated the pulse, an electron at the pass energy giving a signal from the center of the strip. Since each multiplier tube in the array may have a different gain, the gains must be calibrated, or each spectral element must be measured by each channel, accomplished by scanning E_p or the input lens. Electrons emerging from channel plates may be accelerated to a phosphor screen and the readout obtained by a charge-coupled device. Some examples of parallel detection may be found in Hicks et al. (1980), Hansson et al. (1981), and Delwiche et al. (1982).

4.5   Inverse photoemission

For inverse photoemission with high-energy photons, BIS, one often uses an XPS spectrometer run backwards (see Lang and Baer, 1979). An electron gun bombards the sample with electrons in the 1–2 keV range. The peak energy of the electron beam is scanned. The full width at half maximum is about $4k_BT$, where T is the filament temperature. This will be at least 0.2 eV, even for cathodes with low work functions. The emitted photons pass through a monochromator, often the same one used to monochromate the Al- or Mg-Kα lines used in XPS. Thus the photon energy is fixed at 1487 or 1245 eV, hence the term "isochromat spectroscopy." An electron multiplier is the usual detector. If one is not using an existing XPS instrument, there are more possibilities. Since angle resolution is not useful at such high energies, a much larger solid angle may be collected than can be done with the monochromators on most XPS spectrometers, thereby increasing the count rate. This is done by using an array of quartz monochromating crystals, up to 30 or so, for collecting and dispersing the emitted photons. (This may also be done with an XPS spectrometer.) Typical overall bandpasses are about 0.43 eV (Lang and Baer, 1979). The electron beam may damage the sample, e.g., by changing the stoichiometry of compounds by preferential desorption of one type of atom from the surface.

The source for angle-resolved inverse photoemission also is an electron gun. Ideally, it should be able to deliver a collimated beam of electrons of a variable low energy with a narrow energy spread. Biasing the sample to retard the electrons is a possibility, but there will be electron lens effects that will change the cross-section of the beam. Actually, a very narrow bandpass is not helpful due to the large bandpass of the photon detectors normally employed. Several electron guns have been developed for this work (Erdman and Zipf, 1982; Fauster *et al.*, 1983; Stoffel and Johnson, 1984). The currents from these low-energy electron guns are rather small due to space-charge limitations.

The detectors fall into several categories. The simplest has been a bandpass detector, one of two general types. The first to be used, which is still in use, is a Geiger–Müller counter with a CaF_2 window and I_2 as a component of the fill gas. The threshold for photodissociation of I_2 is 9.23 eV. The CaF_2 blocks photons of slightly higher energy than this. The result is a detector with a response about 0.6–0.8 eV wide, peaked at 9.70 eV. A SrF_2 window can narrow this to about 0.4 eV. The other type is an electron multiplier with a high-work-function cathode, e.g., Be–Cu or an alkali halide, and a window with an absorption edge beginning just above the rise in photoresponse of the cathode, e.g., CaF_2 or SrF_2 (Kovacs *et al.*, 1982; Woodruff *et al.*, 1982; Babbe, *et al.*, 1985; Schäfer *et al.*, 1987). The result is a detector with a peak response at about 9.9 eV and a bandpass of about 0.6 eV. The bandpass can be reduced by heating the filter to move the absorption edge to lower energy. Either of these detectors may be used with a large-aperture collection mirror to increase the solid angle captured, for the angle resolution comes from the incident electron beam, not the detected photons. (The mirror reflectance for 9.7 eV photons will be rather low, but a large solid angle for collection makes it beneficial.) Higher resolution is possible with a monochromator to disperse the photons (Chauvet and Baptist, 1981; Fauster *et al.*, 1983; Johnson *et al.*, 1986), but because the cross-section for inverse photoemission is so small, narrow slits cannot be used for even better resolution because of prohibitively small count rates. The spot on the sample illuminated by the incident electrons serves as the entrance slit, imposing a requirement on the electron gun that it deliver a small spot. This geometry also eliminates the need for a mirror to focus the spot on an entrance slit. The advantage of the monochromator is that it allows access to a wide range of photon energies, typically 5–30 eV, resulting in more information. In fact, if the photon energy detected matches the plasmon energy, there is a resonance in the cross-section. The monochromator used may have an exit slit with a single detector, a position-sensitive detector on the focal surface, or several discrete detectors placed on the focal surface. Count rates with the former will be quite low, for the solid angle accepted by the monochromator will be small and the grating efficiency low. When a position-sensitive detector is used, the isochromat mode is not used, for the electron energy is then held constant and the photon spectrum measured. This corresponds to the constant initial state spectrum (CIS) in photoemission. Isochromats could be constructed if data were taken at enough incident electron energies. Typical overall resolution is 0.1 eV at 10 eV or so, but since the bandpass is constant in wavelength, the bandpass in energy increases at higher photon energies.

A clever intermediate between a bandpass detector and a grating monochromator was developed by Smith and colleagues (Childs *et al.*, 1984; Hulbert *et al.*, 1985). A lens made of LiF images the spot on the sample bombarded by the electrons onto a pinhole in front of the detector, an electron multiplier. The LiF lens transmits all photon energies up to about 11.7 eV, but the refractive index is so wavelength-dependent near this cutoff that different wavelengths from the sample are focused at different points along the axis. Thus the pinhole transmits only a small band of energies, about 0.2 eV wide, centered at a photon energy between 9 and 11 eV, somewhat tunable by moving the pinhole along the axis. An occluding disk is placed at the center of the lens to block paraxial rays which would pass straight through to the pinhole.

4.6 Background subtraction

Core-level spectroscopy is often used for compositional analysis. Doing so requires finding the area under a peak without including any area due to inelastically scattered electrons. Because of the analytical applications of XPS, a great deal of effort has gone into the removal of the background in spectra (see, for example, Smith, 1994a). Of more concern in the study of high-temperature superconductors by photoelectron spectroscopy is the fact that the decomposition of a composite core-level spectrum into components cannot be done accurately unless the background is removed. The removal of the background of inelastically-scattered electrons is frequently carried out in angle-integrated photoelectron spectroscopy of valence bands. Such background removal, important though it may be, has been carried out infrequently in angle-resolved photoelectron spectroscopy of valence bands before the recent intensive study of cuprates by this technique. There are two reasons for this. Firstly, we have a poor understanding of the scattering processes that produce the background in angle-resolved valence-band photoelectron spectroscopy, as discussed in Chapter 3. And, as we shall see, the subtraction generally requires identifying a part of a spectrum that consists only of inelastically scattered electrons. Moreover, in the cuprates, some, or most, of what appears as a background of inelastically scattered electrons may, in fact, be part of the primary spectrum, the incoherent part of the spectral function. Secondly, when taking high-resolution data near E_F, the scans would require an inordinately long time to scan all the way to a binding energy large enough that the initial state is below the bottom of the valence band, for the scan must reach a part of the spectrum that consists only of secondary electrons for the background subtraction.

In an XPS core-level spectrum the background usually appears as a region of constant intensity for initial-state binding energies lower than that of the core level, and as a region of larger constant intensity at binding energies greater than that of the core level. The first rational scheme for background subtraction was that of Shirley (1972). (It was applied to the valence XPS spectra of Au; see also Sherwood, 1983). First, if there is any signal apparently originating from above E_F, a spurious signal, its value is subtracted from the entire spectrum. The first, highest kinetic-energy data point in the spectrum, at the foot of a Fermi edge for a metal, or at the beginning of a core-level spectrum after subtraction of the background at lower binding energy, is assumed to consist only of primary electrons. The next point, at deeper binding energy, consists of primaries and secondaries. The number of secondaries is assumed to be proportional to the number of primaries in the first energy bin, with an unknown constant of proportionality. A first guess for this constant may be made by assuming the secondary spectrum grows linearly with binding energy from E_F to the bottom of the conduction band. The counts or current in each succeeding bin consists of a primary contribution and a secondary contribution, with the latter proportional to the sum (integral) of all primaries in the previous bins, with suitable weights if the energy increments are not equal. At some point, representing the bottom of the conduction band, the primaries are assumed to have stopped and the spectrum consists only of secondaries. It can be compared to the area of secondaries from all previous bins and the constant of proportionality re-determined. Then, the secondary spectrum is subtracted. This procedure may have to be iterated. This procedure assumes that no secondary electron can produce another secondary. The spectrum should be flat for initial states deeper than the energy at which the spectrum is declared to be only secondaries, and such spectra often are found. A variation of this method assumes the secondary contribution to an energy bin in the spectrum is proportional to the number of all electrons of higher kinetic energy, both primaries and secondaries. It predicts a particular non-zero slope in the spectrum for the region where only secondaries contribute. Second- or third-order polynomials have been used, with the parameters fitted to regions of the spectra believed to contain no primary spectra (Wertheim and Dicenzo, 1985). In their study of fitting spectra, Joyce et al. (1989) point out the advantage of including the background fit and subtraction along with line-shape analysis of peaks, rather than removing the background before fitting peak shapes. In addition to inelastically scattered electrons, plasmon peaks may appear on the high binding energy side of core level peaks, and these must be removed based on some knowledge of the system under study. Background subtractions based on the electron energy-loss function, $-\text{Im}[1/\tilde{\epsilon}(E, \mathbf{k})]$, as discussed in Chapter 3, have also been used.

Similar techniques are often used for valence-band angle-integrated photo-electron spectra, but the resultant primary spectra are not fitted subsequently in such a detailed way as are peaks from core electrons. Li *et al.* (1993) emphasize that the creation of secondary electrons by other secondaries is an important effect in determining the background in angle-integrated photoelectron spectroscopy with ultraviolet photons. The background subtraction usually is not done as carefully as in XPS when used for analytic purposes, at least in studies to date. However, angle-resolved spectra are rarely analyzed after subtracting a background, largely because no reliable method for background subtraction has been developed. Low-energy electron scattering is not correctly treated by single-scattering approaches. Measurements of $\tilde{\epsilon}(E, \mathbf{k})$ would need to be carried out with very high resolution for application to scattering in high-resolution photoemission work, and the temperature dependence determined to take phonons into consideration. An additional problem arises in high-resolution angle-resolved spectroscopy. Much of the background may arise from elastically scattered electrons. As described in the next section, surface roughness on an atomic scale or surface impurities can elastically scatter escaping electrons, in effect weakening conservation of the parallel component of the electron wave vector. These background electrons originally had a different wave vector than those being collected, but were scattered elastically into the angle acceptance cone of the analyzer. In extreme cases, no peaks may be seen in an angle-resolved EDC, but only a large continuum of such electrons. This probably accounts for the failure to obtain good angle-resolved spectra on many samples of $Bi_2Sr_2CaCu_2O_8$ and $YBa_2Cu_3O_7$ in early studies, and on all samples of Hg- and Tl-cuprates to date.

Angle-integrated inverse photoemission spectra tend to have rather large backgrounds which, of course, are photons produced by electrons which have inelastically scattered before emission. These backgrounds cannot be treated similarly to photoemission backgrounds because there is no region of the spectrum except the lowest energy point that is known to consist only of photons produced by unscattered electrons. However, the background may be modeled along the lines of the discussion of inelastic scattering in the three-step model of photoemission given in Chapter 3. This gives a spectrum of arbitrary amplitude, due to the neglect of scattering matrix elements, but of the proper shape. It is given as a multiple convolution of densities of states for the region above the Fermi level (Dose 1983). As with angle-resolved photoemission, there is not yet a practical scheme for background subtraction in angle-resolved inverse photoemission. Goodman and Henrich (1994) developed a scheme for background subtraction using as its basis the electron energy-loss spectrum. They used it for angle-resolved inverse photoemission of V_2O_5 taken at 9.8 eV, but it has not yet been applied by others or to other materials.

4.7   Sample preparation

Because photoemission is so surface sensitive, it is important to have a clean surface, one that is well crystallized, even for angle-integrated photoemission and core-level studies. Surface purity is mandatory because the surface-impurity valence electrons may contribute to the spectrum, and they may alter the spectrum of the substrate valence electrons if any chemical bonding takes place. Moreover, the photoexcitation cross-section of the impurity may exceed that of the substrate atoms, so they contribute disproportionately. Crystallinity is needed, even in angle-integrated and core-level spectroscopies because the valence-band density of states and the core-level binding energy and screening effects depend on the structure. An amorphous layer on the surface, perhaps the residue of sputtering, may give an altered spectrum.

To obtain a clean surface, the surface is nearly always created in ultrahigh vacuum and maintained there. (We ignore various "capping" techniques used for semiconductors.) At a pressure of 10^{-9} Torr, the equivalent of about a monolayer of residual gas strikes the surface in 1000 s. For a sticking coefficient of unity (often it will be one or two orders of magnitude smaller), one will have 1% of a monolayer of residual gas on the surface in 10 s, an amount that may produce a detectable change in the spectrum. Fortunately, the predominant residual gas in stainless steel chambers at pressures below 10^{-9} Torr is H_2. Generally, all photoemission studies are carried out at pressures in the low 10^{-10} Torr range, and often in the 10^{-11} Torr range. Impurities on the surface may come not only from the ambient vacuum. At room temperature, certainly above it, and possibly below it, impurities may diffuse to the surface from the bulk. For some impurities in some hosts, the system's equilibrium free energy may be lower with an increased surface concentration of the impurity. Such effects can be greatly inhibited by preparing the surface at low temperature and keeping it cold.

Single-crystal samples are generally acknowledged to be the best to use in photoemission studies. They are mandatory for angle-resolved photoemission. If the crystals cleave at all, even badly, cleavage in ultrahigh vacuum is usually the preparation method of choice. For compounds containing rare earths or transition metals, this is usually done at low temperature, as low as 20 K, to reduce changes in surface composition from out-diffusion of impurities, or in the case of high-temperature superconductors from a surface deterioration probably due to the loss of some oxygen to the vacuum. The orientation of the crystals usually has been determined by Laue back-reflection x-ray diffraction before loading into the sample chamber. *In-situ* low-energy electron diffraction (LEED) may be used after cleavage as a check on the orientation and crystallinity of the surface, but usually this would not involve an analysis of spot widths or of current vs. voltage (*I–V*) curves. Final orientation is determined by taking angle-resolved photoemission spectra at a number of detector angular positions and looking for binding energy extrema to locate symmetry lines.

These cleavage surfaces often may not be very flat. There can be surfaces best described as between cleaved and broken. Cleavage may occur along internal cracks or external cracks, thereby exposing a surface that may not be ideal. A cleaved surface may have cleavage steps and there may be facets on the surface that are not parallel to each other. The latter may be discovered by looking at the spread of a laser beam reflected from a sample. Using a surface that is not flat is equivalent to using an increased angular acceptance, hence reduced resolution in the parallel component of wave vector. Observation under a microscope may reveal a number of steps. The scale of these steps, however, is not indicative of surface roughness on the near-atomic scale needed to ensure that in angle-resolved photoemission the component of wave vector parallel to the surface will be conserved and that there is no broadening of the momentum due to surface irregularities. The quality of the surfaces should be studied by scanning tunneling microscopy (STM), but to date, such studies, carried out on the same type of surface as the photoemission studies have rarely been done. These surfaces would be prepared by cleaving in ultrahigh vacuum at a low temperature, typically 20 K. Another problem is that the area scanned by an STM at near-atomic resolution is very small compared with that scanned in photoemission to date. Typical photoemission studies illuminate an area of the order of 100 μm in diameter, while photoelectron microscopy, seldom applied to high-temperature superconductors to date, samples a region of about 1 μm × 1 μm, both far larger than the area an STM can study with resolution adequate for determining whether or not the surface is "atomically flat." An example of an STM study, both imaging and spectroscopy, is the work on $Bi_2Sr_2CaCu_2O_8$ by Shih *et al.* (1991). Atomic resolution can also be achieved with electron microscopy, and it has also been applied to cuprates. Its connection with the samples used in photoemission studies is not clear, however, for the samples used in electron microscopy were prepared by grinding in air. Also, they can be damaged by the electron beam. An example of such work is the study of $Bi_2Sr_2CaCu_2O_8$ by Zhou (1996). The most common technique, other than the angle-resolved photoemission itself, is LEED, which is useful for a quick check of the crystallographic orientation within the cleavage plane. If the diffraction spots are sharp and there is little diffuse background between them, the surface is deemed as good as one can obtain. By this method, an even better surface would not appear much different.

A common way to prepare single-crystal surfaces of elements is to sputter the surface with Ar^+ or Ne^+ ions till the surface is clean, usually as judged by its Auger spectrum, then anneal to restore the crystallinity usually lost by the sputtering process. Annealing may bring impurities to the surface, so this cycle has to be repeated till the surface gives both impurity-free Auger spectra and sharp LEED patterns. This may be the only way to prepare surfaces of single crystals that do not cleave at all, or of surfaces with orientations not achievable by cleaving. A description of methods used for preparing clean surfaces of 74 elements may be found in Musket *et al.* (1982). Sputter-and-anneal cycles can be applied to compounds, but the sputtering usually disproportionates the surface, and the subsequent anneal may not restore the expected stoichiometric surface. Auger and core-level spectroscopy are not only sensitive to surface impurities, but they can also be used to identify those impurities, except for H (both) and Li (Auger). Valence-band photoelectron spectra also show changes due to impurities, but the atomic nature of the impurity cannot be determined unless a peak in the valence-band spectrum has been correlated with a core-level peak in another spectrum, or the same one if the energy range is wide enough. Examples of such peaks seen in valence-band spectra arise from C and O. It must be remembered that the two surface characterization techniques, LEED and Auger spectroscopy, both subject the surface to an electron beam. This beam could damage the surface it is characterizing. This may not be a serious problem in metals, but in insulating samples such damage is not unknown.

Several chemical etches have been developed for some cuprates. They tend to leave Cu–O surfaces, not always the surfaces produced by cleaving. Several XPS studies have been made on these surfaces (Vasquez, 1994a), but such surfaces may not be suitable for valence-band studies by angle-resolved photoemission. The sampling depth with ultraviolet photons is less than in XPS. The surface layer left by the etch, unless removed, may alter the spectrum. Removal of the top layers, e.g., by sputtering, has the problems discussed above. However, as far as is known, there are no published reports on angle-resolved photoelectron spectra from an etched surface of a cuprate.

The quality of the surfaces can have an effect on selection rules in angle-resolved photoemission. For example, Cerrina *et al.* (1984), obtained angle-resolved spectra on cleaved surfaces of GaAs. For good cleaves, the wave-vector component parallel to the surface was conserved, as discussed in Chapter 3. Poorer cleaves, with many steps on the surface, presumably irregularly spaced, gave spectra in which the parallel component of the wave vector was not conserved. Due to the steps, the surface periodicity was broken and the parallel component of the wave vector was no longer a good quantum number. Later, Davis *et al.* (1985) made normal-emission angle-resolved photoemission spectra on the (211) face of Cu. This surface forms a series of periodic steps, with terraces of (111) planes three atoms wide separated by a (100) step one interatomic distance in height. The step period in the (211) plane is 6.01 Å. The spectra indicated that the parallel component of the wave vector was a good quantum number, and the periodic steps did not even produce a significant spectral feature due to umklapp processes, i.e., the addition of a surface reciprocal lattice vector to $k_\parallel$. However, Kevan (1986), in a higher-resolution study, showed that on Cu, randomly dispersed impurities on the surface, in this case a small amount (1–2% of a monolayer) of evaporated K, can add an uncertainty of the order of Λ^{-1} to the parallel component of the wave vector, where Λ is the average mean free path for electron scattering by the impurity. Λ is inversely proportional to the impurity concentration. Earlier work (Tersoff and Kevan, 1983; Kevan, 1983c) showed differences in apparent intrinsic widths for one transition on one crystallographic surface prepared from different crystals of Cu. These differences were attributed to differences in crystal purity, surface impurity concentrations having been estimated to be of the order of 1% of a monolayer. Paniago et al. (1995) reexamined the shapes of the surface-state spectra on Ag and Cu as a function of temperature and angle resolution. They, too, observed a residual temperature-independent contribution to the width, presumably originating from surface defects. (They also showed the influence of angular resolution and electron–phonon scattering on line shapes. They emphasized how difficult it is to obtain the correct line shape from a measured spectrum.) These studies may have important consequences for high-resolution studies of high-temperature superconductors. The metallic cuprate samples are non-stoichiometric, for they contain O vacancies. Moreover, sometimes there are metal ions on the wrong metal-ion site. The extent to which these departures from strict periodicity affect wave-vector uncertainty is unknown at present. The effects seen by Kevan (1986) were observed on a surface state with a narrow lifetime width. 1% of a monolayer coverage is equivalent to 0.1% volume impurity concentration. When $Bi_2Sr_2CaCu_2O_8$ is reduced to $Bi_2Sr_2CaCu_2O_{7.8}$ the O vacancy concentration is 2.5% on the O sublattices, a vacancy concentration of 1.3% considering all sites in the unit cell. This defect concentration eventually may limit the width of peaks in angle-resolved photoemission, but there is no evidence that it yet has limited widths with resolutions used to date, about 10–20 meV in energy and 1° in angle. As we shall remark later, the background, which is due to inelastically scattered electrons in other materials, seems higher in the angle-resolved spectra of the cuprates. Part of this could arise from elastic scattering at defects. However, another point of view is that this high background is intrinsic, the incoherent part of the many-body response of the cuprates to photoexcitation.

No matter how the single-crystal surface has been prepared, its structure may change. Surfaces of covalently bonded materials, e.g., Si, tend to reconstruct, with significant atomic displacement and formation of new interatomic bonds. Surfaces of some metals may reconstruct. Surfaces of some metals and most ionic compounds tend to relax; the surface plane and planes just below it alter their relative spacings a few percent. There may be a kinetic barrier to reconstruction so that a surface prepared by cleaving at low temperature does not reconstruct until it is heated to some higher temperature. Surface reconstruction often is detected easily by the appearance of the LEED pattern, but relaxation does not alter the appearance of the LEED pattern. It must be determined by a detailed comparison of LEED I–V spectra with I–V curves calculated for assumed relaxations.

A final problem associated with single-crystal surfaces (and polycrystalline surfaces as well) of high-temperature superconductors is a change in stoichiometry. As will be reported in Chapter 6, irreversible changes take place when a crystal of Y123, cleaved at 20 K, is warmed to about 100 K for a few minutes in UHV, then cooled back to 20 K. It is suspected that the change is the loss of oxygen to the vacuum. Diffusion of oxygen in to and out of Y123 at room temperature has been measured (Mogilevsky et al., 1994), and oxygen loss has been believed to be the cause of surface reconstructions observed by LEED (Behner et al., 1992). Bi2212 crystals appear not to show such an effect, even upon warming to 300 K. Thus it cannot be assumed that the composition of the region sampled in photoemission is the same as that of the bulk. Cleavage at low temperature could result in the loss of some oxygen, but there is no direct evidence for this. Cleavage at a higher temperature could result in an oxygen loss large enough to convert a surface layer of the sample from a metal to an insulator. Such a change does not require many O atoms to leave the first few layers at the surface. There is some evidence (Chapter 10) that O atoms can be reintroduced to the surface.

The usual sample preparation method for angle-resolved photoelectron spectroscopy of cuprates is to begin with a single crystal. Before use, T_c usually is measured magnetically and the directions of the a- and b-axes determined by Laue back-reflection x-ray diffraction. It is mounted with a thin layer of epoxy on a metal post either before or after the determination of the orientation. A second post or tab is epoxied to the top surface. The lower post is screwed into the cold end of the sample manipulator, usually a closed-cycle He refrigerator. When the sample chamber has been baked and the pressure is in the 10^{-10} Torr range, the sample is cooled to below 40 K, at which point a small movable rod in the sample chamber hits or pushes the top post. The crystal usually cleaves, leaving a clean surface of variable quality. In-situ LEED measurements may be used to check the crystallinity and orientation of the a- and b-axes again. A long-focus microscope may be used to inspect the surface before use. The quality of a laser beam reflected from the cleaved surface may also be used as a check for surface flatness. The samples may be prepared in the chamber used for photoelectron spectroscopy or in a separate chamber from which the sample may be transferred in ultrahigh vacuum.

An alternate cleaving method was used by Schroeder et al. (1993a,b,c). The crystal was epoxied between the two plates of a hinge. The cleavage was effected by advancing a razor blade into the crystal. After cleaving, the hinge could be opened, exposing both cleaved surfaces. This technique was first used to cleave at room temperature, but it can also be carried out at low temperature.

A different method of surface preparation was recently introduced by Ratz *et al.* (1996). They used a "micro milling machine" in which the sample was oriented carefully with respect to the milling machine and the surface scraped off by a diamond blade moving accurately parallel to the (001) surface of the crystal. This was done at a temperature of about 10 K. The angle-resolved valence-band photoelectron spectra of Y123 obtained this way were similar to those this group had obtained earlier by the cleaving technique described in the previous paragraph, but the structures were not as prominent.

For angle-integrated valence-band photoelectron spectroscopy and for many core-level studies, polycrystalline samples have been used. Certainly this describes all the earliest work on high-temperature superconductors. The two commonest sample-preparation techniques are breaking a bulk polycrystalline sample in vacuum, analogous to cleaving, and scraping a new surface on a bulk sample using a diamond file. This, too, is analogous to cleaving. Several problems can arise from this. The polycrystalline samples may tend to break on grain boundaries, rather than across the grains. This may leave the new surface enriched in those impurities which segregate at grain boundaries. If there are several phases present, the break may occur preferentially at or through grains of a second phase, although these presumably are not present in great concentrations if the sample has been screened previously by x-ray diffraction. Cuprate samples cleave easily along their basal planes. In a polycrystalline sample to be subjected to scraping, the grains with the proper orientations may cleave easily and cleanly upon scraping, but those with basal planes nearly normal to the scraped surface may have rather ragged edges.

Many XPS spectrometers are not equipped with low-energy electron diffraction, so a LEED pattern cannot be used as a measure of sample quality. Other criteria have been developed with appear to be necessary, but not sufficient, conditions for obtaining core-level spectra on cuprates that could be free of the effects of spurious phases and altered surface stoichiometry. (It is assumed no impurity peaks are present in the spectrum.) These have been discussed at length by Vasquez (1994a). First and foremost, the XPS spectrum of the valence-band region of a metallic cuprate should show a Fermi edge. Also the O 1s spectrum should be a single peak with a binding energy of 528.0–528.5 eV. Additional peaks at higher binding energies are characteristic of other phases or stoichiometries, often more insulating. Ba, if present, should have a $5p_{3/2}$ peak at 12 eV, and the Cu 2p peak should exhibit satellites. Just a single spin–orbit-split pair of

Cu 2p peaks is indicative of Cu^0 or Cu^+, not Cu^{2+}.

It has been possible to grow thin films of several high-temperature superconductors epitaxially on several substrates. Moreover, the plane of such a film
need not be the basal plane with the c-axis normal to plane of the film. Films
have been grown with the c-axis in the plane of the film. These offer the possibility
of carrying out angle-resolved photoemission on samples whose surfaces are
not basal planes. According to Roy et $al.$ (1994), many of the cuprate films called
epitaxial were not really epitaxial, although a high degree of orientation may
have been achieved.

Films offer several potential advantages besides the different orientation.
They may be doped easily with different metallic elements. They can have much
smoother and flatter surfaces than those of cleaved single crystals. However,
there may be considerable internal strain. Often, cuprate films have been grown
by sputtering or laser ablation. They must be removed from the growth chamber
and introduced into the photoelectron spectrometer; no side-by-side, connected
systems have yet been set up. The surfaces of the films must then be cleaned
before photoelectron studies. Sputtering and annealing are widely used for clean-
ing many materials in $situ$, but this technique often does not yield surfaces of cuprates that exhibit a Fermi edge when one is expected from the bulk
stoichiometry. Presumably there is loss of oxygen during the treatment (Matsui
et $al.$, 1996). Simply exposing the surface to O_2 gas is insufficient for restoring the
oxygen concentration. Sakisaka et $al.$ (1989b) were able to observe a Fermi
edge and dispersing peaks in angle-resolved photoemission spectra from an epitaxial film of $YBa_2Cu_3O_{7-x}$ that had been heated for 20 minutes at 873 K in 100
Torr of O_2, then cooled to 300 K in the oxygen. Treatments involving a microwave plasma of oxygen, i.e., exposure to atomic oxygen (Terada et $al.$, 1994) or
combined exposure to O_2 and photons (Mogilevsky et $al.$, 1994; Aprelev et $al.$,
1994), to be described in Chapter 10, apparently restored the oxygen content,
and Fermi edges appeared in angle-integrated spectra. Marshall et $al.$ (1995)
obtained angle-resolved photoemission spectra on thin films of Bi 2212 produced
by molecular beam epitaxy. The surfaces on which the measurements were
made, however, were produced by cleaving the film samples at low temperature,
the same technique used for bulk single crystals. The data indicated that despite
the large density of defects expected in the film, and a broader transition at T_c,
data similar to that from bulk samples could be obtained. Ma et $al.$ (1995e) and
Hatterer et $al.$ (1996) had similar results upon cleaving a sputtered film of
Bi2212. To date, most data taken on epitaxial films have been from rather thick
basal-plane films that were cleaved in the same manner that single crystals are
cleaved.

4.8 Potential improvements

We assume higher resolution is desirable. We may already have reached the intrinsic widths of some features in the spectra, but we cannot be sure till we see spectra with unchanged widths when we have better instrumental resolution. We certainly can make more meaningful measurements near the Fermi level if better resolution is achieved. Higher-resolution monochromators can be built, and higher-resolution electron analyzers already exist. They cannot yet be used readily, however, for the count rates become unacceptably low. Count rates can be improved with still brighter sources, and with more efficient monochromators. Diffraction gratings are very inefficient. It is likely that better gratings can be made, although there probably will be a trade-off of high reflectivity, perhaps with multilayer coatings, for a more limited wavelength range. Finally, residual thermal broadening should be reduced. Temperatures of 15–20 K are not difficult to achieve in photoelectron spectroscopy. More refrigeration power should lower this a little, but more would be gained by better radiation shielding. The sample now is exposed to a large solid angle of room-temperature radiation to allow photons in and electrons out. It appears possible to cool the front of the input electron lens and to mount a cold shield around it to block some of the room-temperature radiation.

Increasing the energy resolution of the monochromator or analyzer will help achieve better resolution, but improvements in both are more desirable, and at the same time the angle resolution should be improved. The overall energy bandpass depends on the bandpasses of the monochromator, $\Delta h\nu$, and the analyzer, ΔE, approximately as $[(\Delta h\nu)^2 + (\Delta E)^2]^{1/2}$. Reducing both terms is desirable. The finite angular acceptance of the analyzer, $\Delta\theta$, gives an uncertainty in the component of wave vector parallel to the surface of $\Delta \mathbf{k}_{\|}$, which depends on the collection angle. This, in turn, causes an additional energy spread in the spectrum of $\Delta \mathbf{k}_{\|}$ multiplied by the parallel component of $\nabla_k E_i$, which can be appreciable for a highly dispersing band. In 1994–5 several new third-generation synchrotron radiation sources came on line. Several of these already have beam lines that offer unprecedented resolution for XPS from core levels. Vacuum ultraviolet beam lines that offer improved resolution for valence-band studies are under construction, as are beam lines for photoelectron microscopy (to be discussed in Chapter 11). For all of these beam lines, the improved resolution does not come at the expense of count rates, for the undulator sources are brighter than those currently employed.

Higher instrumental resolution will be wasted if the spectral widths are determined by the samples, either as intrinsic (lifetime) widths or apparent widths due to imperfect surface periodicity. Perhaps better annealing can lead to better, i.e., smoother, cleaved surfaces. The role of atomic-scale defects such as vacancies, impurities, and antisite atoms on the apparent resolution is unknown. Clearly the O vacancies are necessary to keep the surface superconducting. Studies of the surfaces of cuprates by scanning microscopies have been carried out but, so far, not on surfaces prepared in the same way as surfaces are prepared for photoelectron spectroscopy. Photoelectron microscopy spatial resolution falls far short of atomic length scales.

Over the years since 1989, the quality of crystals of Bi2212 has improved. Early studies by angle-resolved photoelectron spectroscopy of the anisotropy of the superconducting gap showed no anisotropy, while later work shows a reproducible anisotropy. The earlier crystals had a shorter structural coherence length as determined by x-ray diffraction peak widths than crystals grown several years later (Campuzano, private communication). Recent studies by Skelton *et al.* (1994), Qadri *et al.* (1996), and Osofsky *et al.* (1996) raise the issue of lateral variations in oxygen concentration in "good" $YBa_2Cu_3O_{7-x}$ crystals. These crystals had been characterized by the usual x-ray techniques and they were good by conventional standards. The width of the transition to the superconducting state was small. The x-ray diffraction peaks were "sharp." Qadri *et al.* applied high-resolution x-ray diffraction to measure the c lattice parameter. There is a known correlation between this lattice parameter and the oxygen concentration, $7 - x$. With high resolution, there were multiple peaks in the $\theta - 2\theta$ scans, indicating up to six areas with different lattice parameters, hence different oxygen concentrations. For two good single crystals the averaged values of $7 - x$ were 6.83 and 6.84, but different regions in these crystals had values of $7 - x$ ranging from 6.71 to 7.00 for one crystal and 6.80–6.99 for the other. Skelton *et al.* scanned a 5 μm ×10 μm x-ray spot across the surface of a "good" crystal of $YBa_2Cu_3O_{7-x}$ and measured the c lattice parameter as a function of position using higher-energy x-ray photons, thereby sampling more of the bulk than in the work of Qadri *et al.* A similar inhomogeneous distribution of oxygen content was found. A similar scan on a single crystal of $Nd_{2-x}Ce_xCuO_{4-y}$ was made, revealing an inhomogeneous distribution of Ce. The role of such inhomogeneities in angle-resolved photoelectron spectroscopy is unknown.

Chapter 5

Examples

Introduction

In this chapter we present photoelectron spectra of several materials. We present valence-band data for all of them, and in most cases core-level spectra are discussed. Na is a simple metal, about as simple a metal there is. If the photoelectron spectra of Na cannot be understood it is difficult to be comfortable with interpretations of photoelectron spectra of the cuprates. Cu and Ni are more complicated, having 3d bands near and spanning the Fermi level, respectively. Then we present results on NiO, Cu_2O, and CuO which introduce new effects. The photoelectron spectra of NiO and especially CuO bear some resemblance to those of the cuprates. The studies reported on the three oxides do not represent all of the work on these materials. Studies on related materials, e.g., Cu- and Ni halides, have been useful, for one can see the influence on the photoelectron spectra of changes in ionicity, number, type, and disposition of nearest neighbors, and distances between nearest neighbor transition metal atoms. Studies of a series of 3d transition-metal oxides are also helpful, but we do not discuss these here. Another approach, also not discussed here, is to start with a clean single-crystal surface of a transition metal and let it oxidize in stages in the experimental chamber, taking spectra and LEED patterns at various stages. In some cases, e.g., NiO, the final stage may be several layers of the same oxide that is studied in bulk samples.

5.2 **Sodium**

The three-step model described in Chapter 3 was applied there to an alkali metal, using parameters for Na. The electronic structures of the alkali metals Na, K, and Rb have been considered to be described accurately by the nearly-free-electron model. The photoelectron spectra can be calculated with only one parameter, V_{110}, a Fourier coefficient of the pseudopotential, which may be obtained from theory, from optical data or, most precisely, from Fermi surface measurements such as those obtained from the de Haas–van Alphen effect. Li and Cs may be more complex because Li lacks a p-like core level which would make the pseudopotential quite simple and because Cs has a d band not far above the Fermi level. (K and Rb have empty d bands further above E_F.) We ignore the possibility of a charge-density wave in K, and perhaps Na, but such a phenomenon could give an observable photoemission signature, not yet reliably observed (Overhauser, 1985). Values for the mean free path for inelastic scattering need not be accurately known for our comparison.

Smith and Spicer (1969) measured angle-integrated EDCs on polycrystalline films of Na and K. Their sample chamber was isolated from the monochromator by a LiF window, limiting photon energies to a maximum of 11.6 eV. Their data (EDCs) for Na are shown in Fig. 5.1. The background of inelastically scattered

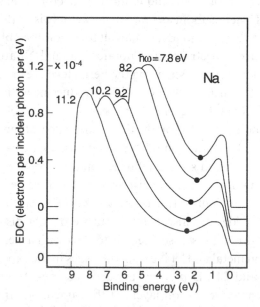

Fig. 5.1 Angle-integrated photoelectron energy distributions for Na at several photon energies. The background of inelastically scattered electrons has not been subtracted. The filled circles mark the minima of the EDC and should be close to the cutoff of the primary distribution, the bottom of the conduction band at the larger photon energies. (Smith and Spicer, 1969)

electrons has not been removed from the spectra. The filled circles indicate a minimum in photocurrent, but without a background subtraction one can say only that they lie close to the bottom of the conduction band. These spectra already give a hint that the conduction band may not be as wide as that of band-structure calculations. There is some agreement with the three-step model calculation, in that the widths of the spectral features change with photon energy in the expected manner, but the shapes are clearly not correct. The calculation predicts a box-like primary spectrum (Fig. 3.13), while the data have approximately triangular shapes. The origin of this discrepancy is not clear. Candidate explanations include an overly simple escape function, the inapplicability of the three-step model, and the inaccurate final-state wave functions, bulk functions having been assumed. It is not useful to pursue this further.

Plummer and colleagues studied Na and K metals by angle-resolved photoemission on epitaxial-layer samples (Jensen and Plummer, 1985; Lyo and Plummer, 1988; Itchkawitz et al., 1990). The data were compared with one-step calculations using the formalism of Mahan (1970), assuming the independent-electron model. One then expects the valence band below the Fermi energy to resemble that of the nearly-free-electron model in all respects. Figure 5.2 (Lyo and Plummer, 1988) shows that this is not the case. The Fermi wave vector is in excellent agreement with the band-structure result (and other experiments) but the bottom of the measured band is 0.6 eV closer to the Fermi level than in calcu-

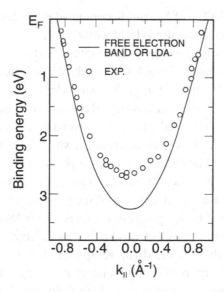

Fig. 5.2 Band dispersion for Na from angle-resolved photoemission (open circles) and the independent-electron calculation of the band structure. $k_\parallel$ is perpendicular to the (110) surface normal. Negligible anisotropy is expected. (Lyo and Plummer, 1988)

lations, i.e., the band calculation gives a band that is too wide, and an effective mass that is too small.

The bandwidth discrepancy is presumably due to the inadequate treatment of the electron–electron Coulomb interactions. Calculations of the real part of the self-energy, $\Sigma_1(\mathbf{k}, E)$, existed at the time, but they accounted for only part of the discrepancy. These measurements stimulated new calculations using several techniques (Shung and Mahan, 1986; Shung et $al.$, 1987; Shung and Mahan, 1988; Surh et $al.$, 1988; Lyo and Plummer, 1988; Northrup et $al.$, 1987, 1989; Shung, 1991). All of them reduced the occupied bandwidth discrepancy in Na and K to 0.1 eV or less, although the calculations do not agree on how this comes about. Northrup et $al.$ find that Σ_1 accounts for the discrepancy. Shung et $al.$ find the real part of the self-energy provides about half of the shift of the energy of the photoelectron peak needed to produce agreement between calculated and measured spectra, but the imaginary part also plays a role in the shift. It causes energy broadening, effectively diminishing the role of wave-vector conservation in direct transitions. It also gives rise to a very short mean free path, about 5 Å for Na (Shung and Mahan, 1986, 1988) which enhances the role of the potential at the surface. These two effects appear to account quantitatively for the observed bandwidth narrowing. The two sets of calculations made different approximations in evaluating the self-energy. These calculations also showed that separation of the photocurrent into a part arising from ther bulk and one from the surface is not possible. The contribution from interference between them is not negligible.

The angle-resolved EDCs of Na and K contained some data points not explicable at the time they were measured (Itchkawitz et $al.$, 1990). They did not correspond to allowed transitions using the calculated band structure, or reasonable distortions of it, and for several years were a subject for concern. Theory (Shung and Mahan, 1986, 1988; Shung et $al.$, 1987) was able to account for the locations of the peaks, but not for their intensities. Recently, the likely explanation for these peaks in the angle-resolved EDCs was given (Wertheim and Riffe, 1995). The peaks were observed in normal emission in which $k_\perp$ was scanned by scanning the photon energy. They occurred in K for photon energies between 20 and 25 eV and 63 and 70 eV, and their initial states appeared to be at E_F, independent of $\mathbf{k}$. Wertheim and Riffe repeated some of the (angle-resolved) measurements in the 17–38 eV region. The K 3p core-level threshold is 18.3 eV, and at higher energies one is exciting both 3p electrons and valence electrons, if the latter transitions can occur. At the threshold for K 3p emission, the apparent valence-band photocurrent arising from initial states at E_F rose by a factor of about 5, then fell at higher photon energies, qualitatively following the expected behavior of the K 3p photoexcitation cross-section. If the 3p electron is excited to a conduction band state above E_F and this state decays when a valence electron fills the 3p hole, giving its energy to the excited electron, the final state is the same as in direct photoemission from the valence band. Interference between the two final states appears not to be important; the 3p excitation dominates. Certainly it is the only process at those photon energies at which there can be no direct transitions from the valence bands. The two-electron process in the autoionization effectively integrates over wave vectors, reducing dispersion effects, as observed (Jensen and Plummer, 1985; Lyo and Plummer, 1988; Itchkawitz et al., 1990). Empty 3d states in the conduction band play a significant role in determining the shapes of the observed spectra. Wertheim and Riffe point out, however, that not all details of their spectra can yet be explained quantitatively; more theoretical work is required.

Core-level photoelectron spectra of the alkali metals have been measured, usually exciting further above threshold than in the experiments on K just cited. The motivation for the core-level studies was not to elucidate the electronic structure of the valence electrons in alkali metals, but to use knowledge of that structure to study the many-body processes taking place when the core hole is created suddenly. The data analysis focused on obtaining values of the asymmetry parameter α, described in Chapter 3, for comparison with related parameters obtained from soft x-ray absorption measurements of the core-level absorption edges and with many-body calculations of the screening of the photohole by the interacting electrons (Wertheim and Citrin, 1978; Ohtaka and Tanabe, 1990).

Angle-resolved inverse photoemission spectra have been measured for Na, (Collins et al., 1988) and angle-integrated spectra for Na, K, Rb, and Cs (Woodruff and Smith, 1990). The detected signals were extremely small and the spectra nearly devoid of structure, expected because dipole matrix elements in the bulk and at the surface are small, as is the joint density of states. Only faint hints of inter-s–p band transitions were found, and there was no sign of d bands above E_F in K, Rb, and Cs. The small signals, high background, and low resolution of inverse photoelectron spectroscopy make it not very useful for the study of alkali metals.

Thus it appears that the photoelectron spectra of nearly-free-electron metals can be accounted for nearly completely by the one-step picture for photoemission, but many-body correlations must be considered in order to obtain quantitative agreement. In addition to Na and K, measurements (Levinson et al., 1983) and calculations including many-body effects (Ma and Shung, 1994) exist for Al.

5.3 Copper

Probably more photoemission studies have been carried out on metallic Cu than any other material. Copper has been the testing ground for new techniques in

photoelectron spectroscopy and for new theoretical ideas. The cuprates and NiO
now are being used, along with Cu, as the testing arena for new understanding
of photoemission. Face-centered cubic Cu has nearly-free electron s–p bands,
and a filled 3d band, the top of which lies about 2.5 eV below the Fermi level.
The interaction between these sets of bands is significant, and the Fermi surface
is distorted from a sphere, contacting the Brillouin zone boundary on the {111}
faces. The states at the Fermi level have considerable d character. Copper was
the subject of the early work by Berglund and Spicer (1964a,b) which led to the
flowering of photoelectron spectroscopy. Angle-resolved photoemission studies
on Cu began early, and have continued. Once the bulk bands were mapped, interest centered on surface states and on photohole and photoelectron lifetimes.
Phonon effects on wave-vector conservation and on photohole lifetimes were studied. Much of the work up to early 1984 has been reviewed by Courths and
Hüfner (1984). We discuss below only those features of the photoelectron spectroscopy that led to an understanding of the electronic structure of bulk Cu, or
to an understanding of the photoemission process itself. We describe only one
surface state on one crystal face, but most of the studies of the widths of the features in angle-resolved spectra. Our understanding of the latter is not yet complete, and it is important for the study of the cuprates by photoelectron
spectroscopy. We do not reference all of the many studies on Cu that have been
made by photoelectron spectroscopy. There have been many since the review of
Courths and Hüfner.

Early angle-integrated photoelectron spectra of Cu showed the 3d band and
the s–p band to have widths in rather good agreement with the energy band calculations at that time. Angle-resolved XPS measurements on polycrystalline films
made by Wagner et al. (1977) and Mehta and Fadley (1977, 1979) were analyzed
by Steiner and Hüfner (1982). The observed angle dependence came from the
effective depth sampled, a smaller depth for collection angles closer to grazing.
Steiner and Hüfner used layer-by-layer densities of states for Cu calculated self-
consistently by Wang and Freeman (unpublished) for a five-layer slab. Two
layers were surface layers and the center layer, representing bulk, had a density
of states in close agreement with that from bulk calculations. The surface layer
has a narrower 3d band because of the reduced number of nearest neighbors.
Steiner and Hüfner fitted the spectra at each angle (Fig. 5.3) with a different linear
combination of the three layer-separated densities of states, the weights determined by the angles and the electron inelastic scattering length. The fit is excellent, according to Fig. 5.3.

Angle-resolved spectra were used to map initial-state bands for comparison
with calculations. A compendium of these results is shown in Fig. 5.4 along with
one set of relativistic energy bands, calculated self-consistently (Eckardt et al.,
1984). These measurements highlight the principal difficulty in band mapping,

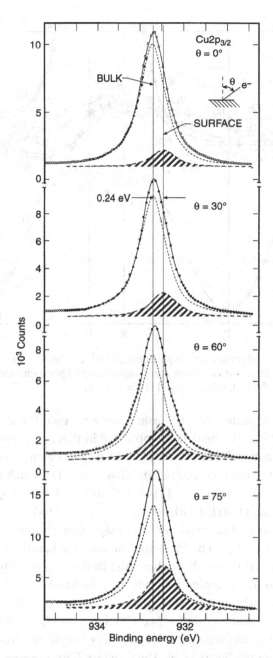

Fig. 5.3 Cu $2p_{3/2}$ XPS spectra taken at several exit angles to vary the relative contribution of the surface layer. The circles are the data points. The solid line is the fit to a sum of bulk and surface contributions, the latter being shown as dashed lines. (Steiner and Hüfner, 1982)

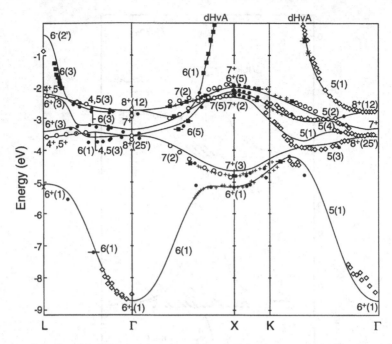

Fig. 5.4 Calculated (solid curves) and experimental (various symbols) band structure of Cu below E_F. The experimental values are from angle-resolved photoemission studies by many authors, cited in the text. (Courths and Hüfner, 1984)

using the correct final states. Several techniques were used for identifying final states. Figure 5.5 shows the final states obtained in these measurements, along with the calculated bands. The data are from Dietz and Eastman (1978b), Dietz and Himpsel (1979), Knapp *et al.* (1979), Nilsson and Dahlbäck (1979), Thiry *et al.* (1979), Grepstad and Slagsvold (1980), Pétroff and Thiry (1980), Courths (1981), Courths *et al.* (1981), Lindroos *et al.* (1982), Lindgren *et al.* (1982), Baalmann *et al.* (1983), Przybylski *et al.* (1983), Courths *et al.* (1983), and Courths and Hüfner (1984). The measured initial-state bands agree extremely well with the calculated bands throughout the Brillouin zone (Fig. 5.4) with a few exceptions, the most prominent of which is the band along Γ–X–K about 4.5 eV below E_F. Here the measured band is too deep by about 0.15 eV, a discrepancy opposite in direction to that for Na. Con Foo *et al.* (1996) determined the Fermi surface in Cu by measuring a large number of angle-resolved CIS spectra with the initial state at the Fermi level. The resultant Fermi surface cross-section followed rather closely that determined by de Haas–van Alphen measurements and LDA band calculations.

One rather interesting result is the observation of a particular surface state on the (100) surface of Cu by Wincott *et al.* (1986). This state, with a minimum binding energy of 2.113 eV, lies in a gap in the bulk 3d bands. This gap, 150 meV wide at the point of minimum surface-state binding energy, does not exist in non-

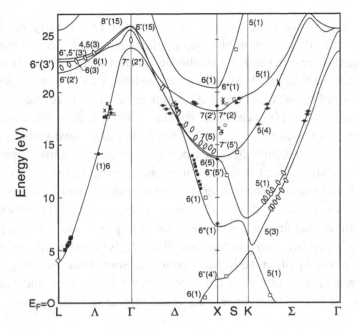

Fig. 5.5 Calculated (solid curves) and experimental (various symbols) band structure of Cu above E_F. The experimental values are from angle-resolved photoemission studies by many authors, cited in the text. (Courths and Hüfner, 1984)

relativistic calculations, but is present in the relativistic calculations because of spin–orbit splitting.

Nilsson and Larsson (1983) examined whether the inclusion of the electron's self-energy could reduce the discrepancy in the binding energies between LDA calculations and the measured bands. They used a self-energy for a free-electron gas that had been calculated for a range of electron densities. The real part of the self-energy was interpolated for use with the muffin-tin region, so it varied with the local electron density as well as electron energy, and the imaginary part was approximated by a constant. For Cu (111) measured with 21.2 eV radiation, the peaks in the spectrum calculated with the method of Pendry *et al.* (1976) with the self-energy correction moved non-rigidly to smaller binding energy by up to 0.1 eV compared with the same calculation without the self-energy. This was in the direction to produce better agreement with experiment, but for two of the three peaks the shift was not large enough to produce actual agreement. After the above work was carried out and summarized, Baalmann *et al.* (1985) measured the dispersion of a final state along the Σ line in the 10–60 eV range of kinetic energies. The band was not parabolic, and in the 15–25 eV region, the final-state energy increased without an increase in wave vector. This occurs when the wave vector is at the X point in the Brillouin zone were there is a 7 eV gap. The measured gap-edge energies are shifted several eV from the calculated gap-edge energies, presumably for lack of a self-energy correction.

Angle-resolved inverse photoemission was used rather early on Cu. As expected the spectra are broad, but they readily illustrate dispersion effects for states above the Fermi level. Low resolution was not a great disadvantage, for the initial states about 10 eV above E_F are at least 1 eV broad, as described below. Woodruff *et al.* (1982) and Altmann *et al.* (1984) reported extensive angle-resolved inverse photoemission studies on (001) and, in the latter work, (110) surfaces on Cu. Jacob *et al.* (1986) worked on these two surfaces and on Cu (111), both at 300 K and higher temperatures. Figure 5.6 shows typical data taken on the (100) surface for electron incidence normal to the surface and for several angles off normal, illustrating dispersion with the parallel component of wave vector. The measured dispersion follows calculated bands rather well (Fig. 5.7). One can see that final states between E_F and $E_F + \Phi$, not active in photoemission, can be reached, and that the band above E_F is the expected continuation of the band below E_F observed in photoemission. Some band mapping of states above E_F was carried out on a Cu (100) surface by Strocov and Starnberg (1995). Strocov *et al.* (1996) have used very-low-energy LEED to obtain final

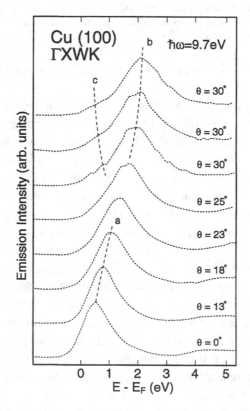

Fig. 5.6   Angle-resolved inverse photoemission spectra for the (100) surface of Cu. (Woodruff *et al.*, 1982)

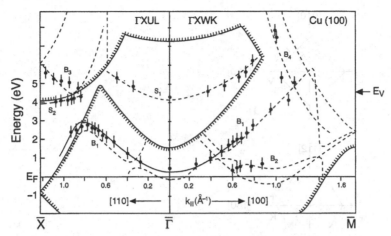

Fig. 5.7 Energy bands above E_F for Cu, obtained by angle-resolved inverse photo-emission spectroscopy. S and B denote surface and bulk bands. Dashed lines represent possible bulk bands. E_v is the vacuum level. (Jacob *et al.*, 1986)

states above E_F for Cu. By fitting these to bands calculated with a complex potential they obtained final-state bands they recommend for use in photoemission.

As the bulk electronic structure became better understood, attention turned to surface states. The (111), (100) and (110) surfaces each displayed one or more bands of surface states. We discuss only the (111) as an example. The Brillouin zone for the (111) surface of an f.c.c. crystal is shown in Fig. 5.8. Bulk bands, projected along a (111) wave vector, intersect this surface zone in most, but not all, of its area. The projected bulk bands exist at all surface wave vectors in the shaded regions. Surface states may form bands in the unshaded regions. They may contribute to a photoelectron spectrum. The surface bands shown have been calculated and observed. They were first identified in angle-integrated spectra for some materials as a peak, unexpected from the calculated band structure, which was more sensitive to adsorbed overlayers than the rest of the spectrum. Angle-resolved spectra were more informative, for they could demonstrate dispersion with surface wave vector, and test for lack of dispersion with wave vector normal to the surface. Such a spectrum is shown in Fig. 5.9. These data demonstrate some unusual features (Kevan, 1983b). The peak widths decrease with depth below the Fermi surface. Photohole lifetimes can be obtained rather easily from surface states since the hole group-velocity is zero and the intrinsic width, Γ, is then the photohole lifetime width, Γ_h. Γ_h should be zero at the Fermi level and increase below it from electron–electron scattering. One source of additional broadening is the finite angle acceptance, which contributed an additional energy broadening proportional to the slope of the dispersion curve, i.e., zero at the Γ point, and largest at E_F. There appears to be yet another source of broadening, scattering due to surface impurities and defects (Kevan, 1986).

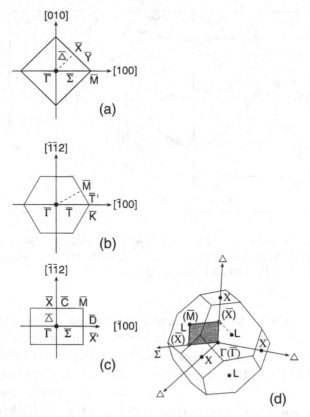

Fig. 5.8 Surface Brillouin zones for three surfaces of an f.c.c. crystal. (a) (001); (b) (111); (c) (110). (d) shows how some symmetry points in the Brillouin zone for the bulk crystal project into one quadrant of the Brillouin zone for the (110) surface.

Bulk and surface states were used for the study of lifetime-and phonon broadening. Early work (Eastman *et al.*, 1978) indicated that the final (bulk) states in an s–p conduction band along the [100] direction in Cu increased in width from 1.2 eV at 10.5 eV above E_F to 1.8 eV at 13.5 eV above. Goldmann *et al.* (1991) measured lifetimes in the s–p bands above E_F by inverse photoemission on (100) and (110) Cu surfaces, going as far as 45 eV above E_F where the width was about 6 eV. Their data, and data from several photoemission studies, could be fit by the simple linear relation $\Gamma_e = 0.13(E - E_F)$ for an electron at energy E. Still further above E_F, the widths fall below the extrapolation of this linear relationship (Himpsel and Eberhardt, 1979).

Additional detailed studies of the effect of temperature on photoemission have been carried out, and the sample of choice has been Cu. In Chapter 3 we showed that increasing temperature caused a loss in the intensity from direct, **k**-conserving transitions and an increase from those that did not conserve **k**, the latter giving a spectrum like a weighted density of states. We also commented that

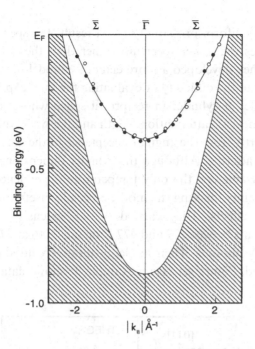

Fig. 5.9 Dispersion, energy vs. wave vector, along [211] in the surface, of a surface state on Cu (111). The shading marks regions where there is a bulk band state with the indicated energy and wave-vector component parallel to the surface. The data were taken with two photon energies, 11.8 and 16.8 eV, filled and open circles, respectively. (Kevan, 1983b)

phonons could contribute to the widths of features in an angle-resolved EDC. Also, but not mentioned there, photoelectron peak energies will shift as thermal expansion and the electron–phonon interaction shift the energy bands in a non-rigid way. Surface states should be even more sensitive to temperature due to a Debye temperature, Θ_D, that is lower than that of the bulk and to larger anharmonic effects. This has been borne out in a series of experiments. Studies of the photoemission of cuprates are not carried out above room temperature; usually they are carried out at 100 K and below. The possible relevance to cuprates of studies on Cu many hundred degrees above room temperature is that phonon emission may play a role in high-energy and -angle resolution studies of cuprates. Phonon emission can occur at low temperatures.

White *et al.* (1987) extended the theory of Shevchik (1977) for the effect of phonons on angle-resolved photoelectron spectra. They found that the spectrum could not be written as the sum of two spectra, one from direct (k-conserving) transitions and one from non-direct (phonon assisted) transitions, which could be separated by measuring at two disparate temperatures. They emphasized what had been noted previously by others, that the non-direct transitions involve rather small changes in wave vector. They are nearly vertical, and non-direct transitions from an initial state $\mathbf{k}_i$ do not occur to final states throughout the

Brillouin zone with uniform probability. As a result, the spectrum of indirect transitions resembles the direct spectrum rather than the broader, smoother density of states. They developed a more elaborate model for phonon-assisted transitions, but due to the difficulty of evaluating the final expressions, worked with a simpler model, a cylinder in reciprocal space whose dimensions were determined by the inelastic attenuation length and the wave-vector uncertainty in the transition, partly from the angular acceptance of the analyzer, and partly from the phonon scattering. Although the cylinder dimensions involve several inexactly known parameters, the only temperature dependence is in the pho-non-induced uncertainty, making the model especially useful for studying tem-perature-dependent effects. They carried out an extensive series of angle-resolved measurements between 77 and 977 K on the (100) and (110) surfaces of Cu, using 40–106 eV photons. Their model qualitatively fitted the temperature dependence observed. Figure 5.10 is an example. It shows data from an earlier

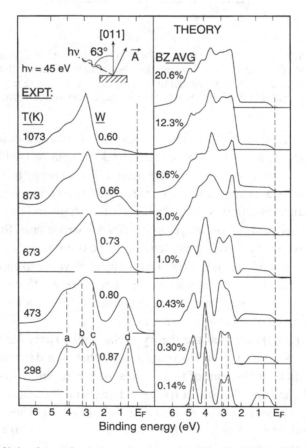

Fig. 5.10 (*Left*) Angle-resolved photoelectron spectra for Cu (110) taken at normal emission at various temperatures, along with calculated spectra (*right*). (White *et al.*, 1987)

paper (Williams *et al.*, 1977) taken on the (011) surface of Cu with 45 eV photons, along with calculated spectra. The low-energy peak weakens and vanishes as the temperature increases. The Debye–Waller factor is labeled W in the figure. The loss of this direct peak is not made up by non-direct transitions because the density of states for such transitions is so small. The direct peaks at larger binding energy are also lost as the temperature rises, but they are replaced by a structure resembling the density of states for the 3d bands, the result of increased non-direct transitions. The calculations involve integrating over a volume in reciprocal space, a cylinder centered on $\mathbf{k}_i$. The dimensions of this cylinder are shown in units of $2\pi/a$. At the highest temperature, this cylinder has a volume of 20.6% of the Brillouin zone. Two important conclusions are that at room temperature there is a significant contribution from non-direct transitions, and a detectable one at 77 K, and that non-direct transitions involving highly dispersing bands can produce shifts in energy with temperature that are due to phonon-assisted transitions, not to shifts in interband energies.

Matzdorf *et al.* (1993b) and Goldmann and Matzdorf (1993) describe the interpretation of the thermal effects, changes in peak positions, widths, and areas, seen in their angle-resolved spectra on Cu (Matzdorf *et al.*, 1993a,b). They studied bulk transitions into s–p final states. In addition to the intensity loss due to the Debye–Waller factor described above and in Chapter 3, they needed at least one additional effect for an explanation of all their observations, scattering by phonons of electrons in to and out of the cone of photoelectrons to be collected after escape. This was an elaboration on the treatment of White *et al.* (1987). In order to reach a point at which numerical modeling was possible, the electron–phonon matrix element was set equal to a constant and an isotropic Debye phonon spectrum was used. The Debye–Waller factor for the loss in intensity of direct transitions was recovered, but it was multiplied by another factor, unity at low temperature, which led to an additional intensity loss linear in T for $T \gg \Theta_D$. This factor produced a better fit to some previously puzzling data.

Aebi *et al.* (1993) used a Cu (100) surface to study the effects of angular resolution on the measured widths of transitions from the s–p bulk bands, and observed a strange effect. Their angular acceptance took three values between a few tenths of a degree to just over one degree. As the angular acceptance was decreased, the Fermi edge in the spectrum diminished and the peak at 1 eV binding energy from photoexcitation out of the s–p band sharpened from 0.7 eV to 0.6 eV. The initial state was then measured off normal incidence, placing $\mathbf{k}_{\parallel}$ in the second surface Brillouin zone. The peak became much larger, using peaks from the 3d band as a reference, and narrower at each angular acceptance, reaching a width of 0.25 eV at the smallest acceptance. The two sets of spectra had the same initial states, and final states differing only by a surface reciprocal lattice vector. These data were explained by Matzdorf *et al.*

(1994) without the need for any new broadening mechanism. They carried out one-step photoemission calculations for the two transitions as a function of $\mathbf{k}_{||}$ in the first and second surface Brillouin zones. In the first zone, as the initial state disperses toward E_F, the peak width increases due to the effect of increased photoelectron broadening. (The photohole width is expected to decrease.) The calculated peaks for the second zone were narrower than those for the first. This was traced to a diminished contribution of the photoelectron width to the total width due to changes in the normal components in the electron and hole group velocities. The contribution of the large final-state width was reduced by a factor of 7 in the second zone due to the different group velocity.

Work on lifetime widths in Cu is continuing, especially studies of the phonon contribution. McDougall *et al.* (1995) studied the s–p-like surface state on Cu (111), the same one studied by Tersoff and Kevan (1983) and Kevan (1983b,c) described in Chapter 4. They worked with the state at the bottom of the surface-state band with a binding energy of 390 meV at 300 K. For surface states the width depends on the initial-state width, not the final-state width. McDougall *et al.* varied the temperature between 30 and 625 K. They used Ar resonance lines, doublets at 11.62 and 11.83 eV, to excite, and with their overall energy resolution of 10–15 meV the two peaks in the spectra did not overlap. The angular resolution was $0.12° \times 1.2°$, leading to a peak count rate of 10 counts per second. They stated that at the beginning of each day they scanned the surface of the sample to find the region that gave the narrowest line width; not all regions were the same. This appears not to be an indication of an inhomogeneous distribution of surface impurities and/or defects which lead to wave-vector uncertainty, for the state they were studying at the minimum of the surface band has $\nabla_k E(\mathbf{k})$ vanishing in the two surface dimensions. The peak width increased linearly with temperature. At room temperature it was narrower than that of Matzdorf *et al.* (1993a,b). They described the width of this s–p surface state as the sum of terms from electron–electron scattering and electron-phonon scattering. The former was estimated from models of the interacting electron gas to be $\Gamma = 2\beta[(\pi k_B T)^2 + (E - E_F)^2]$, with $\beta = 0.015$ eV^{-1}. This gives $\Gamma = 5$ meV for their surface state, with little temperature dependence. This is smaller than their observed 30 meV width from extrapolation to 0 K. The electron–phonon width was written as $\Gamma = 2\pi\lambda k_B T$, although a more proper expression for holes well below E_F is more complicated, depending on the phonon spectrum as well as the hole energy. The measured slope gave an electron–phonon coupling constant of $\lambda = 0.14 \pm 0.02$. They discuss this (surface-state) value with respect to other measurements of λ, averages over the Fermi surface and over regions of it, and theoretical values.

The same Cu (111) surface state was re-studied by Paniago *et al.* (1995) with a view to a better understanding of the effect of instrumental parameters on the observed width.

The temperature dependence of the structures in inverse photoemission peaks has also been studied on the (100), (110), and (111) surfaces of Cu (Jacob *et al.*, 1986; Fauster *et al.*, 1987; Schneider *et al.*, 1990). With the rather low resolution, the only effect seen is the loss of peak intensity with increasing temperature. The loss can be fitted with a Debye–Waller factor, but one with a Debye temperature smaller than that of the bulk, indicative of softer phonons on the surface.

All the work described above concerned data taken by conventional instrumentation. Display analyzers present data in a very different way, and sometimes a different manner of data presentation can lead to new insights. Naumović *et al.* (1993) measured two-dimensional maps of intensity in electron kinetic-energy windows, taking data over almost a complete hemisphere above the sample, and presented them as stereographic projections called "diffractograms." The sample was a (100) surface of Cu, excited with the Kα line of Mg (1253.6 eV) or Si (1740 eV). The data were taken at 16 kinetic-energy windows between 60 and 1740 eV. At these kinetic energies, especially the higher ones, the photoelectron wavelength is less than the interatomic spacing, and one can think of the pattern as the result of single-scattering of a photoelectron by one of the atoms in or near the surface. This allows the data to be analyzed geometrically, and such photoelectron diffraction can give a picture of the atomic arrangement on the surface. At lower energies, multiple scattering becomes important, but geometrical information can be recovered. We mentioned these diffraction effects in Section 3.6, and do not pursue them further here, except to note that Naumović *et al.* (1993) carried out a single-scattering calculation on a cubic cluster with 172 atoms, all in their bulk positions, and were able to get a good qualitative representation of their data, even at the lower kinetic energies. The agreement at lower kinetic energies apparently is because the shorter mean free path for inelastic scattering reduces the contribution of multiple scattering which is expected to grow as the electron wavelength increases.

Aebi *et al.* (1994a) carried out the same type of measurement but at much lower kinetic energies. They excited with the 21.2 eV He I line, and measured the angular distribution of electrons in an energy window chosen to give electrons originating from E_F. They worked with the (100), (111), and (110) surfaces of Cu. Each resultant diffractogram gives a cross-section of the Fermi surface (Fig. 5.11). (At these energies, multiple-scattering dominates, and one does not try to recover geometrical structure, but studies instead the electronic structure.) A similar study by a similar technique was carried out by Stampfl *et al.* (1995), while Thomann *et al.* (1995) obtained similar results using a display analyzer.

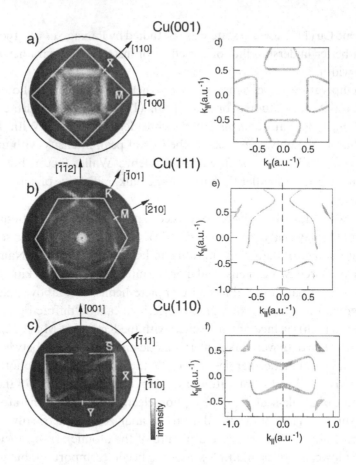

Fig. 5.11 (*Left*) Angular distribution of photoelectrons originating from E_F taken on three surfaces of Cu. The surface Brillouin zone is outlined in white and the brightness of the pattern is a measure of photoelectron current. The expected patterns (*right*) assume free-electron final states. (Aebi *et al.*, 1994a)

Qu *et al.* (1995) used a display analyzer and several photon energies to obtain a data set from which they extracted the full three-dimensional Fermi surface of Cu. Aebi *et al.* have called this method of taking and displaying photoemission data "*E*(**k**)," because it reports photocurrent in a fixed energy window for all values of **k**. In terms of the spectral density function $A(\mathbf{k}, E)$, it keeps E constant and scans $\mathbf{k}_\parallel$ throughout a section of the Brillouin zone. The usual experiments keep $\mathbf{k}_\parallel$ constant and scan E for each angle-resolved EDC.

More recent work along these lines was carried out by Osterwalder *et al.* (1996), with the result that the photoelectron spectrum from the 3d bands in Cu showed the dispersion expected of delocalized Bloch initial states, and also the angle dependence expected from a highly localized picture for the 3d electrons.

A similar duality picture is evident in some of the photoemission studies on NiO and the cuprates. Osterwalder *et al.* measured the angular distribution of the intensities of photoelectrons emitted from (100) and (111) surfaces of Cu, using the He I and II lines. Instead of using the highest resolution possible, they used a wide energy bandpass, 3.6 eV, so that any electron emitted from the 3d bands would be collected, the dispersion of these bands having been well studied previously. These were reported as stereograms with 3600 pixels. They calculated the angular distributions expected for emission from an atom in a cluster, using a single-scattering formalism, even though the single-scattering picture is not expected to be valid for low-energy electrons. (The effect of scattering of the photoelectron by neighbors on the detected angle and energy distribution is sometimes called photoelectron diffraction. It was only mentioned in Chapter 3. It can be used to determine the structure in the near vicinity of the photoexcited atom.) Nonetheless, they achieved very good agreement for both surfaces and both energies in such a modeling when the emission was from a state with d-like symmetry. Agreement was poor when emission from an s- or p state was assumed. The clusters used in the calculation were rather small, 15 Å in diameter and 6 Å deep. Larger clusters would not improve the already good agreement with experiment because the mean free path for inelastic scattering is short.

Osterwalder *et al.* discuss the "duality" observed in photoemission from the 3d bands of Cu. The dispersion of the bands, measured in angle-resolved photoemission taken with resolution adequate to resolve individual bands is evidence that the delocalized Bloch states used in LDA calculations of the electronic structure have considerable validity. The angle distributions of energy-integrated photoemission from the same bands was accounted for by a picture of a photohole localized on one atom, scattering from only a few shells of nearest neighbors. This group had already made a similar observation in Al (Osterwalder *et al.*, 1990), in which the angle distribution of photoelectrons from s–p valence bands was very similar to that from the 2s core level. They suggest that the photoexcitation and measurement processes are responsible for the localization effect. They view the initial state as a sum of Bloch waves. The final state is not simply a plane wave, but a sum of them because of the large uncertainty in wave vector due to inelastic scattering and finite resolution. The integration over a significant fraction of the Brillouin zone leads to a localization in real space, making the initial state resemble a single Wannier function. In addition, there are many-body effects of the electron–electron Coulomb interaction which also lead to spatial localization of photoholes produced in a band (Penn, 1979; Almbladh and Hedin, 1983).

The core levels of Cu that are accessible with photons of energies up to 2 keV are the 3p, 3s, 2p, and 2s. The s levels give a small signal, because of their low, two-fold, degeneracy and because transitions are possible only to p-like final states. The p-level peaks are stronger due to the larger degeneracy and because the transitions are predominantly to d-like final states with their larger degeneracy. The 3p binding energies are 77.3 and 75.1 eV for $j = 1/2$ and 3/2, respectively, so their photoexcitation transitions near threshold overlap and interfere with direct transitions from the valence bands. This resonance was illustrated in Fig. 3.24. This interference has been put to use with the cuprates by monitoring the emission from part of the valence band as the photon energy is scanned through the Cu 3p photoexcitation energy. An increase in valence-band emission indicates Cu character in the valence-band states being sampled.

A weak satellite appears below the valence band in Cu, 15 eV below E_F, where no initial states are expected in the independent-electron model (Iwan *et al.*, 1979). A similar peak, to be discussed in the next section, appears at about 6 eV in Ni. This peak, actually a doublet (11.8, 14.6 eV), increases in area by a factor of about 5 as the photon energy is scanned through the 3p photoexcitation threshold, then the area diminishes. This satellite represents some sort of shakeup event in which the system with the photohole also contains another excitation. It cannot be explained as energy loss of the photoelectron by plasmon excitation. The explanation of the corresponding satellite in Ni, discussed in the next section, requires that the 3d band not be full, so it seems not to apply to Cu. Parlebas *et al.* (1982) were able to provide an explanation along the lines of that for Ni by invoking s–d hybridization in the Cu conduction band. The hybridization imparts some empty 3d character to the states between the Fermi level and the unhybridized 3d band. It is important to understand this satellite, for in many materials, including the cuprates, one frequently finds a peak below the valence-band region of the EDC. Often this peak is due to an impurity on the surface, most commonly O and C, and one must first demonstrate that such a peak is or is not due to a surface impurity. For 3d metals, the satellite should show a resonance at the 3p photoexcitation threshold if it is an intrinsic part of the spectrum. In non-metallic materials, such satellites become the norm and are the result of electronic correlation. In molecules, the satellites often can be related to one-electron excited states. In the case of this satellite in Cu, the excited state of the system after photoemission has two holes in the 3d band, normally full, and the Coulomb repulsion between them is strongly screened by the s electrons (Parlebas *et al.*, 1982). The calculation accounted for the fact that there is a satellite, its energy, approximate width, and the dependence of its intensity on photon energy. It treated s–d hybridization as a wave-vector independent parameter.

One needs to explain not only the resonance, but also the presence of the satellite at photon energies away from resonance. Away from resonance, the satellite is ascribed to the emission of an electron from the 3d band with the simultaneous excitation of a second 3d electron, leaving two 3d holes. This occurs over a wide range of photon energies. When the 3p electron can be excited to the empty part of the 3d band in Ni, another 3d electron can be excited simultaneously, while the 3p hole is filled by a 3d electron, leading to two 3d holes. The final states in the two processes are the same, so there can be interference between the final-state wave functions leading to the resonance. A Fano line shape in the satellite intensity as a function of photon energy results. The problem is that in Cu, the 3d band is nominally full, and the 3p excitation is to an s-like state, not a 3d state. This less localized s state is less likely to be accompanied by the shakeup excitation of a 3d hole. Parlebas *et al.* showed that the s–d hybridization could lead to the same two-hole final state, although the spectral intensity should be weaker.

There is also an Auger process, (3p3d3d) or (MMM), once the photon energy is high enough to create the 3p hole. The spectral peak from the Auger electron can be distinguished from the satellite photoemission peak, for the Auger electron appears at a constant kinetic energy, while the photoelectron peak is at a constant binding energy. The Auger decay is an incoherent process and its intensity simply adds to that of the photoelectrons.

The Cu 2p core levels have been studied. The spin–orbit splitting of 19.8 eV is easily resolved and the stronger and narrower $j = 3/2$ peak at 932.5 eV binding energy is usually selected for line-shape analysis. Additional unresolved structure is expected if the binding energy, electronic screening, or both are different for the surface layer from the bulk. Egelhoff (1984) reported a surface core-level shift of -0.22 ± 0.05 eV (toward lower binding energy) on a (100) surface. The calculations available at the time indicated the surface core-level binding energy should shift by -0.66 eV, leaving a shift of $+0.44$ eV to assign to a screening difference. The sign indicates surface screening of the core hole is less effective than that in the bulk. At higher binding energies, there may be satellites due to the excitation of plasmons, bulk or surface, either intrinsic, excited in the photoabsorption process, or extrinsic, produced during the escape of the electron. To the extent that the conduction electrons are free-electron-like, and the 3d bands are full, the spectra should be as just described. If there are correlation effects, satellites are expected to appear on the core-level spectra.

Citrin *et al.* (1983) measured 2p spectra on polycrystalline Cu and made an extensive analysis of the data, taken at several collection angles to alter the relative contributions of surface and bulk. Their fit is shown in Fig. 5.12. They obtained a surface shift of -0.24 ± 0.02 eV, a lifetime width of 0.60 ± 0.02 eV and a singularity index α of 0.042 ± 0.006. The latter is responsible for the slight asymmetry of the line. The small value of α apparently is a consequence of the dominantly s–p character of much of the Fermi surface; the filled 3d band plays little role (Hüfner and Wertheim, 1975c).

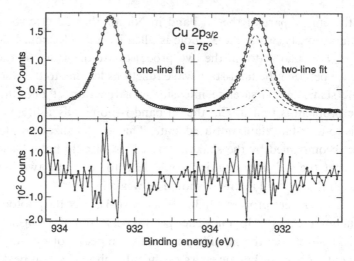

Fig. 5.12   Cu $2p_{3/2}$ XPS spectrum fitted to one and two components. The patterns of the residuals from the least-squares fit show deviations that are more random for the two-component fit. (Citrin *et al.*, 1983)

5.4   Nickel

Nickel, f.c.c. like Cu, is ferromagnetic at room temperature. Its band structure resembles that of Cu except that there are two sets of bands, one for spin-up electrons, one for spin-down. The majority spin 3d band is full, with its top about 0.30 eV below E_F, while the minority spin band is not quite full, containing 0.6 holes per atom. Its top is about 0.33 eV above E_F, according to band calculations (Wang and Callaway, 1977). (Experimental values of the exchange splitting tend to be smaller, about half the calculated 0.6 eV, and they are band- and wave-vector dependent.) This greatly complicates the photoelectron spectrum, for to the extent that spin–orbit coupling is negligible, spin is conserved in the photoexcitation and the observed spectrum is a superposition of the spectrum for each set of spin-separated bands. Because Ni is ferromagnetic and Cu is not, correlation effects play a bigger role in Ni than in Cu, despite the fact that the Cu 3d bands lie deeper than those of Ni and might be expected to have wave functions more localized on the ions. The increased role of correlation is apparent in the photoelectron spectrum of Ni, which is less precisely described by the independent-electron model.

Early angle-integrated UPS and XPS studies on Ni provided a surprise: the 3d bandwidth was about 30% less than that calculated in the independent electron model (Eastman, 1971; Hüfner *et al.*, 1972; Hüfner *et al.*, 1973; Höchst *et al.*,

1977). This discrepancy held up as more self-consistent band calculations were carried out and more photoelectron spectra taken. Angle-resolved spectra (Himpsel *et al.*, 1979; Eberhardt and Plummer, 1980) gave a more precise picture of the discrepancy. The 3d band was about 30% narrower than in self-consistent, independent-electron calculations, although the measured Fermi wave vectors were in agreement with the calculated bands. The exchange splitting was determined by measuring the band splitting at several symmetry points (Eberhardt and Plummer, 1980). Values measured at room temperature ranged from 0.26 to 0.49 eV, all smaller than those of the calculations. At the X point in the Brillouin zone, the exchange splitting was only 0.17 eV (Heimann *et al.*, 1981). Aebi *et al.* (1996) have mapped the Fermi surface intersection with a (110) plane using their angle-scanning photoelectron spectrometer. At room temperature (0.48 T_C, where T_C is the Curie temperature) the observed intersection agrees well with band calculations for $T = 0$, and (low-temperature) de Haas–van Alphen data. Spin assignments for the sections of the Fermi surface can be made from band calculations when those sections are well separated in **k**-space. Data taken above the Curie temperature, at about 1.1 T_C, show that some of the majority- and minority-spin features have coalesced, while some other features ascribed to exchange splitting do not move.

Measurements with larger photon energies and a larger range of electron kinetic energies (Guillot *et al.*, 1977) revealed a satellite peak which occurred at an apparent binding energy apparently impossible to reconcile with independent-electron calculations, about 6–7 eV below E_F. The strength of the satellite increased then decreased as the exciting photon energy passed through the 3p core-level photoexcitation threshold, about 63 eV, from below (Fig. 5.13). The integrated area as a function of photon energy (not the shape in the EDC) could be fitted with a Fano line-shape (Fig. 5.14). This satellite had previously been observed in XPS (Hüfner and Wertheim, 1975a; Kemeny and Shevchik, 1975) and UPS (Smith *et al.*, 1977; Tibbetts and Egelhoff, 1978). The satellite represented a discrete excited state of the ionized system, and Guillot *et al.* proposed, as had Hüfner and Wertheim (1975a), that the final state had two 3d holes. The process proposed was

$$3p^6 3d^9 3s \rightarrow 3p^5 3d^{10} 4s \rightarrow 3p^6 3d^8 4s + \epsilon.$$

The first step is photoexcitation and the second, autoionization via Coulomb interaction. The system is left with two 3d holes on the same atom, as suggested by Hüfner and Wertheim (1975a). Penn (1979) carried out a calculation which reinforced the above picture. The two holes were, however, itinerant, residing at and near the bottom of the 3d band, but correlated by their (screened) Coulomb interaction so that as they moved, the probability that they were at one site was appreciable. The Coulomb repulsive energy of the two holes raises the system

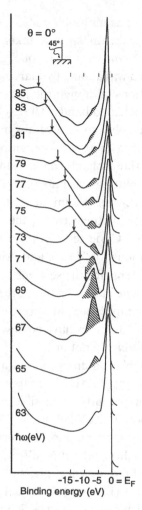

Fig. 5.13 Photoemission spectra for Ni (100) with photon excitation passing through the 3p → 3d resonance. The dashed region is the "6-eV" satellite. The arrows mark the (3p,3d,3d) Auger peak. (Guillot *et al.*, 1977)

energy, thereby lowering the photoelectron kinetic energy, giving a peak in the EDC below the bottom of the 3d band.

The valence-band satellite had another possible explanation, one based on the independent-electron approximation. Kanski *et al.* (1980) proposed that at high photon energies a transition was possible from near the bottom of the 3d band to an empty band far above the Fermi level. This final-state band was rather flat near the zone boundary, giving it a large density of states, and with a large photo-electron lifetime width at such high energies, it could be accessed with a range of photon energies. They calculated the photoelectron spectrum and found a peak

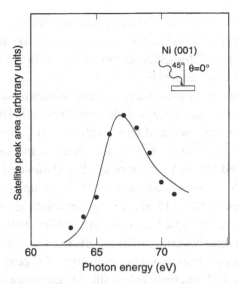

Fig. 5.14 The area of the 6-eV satellite spectrum (shaded region in Fig. 5.14) vs. photon energy. The points are experimental values and the line is a fit to the Fano line-shape function with $\Gamma = 2\,\text{eV}$, $q = 2.5$, and $E_0 = 66\,\text{eV}$. (Guillot *et al.*, 1977)

resembling the satellite, a peak which appeared to resonate at a photon energy of 66 eV because of the large density of final states at this energy. The calculated resonance spectrum was in good agreement with that measured by Guillot *et al.* (1977). It is not easy to determine which picture for the satellite, the one-electron process of Kanski *et al.* or the two-hole excitation, is correct just from the observed resonance spectra. The fact that a similar satellite appeared on core-level photoemission spectra (see below) convinced many that the two-hole final state was the cause of the satellite. However, more convincing evidence came from differences in the predicted spin polarization of the electrons in the two explanations. Subsequent experiments favored the two-hole picture (see Kotani, 1987 for references).

The 2p core-level spectra of Ni also show the 6-eV satellite (Kemeny and Shevchik, 1975; Hüfner and Wertheim, 1975b), as do the 3s and 3p spectra. A calculation similar to that of Penn was carried out by Tersoff *et al.* (1979) for this core-level satellite. The main line is asymmetric and a value of 0.24 was found for α, the asymmetry parameter, for the 2p, 3s, and 3p lines of Ni by Hüfner and Wertheim (1975b). The rather large value of α was attributed to the screening role of the 3d states at the Fermi level (Hüfner and Wertheim, 1975c). The main peak of the Ni 3s core spectrum is not a single peak, but a doublet with a splitting of 1.8 eV (Hüfner and Wertheim, 1972). (The splitting was not resolved, but resulted from a fitting. However the equivalent, but larger, splitting in magnetic compounds of Ni was resolved.) This splitting is the multiplet splitting of the core photohole with the incomplete 3d shell. Recently Nilsson *et al.* (1993) determined the surface shifts of both the $2p_{3/2}$ main line and its satellite to be -0.30 and -0.7 eV, respectively. This difference must be attributed to different effects of the surface on final-state screening.

Eberhardt and Plummer (1980) observed three additional satellites in Ni when the photon energy was at the 3p photoexcitation threshold. These were at 113.4, 17.8, and 22 eV. They were assigned tentatively to various excited states of the Ni^+ ion, an interpretation suggested by the then-recent observation that Ni phthalocyanine showed the 6-eV resonant satellite (Iwan and Koch, 1979), and in this molecule, the Ni atom is quite isolated. Björneholm *et al.* (1990) reported valence-band satellites in Ni at 13 and 18 eV, similar to those just mentioned, and attributed them to three-hole final states, i.e., to $3d^7$ final states. Sakisaka *et al.* (1987) made an extensive study of the 6-eV satellite and reported that its apparent binding energy varies with photon energy. The peak shifts from about 6.0 eV using photons with energies below the 3d resonance to about 6.8 eV at resonance. This suggests the satellite has an unresolved composite structure with a weak component undergoing the strongest resonance. The principal multiplets of the d^8 configuration are the ^1G and ^3F. By placing these at 5.0 and 7.3 eV, respectively, and placing the other three weaker components in this region, they could simulate the satellite. The behavior with photon energy then arises because the singlet : triplet intensity ratio, according to calculations (Davis and Feldkamp, 1979a,b), varies from 0.5 off resonance to 2.3 on resonance. This enhances the 7.3 eV component on resonance and shifts the peak of the composite structure to higher binding energy.

Theoretical studies of the electronic structure of Ni were extended beyond the independent-electron model to gain a better understanding of the narrowing of the 3d band and of the occurrence of the satellite. Also, the exchange splitting found by photoemission was about half that from band calculations, and that required an explanation. The satellite calculations were mentioned above. These were followed by more extensive work. Liebsch (1979a,b) calculated the **k**-averaged spectral density function $A_\sigma(E)$ for each spin σ, and related the results to free-atom parameters. He accounted qualitatively (semiquantitatively) for the narrowing of the 3d spectrum, the short photohole lifetime at the bottom of the 3d band (Eberhardt and Plummer, 1980), and the appearance of the satellite. In fact, a second satellite 2 eV deeper was predicted. However, the band narrowing and the satellite energy could not be fitted simultaneously to experiment with one value of the intra-atomic Coulomb integral (see also Tréglia *et al.*, 1980). The low-density approximation for the 3d holes used in this work was removed subsequently (Liebsch, 1981), giving increased hole–hole and electron–electron correlation. It improved agreement with experiment in band narrowing and in the energy of the satellite. Davis and Feldkamp (1980) also addressed the discrepancies between the first papers of Liebsch (1979a,b) and experiment, obtaining fits similar to those in a later paper of Liebsch (1981). They included screening of the Coulomb interaction by the s electrons, using a simple density of states for them, with no s–d hybridization. Jordan (1989) carried out a one-step calculation of the emission from the (110) surface of Ni using a self-energy correction, and achieved rather good agreement with experiment, including the small exchange splitting between a pair of bands at the X point (Heimann *et al.* 1981). Unger *et al.* (1994) calculated quasiparticle bands for Ni and angle-integrated valence-band photoemission and inverse photoemission spectra. As did Liebsch (1981), they found it difficult to obtain good agreement with experiment in both the amount of valence-band narrowing and the energy position of the satellite. They attribute this to inadequacy of the simplified Hamiltonian, rather than to any subsequent approximations.

Jo *et al.* (1983) carried out more extensive calculations of the resonant photoemission spectrum (emitted current vs. photon energy) of the valence-band satellite of Ni, including s–d hybridization. They also calculated the photon energy dependence of the Auger electron emission caused by the 3p photohole. The latter process is the only one expected if there is no coherence between the two photoemission processes at the energy of 3p photoexcitation. With a suitable choice of the Fano asymmetry parameter q, a ratio of dipole transition matrix elements (see Chapter 3), they could fit both the resonant photoemission and Auger spectral line shapes. Further discussion of the theory can be found in Kotani and Toyozawa (1979), Kotani (1987), and Allen (1992), the last two giving comprehensive references to experimental work as well.

Studies of the satellites continued to be made, and the interpretation of them has evolved somewhat from the above. Van der Laan *et al.* (1992) measured the on- and off-resonance spectra from the 3s, 3p, and valence electrons in Ni. They fitted their data to an elaborate (10 sets of terms) cluster model which included atomic Coulomb, exchange, and spin–orbit interactions on the central atom, hybridization with bulk (ligand) states, a molecular field for spin polarization, as well as Coulomb interactions. The initial state was taken as a linear combination of the three states $|3d^8\rangle$, $|3d^9\underline{L}\rangle$, and $|3d^{10}\underline{L}^2\rangle$, where $\underline{L}$ stands for a hole on a ligand site, in this case a Ni site adjacent to the photoexcited Ni atom. The final state for valence-band (and its satellite) photoemission is a linear combination of $|d^7\mathbf{k}\rangle$, $|d^8\underline{L}\mathbf{k}\rangle$, and $|d^9\underline{L}^2\mathbf{k}\rangle$, where $\mathbf{k}$ represents the photoelectron. The symbol d^n stands for all the multiplets of that configuration. The coherent sum of direct photoemission and resonance photoemission was calculated by fitting six parameters. All other parameters were fixed by atomic calculations, e.g., of Slater integrals. The calculations were then repeated for the 3s and 3p emission, i.e., the final states had a core hole which interacted with the 3d electrons. The calculated off-resonance valence-band spectrum had a main peak with dominant

$|d^9\underline{L}^2\rangle$ character with a very weak $|d^8\underline{L}\rangle$ 6-eV satellite. At resonance the satellite peak grew considerably, but also the main peak picked up a little $|d^8\underline{L}\rangle$ character. The 15-eV satellite, from the $|d^7\rangle$ component of the final state, was very weak in both the calculation and in the measured spectrum. The core-level spectra were much more complex, with numerous satellites at resonance, most of which were reproduced in the calculation.

Chen and Falicov (1989) carried out an exact many-body calculation on a four-atom Ni cell with periodic boundary conditions, i.e., two (100) layers of Ni. Chen (1992a,b) extended the calculation by connecting layer states on one side of the slab to tight-binding bulk states. This way surface-specific effects could be identified. In calculations on these periodic clusters, single-particle and many-body effects are given equal importance. The photoelectron spectrum in the valence-band region was calculated, and satellites appeared at about 6 and 15 eV below E_F. These were attributed to two-hole final states as above, but there was also some contribution of two-hole final states in the lower half of the valence-band region of the spectrum.

Work by López et al. (1994a,b, 1995) measured the intensities of the Auger electron from the decay of the 2p and 3p photoholes and the satellite while scanning the photon energy through the $2p_{3/2}$ and $3p_{5/2}$ peaks. They concluded that at resonance the satellite gained its strength as a result of the incoherent superposition of the satellite and the Auger peak. The valence-band photoemission, on the other hand, increased far more at resonance, so a coherent superposition of final states had occurred. However, other recent work (Qvarford et al., 1995) on the corresponding resonance on Cu in cuprates indicates that the original view of this resonance is correct. Weinelt et al. (1997) carried out a reexamination of the resonance from the 2p photoexcitation in Ni. By varying the polarization, they could align the electric field parallel or perpendicular to the surface. The latter enhances the relative contribution of the direct photoemission process. The results were rather complex. Resonant photoemission does indeed occur, but so does incoherent (Auger) emission. The relative importance of the two depends on geometry and on photon energy. The intensities of features in the valence-band and 6-eV satellite emission spectra varied with photon energy, but they could be fitted only approximately by a single Fano line shape each. The reason is that some coherence is lost at the higher photon energies as inelastic scattering events become significant. The values of q are quite different for the valence-band resonance and the satellite resonance. They are also different for the L_3 and L_2 spectra, and they depend on the polarization of the exciting radiation.

Woodruff et al. (1982) obtained angle-resolved inverse photoelectron spectra (9.7 eV photon energy) on Ni (001) and (110) surfaces. Most of the data could be explained by the band structure and the three-step model, including dipole matrix elements, as for Cu. This includes transitions from s–p bands into s–p bands and into d bands. The latter transitions were weak, in agreement with the calculation, but in disagreement with earlier, effectively angle-integrated, inverse photoemission spectra (Dose *et al.*, 1981). Integrating the calculated direct-transition spectrum over the entire Brillouin zone gave a close approximation to the angle-integrated spectrum. Spectra taken at normal incidence on the (110) surface could not be explained by direct transitions.

Since the above work was carried out, spin-polarized photoemission, first angle-integrated, then angle-resolved, has been employed repeatedly on Ni, the former on core levels as well as valence bands. This gives additional information on the magnetism of Ni, and confirms further the nature of the satellites. Studies of exchange splitting as a function of temperature have been carried out. Actually, high-resolution photoelectron spectroscopy can resolve the exchange splitting, as mentioned above, and as found in recent work (Kreutz *et al.*, 1995), but spin measurement provides reassurance for the assignment of the structures. A description of the results would take us too far from our purpose. There has also been considerable recent effort measuring and interpreting the linear and circular dichroism of the absorption from core levels in Ni.

5.5   Nickel oxide

Nickel oxide, NiO, is a cubic insulator which is antiferromagnetic below 520 K. In the independent-electron approximation it should be a metal because of the partially filled Ni 3d band. It and several other 3d metal oxides have long presented a problem for band theory, for they all should be metals while, in fact, some are insulators, often with magnetic order. Band calculations for paramagnetic NiO predict a metal. A spin-polarized band calculation (Terakura *et al.*, 1984) results in an antiferromagnetic insulator with a band gap of about 0.4 eV, about a factor of 10 smaller than the experimental gap. Self-interaction corrections bring the calculated gap up to 2.45 eV (Svane and Gunnarsson, 1990). One accounting for the properties of 3d transition-metal oxides and halides is based on localized electronic states. Band theory was believed to be inapplicable and these materials were at first thought to be Mott insulators, so called because highly localized electrons do not overlap with electrons on neighboring atoms to give any appreciable band dispersion. However, for NiO a Mott insulator would require that both the top of the valence band and the bottom of the conduction band both be based on Ni 3d states. Fujimori and Minami (1984), Sawatzky and Allen (1984), and Zaanen *et al.* (1985) were able to show, based on a systematic analysis of photoelectron and optical spectra, that NiO is more properly described as a "charge-transfer" insulator. In between the two Ni 3d

Hubbard bands with their gap of the order of U, the multiplet-averaged 3d–3d Coulomb energy, there is a filled O 2p band. The top of the valence band then is largely O 2p in character, but the conduction band bottom is mostly Ni 3d-like. An optical transition across the band gap displaces an electron from an O site to a neighboring Ni. Later, band theory was modified to include correlation effects better, the LDA + U method (Anisimov *et al.*, 1991). This gave a better accounting of the electronic structure of NiO, but the resultant electronic structure has not yet accounted for all the spectroscopic results on NiO, while a cluster model, emphasizing localized electronic states does somewhat better, with a reasonable choice of a few parameters. The LDA + U calculation of Anisimov et al. (1991) gave a charge-transfer gap of 5.7 eV. The valence-band maximum had considerable O 2p character while the bottom of the conduction band came from Ni 3d states, in accord with the charge-transfer picture. Norman and Freeman (1986) carried out a supercell calculation of the ground state of NiO, and of NiO with a hole and with an electron. From these they obtained a value of U of 7.9 eV, and a gap of 4.3 eV. They concluded that NiO was a charge-transfer insulator. Their analysis of the charge distibution led to agreement with Sawatzky and Allen (1984) that the photoemission peak nearest the Fermi level was from a $d^8\underline{L}$ final state, and the satellite about 7 eV deeper from a d^7 final state (see below). A recent calculation (Mizokawa and Fujimori, 1996) of the electronic structure of NiO added a self-energy correction to a Hartree–Fock band calculation. An earlier Hartree–Fock calculation without such a correction yielded a band gap that was too large (Towler *et al.*, 1994). The Hartree–Fock bands with the self-energy correction gave good agreement with the observed band gap. The calculated valence-band photoemission spectrum, including the satellite, was in qualitative agreement with experiment, but there were some differences in the widths, intensities, and 3d contributions to parts of the spectrum.

Band theory does not adequately account for the change in Coulomb energy when an electron is transferred to or from a transition-metal site. The Mott insulator, with its localized states, is more precisely described by the Hubbard model, a model with a rich variety of behavior as the electron count is varied. The Hubbard model must be augmented, however, to account for the O 2p states. Thus the Ni 3d valence band is the lower Hubbard band, but the ligand O 2p band is interposed between it and the bottom of the empty upper Hubbard band, a Ni 3d band. ("Band" here is used loosely, for the model usually supposes localized levels, and the broadening due to the periodic lattice is added empirically via a hopping integral.) Application of the Hubbard model to a small cluster of atoms has produced reasonable agreement with many aspects of the photoelectron spectra of NiO, especially the Ni 2p core-level spectrum.

Both the band model with corrections and the localized model with account taken of several shells of atoms around the central atom now give a reasonably good account of valence-band spectra in NiO. Neither is yet the final model, for there are problems with doping with donors and acceptors. Hüfner (1994), has presented a superb review of the electronic structure of NiO (along with those of other transition-metal oxides, halides, and the cuprate superconductors), especially with regard to photoelectron and other spectroscopies. We strongly urge the reader to read this review, for it presents a very clear exposition of the status in 1994 of our understanding of the role of electronic correlation in the electron spectroscopy and other properties of transition-metal compounds.

Many early studies of NiO were made on the surfaces of Ni foils that had been oxidized in the laboratory or in the photoelectron spectrometer. Such films could be epitaxial if grown on (100) faces of Ni single crystals. These have the advantage that any charge produced in the thin insulating oxide layer can be neutralized rapidly by charge flow from the underlying metal. Wertheim and Hüfner (1972) reported XPS measurements on the valence bands and on several core levels, O 1s and 2s and Ni 2p. They showed that the Ni 3d states dominated the upper regions of the valence band and O 2p states the more deeply bound regions. There was an apparent valence-band satellite at about 9 eV binding energy and several satellites with the Ni 2p peak. Eastman and Freeouf (1975) measured valence-band spectra on a single crystal of NiO, using a range of photon energies, 5–90 eV. From the expected dependence of the photoexcitation cross-sections they determined the individual contributions of the Ni 3d and O 2p states to the valence-band spectrum. At all these energies the O 2p part of the spectrum was larger than in XPS spectra because of the larger O 2p cross section at lower photon energies. The top of the valence band was assigned to only Ni 3d states, the bottom (3–6 eV) to pure O 2p states, and the band center to both. The width of the pure Ni 3d region was explained by crystal-field splitting of the localized initial and final states of a cluster.

Inverse photoelectron spectra (9.7 eV photon energy) were measured by Scheidt et al. (1981) in the course of a study of the oxidation of Ni by inverse photoelectron and appearance-potential spectroscopies. The principal structure was a peak 4 eV above the Fermi level. The rise to this peak begins about 3 eV above E_F.

Oh et al. (1982) and Thuler et al. (1983) studied the Ni 3p resonance in NiO using single crystal (100) surfaces produced by cleaving in vacuum. The former reported no charging effects, but the latter reported shifts in spectral peaks of up to 5 eV due to charging. They measured at several values of incident photon flux and extrapolated their spectral peak positions to zero flux. (Heating the sample to increase its conductivity led to oxygen loss at the surface and an altered spectrum.) Both groups measured EDCs at several photon energies passing through the resonance. Both groups also measured CIS spectra or constructed them from EDCs, and the latter group reported a low-energy CFS spectrum which gave an absorption-like spectrum near the 3p photoexcitation threshold. Upon scanning through the 3p resonance, at about 67 eV, the contribution of the upper part of the valence band goes through a minimum, while valence-band satellites at about 8 and 10 eV (Oh et al., 1982) or 9 and 22 eV (Thuler et al., 1983) show an enhancement even larger than that found in Ni metal. The lower part of the valence band, of O 2p character, shows almost no change in intensity as the photon energy scans through the resonance. Thuler et al. also mention some unpublished angle-resolved spectra which showed significant dispersion in the parts of the valence band attributed to O 2p states, and no dispersion as large as 0.3 eV in the regions assigned to Ni 3d states.

Sawatzky and Allen (1984) measured XPS and BIS spectra on a single-crystal sample cleaved in vacuum. The insulating sample charged under the electron beam so the BIS spectra were taken for several incident currents and extrapolated to zero current. The BIS spectrum showed a large peak at 4 eV, in agreement with that of Scheidt et al. By plotting the valence-band XPS and the BIS spectra on the same energy scale, an energy gap was made evident, but its value depends on how one interprets the shapes of the spectra on either side of the gap, and in any case, is affected by the instrumental resolution of 0.6 eV. Their best estimate is 4.3 eV. A fit to a model for a $(NiO_6)^{10-}$ cluster resulted in a Ni 3d–3d Coulomb repulsive energy of 7–9 eV, and the picture of a band of empty O 2p states in the Ni 3d Hubbard gap.

Fujimori and Minami (1984) proposed a different interpretation of the satellite from their detailed calculation of the states of a $(NiO_6)^{10-}$ cluster before and after electron removal and addition, fitting their calculation to photoemission and inverse photoemission spectra. The ground state was taken to be the lowest-energy linear combination $a_0|d^8\rangle + a_1|d^9\underline{L}\rangle$, where the ligand orbitals involve both O 2p state and Ni 4s and 4p states. These basis states differ in energy by Δ and are coupled by a transfer integral. The final states for photoemission are $b_0|d^7\rangle + b_1|d^8\underline{L}\rangle + b_2|d^9\underline{L}^2\rangle$, each basis state with a different energy and each mixed with its neighbor in the list by one of two transfer integrals. States $c_0|d^9\rangle + c_1|d^{10}\underline{L}\rangle$ were used for final states in inverse photoemission. The 3d photoemission strength into a final state is proportional to $|a_0b_0T_0 + a_1b_2T_1|^2$, where the T_is are transition matrix elements, and there are three sets of b_is, found from a 3×3 Hamiltonian matrix, while the a_is come from diagonalizing a 2×2 matrix. However, the five degenerate 3d states are split into two levels by the cubic crystal field and the d^7 and d^8 states have significant multiplet splitting. Including these, and considering in more detail the symmetry-adapted O 2p orbitals, increased the sizes of the Hamiltonian matrices, up to 17×17. Some off-diagonal matrix elements were calculated, then scaled to fit experiment. Diagonal matrix elements were evaluated from atomic data or treated as fitting parameters. Transition matrix elements were derived from fractional parentage coefficients.

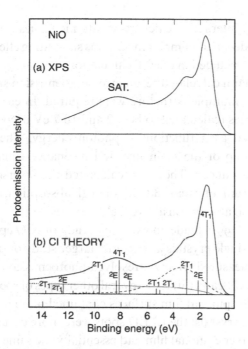

Fig. 5.15 (b) Calculated valence-band spectrum for a NiO_6 cluster (vertical lines). This was convolved with a broadening function (see text) and the dashed spectrum from the O 2p valence band and the (dotted) background of inelastically scattered electrons added in to give the solid line. The experimental XPS spectrum is shown for comparison in (a). The peak at 8 eV is what is usually called the satellite. (Fujimori and Minami, 1984)

Figure 5.15(b) shows the calculated spectrum as vertical lines and as a curve after convolution with a Lorentzian; the experimental XPS spectrum is shown in Fig. 5.15(a). The spectrum calculated with the assumption that the final state is simply $|d^7\rangle$ with crystal field and multiplet splitting is not shown. It accounts rather well for the main peak and its shoulder. The full configuration-interaction calculation, the spectrum from which is shown in Fig. 5.15(a), retains that fit from its $|d^7\rangle$ component, but also gives a good account of the rest of the spectrum, the main satellite and the featureless spectrum in between. (The dashed line indicates a Gaussian added to simulate the photoemission from the O 2p states not treated in the calculation.) However, the detailed final states are not the same after the configuration interaction as before. Before, the $|d^7\rangle$ final states gave the main peak, assigning the principal satellite to $|d^8\underline{L}\rangle$. Analysis of the individual final states contributing to the spectrum in Fig. 5.15(b) shows that the main peak has its major strength from $|d^8\underline{L}\rangle$ final states, and the satellite arises nearly equally from $|d^7\rangle$ and $|d^8\underline{L}\rangle$. Contributions from $|d^9\underline{L}^2\rangle$ appear throughout the spectrum. The Racah coefficients used in the simpler fit were reduced from those for the free Ni ions in order to produce a fit to the XPS spectrum. In the full configuration-interaction calculation, atomic values were used, indicating the reduction needed in the simpler model was simulating effects of configuration mixing explicitly included in the full calculation.

Fujimori and Minami calculated resonance photoemission spectra, but some simplifications of the multiplet structure were required. Intensities of the main line and of the satellites (calculated to be 7.2 and 10.7 eV deeper than the main peak) were calculated as a function of photon energy. The experimentally observed resonance dip of the main line and resonance enhancement of the satellite could be reproduced. They also calculated the inverse photoemission spectrum, the spectrum of weak 3d–3d optical absorption, and the strong charge-transfer optical absorption above 4 eV.

Kuhlenbeck *et al.* (1991) made an extensive study of NiO epitaxial layers on Ni and cleaved NiO single crystals as part of a larger study of the adsorption of NO on NiO. They measured (i) angle-resolved photoemission over a range of photon energies, (ii) XPS, and (iii) took several other types of spectra. LEED patterns showed that the epitaxial film surfaces contained more defects than those of cleaved NiO single crystals; the LEED spots were more diffuse. They showed that a 4- or 5-monolayer epitaxial film had essentially the same electronic structure as a cleaved bulk single crystal but the relative intensities and widths of some of the spectral peaks differed between the two types of sample. The measured bands based on O states followed the bands calculated with the LDA, but the bands based on Ni were much flatter (see the following paragraphs). Defects in the epitaxial film or introduced in the NiO crystal surface by polishing caused some weak, featureless additional structure in the valence-band spectra located between the structure from the states of least binding energy and the Fermi level.

Shen *et al.* (1991a) carried out a very extensive study of the valence bands of NiO by angle-resolved photoemission. This provided considerable insight into the electronic structure of NiO. They measured a series of spectra at different photon energies, collecting normally emitted electrons from the (100) cleavage face. An example is provided in Fig. 5.16. These provided spectra of E vs. $\mathbf{k}$ with $\mathbf{k}$ along the $\langle 100 \rangle$ direction (Fig. 5.17). They also scanned off-normal by varying the collection angle in such a way as to scan the parallel component of $\mathbf{k}$ along the $\langle 100 \rangle$ and $\langle 110 \rangle$ directions in the surface. In order to keep the perpendicular component of $\mathbf{k}$ constant during these scans, the photon energy was changed as the angle scanned. These data and the resultant band map are shown in Figs. 5.18 and 5.19. In order to map initial states, the final states were assumed to be free-electron-like, with an effective mass of $0.95m_0$ and an inner potential of -8 eV, referenced to the Fermi level.

These experimental band maps agree rather well with the LDA band structure these authors calculated for an assumed antiferromagnetic state, but only for those bands that are based primarily on the O 2p states. The band states based

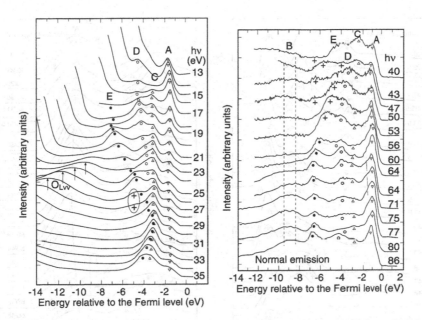

Fig. 5.16 Angle-resolved EDCs taken at normal emission from the (100) face of NiO at several photon energies. (Shen *et al.*, 1991a)

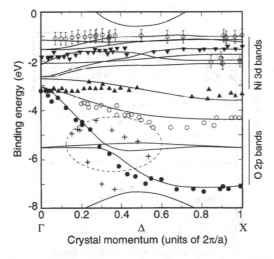

Fig. 5.17 Comparison of the experimental band dispersion obtained from Fig. 5.16 and similar data (various symbols) with LDA calculations for antiferromagnetic NiO. The effects within the dashed oval may result from band folding due to magnetic ordering. (Shen *et al.*, 1991a)

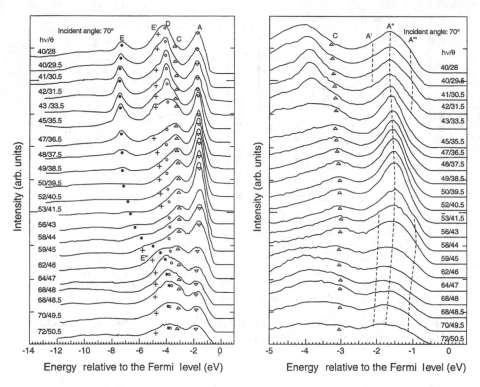

Fig. 5.18 Angle-resolved EDCs taken at various angles from the normal to the (100) face of NiO at several photon energies. The photon energy was scanned with the angle to keep $k_\perp$ fixed. (Shen *et al.*, 1991a)

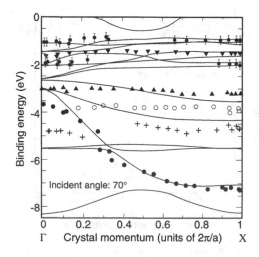

Fig. 5.19 Comparison of the experimental band dispersion obtained from Fig. 5.18 and similar data (various symbols) with LDA calculations for antiferromagnetic NiO. (Shen *et al.*, 1991a)

largely on Ni 3d atomic states do not agree as well with the bands calculated in the LDA. The agreement is better for the bands calculated with an assumed anti-ferromagnetic NiO than for those with paramagnetic NiO. (The latter are not shown in Figs. 5.17 and 5.19.) However, the experimental 3d-based bands exhibit dispersion, more dispersion than cluster models give (which often is zero) but, for some bands, definitely less than that of the LDA calculations. Other discrepancies with the LDA bands are that calculated bands just below E_F (see Figs. 5.17 and 5.19) do not appear in the data and, of course, the satellites observed are not in the LDA calculation. An important characteristic of the angle-resolved spectra of Shen *et al.* is that there are features best accounted for by the LDA band calculations and other features best accounted for by a localized picture; if not a NiO_6 cluster, then a larger cluster.

If NiO is treated as a cluster, the charge-transfer modification of the Hubbard model is used. In between the filled and empty Ni 3d "bands" there is a filled O 2p band, and the observed optical gap is between the top of the O band and the bottom of the empty Ni 3d band. But the valence-band spectra of Shen *et al.*, and earlier work, all show final states at the lowest binding energy, apparently at the top of the valence band, that are mostly Ni 3d in character, not O 2p. The filled O 2p bands lie deeper than the Ni 3d. The charge-transfer picture just described omitted Ni 3d–O 2p hybridization, which can change the picture, and which was taken into account in the cluster calculations. Some of the O 2p states in the Hubbard gap are pushed up by hybridization, and some are pushed down. The latter are responsible for the dispersing O 2p bands seen at binding energies greater than the flatter Ni 3d bands, which have been pulled up by the hybridization. The lower Hubbard band is considered to correspond to the satellite below the valence band. The wave-function character of the states at the top of the valence band are not well described by LDA + U calculations. Wei and Qi (1994) obtained mostly Ni 3d states at the top of the valence band in their LDA calculation. However, the LDA + U version has the 3d states lowered considerably, and the gap widened to 3.7 eV (Anisimov *et al.*, 1993). The top of the valence band is now a hybrid of Ni 3d and O 2p with approximately equal weights, in qualitative agreement with the above statements about the cluster picture. Mizokawa and Fujimori's (1996) Hartree–Fock band calculation with and without a self-energy correction give a photoemission spectrum with mostly O 2p states at the top of the valence band.

Shen *et al.* also examined the doublet valence-band satellite by angle-resolved photoemission. The two overlapping peaks at about 9 and 10.5 eV disperse with wave vector over a range of about 0.5 eV. However it is not certain that this is due to a single broad peak dispersing or to changes in intensities of unresolved peaks due to several multiplet final states. The authors favor the dispersion picture, however, after due consideration. Resonance effects are evident when the photon energy passes through the Ni 3p excitation energy, and they are larger in the satellite that disperses less.

Several calculations were made on NiO to try to obtain a better explanation of all of the features seen in the angle-resolved photoemission spectra. Since the cluster models required several final-state configurations to come close to experimental spectra and to obtain the satellite, such configurations should be incorporated into the LDA + U method. This was done in an approximate way by using different transfer integrals for hybridization with the oxygens for each of the three possible three-hole final states $|3d^7\rangle$. The resultant calculated valence-band XPS and BIS spectra then fit measured angle-integrated spectra very well, especially the BIS. The calculated valence-band spectrum exhibits a satellite at about 8 eV below the band edge (Anisimov *et al.*, 1993).

Later, Anisimov *et al.* (1994) (also Anisimov, 1995) used the Anderson impurity model to describe the spectra of NiO, taking their parameters not from experiment, but from the LDA + U results. The wave functions were of the configuration-interaction type used in most cluster calculations. They calculated the valence-band photoemission spectrum for 3d holes and for O 2p holes. The former was compared with XPS valence-band spectra and a linear combination of the two was compared with angle-integrated photoelectron spectra taken with the He II line. (Dipole matrix elements were not evaluated.) Agreement was considered good, as was agreement between calculated and measured O 1s x-ray emission spectra.

Bala *et al.* (1994) state that neither the band picture nor the cluster picture is correct for NiO, although the cluster picture is the better of just these two. Instead, they propose that a polaron picture is more correct, in which the O 2p hole charge interacts only with the spins of the Ni 3d electrons, which are treated as spin waves. They applied this picture to a two-dimensional NiO lattice, making use of several parameters obtained from fitting angle-integrated photoemission spectra to a cluster model for NiO. They began with a set of bands for the O 2p hole, which then flattened considerably when the hole–spin interaction was turned on. The agreement between the four bands in the calculated O 2p photohole spectral function and four of the bands mapped by Shen *et al.* (1991a) was characterized as semiquantitative, as can be seen in Fig. 5.20.

Manghi *et al.* (1994) used their new method of treating the Hubbard Hamiltonian by a configuration-interaction expansion using configurations with different numbers of electron–hole pairs. They began with an LDA band calculation for paramagnetic NiO, obtaining the usual result, metallic NiO. They then used their formalism with $U = 10$ eV to calculate the energy and wave-vector-dependent self-energy and the spectral density function $A(\mathbf{k}, E)$. This went beyond mean-field theory in treating the correlations on the Ni sites. The resultant density of states has a gap of about 4.5 eV and the top of the valence

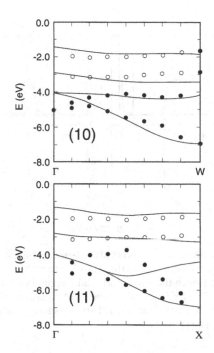

Fig. 5.20 Calculated spectral density function peak positions (open and filled circles) for O 2p holes interacting with magnons in a planar model for antiferromagnetic NiO along the [10] and [11] directions in the square Brillouin zone, and (curves) the corresponding "band" dispersion (Shen *et al.*, 1991a) from angle-resolved photoemission (Bala *et al.*, 1994).

band is mostly O 2p-like. The spectral density function for a fixed value of **k** may have several peaks of differing widths, including peaks interpretable as satellites (Fig. 5.21). Finally, the quasiparticle band structure was presented (Fig. 5.22) for comparison with the spectra of Shen *et al.* There is rather good agreement for many of the bands, both flat ones, primarily of Ni 3d character, and more dispersing bands based largely on O 2p states. The authors anticipated that the inclusion of antiferromagnetism would not alter the results significantly. Takahashi and Igarashi (1996a,b) obtained similar good agreement with band dispersion and the satellite by calculating the self-energy for quasiparticle bands that had been treated in the Hartree–Fock approximation instead of the LDA.

The LDA + U method has also been used to calculate the angle-integrated valence-band photoelectron spectrum, the BIS spectrum, and the Ni 2p XPS spectrum in NiO with considerable success (Anisimov, 1995). The Ni atom was treated as an impurity in the continuum of O 2p states which were obtained from an LDA + U calculation. All of the hybridization parameters, taken as fitting parameters in cluster calculations, were evaluated from the LDA wave functions. The calculated spectra show satellites, not present in LDA calculations.

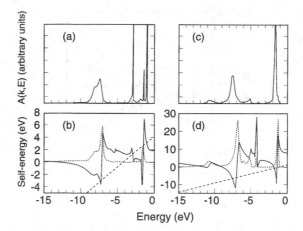

Fig. 5.21   (a,c) Spectral density function and (b,d) self-energy for two states with **k** half-way between Γ and X obtained by correcting the LDA bands in NiO for important many-body effects. (b,d) show both the real (solid) and imaginary (dashed) parts of the self-energy. (b) is for a band 4.25 eV below E_F in the LDA calculation having mixed O 2p–Ni 3d character. The band for the right side is 1.08 eV below E_F and is mostly Ni 3d in character. (Manghi *et al.*, 1994)

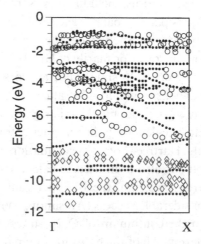

Fig. 5.22   Calculated quasiparticle dispersion (filled circles) and the corresponding "band" dispersion (open symbols) (Shen *et al.*, 1991a) from angle-resolved photoemission. (Manghi *et al.*, 1994)

Aryasetiawan and Gunnarsson (1995) and Aryasetiawan and Karlsson (1996) calculated the self-energy and spectral density function of NiO starting from an LDA calculation and using the GW approximation. This gave a reasonable value for the gap, and the resultant quasiparticle dispersion was in better agreement with the data of Shen *et al.* than the LDA bands, especially that of the deeper O 2p bands. However, the top of the valence band was determined to be largely of Ni 3d character rather than the expected O 2p, and no satellites were produced. The 8-eV satellite appeared in their later calculation when they began with an LDA band structure, but upon iteration to self-consistency the calculated satellite became very weak, appearing as two satellites at 7 and 13 eV.

Resonance photoemission of valence-band electrons upon photoexcitation of Ni 3p, Ni 2p, and O 1s electrons has been studied by Tjernberg *et al.* (1996). They used single crystals of NiO. Their angle-resolved spectra with Ni 3p excitation were presumably taken normal to the (100) cleaved surface. The other spectra were taken in an angle-integrated mode. Different parts of the valence-band spectrum exhibit resonance depending on their atomic character, and the resonance intensity depends on the degree of localization of the valence-band states. Photoexcitation through the Ni 3p edge produces resonances at 1.5, 4.5 and 9.5 eV in the valence band, but the photon energy dependence of them is not identical. (The first peak in the valence band, the largest peak, is at 2 eV below their reference Fermi level.) The 9.5 eV peak is usually called a satellite, and its 3p resonance, with its Fano line shape in a plot of intensity vs. photon energy, has already been discussed. The Ni 2p resonances occur at 3.5, 5.0, and 9.5 eV, with the latter exhibiting the largest resonance. These show far smaller, essentially no, interference effects because the transition dipole matrix element for direct valence-band excitation at these energies is much smaller than that for core hole excitation. The resonances from the O 1s level occur in the 9.5-eV satellite and in satellites at 20 and 40 eV binding energy when exciting in the first O 1s peak. Excitation above this peak causes the 9.5-eV satellite intensity to weaken, while the other two remain strong. These two peaks do not have binding energies that remain constant as the photon energy is varied. Instead, their kinetic energies are constant; they are Auger peaks. There is no resonance of the oxygen-derived valence-band states at the O 1s edge because the O-based valence-band states are sufficiently delocalized that their overlap with an O 1s core hole is negligible.

The resonances are not as simple as just described. Each core level gives rise to a structured absorption spectrum, not a single peak followed by a monotonic decay. Photoexcitation in the region of some of the above-edge absorption structures gives a somewhat altered valence-band emission spectrum (see Tjernberg *et al.*, 1996). Presumably unresolved multiplet structure plays a significant role here.

The Ni 2p core-level spectrum has been discussed extensively in the literature. In addition to the satellite of the $2p_{3/2}$ peak, which required an explanation, the main $2p_{3/2}$ peak itself is actually a doublet with a splitting of about 2 eV. See, e.g., Uhlenbrock et $al.$ (1992) for recent data taken on a single crystal. van Veenendaal and Sawatzky (1993) calculated the $2p_{3/2}$ spectrum for the commonly used $(NiO_6)^{10-}$ cluster and for a larger cluster, one with other Ni atoms, $(Ni_7O_{36})^{58-}$. In order to make this a more tractable problem, they reduced the number of holes they treated, thereby reducing the number of hybridization terms. Their principal interest was to determine the effect of the additional Ni next-nearest neighbors on the photoexcited Ni. They found a prominent peak about 2 eV deeper than the main peak, in agreement with experiment. The final state for this peak had a screening electron that did not come from a neighboring O orbital, an $|\underline{L}\rangle$ final state, but rather from a neighboring $(NiO_6)^{10-}$ unit, $|\underline{L}'\rangle$. The final state for the photoexcited Ni giving this peak is then $|cd^9\underline{L}'\rangle$. The satellite structure is also altered somewhat (less dramatically) by the presence of other $(NiO_6)^{10-}$ units. The role of the more distant neighbors to the photoexcited Ni ion in the origin of the first satellite on the Ni $2p_{3/2}$ line was later demonstrated by Alders et $al.$ (1996). They grew a monolayer of NiO on a (100) surface of MnO, and measured the Ni 2p spectrum with good resolution. They also grew additional monolayers, up to 20, and studied the line shapes as a function of thickness. The position of the O 1s peak was used to correct for charging effects. The satellite was very weak or absent in the sample with 1 ML of NiO, growing relatively more prominent until the line shape saturated between 10 and 15 ML, in accord with expectation from the model calculation of van Veenendaal and Sawatzky.

It is difficult to dope NiO. The only known dopant to date is Li, which substitutes for Ni, giving a p-type sample. Nickel vacancies, often naturally occurring, also give a p-type sample. Annealing in a reducing atmosphere to produce oxygen vacancies results in an n-type sample. The atomistic nature of the donor and acceptor centers is not well known. The one-electron model used for doping semiconductors such as Si is invalid. Inverse photoemission and valence-band photoemission on an n-type or a p-type doped sample would, when combined, show the same band gap as an intrinsic sample. Such is not the case when correlation effects are large. Reinert et $al.$ (1995) carried out a study by photoelectron, inverse photoelectron, and electron energy loss spectroscopies on p- and n-doped NiO. Hole doping by Li substitution causes a shift of about 0.4 eV to lower apparent binding energy in the O 1s and Ni 2p core-level spectra, and in the valence-band maximum. The lowest peak in the inverse photoemission spectrum shifts about 0.8 eV in the same direction, i.e., further above E_F. This peak is about 1.4 eV below the valence band edge (Hüfner et $al.$, 1992). (These shifts are at saturation, hole concentrations between 5 and 20%. More than half of the shift has occurred for the lowest doping level used, 2%.) In the one-electron picture these data would mean that the charge-transfer gap increased by about 0.4 eV upon doping. Electron doping by vacuum annealing causes a shift of the Ni 2p core spectrum of about 0.75 eV to higher binding energies, a shift of opposite sign to that found upon hole doping. At larger defect concentrations a new peak appeared at lower binding energies, whose intensity correlated with the loss of intensity of the O 1s peak. This peak was attributed to neutral Ni centers in the ionic crystal. The lowest energy peak in the valence band also shifts 0.75 eV to greater binding energies, while the lowest peak in the inverse photo-emission spectrum shifts to higher energy. Both types of valence-band spectra show new peaks nearer E_F at high defect concentrations, 0.6 eV on either side of E_F. Reinert *et al.* (1995) interpret their spectra as indicating that either type of doping introduces new states near the middle of the charge-transfer gap. They lie 1.2 eV above the top of the valence-band edge and are empty in the case of Li doping. They lie 0.6 and 1.7 eV above it for O-vacancy doping, and the 0.6 eV level is occupied. These are shown schematically in Fig. 5.23. This figure also shows the shifts in the location of the Fermi level.

5.6   Cuprous oxide

Cuprous oxide, Cu_2O, is a cubic semiconductor with a band gap of 2.17 eV. The Cu atoms have a nominal configuration of $3d^{10}$, so Cu_2O has more in common with Cu itself than with NiO and the cuprates, both of which have 3d holes. Each O atom in Cu_2O has four nearest Cu neighbors tetrahedrally disposed, while each Cu has two O nearest neighbors, forming O–Cu–O linear chains along the cube body diagonals.

Self-consistent band calculations were carried out by Kleinman and Mednick (1980). The calculated band gap was 1.07 eV, well below the measured one, a discrepancy common to band calculations on insulators and semiconductors. Robertson (1983) produced a tight-binding band structure, adjusting some parameters to give better agreement with experiment than the self-consistent calculation while still retaining many of the features of the previous results.

Valence-band photoelectron spectra, including resonance spectra were obtained by Thuler *et al.* (1982), along with similar spectra for Cu and CuO. They compared their spectra taken at photon energies around 75 eV with band calculations, and found reasonable qualitative agreement, but the O 2p bands were about 2 eV deeper than in the calculations. Ghijsen *et al.* (1988) obtained valence-band spectra with He I and II lines, and with the Al-Kα line. Both these studies were carried out on polycrystalline samples made by suitable exposure of Cu to oxygen and heat treatment. Ghijsen *et al.* compared their spectra to

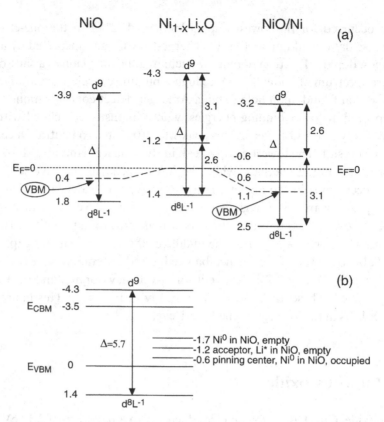

Fig. 5.23 (a) Energy-level scheme for intrinsic NiO (*left*), hole-doped NiO (*center*) and electron-doped NiO (*right*), drawn with Fermi levels aligned, and all energies measured in eV with respect to E_F. The dashed line labelled VBM is the valence-band edge, while the solid line below it represents a peak in the valence-band photoemission spectrum. The uppermost solid line represents a peak in the inverse photoemission spectrum. The conduction-band edge is not shown. Solid lines in the gaps represent donor or acceptor levels, which are all placed in (b). (Reinert *et al.*, 1995)

the calculated density of states, making use of the different photon energies used to separate the Cu 3d contributions from the O 2p. The He I spectrum will have the largest relative contribution from the O 2p states, and the Al-Kα the least. The spectrum obtained with He II radiation should contain both, and indeed, except for shifts in energy of up to 1 eV, is in good agreement with a broadened density of states from band calculations. The Cu 3d-based bands contribute between 1 and 4 eV binding energy and the O 2p binding energies range from 6 to 7 eV.

There is a weak satellite in Cu_2O at 15.3 eV binding energy, below the valence band (Thuler *et al.*, 1982) previously seen by Wertheim and Hüfner (1972). It shows resonant enhancement around 76 eV photon energy, the threshold for Cu

3p photoexcitation. It is similar in energy, strength, and width to a satellite in metallic Cu at 14.9 eV. (Cu also has a weaker 11.9-eV satellite.) Both Cu satellites have been attributed to a two-hole final state, d^8 configuration, as in Ni, but produced differently, by s–d valence-band hybridization in Cu. The d^8 states are split by multiplet structure, giving two observable peaks in Cu, but only one peak is detectable in Cu_2O. The resonance spectra (satellite intensity vs. photon energy) for Cu and Cu_2O are almost identical, except for a shift of 0.85 eV to higher energy for Cu_2O. XPS data show equal apparent binding energies for these two materials. The 0.85 eV shift must result from differences in the final states. The satellites in the two materials lie below the peaks in the 3d part of the valence-band spectra by amounts that differ by 0.6 eV, that for Cu_2O being the larger. From these shifts, Thuler et al. estimate the d–d correlation energies for the d^8 satellite to be 8.2 eV for Cu and 7.9 eV in Cu_2O.

The same satellite was studied by Tjeng et al. (1992) who showed that it resonated when the Cu 2p photoexcitation threshold was crossed, but not when the O 1s threshold was crossed. There was a very weak O 1s resonance in the 10–15 eV region of binding energy and a large resonance at 21 eV, the latter presumably from the deepest O 2p states. The valence region showed no sign of a resonance for either core level.

Cu core-level XPS spectra were obtained by Kim (1974) and by Fiermans et al. (1975). Ghijsen et al. (1988) re-investigated the Cu 2p spectra and measured the O 2s spectra. No satellites were reliably detected with the Cu 2p peaks. Ghijsen et al. showed that this was consistent with the picture for the two-hole final-state satellite in Ni; with a 3d shell that is initially full in Cu_2O the satellite must have one s–p hole, resulting in a much broader and weaker satellite. The O 1s spectrum was a single peak, with a very weak peak whose origin was not discussed. From an analysis of their valence-band and core-level data, Ghijsen et al. obtained Coulomb energies, U, of 9.3 eV for d electrons on Cu, and 4.6 eV for p electrons on O in Cu_2O, the former in reasonable agreement with the value of Thuler et al.

Ghijsen et al. also measured BIS spectra for Cu_2O, detecting 1487 eV photons. Combining the BIS spectrum with the valence-band spectrum taken with XPS (the Fermi levels were known in each spectrum) resulted in a band gap of 2.4 eV, in good agreement with the value determined optically. The structures in the density of states from the band calculation were in good agreement with those in the BIS spectrum over a 15 eV range above E_F, provided the density of states was shifted by 1.2 eV. Thus it appears, as Ghijsen et al. conclude, that Cu_2O is quite well described by the independent-electron model, at least as well as metallic Cu is so described. Some discrepancies exist, and more are liable to appear as more data are taken, but there are no major unexplained structures, and no unexpected discrepancies in the energies of structures.

Schulz and Cox (1991) carried out LEED, XPS, and angle-integrated UPS studies on the (111) and (100) surfaces of single-crystal Cu_2O. The (111) surface plane is composed entirely of O ions, with an all-Cu plane below, and a plane of O below that. This triad of planes then repeats on into the crystal. The (100) surface is composed only of oxygen ions with a plane of Cu below it. This pair repeats into the bulk. The surfaces were polished *ex situ*, then cleaned in the experimental chamber by ion bombardment, followed by an anneal. XPS showed rather stable Cu:O ratios with annealing temperature for the (111) surface, but higher-temperature anneals caused this ratio to fall for the (100) surface. An anneal at 1000 K was believed to yield a stoichiometric surface. The spectra taken with vacuum ultraviolet photons were similar to those obtained by others on polycrystalline samples, except for a peak at 9.4 eV whose intensity varied non-monotonically with annealing temperature, especially on the (100) surface. The origin of this peak is not yet clear.

5.7   Cupric oxide

Cupric oxide, CuO, is a monoclinic semiconductor with a bandgap of about 1.4 eV. The copper atoms have a nominal $3d^9$ configuration, and LDA band calculations predict it to be a metal. The Cu atoms have four nearest-neighbor oxygen atoms, all five atoms lying in a planar rectangle. LDA band calculations (Ching *et al.*, 1989) indeed produced the bands of a metal, but a gap of 1.4 eV results when self-interaction corrections are made (Svane and Gunnarsson, 1990).

As with Cu_2O, photoemission studies on CuO have been carried out with polycrystalline samples, sometimes pressed powders of CuO, and more frequently CuO produced by suitable exposure of Cu to oxygen in the sample chamber. The latter results in a thin film less susceptible to charging of the surface. One problem with such samples of oxides is the possible creation of defects (e.g., by loss of oxygen) by the incident photon or electron beam, or by energetic photoelectrons and secondaries.

Valence-band spectra were reported by Wertheim and Hüfner (1972) in XPS, and by Yarmoff *et al.* (1987) using 106 eV photons. The valence-band region showed two peaks, nominally from the Cu 3d-derived bands or states peaking at 3.3 eV, and a weaker one from the O 2p bands, peaking at 6.4 eV binding energy. There were satellites at 10.3 and 13.3 eV. Thuler *et al.* (1982) measured valence-band spectra with photon energies near that of the Cu 3p resonance. The valence-band spectrum resembled that of Cu_2O, but the O 2p peak overlaps the Cu 3d peak more than in Cu_2O. There were two strong satellites, at 10.5 and 12.9 eV, and a weak one at 16.6 eV. Ghijsen *et al.* (1988) measured valence-band spectra with three photon energies, as described above for Cu_2O. There were three overlapping structures in the valence bands, all of which contained some Cu 3d character. These covered a region 7–8 eV wide. The states of primarily O 2p character ranged from 1 to 6 eV binding energy, overlapping the three peaks with Cu 3d character. There was rather good agreement between the spectrum taken with the He II line, exciting both O 2p and Cu 3d states, and the broadened density of states from the band calculation. There were also two satellites at 10.0 and 12.4 eV. The satellites cannot be explained by transitions from a calculated valence band.

Thuler et al. (1982) measured the resonance curve of the satellites and made a comparison with those of Cu and Cu_2O. The resonance shifted 1.4 eV to lower energy for CuO, with metallic Cu as the reference. The final state believed to explain this resonance in Cu and Cu_2O could not be used, for it led to an inconsistency in a valence-band binding energy. Instead, they proposed the same $3d^8$ two-hole excited state used to explain the satellite in Ni to explain the Cu core-level satellites in CuO.

The 12.4 eV satellite was studied by Tjeng et al. (1991, 1992) who showed that it resonated when the Cu 2p photoexcitation threshold was crossed, but not when the O 1s threshold was crossed. There was a large resonance at 16 eV, the latter presumably from the deepest O 2p states. The valence region showed no sign of a resonance for either core level, but the nearly structureless region between 5 and 15 eV showed a slight enhancement upon photoexcitation of Cu 2p and O 1s core levels.

Viescas et al. (1988) made inverse photoemission measurements on CuO with 16 eV electrons. Their spectrum covers only the first 4.5 eV above E_F, and in this region the only structure is a weak peak at 1.5 eV. Ghijsen et al. (1988) took a BIS spectrum of CuO, detecting 1487 eV photons. This was combined with the XPS spectrum of the valence band to give a spectral function. The gap in the combined spectrum was 1.4 eV in good agreement with the optical gap. The structures in the density of states from the band calculation were in good agreement with those in the BIS spectrum over a 20 eV range above E_F, provided the density of states was shifted by 0.7 eV.

Core-level XPS spectra have been reported. There is a satellite in the 5–10 eV region in the spectrum of each Cu core level in CuO (Kim, 1974), as well as in the valence-band spectrum. The 2p satellites are very prominent, nearly as intense as the main lines. The main lines are two well-separated spin–orbit components, each of which has its own satellite and each of which is broader than the corresponding line in Cu_2O (Fiermans et al., 1975; Mosser et al., 1991). In addition, the lesser bound of the two $2p_{3/2}$ peaks has a slight shoulder, one that is more prominent in some cuprate superconductors (Parmigiani et al., 1992). This shoulder has been assigned to final states involving holes on ligands of the Cu. These hole states are, in principle, not present for a square-planar array of

O ligands. They appear weakly because of the presence of the two out-of-plane O ligands in CuO, ligands slightly further away than those in the square. However an alternate explanation of this shoulder, multiplet structure, was given by Parlebas et al. (1995). This will be described below.

Shen et al. (1987), Ghijsen et al. (1988) and Eskes et al. (1990) modeled CuO with a cluster. Eskes et al. varied parameters in the Hamiltonian for the cluster and produced valence-band photoemission spectra for comparison with experiment. Shen et al. and Ghijsen et al. used Cu $2p_{3/2}$ photoemission spectra to help determine their parameters. The cluster of Eskes et al., similar to that of Ghijsen et al., and its ground state were discussed in Chapter 2. For Eskes et al. the final-state configurations were $|d^8\rangle$, $|\ d^9\underline{L}\rangle$, and $|d^{10}\underline{L}^2\rangle$ for photoemission and $|d^{10}\rangle$ for inverse photoemission. The three parameters used in obtaining the ground state were used for the excited states. These gave a charge-transfer band gap of 1.8 eV, larger than the experimental 1.4 eV. Two of the three Racah parameters for the multiplet structure of the d^8 component of the final state in photoemission were taken from fits to optical spectra, and one was adjustable. The spectral density was calculated in the sudden approximation and each line in the series was given a Lorentzian broadening for comparison with experiment. The spectral density was broken into its 3d and 2p components for more detailed interpretation. The ratio O 2p:Cu 3d photoexcitation cross-sections at each of the photon energies used in the experiment, 21.2, 40.8, and 1487 eV, is known from calculations (Yeh and Lindau, 1985). At the two lower photon energies both components contribute, while at 1487 eV the spectrum is dominated by the 3d component. Figure 5.24 shows the measured spectra at the three energies, the calculated spectral density function, and the calculated total spectral density function from d holes only, both as lines and as the sum of broadened lines. Structures A, B, and C have their largest contributions from $d^9\underline{L}\rangle$ final states, D_1 and D_2 contain contributions from $|d^{10}\underline{L}^2\rangle$ final states, while the $|d^8\rangle$ final states dominate in structures E and F and contribute to D_1 and D_2. The higher apparent binding energy of the $|d^8\rangle$ final states relative to that of $|d^9\underline{L}\rangle$ is because the Cu 3d–3d Coulomb energy is larger than the O 2p–Cu 3d charge transfer energy. Peak F, not apparent in the measured spectra of Fig. 5.24, appears in resonance photoemission spectra. Eskes et al. compared their cluster Hamiltonian parameters to those of Ghijsen et al. and Shen et al., pointing out some differences, although the fits to the spectra are qualitatively about the same.

van Veenendaal et al. (1993) modeled CuO with several planar clusters: CuO_4, as just discussed; Cu_2O_7; and Cu_3O_{10}. One of the problems they were investigating is the rather large width of the Cu $2p_{3/2}$ peak in CuO (Ghijsen et al., 1988). They had shown earlier that the use of a cluster larger than NiO_6 produced a satellite, as observed, on the corresponding Ni 2p line in NiO. A similar effect might produce an unresolved satellite in CuO, giving the unusually broad line.

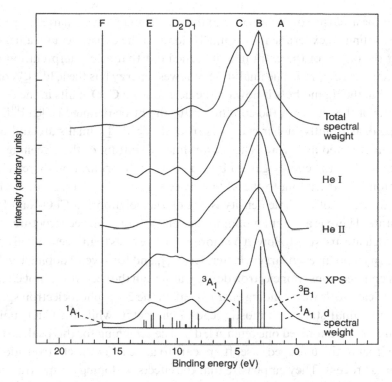

Fig. 5.24 Calculated valence-band photoemission spectrum for a CuO_4 cluster including broadening for comparison with experiment (top curve), experimental spectra taken at three photon energies (middle) and the calculated Cu 3d contribution to the spectrum (vertical lines below the bottom curve), and the resultant broadened spectrum. (Eskes *et al.*, 1990)

If this peak is largely $|\underline{c}d^{10}\underline{L}\rangle$ in character, the width might come from the O 2p band width, but not all compounds of Cu with neighboring O exhibit such a broad peak. Also, the parameters in the CuO_4 and Cu_2O_7 cluster models that gave the best fit to the valence-band photoelectron spectrum were not the same as those needed for a good fit to the 2p core-level spectrum. To keep the calculation more tractable, a reduced basis set was used, Cu $3d_{x^2-y^2}$ and O 2p bonding orbitals, deemed adequate for core-level spectra, but not for valence-band spectra, which were not calculated. The valence states were full, except for one hole per Cu site, or, to simulate hole doping, one hole per Cu site plus one hole per cluster. The calculated core-level spectrum for the CuO_4 cluster with one valence hole gives two peaks. The main line (lower apparent binding energy, lowest total energy) consists chiefly of the $|3d^{10}\underline{L}\rangle$ state, and the satellite is mostly $|3d^9\rangle$. A cluster with additional oxygen atoms, CuO_{10}, does not give a calculated spectrum that is much different, but the Cu_7O_{10} cluster give a spectrum with an extra, rather strong line on the high-energy side of the main peak. Moreover, the ratio of the intensity of the satellite to that of the main line, a parameter not used in any fitting to experiment, was much closer to the experimental value. In examining the origin of the extra line, it turned out that a re-interpretation of the main line was required. The final state of lowest energy has the $|\underline{c}3d^{10}L\rangle$ configuration, but the ligand hole is now on the neighboring CuO_4 units in the cluster. In terms of the site with the core hole, this might be denoted as $|\underline{c}3d^{10}L'\rangle$. The new peak at slightly higher energy is from the $|\underline{c}3d^{10}L\rangle$ final state and the satellite is interpreted as before. The reason for the shifting of the screening to the next unit cell is the energy gained by Cu 3d–O 2p hybridization there. This hybridization is that leading to the Zhang–Rice singlet. However, it should be emphasized that all of the hole density is not on the neighboring CuO_4 unit for this final state. There is a lesser amount in the central, core-ionized structure.

These calculations were followed by more extensive ones (van Veenendaal and Sawatzky, 1994) on an even larger cluster, Cu_4O_{13}, and for several doping levels, both holes and electrons. For electron doping, although the spectrum is not changed a great deal, the lowest binding energy peak in the $2p_{3/2}$ photoelectron spectrum is from a final state with mostly $|c3d^{10}\rangle$ character, while the CuO_4 units neighboring the core-ionized one remain $|3d^9\rangle$. The screening for this peak is different than that in the undoped case. It appears to arise from the electron added in the doping process. They emphasize that the effects of doping on spectra cannot be simulated by using Cu^+ or Cu^{3+} sites in addition to Cu^{2+} sites. Consideration of more than one CuO_4 unit in a cluster should also alter the valence-band photoelectron spectrum from that of a single CuO_4 unit.

Parlebas et al. (1995) used the same CuO_4 planar cluster model in a calculation of the Cu 2p XPS spectrum. This incorporated hybridization between the Cu 3d and O 2p states, Coulomb repulsion on the Cu site, and spin–orbit splitting. The calculated spectrum is shown in Fig. 5.25, along with a measured spectrum. The large number of lines in any one peak indicates multiplet structure. The calculated lines were convolved with a Gaussian function and a Lorentzian function, each of 1.2 eV full width at half maximum, to produce the calculated spectrum for comparison with experiment. The Lorentzian was to simulate lifetime broadening and the Gaussian represented instrumental resolution. The main peak and satellite represent $|\underline{c}d^{10}L\rangle$ and $|\underline{c}d^9\rangle$ final states, respectively, as in the cases discussed above. The width of the satellite is attributed to multiplet splitting. The authors point out that the calculated shape of this satellite is sensitive to the local electronic structure. It changes when the hybridization anisotropy or the 3d crystal field splitting ($10Dq$) change. Note in Fig. 5.25 that the width of the experimental main line is greater than that of the calculation. This was previously attributed to screening by atoms more distant than those in the small cluster. Parlebas et al. state that the same parameters used to produce the fit of Fig. 5.25 gave successful fits to valence-band XPS spectra and x-ray emission spectra. An

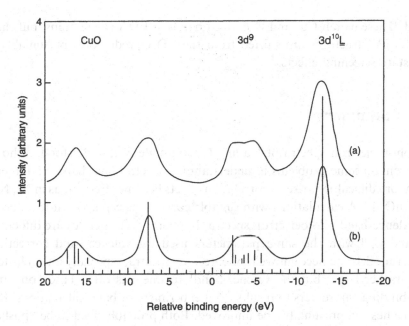

Fig. 5.25 Calculated Cu 2p XPS spectrum (vertical lines) and the resultant spectrum after broadening, and the measured spectrum (upper curve). Parlebas *et al.*, (1995)

earlier calculation (Okada and Kotani, 1990) of the Cu 2p XPS spectrum also gave a good fit, but the same parameters may not yield a good a fit to valence-band spectra.

Bocquet and Fujimori (1996) have calculated the 2p photoelectron spectra for all the divalent and trivalent transition metals located at the center of a cluster of six octahedrally disposed ligands. They used an extensive set of orbitals (configuration interaction) and calculated most of the electrostatic and exchange integrals. The three parameters Δ, U and t, defined previously, were left as fitting parameters. Measured $2p_{3/2}$ satellite intensities and splittings could be used to select values of parameters from universal plots in this paper. The calculations cover Ni^{2+} and Cu^{2+}. Hence they can be used for NiO, and CuO and cuprates, although the geometry of the last two is not ideally octahedral. Obviously the effects of relaxation by more distant ligands, discussed above, will not be described, but these are lesser effects than replacing one set of ligands by another.

Inverse photoemission spectra were made on polycrystalline samples by Wagener *et al.* (1990c), using both 15–35 eV photons and 1487 eV photons. A shoulder in the spectra appeared about 1 eV above threshold, the conduction-band minimum, which was assigned to empty O 2p states. A second structure, a broad plateau about 2.4 eV above the conduction-band minimum, was assigned to empty Cu 3d states; a final state in inverse photoemission of Cu $3d^{10}$. Structures at 7, 10.8, and 18 eV above the conduction-band edge also were also found. These match Cu- and O-derived bands in LDA calculations, but the 1-and 2.4-eV structures are shifted from the LDA predictions, presumably by final-state screening effects.

5.8   Summary

The photoelectron spectra of Na and Cu are quite well understood, although there remain some problems in understanding spectral line shapes and intensities. More difficulties arise as correlation effects become stronger, as in Ni, NiO, and CuO. LDA calculations with suitable corrections can account for much of the valence-band photoelectron spectra, but core-level satellites are difficult to describe well with the same parameters used for valence-band corrections. Cluster models can account well for core-level spectra, especially if the cluster is made rather large, but the valence-band treatment is often based on fitting tight-binding parameters to results from experiment or band calculations. Both approaches can presumably be improved. Both probably need to be "pushed" further.

Chapter 6

Early photoelectron studies of cuprates

6.1   **Introduction**

As soon as the cuprates were discovered, measurements of all types were made on the samples then at hand. Photoelectron spectroscopy was employed early to learn something of the electronic structure of these materials. As there were many more XPS spectrometers than ultraviolet, many XPS papers appeared in the first year of the "cuprate era." One of the early goals was to seek evidence of Cu^{3+} ions in those cuprates that were superconducting, thereby demonstrating that the holes introduced by doping resided, at least partially, on the Cu sites. Cu 2p spectra existed for a series of compounds in which the Cu ions had a known valence, mostly +1 and +2, but at least one, $NaCuO_2$, in which it was trivalent. The Cu 2p apparent binding energy increased as the valence increased, so the binding energy could be used to determine the valence to about one significant figure. (The shift is not linear with valence. The binding energy difference between Cu^{2+} and Cu^{3+} is rather small (Karlsson *et al.*, 1992).) The search was not successful. Core-level spectra from all the other elements in the cuprates were studied, not just the Cu 2p. Valence-band spectra were measured for comparison with densities of states from band calculations. Considerable agreement was found, but not enough to say definitively that LDA calculations did or did not give an accurate picture of the electronic structure of the cuprates. Another goal of the early work was to detect a Fermi edge in the valence-band spectrum in order to show that the cuprates were or were not similar to other conductors. This goal, too, proved difficult to reach. Other spectroscopies were employed at the same time. The empty states just above E_F were probed by electron energy loss spectroscopy and x-ray absorption spectroscopy, and the filled states just below E_F were probed by x-ray fluorescence. All these techniques use core levels on specific atoms, so the degree of localization of the states near E_F on those atoms could be assessed. In fact, the equivalent of x-ray absorption spectra can be taken in photoemission by total or partial yield spectroscopy, as discussed in Chapter 3. Some electron energy loss spectra can be found in the spectra of inelastically scattered electrons in the XPS spectra of core levels, but these are rather weak, and only plasmon losses may be identified reliably.

Unfortunately, many of the samples used in the early investigations were polycrystalline ceramics. The surfaces studied by XPS were prepared in vacuum or ultrahigh vacuum by breaking or, more commonly, by scraping, both being carried out at room temperature. The samples usually had been characterized by x-ray diffraction and found to be mostly (> 90%) of the desired phase. Often the sharpness of the transition to superconductivity had been determined on the same sample, or on one prepared as identically as possible. The transition temperature studies average over the entire volume of the sample while the x-ray powder patterns sample depths of a μm or so. The XPS studies of the Cu 2p levels sample a depth of only around 60 Å (about three mean free path lengths), and this may differ from the bulk for several reasons. The cleaving or scraping may break the samples along grain boundaries rather than across the grains. The material on the surface may not be the desired phase. Although it represents a minority constituent of the entire sample, the wrong phase could dominate on the surface if the samples break along grain boundaries. Sometimes these phases are carbonates, which could be identified from the peak due to the C 1s level. More importantly, in retrospect, at room temperature the surfaces of many of the cuprates are not stable in UHV, and oxygen may be lost. It cannot be replaced simply by backfilling the chamber with O_2 gas and annealing the sample. (We will discuss a few studies on oxygen loss and restoration in Chapter 10.) The result is that the surface regions may not have the composition of the bulk. In a number of cases they are not even metallic, but are insulating. This can be inferred in several ways. One is that the core-level spectra of all the component elements may be shifted to higher binding energy than expected by the positive charge left on the insulating surfaces by the escaping photoelectrons. There may be multiple peaks from a core level, more peaks than can be explained by the number of inequivalent sites expected from the structure, from surface effects, and from the effects of atomic structure. Moreover, the relative intensities of these peaks may be sample dependent. Another sign of a problematic sample is that in the valence-band spectral region a peak at about 8–9 eV binding energy appears with sample-to-sample differences in its intensity relative to that of the valence bands. It was sometimes attributed to an impurity, but it now appears that it is a satellite of the Cu–O valence-band spectrum in more insulating cuprates, and the variable intensity arose from varying amounts of an insulating phase on the surface. Finally, the XPS spectrum of the valence band rarely showed any intensity at the Fermi level. We know from more recent ultraviolet photoemission studies that the intensity at the Fermi level is small, and that these states have a smaller photoexcitation cross-section at 1.487 keV than in the vacuum ultraviolet. Still, states at the Fermi level were seen in XPS of a few samples, aided by the increased escape depth 1.487 keV photons provide over vacuum-ultraviolet photons, i.e., some signals came from below the insulating layer. X-ray absorption and emission and electron energy loss spectroscopies probe greater depths than photoemission, so they are less sensitive to the effects just described.

The XPS studies of the elements other than Cu and O have been rather puzzling. For a given core level there may be several peaks. One has to decide whether these arise from inequivalent atomic sites in the bulk, a surface site and a bulk site, final-state multiplet structures, or several of these. As long as the atom does not have a partially filled valence shell, multiplet structure tends to be simple. In such cases there may be only a spin–orbit-split pair of lines. For deep core levels, these are well separated compared with the line widths. Picking the strongest such line, and determining that it is not from the surface by measuring spectra at several take-off angles or adsorbing different atoms on the surface, its position is characterized by a binding energy. Often this is compared with the binding energy of the same level in a reference material. For example for Ba in Y123 one might use Ba metal or BaO. In measuring the binding energy in a sample, a zero-energy reference is needed. For metallic samples, the Fermi energy could be used, but it may not be a prominent feature in the spectrum. The Fermi energy of a good metal, e.g., Au, in contact with the sample may be used, for the Fermi levels should be the same. Sometimes instead a small spot of metal, e.g., Au, may be evaporated on part of the sample surface for a reference. It has been customary to use the binding energy of the sharp Au $4f_{7/2}$ peak as a secondary reference. The C 1s peak has been used occasionally, but this can be risky, for the binding energy of carbon depends on its chemical state and, for the carbon left on the surface from undesorbed hydrocarbons, this can depend on the unknown valence. At any rate, binding energy shifts have been reported for most elements in the cuprates, but their interpretation sometimes remains unclear.

A positive binding-energy shift is taken to mean a larger binding energy for the core level in the material under study than it has in the reference material. Unfortunately, this shift can have several causes, all of which may be relevant. In an ionic material the cation has a positive charge due to the loss of one or more electrons. This increases the binding energy of all remaining electrons on that ion. Thus a positive core-level binding shift indicates an increase in oxidation state. The anion core levels shift in the opposite direction. In addition, there is the Madelung (electrostatic) potential from the ions on the rest of the lattice. The cation site is a site of negative electrostatic potential, which decreases the core-level binding energies. The opposite is true for anion core levels. The situation is modified in the case of partially covalent bonding, for then the shift is not as large because the valence electron has been displaced rather than removed. These are both initial state effects. In addition, the electrons surrounding the core-ionized atom can polarize around the site of that atom, screening the hole and reducing the system energy. This increases the energy of the emitted electron. This continuum polarization picture holds for a metal, but insulators can polarize as well, with transfer of an electron from a ligand to the core-ionized atom or ion. This shifts the binding energy to smaller values. Both initial-state effects and final-state screening can occur at once. The best way to understand binding-energy shifts is to carry out total energy calculations for a large enough cluster in its ground state and in a core-ionized state for both the material at hand and the reference material. However the reference material is usually a metal and the other material may not be, so there may be different effects on each from any approximations made.

Vasquez (1994b) has reviewed the situation for the core levels of Ca, Sr, and Ba in cuprates. The spectra are not very well understood. For example, the binding energy of the Ba $3d_{5/2}$ level is less, by up to 3 eV, in Ba-containing cuprates than in Ba metal. (The shift is only about 1 eV to lower energy for BaO.) Sr 3d and Ca 2p core levels also shift to lower binding energies in the cuprates. Shifts of these levels for all three atoms to higher binding energies, often observed, are now known to be due to contaminating compounds. The negative sign of the shift in the cuprates is not understood, for the arguments on ionicity and screening predict a shift to higher binding energy. Vasquez discusses the arguments for and against each of the mechanisms by which an alkali-metal core-electron binding energy can shift in a cuprate. At the moment, the most promising reason for the negative binding energy shift of the Ba $3d_{5/2}$ level seems to be the effect of the Madelung term mentioned above, with suitable corrections for non-spherical ions and covalency. One approach along these lines is that of Parmigiani *et al.* (1991b). Brundle and Fowler (1993) also discuss this shift.

The Ca 2p spectra for Bi2212 show two peaks attributed to the bulk, where inequivalent sites are not expected. The weaker of the two peaks has been assigned to Ca on Sr sites. The asymmetry or splitting in the Sr 3d line has been attributed to some Sr on Ca sites. X-ray and neutron diffraction studies also indicate the exchange of some Ca and Sr between sites.

The O 1s core level might seem more difficult to study for there are several inequivalent O sites in cuprates. Several early studies indeed showed more than one peak, but as Bi2212 single crystals became better, eventually only one O 1s peak was found (Meyer *et al.*, 1988b). The extra peaks observed earlier, usually on Y123, were spurious (Srinavasan *et al.*, 1993; Brundle and Fowler, 1993). The single peak of Meyer *et al.*, had a shape which strongly suggested some underlying structure (Meyer *et al.*, 1988b). This became more apparent when the resolution was improved to 0.35 eV (Parmigiani, 1991a). These authors also varied the collection angle in their angle-resolving spectrometer to vary the surface-to-bulk relative sensitivity. The strongest component was assigned to O in the surface Bi–O plane. The next strongest, at slightly lower binding energy, was assigned to oxygen in the CuO_2 planes, and a separate peak at 531 eV could be varied in relative intensity by annealing in O_2 or Ar. It was attributed to surface oxygen and to oxygen intercalated below the surface Bi–O plane, although additional work is needed to assign more specific locations.

The O 1s peak in cuprates (and BaO) is at a lower binding energy than the 1s peak in many ionic oxides. Barr and Brundle (1992) have discussed the meaning of this, pointing out that the O 1s binding energy is lower in cuprates than in CuO and Cu_2O. They also noted that the O 2p contributions to the valence band tend to have lower binding energies in cuprates and BaO. They suggested that the latter is caused by the location of low-lying cation core levels, e.g., Ba 5p, which tend to push the O 2p valence electron energies higher. This was stated to tend to decrease the O 1s binding energy as well. In their picture, the Ba ions are highly ionic, and the corresponding anions are the CuO_2 planes, which are internally bonded covalently, but carry an overall negative charge.

Balzarotti *et al.* (1991) cleaved Bi2212 crystals at 20 K and measured the O 1s spectrum as a function of time at this temperature. The spectrum immediately after cleaving had components at 528.8, 531.5, and 533.5 eV, with the latter the weakest. Three hours after the cleave the 533.5 eV component had become the largest and five hours after the cleave it was still larger. Warming to 300 K restored the spectrum to one similar to the original spectrum, and keeping the sample overnight at 300 K in UHV led to no additional changes in the spectrum. The 533.5 eV peak was clearly the result of a condensate at low temperature, probably CO. Changes in the valence-band spectra occurred as well, but they were relatively much smaller.

The Cu 2p core levels have been extensively studied, along with those of Cu_2O and CuO (Chapter 5). The reason is that they show a strong satellite originating in many-body (correlation) effects involving the incomplete 3d shell and ligand valence electrons. If the Cu 3d levels are full, as in Cu and Cu_2O, there is no satellite, or it is extremely weak. The satellite energy and relative intensity have been used to fix parameters in several cluster calculations. When the minimum cluster, $(CuO_4)^{6-}$, is used, different parameters result from data from CuO and a cuprate. The satellite structure is similar in cuprates with Cu surrounded by four oxygen atoms in the plane, but it changes somewhat in cuprates with apical oxygens.

Figure 6.1 shows the Cu $2p_{3/2}$ spectrum of CuO and Bi2212 taken with 0.285 eV resolution (Parmigiani *et al.*, 1992). The main line in Bi2212 has a small feature, B, assigned to the presence of apical O atoms not present in CuO. They showed that spectra taken on two other cuprates, one with and one without apical oxygen atoms, confirm this interpretation. As discussed in Section 5.6, Parlebas *et al.* (1995) found the characteristic shape of the Cu $2p_{3/2}$ main line arises from unresolved multiplet structure which depends on the anisotropy of the Cu 3d–O 2p hybridization. The anisotropy will change when apical oxygen atoms are added to a square-planar CuO_4 cluster.

Parmigiani and Sangaletti (1994) have tabulated 36 values of the energy separation of the satellite from the main line and the relative intensity ratio for the Cu $2p_{3/2}$ lines in six cuprates, as well as four pairs of similar data for CuO. The energy separations appear nearly independent of the material, with a possible trend that CuO and Nd_2CuO_4 have slightly higher separations. The scatter in the intensity ratios is much larger, even for one cuprate. For example, the three measurements of this ratio reported for Y123 are 0.24, 0.33 and 0.35. These two measured quantities enter into the determination of several parameters in the CuO_4 cluster models used to explain the valence band and Cu 2p photoelectron spectra, as discussed by Shen *et al.* (1988). They used their own measured values for Y123, an intensity ratio of 0.35 and a satellite separation of 8.8 eV. The use of 0.24 and 8.8 eV (Fujimori *et al.*, 1987) would have altered the resultant parameters.

There have been several studies of trends in the Cu 2p satellite-to-main-line ratio and satellite separation from the main line as a function of doping in Y123 (Rao *et al.*, 1990; Yeh *et al.*, 1990; Santra *et al.*, 1991; Rao *et al.*, 1992; Yeh,

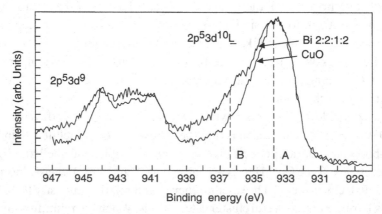

Fig. 6.1   XPS spectrum of the Cu $2p_{3/2}$ core level in CuO and Bi2212 normalized to the peak height of the "main line." Component A occurs in CuO and at least one other cuprate without apical oxygen atoms, while component B is present in two cuprates, including Bi2212, with apical oxygen atoms. (Parmigiani *et al.*, 1992)

1992). These can be related to changes in the parameters of the CuO_4 cluster models. In particular, Rao *et al.* related the intensity ratio directly to the charge-transfer energy Δ. The measured ratios and shifts varied monotonically with composition, from which hole concentrations were inferred, or sometimes measured, but the sign of the monotonic shift with hole concentration did not always agree between groups.

As a result of some of the early XPS studies, the question of how insulators could be superconducting occasionally was discussed seriously. However, we have chosen not to document here every one of the earlier XPS (and UPS) studies. Some of these early papers are best left unread by persons beginning research on cuprates. Several reviews of photoemission studies on cuprates exist, some of which cover more thoroughly the early literature (Fuggle *et al.*, 1988; Lindberg *et al.*, 1990a; Meyer and Weaver, 1990; Vasquez, 1994a,b; Parmigiani and Sangaletti, 1994; Nagoshi *et al.*, 1994; Veal and Gu, 1994; Loeser *et al.*, 1994; Sakisaka, 1994; and Shen and Dessau, 1995). Al Shamma and Fuggle (1990) published a bibliography of 505 papers dealing with electron and x-ray spectroscopy of cuprates, most of them on the spectroscopy of core levels. In the rest of this chapter we focus on work published primarily before the start of 1991.

List *et al.* (1988) first reported the loss of oxygen to the vacuum from single crystals of $EuBa_2Cu_3O_7$. They cleaved the crystals in ultrahigh vacuum at 20 K and measured valence-band spectra by angle-integrated photoelectron spectroscopy. A clear signal was found at the Fermi level, indicating the surface region was metallic. The spectra remained unchanged for at least 8 hours at 20 K. Warming the surface to 80 K left the Fermi-edge part of the spectrum unchanged, but slight changes were noted elsewhere in the spectrum. Warming to 300 K produced more spectral changes and after 8 hours at 800 K the Fermi edge disappeared. There were also changes elsewhere in the spectrum, and all the features shifted to higher binding energy by about 0.5 eV. A 9-eV satellite peak grew. Cooling to 20 K did not restore the original spectrum. The Ba 4d core-level spectrum also changed irreversibly during this temperature cycle, from its original spectrum to one resembling that of a different, poorly oxygenated sample. The 0.5 eV shift is what is expected if an insulating layer, such as the loss of oxygen could produce, were present on the surface. It would charge positively due to the poorly compensated loss of photoelectrons, thereby shifting all initial states to greater binding energy. The lack of a Fermi edge in the spectrum and the appearance of the satellite are also indicative of an insulating surface layer.

XPS studies (Fowler *et al.*, 1990) showed a Fermi level in Y123 at 300 K. These suggested that the crystals selected for the XPS study were more stable than those used by List *et al.* However, the escape length of valence-band photoelectrons in XPS is greater than that in ultraviolet studies, and this may contribute to the difference. Subsequent work by other techniques has verified that oxygen is indeed lost, even at low temperatures. Ogawa *et al.* (1991) cleaved single crystals of Y123 at 300 K, and concluded from Auger spectroscopy that all the oxygen in the surface chains was lost to the vacuum. Edwards *et al.* (1992) cleaved Y123 crystals at low temperature and studied the surface by STM. Upon heating above 70 K, atomic resolution was irreversibly lost. This was attributed to the loss of oxygen. Both these studies also showed that the cleavage occurred between the Ba–O and Cu–O chain layers. As a result of these studies, most work today is carried out on single crystals and, increasingly, epitaxial films which are treated, i.e., cleaved, as single crystals.

6.2   **Bi2212**

List *et al.*, (1989) measured the height of the step at the Fermi edge as a function of photon energy. They fitted it to the energy dependence of the photoexcitation cross-section for a linear combination of Cu 3d and O 2p states, finding the states near E_F to have about 65% O 2p character and 35% Cu 3d character.

Lindberg *et al.* (1989c) analyzed core-level spectra taken on single crystals of Bi2212 cleaved *in situ*. The photocurrent from a core level on an atom at depth z is proportional to $\exp[-z/\Lambda(E_f)]\sigma(h\nu)T(E)$, where Λ is the mean free path for escape without inelastic scattering, σ the photoexcitation cross-section, and T the analyzer transmission. This was applied to core levels on all five types of atoms. When summing over layers, the first factor gave a type of structure factor. Literature values of Λ ("universal curve") and σ (calculated) were used, and $T(E)$ came from an analysis of the analyzer. By summing over atomic layers, one could obtain the expected relative intensity from each core level. The O 1s level was used for normalization. The results were quite consistent with the expected structure of Bi2212 when the top layer is a Bi–O plane. They also indicated that the surface did not become enriched in any one element to a detectable degree, and that there was not much intermixing of Ca and Sr.

A significant demonstration of the effect of superconductivity on the photoelectron spectrum of a cuprate was the work of Imer *et al.*, (1989a). They used the He I and He II lines to obtain angle-integrated spectra with 20 meV resolution of a Bi2212 single crystal and a sample consisting of a few grains. They produced a new sample surface every 20 minutes during a 3-hour run by scraping with a tungsten brush or peeling the single crystal with a sharp blade. A comparison of spectra taken at 15 K and 105 K clearly show a loss of intensity at the Fermi energy and a pileup just below it, shown in Fig. 6.2. By comparing their spectrum with that expected from a BCS spectral function they obtained a value of the

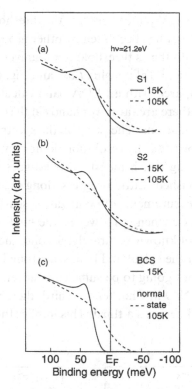

Fig. 6.2 Photoelectron energy distribution curves from two samples of Bi2212 at (a) 15 and (b) 105 K. (c) is a Fermi Dirac function for 105 K and a BCS spectral density for 15 K with $\Delta = 30$ meV, both convolved with the instrumental function. (Imer *et al.*,1989a)

order parameter Δ of 30 meV $\pm$ 5 meV. Sakisaka (1989) pointed out that the failure to have a pileup below E_F in the spectra taken at 105 K could be due to thermal broadening and a reduced intensity from a Debye–Waller factor, but subsequent work has shown this is not the case. Other criticism of this work (Chang *et al.*, 1989c) was rejected (Imer *et al.*, 1989b). In fact, in a later *tour de force*, this same group (Grioni *et al.*, 1991) carried out a similar experiment on a better-understood superconductor, Nb_3Al. Here, with about 13 meV resolution they were able to see spectra similar to their earlier spectra on Bi2212, even though the expected value of Δ was only 3.4 meV. The fit to their data gave 1.65 meV. They attributed the smaller value to possible weakened superconductivity in the surface layers, the only layers photoemission probes.

Many of the early angle-resolved studies of cuprates were carried out at rather low resolution, a spectral band pass of 200 meV and more. This is often adequate for examining the entire valence-band region, but not adequate for work near the Fermi energy. The first band mapping was that of Takahashi *et al.* (1988, 1989) on Bi2212. They showed that the intensities of the valence-band spectra were enhanced at about 18 eV photon energy, the threshold for O 2s excitation. (Although this enhancement has been seen by other groups, e.g., Manzke *et al.* (1989c), its connection with the 2s threshold is not universally accepted.) Figure 6.3 shows some of the data of Takahashi *et al.*, and Fig. 6.4 shows the result of this band mapping, compared with an LDA band calculation. There are three features to notice. First, there are far fewer bands in the experimental electronic structure than in the calculations. The peaks in the spectra are broad. Lowering the temperature and improving the resolution do not reveal many new features in this diagram. Identifying which measured band goes with which calculated one is not easily accomplished. Little progress along these lines has been made since 1989. It requires measurements of polarization sensitivity and calculations of the momentum matrix elements. However, the final states are about 15 eV above E_F and they are not known accurately. Second, the measured bands tend to be flatter than most of the calculated bands, a strong hint that LDA calculations by themselves are not going to be sufficiently accurate. (See the discussion of the band mapping of NiO in Chapter 5.) Third, there are two Fermi-surface crossings in Fig. 6.4, and perhaps a third. This implies there is a Fermi surface,

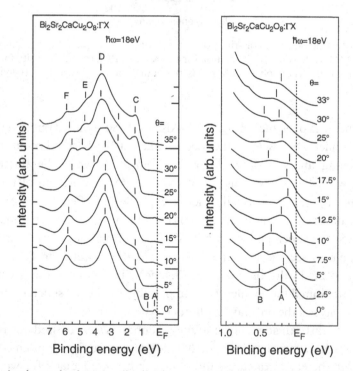

Fig. 6.3 Angle-resolved energy distribution curves taken on Bi2212 with 18 eV photons. As the angle from the surface normal increases, the component of the wave vector along the Γ–X direction in the Brillouin zone increases. (Takahashi *et al.*, 1989)

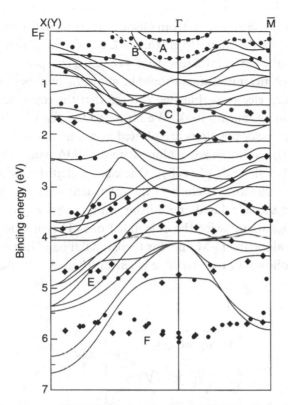

Fig. 6.4 Band structure of Bi2212 calculated by the LDA (Massidda *et al.*, 1988) (solid curves) and experimental band mapping points determined with 18 and 40 eV photons (filled circles and diamonds, respectively) from the spectra of Fig. 6.3 and other spectra. The letters denote the peaks in the spectra. (Takahashi *et al.*, 1989)

not an obvious property of cuprates at that time. However, the number of crossings of E_F along the Γ–X and Γ–$\overline{M}$ lines in Fig. 6.4 is not in agreement with other work, both contemporaneous (Minami *et al.*, 1989 and Olson *et al.*, 1989) and more recent (Chapter 7). This is probably the result of differences in sample quality. Visual appearance of a cleaved surface, the appearance of a LEED pattern, and the sharpness of the transition to superconductivity cannot always be relied upon to indicate surface quality.

The first results on Bi2212 photocurrent or the first that were widely accepted were arguably the angle-resolved spectra of Olson *et al.* (1989), Manzke *et al.* (1989a,b,c) and Claessen *et al.* (1989) in which several points on the Fermi surface were determined and a gap due to superconductivity was not only detected, but measured. The material was a single crystal of Bi2212. This work was carried out at improved resolution, but was limited to studying the first 500 meV or so below E_F. Additional papers resulting from the same set of measurements followed immediately (Mante *et al.*, 1990, Olson *et al.*, 1990a,b).

Figure 6.5 shows several spectra from Olson *et al.* (1990a) taken with 28 meV overall resolution. As the angle from the surface normal is scanned, a peak moves from below the Fermi level toward it, becoming sharper as it approaches E_F. At larger angles, the initial state should be above E_F and there should be no spectrum of primary electrons, but one can see that the top spectrum has an appreciable magnitude, but no peak. We return to this feature shortly. In the inset to Fig. 6.5 the peak positions are plotted as a function of wave vector in the basal plane of the Brillouin zone, along with the LDA band (Krakauer and Pickett, 1988a,b; Massidda *et al.*, 1989). One can see that the measured Fermi wave vector in this direction is close to the calculated one, but the measured band is flatter than the calculated one. Fitting with parabolas gives an effective mass of about $2m_0$, about twice the calculated one. The situation was much more complicated along the Γ–$\overline{M}$ line, for there is a flat band just below E_F and it was difficult to follow it or to determine where or whether it crossed E_F.

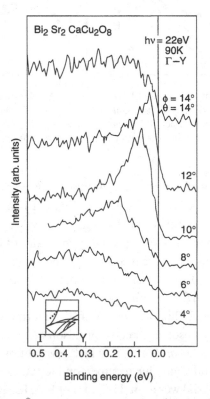

Fig. 6.5   Angle-resolved energy distribution curves for Bi2212 at 90 K taken with 22 eV photons. The angles are measured from the normal to the *a–b* plane surface. The inset shows the resultant binding energy vs. wave vector (points) and the LDA calculated bands of Massidda *et al.*, 1988. (Olson *et al.*, 1990a)

The peaks in Fig. 6.5 are obviously not symmetric. Olson *et al.* (1990a) fitted them with an asymmetric Lorentzian, one whose width depended on the energy below E_F as $| E- E_F|^p$ with $p = 1$ or 2. A value of $p = 2$ was expected for a Fermi liquid, but several models extant at that time (and still in existence, albeit modified) predicted $p = 1$. The fit was far better with $p = 1$ than with $p = 2$. Moreover, the coefficient was rather large: $\Gamma = 0.6[E - E_F|$. Smith *et al.* (1993) later pointed out how difficult it can be to extract the correct value of a lifetime-limited width from an experimental spectrum. In any case, it is clear that these spectra do not get close enough to E_F to allow a meaningful comparison with Fermi liquid theory. Measurements much closer to E_F must be made with higher resolution, and with all the conditions emphasized by Smith *et al.* met. However, it is interesting to note that when Santoni and Himpsel (1991) later re-analyzed photoemission data for metallic Fe, the width they obtained was also $0.6|E - E_F|$, again over a range below E_F too wide for comparison with Fermi liquid theory. The states involved were almost purely Fe 3d states, while those near the Fermi surface in cuprates are only partially Cu 3d in character.

Olson *et al.* (1989, 1990b) also obtained a value for the superconducting gap of 24 meV $\pm$ 5 meV. It did not depend on wave vector in the basal plane of the Brillouin zone (Olson *et al.*, 1990b). The latter result did not survive, for later measurements showed a distinct anisotropy, discussed in the next chapter. The isotropic gap originally observed evidently resulted from surfaces in some way inferior to the best obtained later, but the actual reason is not known. Poor surfaces (bad cleaves) do not show any structure in angle-resolved photoelectron spectra. Manzke *et al.* (1989a) obtained $\Delta = 30$ meV $\pm$ 5 meV.

We return to the large background in Fig. 6.5. The peaks near the Fermi energy in Bi2212 angle-resolved ultraviolet photoemission spectra are only about twice as large as the background 300–500 meV below the Fermi energy. This ratio has remained about the same for spectra taken by several groups on crystals grown by several groups over a period of seven years. The peaks in angle-resolved photoelectron spectra taken on most other materials do not display such a large background. Figure 6.6 shows spectra taken under comparable conditions for Bi2212 and $TiTe_2$, illustrating the difference in apparent backgrounds. Over the past seven years, the visual quality of the cleaved Bi2212 surfaces has improved. Improved crystals resulted in the observation of several new features in the photoelectron spectra, but the background level has not changed. The background in the photoemission spectra of most materials has been interpreted as due to inelastically scattered electrons, but for the cuprates it appears to be different. One interpretation of this background is that for the cuprates this is not a background of electrons inelastically scattered by inter-electron Coulomb potentials, but rather an intrinsic feature of the photoelectron spectrum due to leaving the system in a continuum of excited states. An example of

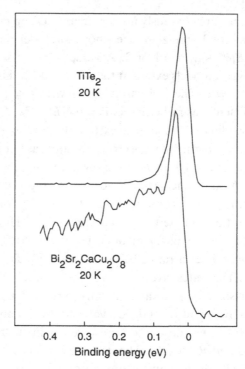

TiTe$_2$
20 K

Bi$_2$Sr$_2$CaCu$_2$O$_8$
20 K

0.4   0.3   0.2   0.1   0

Binding energy (eV)

Fig. 6.6 Angle-resolved energy distribution curves for Bi2212 at 90 K and TiTe$_2$ taken under nearly identical conditions of angle and energy resolution and photon flux. The ordinates are in arbitrary units, but count rate is about 20 times larger for TiTe$_2$ than for Bi2212.

the latter (Xiao and Li, 1994) will be described in the next chapter. Another interpretation is that the surface is intrinsically defective due to the oxygen vacancies introduced by doping to produce the metallic superconductor from the insulating parent compound. There is, as yet, no experimental proof for the origin of the large background found in cuprate spectra. Figure 6.6 also illustrates how weak a feature one is dealing with in studying the near-Fermi edge photoemission in cuprates. Under comparable conditions, the count rate for the TiTe$_2$ spectrum is about 20 times that of the Bi2212 spectrum. At larger binding energies, the count rate for Bi2212 increases considerably, but this region is of less interest.

An improved band mapping in the region of overlapping transitions (as in Fig. 6.3) was carried out by Böttner *et al.* (1990). Of their three samples, the data from the best was analyzed, Bi2212 doped with Pb to a Pb:Bi ratio of 0.28. Because of the large number of spectra to be taken, the resolution was reduced to 200 meV, allowing faster scans. Instead of locating peaks in the raw spectra, they fitted the spectra. The method of fitting is rational, and well suited for a background of inelastically scattered electrons. Its applicability to cuprate spectra is not yet known to be valid, but no better method of background subtraction is known. They first shut the pass window on their display analyzer and subtracted the resultant spectrum from the actual spectra. This removes electrons produced in the analyzer, e.g., by reflected or scattered radiation. Then they subtracted a background due to inelastically scattered electrons by assuming at binding energy E, the current of such electrons is proportional to the current of primaries at all lower binding energies. The remaining spectra were then fitted by five structures with a Doniach-Šunjić line shape in which the width parameter Γ for each was not constant, but increased linearly with binding energy. This fit was not used in the first 2 eV. The energy of the singularity from the fit was used as the peak energy. (Maetz *et al.* (1982) have discussed the use of the Doniach–Šunjić line shape for tight-binding bands in addition to the localized states for which it was proposed.)

The energies of the fitted peaks were plotted against wave vector to obtain a band map. However it is not obvious when to add or subtract a reciprocal lattice vector, i.e., when to place a point in the first, second or third Brillouin zone. Plots were made in the reduced and extended-zone schemes. The latter proved to give somewhat smoother curves, and hints of band crossings not evident in the former. Finally, the intensities of each peak were calculated from the fits, and plotted as a function of wave vector. These formed smooth curves in the extended zone plots and helped identify band crossings. The final results of the analysis are presented in Fig. 6.7, along with the LDA bands of Massidda *et al.* (1988). The first and second Brillouin zones are shown. (The data points in the two panels are different.) The glaring discrepancy at and below 6 eV, also seen by Takahashi *et al.* (1988, 1989) and Minami *et al.* (1989), was tentatively assigned to the effect of excess oxygen, obviously not considered in the LDA calculation. The other experimental bands lie in regions densely populated with calculated bands. As an aid in identification, estimates of the expected intensity were made by calculating $\cos^2 \delta$, where $\delta = (\mathbf{A}, \mathbf{k})$. This comes from the classical picture of absorption of radiation. $\mathbf{A}$ came from the angle of incidence and polarization, without consideration of refraction by the sample. In each case there are LDA bands near the measured ones that can give adequate photocurrent for the geometry used, but actual identification is not possible. The calculated bands often are separated by an energy too small to resolve. Higher resolution may not yield better spectra due to the intrinsically broad photoelectron peaks.

Inverse photoemission data complement the photoemission spectra. Claessen *et al.* (1989) measured angle-resolved inverse photoemission spectra on a Bi2212 crystal cleaved in ultrahigh vacuum, detecting 9.9 eV photons with 0.68 eV resolution. They found two peaks (2.9, 5 eV) which did not disperse as the wave vector was scanned along the Γ–X line. Dispersive peaks were expected from the LDA calculation. They also found a weak enhancement at E_F near the value of $\mathbf{k}$ where a band crosses the Fermi level in the photoelectron spectra. Similar

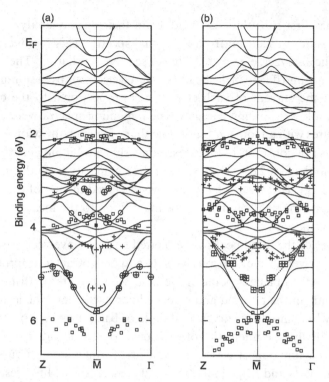

Fig. 6.7 Experimental band structure for Bi2212 obtained from band mapping in two Brillouin zones. (a) Points assigned to the first Brillouin zone. (b) Points believed to have been measured in the second zone. They have been mapped into the first zone. The solid lines are from LDA calculations (Massidda *et al.*, 1988). (Böttner *et al.*, 1990)

results were reported by Drube *et al.* (1989) and Wagener *et al.* (1989). Both groups used a monochromator to disperse the emitted photons. They scanned through the O 2s threshold region and observed an enhancement of the Fermi edge, which they attributed to the O 2p nature of the edge. At resonance the incident electron excited O 2s electrons to the O 2p states at E_F, from which they decayed radiatively, producing an additional photon flux. This is the same resonance reported in photoemission by Takahashi *et al.*, mentioned above.

Wagener *et al.* (1990b) extended this work by studying single crystals of Bi2212 and $Bi_2Sr_2CuO_6$ at 300 K and 60 K. The angle-resolved spectrometer detected photons in the 10–44 eV range with a bandpass varying between 0.3 and 1.0 eV. The 18 eV resonance for states in the first 2 eV above E_F was clearly delineated, and assigned to the O 2s resonance. The incident electron creates a hole in the O 2s shell which is filled by a radiative transition from the O 2p band. Upon scanning off the sample normal, the emission of photons by electrons decaying into states in the first 2 eV above E_F increased monotonically with increasing angle. This effect is largest in the spectrum taken with 18 eV photons.

The latter is taken to mean the final states are O 2p-like, and the former, that they are of $2p_x$ and $2p_y$ character. Most of the spectra were taken at 60 K. Reproducible differences were seen between spectra taken at 60 K and 300 K, shifts of edges and changes in intensity that were too large to attribute to the superconductivity at 60 K. Rather, these were assigned tentatively to charge transfer upon a structural transition at 210 K.

Bernhoff *et al.* (1990) also carried out angle-resolved inverse photoemission measurements with 0.35 eV resolution on Bi2212. These authors pointed out how difficult such measurements are. The very low count rates require long data-collection times and the incident electron beam causes surface deterioration in about 48 hours. In the first 2 eV above E_F they observed four dispersing peaks, one of which split into a pair of peaks. These were plotted on calculated band structures, as shown in Fig. 6.8. It is clear that there are similarities, but there certainly is not close agreement between the LDA bands and the data. Especially, there are energy shifts of up to about 0.5 eV, and one or more bands that are difficult to match to a calculated band. The measured bands are flatter, as are bands just below E_F mapped from photoemission data. However, the

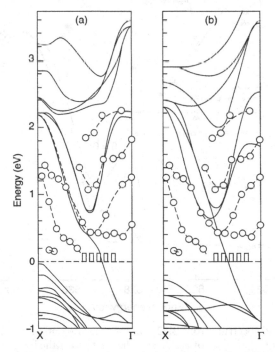

Fig. 6.8 Experimental band structure (open circles) above E_F for Bi2212 determined from angle-resolved inverse photoemission plotted on two calculated bands, (a) those of Krakauer *et al.* (1988) and (b) of Massidda *et al.* (1988). The rectangles indicate regions of enhanced intensity at the Fermi level. (Bernhoff *et al.*, 1990)

band approaching the Fermi level from above matches reasonably well with a band approaching E_F from below in angle-resolved photoelectron spectra.

The region just above E_F can be studied also by x-ray absorption spectroscopy and electron energy-loss spectroscopy (Nagoshi et al., 1994; Fink et al., 1994). Transitions from a core level to just above the Fermi level occur at each core threshold, provided the dipole matrix elements are not too small. Quantitative x-ray absorption spectra may be obtained by transmission of x-rays through a thin sample. The spectral shapes may be studied by electron energy-loss spectroscopy (which can also give quantitative information) and by photoelectron total yield or partial yield spectra, the latter collecting low-energy secondary electrons, which constitute most of the total yield. If polarized radiation and single crystals are used, one can obtain symmetry information on the states just above E_F. A number of such studies were carried out shortly after cuprates were discovered, and when single crystals became available, polarized radiation was used. Figure 6.9 shows the dependence on angle of incidence of the partial yield spectrum in

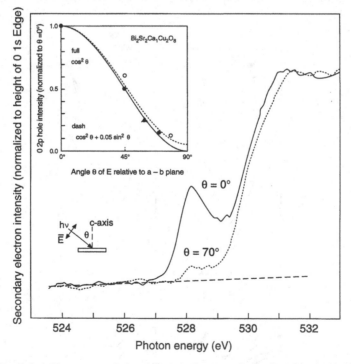

Fig. 6.9 Partial yield spectrum from Bi2212 at the O 1s absorption threshold with linearly polarized radiation at two angles of incidence. The inset shows the integrated intensity of the first peak as a function of angle, fitted with the $\cos^2 \theta$ dependence expected if the final states were only O $2p_x$ and $2p_y$ states (solid curve). The dashed curve allows for 5% $2p_z$ final states, imperfect polarization, or sample disorder. Different data-point symbols denote different samples. (Himpsel et al., 1988)

the region of the O 1s threshold from a single crystal of Bi2212 measured by Himpsel *et al.* (1988) with linearly polarized radiation. Spectra were measured at several angles (inset) and the angle-dependence of the peak intensity was very close to that predicted for absorption to O $2p_x$ and O $2p_y$ states (lobes in the *a–b* plane) at and above E_F. Transitions to these states are most intense when the incident electric field is parallel to the bond axis, i.e., at normal incidence. An admixture of more than about 5% $2p_z$ states would not fit the data.

At the same time, photoelectron spectroscopy was used to study other aspects of the electronic structure of Bi2212. For example, Wells *et al.* (1989) worked with single crystals of La-doped Bi2212. They measured angle-resolved spectra, but instead of band mapping, they applied polarization selection rules to determine the symmetry of the initial states by collecting electrons in a mirror plane of the surface (Hermanson, 1977). The analysis assumed the nearly tetragonal structure is actually tetragonal, as did the band structure calculation. They concluded that the states near the bottom of the valence band are mainly O $2p_z$ states, and those near the top (0–2 eV binding energy) are largely O $2p_x$ and O $2p_y$ in character. (See also the comments in Johnson (1990) and Wells *et al.* (1990a).) Since they were working with 16–20 eV photons, the photoexcitation cross-sections were dominated by the O 2p states. Lindberg *et al.* (1989a) studied the same crystal and showed that there was little or no dispersion of any of the valence-band peaks with wave vector normal to the surface, but there was dispersion with wave vector in the plane of the surface.

Wells *et al.* (1990b) studied the Bi–O-derived parts of the electronic structure near E_F. When scanning from Γ to X or Y, the states near the Fermi surface crossing are derived primarily from Cu 3d and O 2p states from atoms in the CuO_2 planes. Along Γ–$\overline{M}$, the states near E_F, if not derived from states based on the Bi–O planes, at least have more Bi–O character than those near E_F along Γ–X and Γ–Y. A Bi–O plane is the cleavage plane. They scanned along these two directions in a sample cleaved from a crystal that had been annealed in 12 atmospheres of O_2, (T_c = 80 K). They observed spectra similar to those of Olson *et al.* (1990b), a Fermi edge, or for the Γ–$\overline{M}$ line, something close to it, which showed a "pileup" upon cooling to below T_c. Then a small amount of Au was evaporated on the surface, equivalent to 0.5 Å. This reduced the amplitude of the spectra along Γ–X a little, but the pileup appeared upon cooling. This is shown in Fig. 6.10. The states sampled, in the CuO_2 plane, were little affected by the Au on the surface and remained superconducting. Along Γ–$\overline{M}$, however, the effect of the Au was to reduce the spectrum considerably more, the Fermi edge was very weak or non-existent, and there was no pileup in the spectrum below T_c. The Bi–O surface plane was metallic and superconducting before the deposition of the Au, but the Au disrupted it and destroyed its metallicity. These states could also be changed from metallic to non-metallic by removing

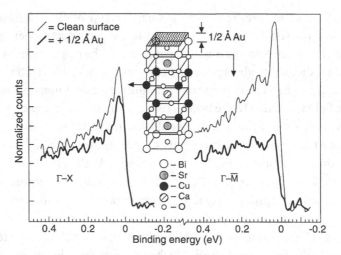

Fig. 6.10 Angle-resolved photoemission spectra of Bi2212 at 20 K before (thin line) and after (thick line) the deposition of 0.5 Å equivalent of Au on the surface. The spectra along Γ–X arise primarily from a CuO$_2$ plane, while those for Γ–$\overline{M}$ have a significant component with Bi–O character. (Wells *et al.*, 1990b)

some of the oxygen. This was done by cleaving an as-grown crystal ($T_c = 89$ K). For this surface there was no Fermi edge, or near-Fermi edge along Γ–$\overline{M}$. Note also that in this experiment, the random sprinkling of Au atoms on the surface did not have a noticeable effect on wave-vector conservation during photoemission nor on the "inelastic background."

The related compound Bi$_2$Sr$_2$CuO$_6$ was studied by Lindberg *et al.* (1989b) by angle-integrated photoelectron spectroscopy on single crystals cleaved *in situ*. Bi$_2$Sr$_2$CuO$_6$ has one CuO$_2$ layer per unit cell, while Bi2212 has two. The O 1s spectrum was the single peak characteristic of a good surface, and the Cu 2p spectrum had a strong satellite similar to that of other cuprates. The valence-band spectrum was similar to that of Bi2212 and there was a Fermi edge characteristic of a metal.

6.3   Y123

As with early photoelectron spectra from Bi2212, the first photoelectron spectra from Y123 were taken on polycrystalline samples. When single crystals were produced they were twinned; in some regions the *b*-axis was nearly parallel to the *a*-axis in other regions, precluding the secure identification of bands arising from the Cu–O chains. Nonetheless, these twinned crystals proved very useful.

In an early study of angle-resolved photoemission from Y123, Sakisaka *et al.* (1989a,b) cited 12 earlier publications on photoelectron spectra on Y123, none of which report reliably a signal at the Fermi edge. Their own work, carried out at 100–200 meV resolution on epitaxial films, clearly showed dispersing bands passing through the Fermi level. They annealed the films in O_2 in the experimental chamber, let the samples cool to room temperature, then pumped the gas out. All measurements were made at room temperature. A definite Fermi edge was observed, as well as several bands dispersing in the first 1 eV below the Fermi energy. The peaks at greater binding energy were broad, like those in Fig. 6.3. However, there was a definite satellite at 8.5 eV which grew and shifted to about 9 eV when the sample was annealed in vacuum.

Angle-integrated spectra of the valence bands were obtained by Arko *et al.* (1989) from single crystals cleaved at 20 K, using a resolution of 100–200 meV. By varying the photon energy, bands could be seen to disperse, along with variations in the height of the Fermi edge, but no band mapping could be carried out. The 9 eV peak was not present, and because it was not, a small satellite peak at 10 eV could be seen. This peak, along with a peak at 12 eV previously found, underwent a resonance as the photon energy went through the 74 eV threshold for Cu 3p photoexcitation. The valence-band spectra showed no strong evidence for a resonance. The resonant satellites are similar to those found in Ni and CuO discussed in Chapter 5. Arko *et al.* also looked for resonances at the O 2s threshold. A deteriorated Eu123 surface showed the 9 eV peak. Its resonance occurred at 22 eV, while the Fermi-edge structure had a resonance at 18 eV. The authors urged caution in trying to determine Cu- and O-character of various parts of the valence-band spectrum from their resonance behavior. In addition, they measured spectra over a wide range of photon energies. At 100 eV or so, the O 2p photoexcitation cross-section is expected to diminish by about a factor of almost ten more than the Cu 3d cross-section, when compared with their relative sizes at 20 eV. The spectra just below E_F did not fall by this much at higher energies, indicating considerable Cu 3d character just below E_F.

Later, Eisaki *et al.* (1990) made similar measurements on Bi2212, metallic, but not superconducting $Bi_2Sr_2CuO_6$, and $Bi_2Sr_2CoO_{6+x}$. The latter is a semiconductor. The main valence-band features were nearly the same in all three materials, but the region just below E_F varied. The intensity at E_F was about twice as large for Bi2212 as for $Bi_2Sr_2CoO_{6+x}$, roughly in the 2:1 ratio of the number of CuO_2 planes. This strongly suggests the states near E_F are localized around these planes. By going to a photon energy of 120 eV, they again found about a 2:1 ratio in the intensities at E_F. (The Bi–O planes contribute negligibly at this energy because of small photoexcitation cross-sections.) Thus, in agreement with Arko *et al.*, they find considerable Cu 3d character in the photoelectron spectrum near E_F. They then measured constant initial-state spectra for electrons from various regions of the valence band, varying the photon energy through the Cu 3p resonance. As in the work of Arko *et al.*, a 12.7-eV satellite showed a weak resonance, but the region near E_F did not. They used a simple cluster model to show that it was possible to have a Cu 3d–O 2p hybrid ground state that had most of the initial hole density on the O site, 95% in their model, and still have significant Cu 3d character in the photoemission spectrum, 40–50% in their model.

Veal *et al.* (1989) took EDCs with 250 meV resolution on single crystals of YBa$_2$Cu$_3$O$_x$ with x varying between 6.3 and 6.9. The step at the Fermi edge was very small in the sample with $x = 6.3$. It grew as x increased, and for $x = 6.9$ a Fermi edge that was sharp, considering the resolution, was found. This was the first evidence from photoelectron spectroscopy for a metallic sample of Y123.

The valence bands were studied with 100 meV resolution by Campuzano *et al.* (1990a,b) on twinned single crystals cleaved at 8 K. They used an angle-resolving analyzer (1.4° angle resolution) and sought Fermi-level crossings as the parallel component of wave vector was varied. Their Fermi surface is shown in Fig. 6.11. The calculated Fermi surface (Yu *et al.*, 1987) is shown twice, representing the Fermi surface of a twinned crystal for which Γ–X and Γ–Y spectra appear together. It is clear that there is considerable agreement with the LDA calculation, but the parts of the Fermi surface due to bands derived from states on the Cu–O chains (curves parallel to Γ–X or Γ–Y) were not definitively identified.

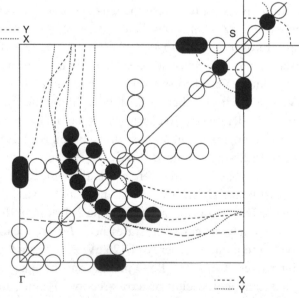

Fig. 6.11 Experimental (Campuzano *et al.*, 1990a) and theoretical (LDA – Yu *et al.*, 1987) Fermi surface for twinned Y123. Open circles represent spectra with no Fermi-surface crossing. Filled symbols represent observed Fermi-surface crossings. Both the experimental points and the calculated bands have been reflected about the diagonal of the zone.

Cu $2p_{3/2}$ spectra from Y123 were measured by several groups (Fujimori et $al.$, 1987; Shen et $al.$, 1987; Gourieux et $al.$, 1988; Steiner et $al.$, 1988), although when the O 1s spectra were presented, there was some strength in the 531-eV component. Some of these authors interpreted their data with the CuO_4 cluster model described in Chapter 5. Seino et $al.$ (1990) pointed out that such a model neglected the Cu atoms in the chains which also contribute to the XPS spectrum. They used an expanded cluster model to calculate the Cu $2p_{3/2}$ photoelectron spectrum and the polarized soft x-ray absorption spectrum from the Cu 2p and O 1s levels, five spectra in all. Their model consisted of the square-planar CuO_4 cluster, an apical O atom above the central Cu, a chain Cu atom above the apical O, and the apical O above the chain Cu. In addition, two chain O atoms were associated with the chain Cu. The chain CuO_4 unit above the CuO_4 cluster has a different geometry and was treated differently. They used an Anderson model for the CuO_4 cluster in the CuO_2 plane and a cluster model for the chain CuO_4 cluster. The former models the dispersion of the O 2p bands in a simple way. The two parts of the model are coupled by hybridization between the planar Cu atom and the apical O above it, although this coupling was considered not very important and was omitted for much of the numerical work. Both $YBa_2Cu_3O_7$ and $YBa_2Cu_3O_{6.5}$ were treated, the latter by omitting one chain O atom. The ground states were calculated, along with suitable excited states for each spectrum. Reasonable agreement for all five spectra could be found with reasonable values of the fitting parameters. However Bi2212 has a peak in its polarized ($E\|c$) O 1s absorption spectrum which this two-cluster model would assign to apical oxygen atoms. Bi2212 has no apical oxygen atoms. It is not clear whether distortions of the square planar CuO_4 cluster can give rise to such a peak.

Weaver et $al.$ (1988) obtained XPS spectra on polycrystalline Y123 and on cleaved single crystals of $RBa_2Cu_3O_x$ with R = Nd and Ga. The rare earths provided their usual intense and characteristic spectra. Some of these spectra, e.g., the 5p between 20 and 30 eV, are rarely used, for they are fairly broad, show some multiplet structure, and are not very sensitive to bonding in the crystal. The 4f spectrum completely (Nd) or partially (Gd) overlaps much of the valence-band spectra in these compounds. The O 1s spectra showed a rather broad main peak at about 529 eV with a second structure at 531 eV, the latter due to contamination. (All measurements were taken at room temperature.) The main peak was decomposed into two peaks. The weaker peak at lower binding energy was assigned to O atoms in the Cu–O chains and the larger component to O atoms in the CuO_2 planes. A third component was needed to fit the spectra, one located between the two-component main peak and the 531-eV peak. This was assigned as a satellite from a final state of an in-plane O with a $3d^{10}\underline{L}$ configuration on a neighboring Cu instead of the $3d^9$ of the main line. In replacing Y by Nd or Gd, the main line components remained unchanged, and the satellite

became slightly more intense than in Y123. However Brundle and Fowler ("ã)
later made the peak assignments differently. The largest peak (528.1 eV) was
assigned to O in the CuO_2 planes, the next largest peak (527.4 eV) to Ba–O
plane oxygen, and the weakest peak (529.2 eV) to Cu–O chain oxygen.

Wagener *et al.* (1990a) made angle-resolved inverse photoemission measurements of single crystals of Y123 cleaved at 60 K. The resolution was 0.3–0.5 eV,
detecting photons in the energy range of 10–23 eV. The angle of incidence of the
electrons could be varied. Varying the incident electron energy varied the value
of $k_\perp$. With normally incident electrons a Fermi edge was observed and structure
in the empty conduction bands was found at 0.5, 1, 1.8, and 3 eV above E_F. It
did not disperse with $k_\perp$. The structure in the 0–2 eV range was assigned to
empty states based on states on the CuO_2 planes, in accord with LDA band calculations. They resonated when the photon energy passed through 18 eV,
approximately the energy of the O 2s photoabsorption. Spectra taken with electrons incident off normal, giving a component of $k_\parallel$ along [100], show some dispersing bands from 0 to 0.5 and from 1.5 eV. These were assigned to Cu–O
chain bands.

6.4 $R_{2-x}Ce_xCuO_4$

$R_{2-x}Ce_xCuO_4$ with R standing for a rare earth, usually Nd, is believed to be electron-doped, i.e., n-type, superconducting cuprate for suitable values of x. With
Nd, T_c peaks at about 24 K for $x = 0.15$. These materials are superconducting
over only a narrow range of x, from 0.14 to 0.18. These cuprates have been studied frequently to see if there is electron–hole symmetry in the cuprates, and
because the magnetic moment on the rare earth, normally destructive to superconductivity, seems to be harmless here. Early work on scraped polycrystalline
samples (Grassmann *et al.*, 1990), which showed a single O 1s spectral peak and
no 9 eV peak below the valence band, gave valence-band spectra with no intensity
at the Fermi level. Resonance studies using the rare earth (Pr, Nd, Sm) 4d resonance showed there was some rare-earth 4f character in the valence-band region,
but beginning 1 eV below E_F and extending to the bottom of the valence band.

Weaver *et al.* (1988) obtained XPS spectra from cleaved single crystals of
R_2CuO_4 with R = Pr, Eu, and Gd. The O 1s spectra all showed a broad main
peak at about 528.5 eV and a peak at 531 eV assigned to contamination. The
main peak was decomposed into two components of nearly equal size, assigned
to O in the CuO_2 planes and to off-plane O, bound to the rare earth. The in-
plane O component was the less-bound peak, not the greater-bound peak as in
Y123. There was a weaker third component, a satellite assigned to a final state
of an in-plane O with a $3d^{10}\underline{L}$ configuration, as described above for Y123. The

rare-earth 4f spectra significantly overlap much of the valence-band spectra in these compounds. In the case of Pr_2CuO_4, Weaver et al. took resonance photoemission spectra at the Pr 4d threshold and were able to show, with better resolution than in XPS spectra, that there were two 4f contributions to the valence-band spectra, at 1.2 and 4.4 eV below E_F. Such double peaks from rare earths have been widely studied, but it will take us too far afield to go into a detailed description of them. Several chapters of a book (Gschneidner et al., 1987) have been devoted to their description and explanation. We are not yet at the stage where an analysis of their line shapes can tell us much about the electronic structure of the cuprates.

Gunnarsson et al. (1990a) measured the spectrum of ceramic samples of Nd_2CuO_4 on and off the resonance at the Cu 3p photoexcitation threshold. The satellites at 9 and 12 eV went through a resonance, indicating the roles of Cu 3d states in these satellites. These authors also used an Anderson impurity model to calculate the expected valence-band spectrum on and off resonance, including the effects of photon polarization. One result of the calculation, not found in the measured spectra, was a well-resolved peak just below E_F due to a singlet final state, sometime called the "Zhang–Rice" singlet. A similar peak was calculated by the same model for CuO, and such a peak appears in the spectrum. Except for this peak, the calculated valence-band spectra for CuO and Nd_2CuO_4 are similar. Gunnarsson et al. (1990b) discussed this discrepancy, for the CuO_4 units should be very similar in the two materials. Clearly the differences in the more distant atoms play a role, one that is not adequately described by changes in parameters in the Anderson impurity model.

Sakisaka et al. (1990a,b) took angle-resolved spectra with 100–200meV resolution in epitaxial films of $Nd_{1.83}Ce_{0.17}CuO_4$. The films were annealed in O_2 in the vacuum chamber. There was a small peak at 9.7 eV the authors suggest might have been from C contamination although, in retrospect, it might be assigned to the 9 eV peak frequently seen in cuprates. A clear Fermi edge and six dispersing bands could be seen. The peak closest to E_F showed a resonance at the Cu 3p excitation threshold, indicating considerable Cu 3d character at or near E_F. Its dispersion had a shape roughly agreeing with that of a band in an LDA calculation, but the energy changes were about a factor of ten smaller than in the calculation. It appeared not to cross the Fermi level in the directions they scanned. They ruled out much Cu 4s and O 2p character to this band by comparing the dependence of its intensity on photon energy with that expected from calculations (Yeh and Lindau, 1985). A Fermi edge was also observed in XPS by Suzuki et al. (1990) and by Vasquez et al. (1991) Suzuki et al. also showed that scraping a surface initially produced by breaking caused the XPS core-level spectra to deteriorate. Vasquez et al. etched epitaxial NCCO films with HCl and were able to produce a surface that showed only a trace of a high binding-energy O 1s component. Samples with non-zero Ce concentrations showed a Fermi edge in XPS that grew with increasing Ce concentration.

Namatame *et al.* (1990) studied polycrystalline samples of $Nd_{2-x}Ce_x CuO_4$ with $x = 0.05$ to 0.20 by angle-integrated photoemission using synchrotron radiation. The samples were fully oxygenated and not superconducting. The surfaces were scraped in vacuum at about 80 K. The resonances from Cu 3p, Nd 4d, and Ce 4d states were used to determine valence-band wave-function character. The valence band extends from about 1 to 8 eV binding energy with a satellite at 13.3 eV. The satellite shows a resonance at the Cu 3p photoexcitation threshold. A second satellite, at 15.7 eV, shows a resonance at slightly higher photon energy. The main valence band did not show a Cu resonance. The Nd 4d resonance revealed Nd 4f states throughout the entire conduction band, while the corresponding Ce resonance showed Ce 4f states to be localized in the uppermost 3 or 4 eV of the valence band, the width indicating considerable hybridization with O 2p states. By a comparison with $La_{2-x}Sr_xCuO_4$ they were able to show that the Fermi level of their samples was located in the band gap of the insulating samples.

Fujimori *et al.* (1990) used similar samples, but annealed in O_2 and Ar or CO_2 to reduce them to samples which were superconducting. They used He I, He II and Al-Kα lines to study sample surfaces scraped at about 80 K. They used only data from surfaces that showed a single O 1s peak at 529 eV binding energy and no 9–10 eV "valence band" peak. Core-level spectra showed not only that Nd is indeed replaced by Ce but that the valences were, in fact, Nd^{3+} and Ce^{4+}. Despite superconductivity, there was no photoemission strength from the Fermi level. The photoelectron spectra began at about 1 eV binding energy. The conduction electrons introduced by replacing up to 10% of the Nd by Ce did not appear in the spectra, nor did they cause a detectable shift in the Fermi level possibly because of sample conditions at the surface. Doping with Ce did, however, cause a decrease in the intensity ratio of the Cu $2p_{3/2}$ satellite to the main $2p_{3/2}$ line, at a rate consistent with that expected if all electrons introduced by doping were in Cu 3d states, although there was no evidence of a new, shifted Cu 2p core-level peak from a Cu $3d^{10}$ configuration.

Hwu *et al.* (1990) reported the observation of such a shifted Cu 2p core-level peak as Ce was added to NCCO. In EDCs a weak satellite around 16 eV has been assigned as a satellite on the Cu^{2+} ion in a cluster model. The corresponding satellite on a Cu^+ ion in such a cluster is not visible in EDCs. CIS spectra taken on a series of Ce-doped NCCO polycrystalline samples showed a clear double-peaked resonance for a satellite with 17.6-eV binding energy in all samples, including $x = 0$, and a second double-peaked resonance for an initial state energy of 20 eV for those samples with x in the range 0.05 to 0.20. The latter was assigned to the Cu^+ initial state. Suzuki *et al.* (1990) also determined in increase in the fraction of Cu^+ sites with Ce doping by the decrease in the relative intensity of the satellite on the Cu $2p_{3/2}$ line with Ce doping.

Reihl *et al.* (1990) did observe a shift in the Fermi level with Ce doping of NCCO. They used a range of polycrystalline samples that had been reduced. They used photoemission with the He I and II lines and inverse photoemission ($\hbar\omega = 9.5$ eV) on fractured surfaces. There were significant peaks at 10 eV, peaks generally associated with other oxide phases. As the Ce concentration was increased, the main peak shifted slightly to larger binding energy, interpreted as a shift of the Fermi level due to electron addition. The shift was 0.3 eV for $x = 0.2$. There was no intensity at E_F in either type of spectrum.

Allen *et al.* (1990) carried out angle-integrated resonance photoemission studies on polycrystalline samples of $Nd_{2-x}Ce_xCuO_{4-y}$ with $x = 0$ and 0.15 and a single crystal with $x = 0.15$. The single crystal was cleaved and measured at 20 K. EDCs were taken for various photon energies around the Cu 3p and Nd 4d resonances of the ceramic samples, after showing that EDCs of the ceramic sample with $x = 0.15$ were nearly identical in shape to those of the single crystal. The 12 eV valence-band satellite showed a clear resonance at the Cu 3p threshold. The entire valence band showed a weak resonance around the Nd 4d threshold. For $x = 0$, there was no Fermi edge; the EDC began its rise at about 0.5 eV binding energy. This region gained a small intensity when x was 0.15, a metallic (and superconducting) sample. A comparison of the EDC for $x = 0.15$ sample with that of $La_{1.85}Sr_{0.15}CuO_{4-y}$ after aligning the energies of the main peak showed that both had the same Fermi level. The EDCs of the two ceramic samples were normalized (by a method of debatable validity) to obtain the EDC of the additional electrons introduced by Ce doping. Then the magnitude of this was plotted as a function of photon energy. The plot varies with photon energy as expected from the Cu 3d and O 2p photoexcitation cross-sections. It also showed a small resonance at the Cu 3p threshold, but no resonance at the Nd 4d threshold. (States further into the valence band show such a Nd resonance.) The authors conclude that the electrons introduced by Ce doping occupy new states in the charge-transfer gap above the top of the valence band of the insulating sample and that E_F is the same for both electron and hole doping. The newly occupied is region about 0.5 eV wide (as determined with 0.35 eV resolution).

6.5   Other cuprates

A Fermi edge was detected by photoemission in $Y_{1-x}Ca_xBa_2Cu_4O_y$ by Itti *et al.* (1990a) using the He I line and polycrystalline samples. This cuprate is similar to Y123 except that there are two sets of Cu–O chains. It is much more stable. As x

increased from 0 to 0.1 the size of the Fermi edge diminished. Under the same experimental conditions no Fermi edge was seen in Y123.

$Tl_2Ba_2Ca_2Cu_3O_{10}$ was studied by Meyer *et al.* (1989) via XPS and inverse photoemission spectroscopy. The samples were both sintered ceramics and single crystals, both fractured and measured at 300 K in ultrahigh vacuum. The XPS spectra were taken with a resolution of 0.6 eV. The valence-band XPS spectrum consisted of a single broad peak with a shoulder, together about as broad as that expected from LDA band calculations (Marksteiner *et al.*, 1989), but its peak was shifted by about 1.5 eV to higher binding energy. Meyer *et al.* stated there was evidence for intensity at the Fermi level. However, there was a strong peak at 531 eV in the O 1s core-level spectrum, with perhaps half the intensity of the 529-eV main line. The inverse photoemission spectrum identified several empty states in the first 16 eV above E_F, each of which could be assigned to a specific subshell on one of the constituent atoms. The spectrum of emitted photons was dispersed by a monochromator and detected with a position-sensitive detector. At a photon energy of about 18 eV, the (controversial) threshold for O 2s excitation, there is a resonance in the emitted photons. This resonance was apparent for electron final states in the range 0–4 eV above E_F, signifying O 2p character there.

Work on $La_{2-x}Sr_xCuO_4$, the original high-temperature superconductor (although Ba was used instead of Sr), began early, always on sintered polycrystalline samples. For example valence-band spectra taken by Shen *et al.* (1987) were on surfaces that showed a composite peak with a high binding-energy component for the O 1s spectrum. Surfaces scraped in vacuum at room temperature often gave spectra that changed with time. Later work with single crystals cleaved at a low temperature in ultrahigh vacuum showed that even under these conditions the surfaces were not stable. For example Mikheeva *et al.* (1993) cleaved single crystals of La_2CuO_{4+x} at 30 K and kept the surfaces at this temperature while measuring valence-band spectra. They obtained angle-resolved EDCs with 120 meV resolution and observed a number of broad overlapping peaks in the valence-band region, 0–6 eV binding energy. After a period of 2–3 hours at 30 K a new broad peak appeared at about 5 eV binding energy. Its contribution to the EDC increased when the collection angle was moved away from the normal, implying a surface origin for this feature. Common surface condensates, water and CO, produce new peaks of known binding energies, not 5 eV. Mikheeva *et al.* suggested a possible surface reconstruction, even at such a low temperature.

6.6   Summary

We have arbitrarily limited the discussion in this chapter to work published before 1991, with a few exceptions. Angle-integrated photoemission spectra and inverse photoemission spectra were compared with the density of states from LDA calculations with some success. Some of the differences were intrinsic and some were caused by problems with the sample surfaces. The resolution usually was about 100 meV at best. The spectra showed a number of broad peaks, but far fewer than might be expected from band calculations. Since approximately 1990 there have been improvements in the quality of the crystals grown of many of the cuprates. This, and improvements in resolution (or simply the use of higher resolution which was available earlier) resulted in spectra that revealed a great deal more about the cuprates, spectra which made photoelectron spectroscopy one of the premier techniques for the study of the electronic structure of the cuprates. These high-resolution spectra will be the subject of Chapters 7 and 8. They cover only the first 1 eV or so below E_F. Higher resolution does not benefit spectra at greater binding energies in the case of the cuprates. The spectra are inherently broad, presumably because of very short photohole lifetimes.

Many of the early XPS spectra have yet to be improved upon if they were taken on samples showing a Fermi edge. Resolution improvements have not been great, and XPS is less sensitive to the quality of the sample surface than is angle-resolved ultraviolet photoelectron spectroscopy. Similarly, there have been few measurements of inverse photoelectron spectra of the cuprates more recent than about 1990.

Chapter 7

Bi2212 and other Bi-cuprates

Once at least a few points on the Fermi surface of Bi2212 had been mapped by photoemission, there was considerable activity to complete the Fermi surface mapping and map the bands just below E_F. Further studies of the superconducting order parameter were also carried out. Fortunately at about this time, 1990 and later, better single crystals became available. In the following we describe the photoemission studies made to date on Bi2212. We begin with the normal state, then discuss the superconducting state. Unless stated otherwise all work to be described in these sections was carried out on crystals near optimal doping. The third section in this chapter deals with variations in doping. There have been reviews of the study of the valence bands of Bi2212 and related compounds by photoelectron spectroscopy: Loeser *et al.* (1994) and Shen and Dessau (1995).

As mentioned previously, Bi2212 has an orthorhombic structure but it is frequently approximated by a tetragonal one. The unit vectors in the first Brillouin zone for the tetragonal lattice are rotated 45° with respect to those of the orthorhombic structure. The Γ–X axis is parallel to the Cu–O bond in the Brillouin zone for the orthorhombic structure (see Fig. 2.10(d)), while the Γ–X axis is at 45° to the Cu–O bonds for the tetragonal structure (see Fig 2.10(c)). In the latter case, the Γ–$\overline{\mathrm{M}}$ line is parallel to the bond. It has become fashionable in recent years to drop the symmetry notation in labeling the tetragonal Brillouin zone and use the points $(0, 0)$ (Γ), $(\pi, 0)$ $(\overline{\mathrm{M}})$, $(0, \pi)$ $(\overline{\mathrm{M}})$, and (π,π) (Y) for the corners of one quarter of the zone, with the bond axis parallel to $(0, 0)$–$(\pi, 0)$, (see Fig. 2.10(c)).

Additional conceptual problems may be avoided if the reader also remembers that for the Brillouin zone for the tetragonal lattice, moving away from Γ on the

243

k_x-k_y plane brings one onto the top surface of the next Brillouin zone in the extended-zone scheme, not the central plane (see Fig. 2.7). The extended-zone scheme is frequently used, as is the smallest unit needed to present data, the irreducible wedge of the first Brillouin zone. Also, the Γ point may not always be placed at the center of a figure.

If the material is considered two-dimensional, the three-dimensional Brillouin zone collapses into a two-dimensional one. For the tetragonal and orthorhombic lattices, the point at the center of the top face, Z, becomes equivalent to the point at the center of the median plane, Γ. The orthorhombic median plane is a rectangle with the corners cut off, but in two dimensions, the Brillouin zone is a rectangle with its corners intact. The symbols for symmetry points in a surface Brillouin zone often are the same letters used for symmetry points in the Brillouin zone for the bulk, but written with a bar over them.

7.1   Bi2212 above T_c

7.1.1   Valence bands

After the initial mapping of several points on the Fermi surface, much more extensive mapping was carried out. These measurements required a large number of scans, for even though the crystal orientation was determined by x-ray diffraction, and perhaps *in-situ* electron diffraction, the direction of the sample normal with respect to the analyzer goniometer axes was known only to within a few degrees. Test scans, looking for extrema and symmetry about the normal were required, as were intermittent repetitions of scans in order to check for sample deterioration. Intermittent scans of the Fermi edge of a reference metal also were required.

Bi2212 has several complications. One is the orthorhombic structure. The distortion along the b-axis lowers the symmetry from tetragonal to orthorhombic. If the distortion were small enough, its effects might be treated as a perturbation on the electronic structure of the tetragonal crystal, but it does not appear that it is small enough. Oxygen doping of Bi2212 occurs by adding or removing oxygen atoms from the pairs of Bi–O layers (Pham *et al.*, 1993). This adds two complications. One is that the displacements of the Bi atoms, important in the superlattice, may change. The other is that the states near the $\overline{\text{M}}$ point have some amplitude derived from Bi–O states on the planes. One could hope that the rigid-band model might be valid qualitatively for the valence-band states derived from Cu–O states upon doping, but not those based on the Bi–O planes whose composition and charge change with doping.

Shen *et al.* (1993a) and Dessau *et al.* (1993a) carried out extensive band mapping near the Fermi level of Bi2212. Their resolution was 45 meV, reduced in order to increase the count rate for taking a large number of spectra, with an angular resolution of 1°. Several sets of their spectra are shown in Fig. 7.1. Upon increasing the collection angle, the peak approaching the Fermi energy sharpens, then drops as the initial state crosses the Fermi level. The crossing is assumed to occur when the peak above background falls to one-half its height. This is not always an easy criterion to apply, because the falloff with angle may not be rapid, and there is not an accepted shape for the background. (One method of subtracting the background is to subtract a spectrum taken with the initial state above E_F, with the expectation that this spectrum consists only of background. This spectrum, just a step rising at E_F, may have to be re-normalized. As with all other background subtraction schemes for cuprates, it has no justification in terms of microscopic processes.) The Fermi-level crossing points are shown in Fig. 7.2(a) as solid circles. Shaded circles indicate spectra in which the initial state is very near E_F, indistinguishable from it at the resolution used. This indicates that there is a flat band near the Fermi level. Figure 7.1 (c) is a scan along such a flat band. Note that along Γ–X and Γ–Y there is only one band crossing the Fermi level, but just off these lines, there are two crossings, one from each part of the Fermi surface. Two crossings are expected because each of the two CuO_2 planes per unit cell provides an antibonding hybrid Cu $3d_{x^2-y^2}$–O $2p_x,2p_y$ band. The degeneracy of these bands is lifted by any interaction between them, resulting in a bonding band and an antibonding band. The splitting along Γ–X, if any, is not resolved. We return to this shortly.

There is an apparent lack of symmetry in Fig. 7.2(a). The points along the horizontal line Γ–$\overline{M}$–Z are not the same as along the vertical Γ–$\overline{M}$–Z line. However, in the experiment these lines are not equivalent, for the radiation was polarized. The electric field was parallel to **k** along the horizontal line and

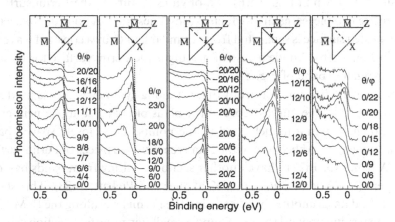

Fig. 7.1 Angle-resolved EDCs from Bi2212 taken at 100 K on a crystal with $T_c = 85$ K. The scans were taken for several values of $\mathbf{k}_\parallel$ along symmetry lines as indicated. (θ and ϕ are emission angles from the sample normal.) (Dessau *et al.*, 1993a)

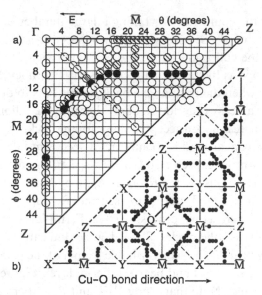

Fig. 7.2 (a) Bi2212 Fermi surface determined from the spectra of Fig. 7.1 and additional similar data. Filled circles represent Fermi-surface crossings in the spectra. Shaded circles represent spectra with peaks near E_F, but which were more difficult to identify as crossing E_F from nearby scans. (b) Fermi surface in the extended-zone scheme obtained from (a) by using symmetry to obtain additional points. (Dessau *et al.*, 1993a)

perpendicular to **k** when the vertical line was scanned. Spectra taken after rotation the crystal by 90° about its normal verify this. This is a result of very strong polarization selection rules. Relative heights of peaks may be very polarization sensitive, as above, and also change considerably as the photon energy is varied.

The points of Fig. 7.2(a) have been mapped into the extended-zone scheme Fermi surface shown in Fig. 7.2(b). An obvious feature of the Fermi surface is that there are two pieces, each of which exhibit "nesting." A considerable length of such a nested surface is separated from a similar length by a constant wave vector, **Q**, in Fig. 7.2(b). An ideal case of nesting is the Fermi surface for the half-filled tight-binding band of Fig. 2.8(d).

All of the band information from Fig. 7.1 along symmetry lines is plotted as points in Fig. 7.3. The curves represent a plausible band, but are neither the LDA bands nor from a model fitted to the points. Two closely spaced bands were drawn along X–$\overline{M}$ because of the occurrence of two pieces of Fermi surface. The LDA bands for Bi2212 have another section resulting from bands based on Bi–O layers (Massidda *et al.*, 1988; Krakauer and Pickett, 1988). This should appear in the data as another band crossing the Fermi level along the Γ–$\overline{M}$ direction, but dispersing upward toward Γ, and a small Fermi surface section centered on $\overline{M}$. The band observed disperses downward toward Γ and is based on states from the CuO$_2$ planes. The missing piece of Fermi surface may not be present

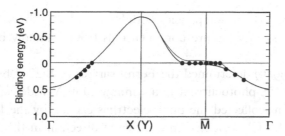

Fig. 7.3 Experimental bands obtained from the data of Fig. 7.1 (solid circles) and a possible band passing through them. (Dessau *et al.*, 1993a)

because the Bi–O based band may actually lie above E_F or the dipole matrix elements for photoemission from it may be too weak. The points and the curves in Fig. 7.3 are believed to represent states derived from states in the CuO_2 planes, with some mixing in of Bi–O character. A rather flat band along X–$\overline{M}$ is found in LDA calculations, but the measured one is much flatter, usually taken as a sign of correlation effects. This flat region turns out to be a cut through an extended van Hove critical point, a line of saddle points. Such an extended critical point was first found in Y123 by Gofron *et al.* (1993), but it has been found in several classes of cuprates, always just below E_F in optimally doped p-type cuprates, and further below E_F in n-type cuprates. The LDA-calculated bands tend to have a single critical point along this line, or a few such points, related by symmetry, just off this line, rather than a line of critical points, the extended critical point. This flat band will be discussed further in the next chapter.

Ma *et al.* (1995b) mapped the region of the extended critical point in more detail, scanning along Γ to $\overline{M}$ then continuing into the next zone. The crystal used was annealed in oxygen, and hence was overdoped. They found a region of zero or little dispersion (20 meV or less) along this line. An unexpected result is that it begins near $\overline{M}$ and continues a large distance, about 40% of the distance from $\overline{M}$ to Z. Thus it is not symmetric about the $\overline{M}$ point as expected for a two-dimensional Brillouin zone. They made four scans along lines in reciprocal space passing through the Γ–$\overline{M}$–Z line, but perpendicular to it. The scan along X–$\overline{M}$–Y was asymmetric. The band rose away from the Γ–$\overline{M}$ line toward X, crossing the Fermi level near $\overline{M}$, while it fell monotonically toward Y, then rose rapidly to cross the Fermi level much further from $\overline{M}$ than the crossing toward X. The other cuts through reciprocal space were made parallel to the Γ–$\overline{M}$ line and they all showed a minimum on this line, dispersing upward away from it, but asymmetrically about the Γ–$\overline{M}$ line. This dispersion is that expected for a line of saddle points. The asymmetry observed is consistent with the symmetry of the orthorhombic lattice. It may or may not be enhanced by the superlattice. Ma *et al.* report several discrepancies with the spectra of Dessau *et al.*, discussed above, the origins of which, beyond the difference in oxygen stoichiometry, are not clear. There are also discrepancies with later work by others. Sample misalignment is a possibility, as are complications from "shadow bands," to be described next.

Aebi *et al.* (1994b) determined the Fermi surface of Bi2212 by taking 6000 measurements of the photocurrent in a window 10 meV wide centered at E_F. Each measurement collected the photoelectrons excited by the He I line in a cone of full width of 1° oriented in a different direction in the 2π steradians above the sample. Background subtraction is very important here. The resultant currents were presented as a gray-scale display on a stereogram. As a check, no use was made of symmetry, so the symmetry in the data was measured, not assumed. The samples were cleaved single crystals, and once a good cleave had been obtained, judged by the measurement of dispersing bands, it remained stable for several weeks. Figure 7.4 shows their result. The most intense features correspond closely to the Fermi surface as calculated by the LDA (Massidda *et al.*, 1988). Agreement with the Fermi surface of Dessau *et al.* (1993a) is not as good, perhaps because of the coarser mesh in **k**-space used in the previous work and the difficulty in determining Fermi surface crossings in the conventional angle-resolved energy distribution curves. The discrepancies with LDA are in the vicinity of the $\overline{M}$ point, a region of flat bands. The weaker features discovered by Aebi *et al.* are now called shadow bands. They appear to be replicas of the Fermi surface displaced along k_x by the Γ–X wave vector or along k_y by the Γ–Y wave vector. These wave vectors, $\mathbf{Q} = (\pi/a, \pi/a)$, are the wave vectors of the

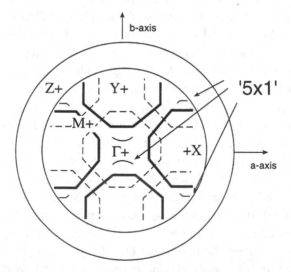

Fig. 7.4 Stereographic projection of lines in reciprocal space at which photocurrents with initial states at or near E_F were found in Bi2212 at room temperature. The heavy lines mark the regions of largest photocurrent. The dashed lines are the shadow bands. The thin lines marked "5 × 1" are caused by the superlattice. (Aebi *et al.*, 1994b)

antiferromagnetic superlattice in the insulating parent cuprate. Aebi *et al.* measured the dispersion of one band giving a normal Fermi-surface crossing, and one of the shadow bands, with the result that the direction of the former dispersion agreed with the LDA prediction, and the sense of the shadow band dispersion with that of an LDA band translated by a vector in **k**-space. They proposed this could be due to a c(2 × 2) superlattice, such as could be produced by (local, dynamic) antiferromagnetic ordering of the Cu spins. (There is no long-range antiferromagnetic order at the temperature of the measurements, 300 K.) Such ordering and, indeed, shadow bands, had been predicted earlier by Kampf and Schrieffer (1990). (See also Schrieffer and Kampf, 1995). The static antiferromagnetism of the square CuO_2 plane with a half-filled valence band has unit vectors twice as large as those of the lattice unit cell because only alternate Cu atoms are equivalent when the spins order. The Brillouin zone dimensions then are half as large, and in folding the electronic structure into the new first zone, new bands appear. These bands will also be induced by antiferromagnetic fluctuations in lattices that are not statically antiferromagnetic. Their strength in photoemission spectra could be high if the fluctuations are large and the coupling to them is strong. The picture of Kampf and Schrieffer was based on the possible coupling by strong antiferromagnetic fluctuations of states **k** and **k** − **Q**, the wave vector characterizing the periodicity of the antiferromagnetic fluctuation, where $\mathbf{Q} = (\pi/a, \pi/a)$ and $|\mathbf{k} - \mathbf{Q}| < \xi^{-1}$, where ξ is the antiferromagnetic fluctuation correlation length. This coupling remaps bands by **Q** into the first Brillouin zone, as just mentioned. The explanation for shadow bands suggested by Aebi *et al.* in terms of antiferromagnetic correlations was criticized (Chakravarty, 1995), partly on the grounds that the spin fluctuations would be too energetic. A surface reconstruction was suggested as a possible alternative. Aebi *et al.* responded to this, (Aebi *et al.*, 1995a) but there was no clear resolution of the question. In addition to the shadow bands, there were some rather faint bands crossing the Fermi level that exhibited two-fold rather than near-four-fold symmetry. These are due to the well-known superlattice along the *b*-axis, the so-called 5 × 1 reconstruction. We return to this reconstruction later.

Haas *et al.* (1995) analyzed the Kampf–Schrieffer model with a one-band Hubbard Hamiltonian and a *t*–*J* Hamiltonian for a small two-dimensional cluster. They calculated the spectral density function for several values of band filling and coupling strength, U/t. Because the clusters were so small, only a few wave vectors in the Brillouin zone could be sampled. They looked for structure in the spectral density function just below E_F when k was larger than k_F. They found such a peak with a strength that could be measured for underdoping with holes (but not for overdoping) when the coupling was strong; $U/t = 8$ and $J/t = 0.4$ were used. The spin correlation lengths did not have to be long. Haas *et al.* suggest that the peaks just discussed may be the shadow band peaks, but they point out that their calculations show these peaks to be weaker near one of the conditions of the experiments, near-optimal doping.

Other explanations of the shadow bands have been proposed. Langer *et al.* (1995) pointed out that the model of Kampf and Schrieffer used an assumed spin susceptibility that led to a long magnetic correlation length, longer than obtained from neutron scattering measurements. Langer *et al.* used a new numerical method for calculating the self-energy for a one-band Hubbard Hamiltonian for several doping levels. The spectral density function had small peaks dispersing through E_F in addition to the main peaks dispersing though E_F. The small peaks were separated by $\mathbf{Q}$ from the main peaks and were identified as the shadow bands. They result from the coupling of states $\mathbf{k}$ and $\mathbf{k} - \mathbf{Q}$ by antiferromagnetic fluctuations. The coupling is large if the $\mathbf{Q}$ component of the magnetic susceptibility is large. They also calculated ξ and obtained a small value, in agreement with experiment. With smaller coupling, the shadow bands would be weaker, but they become more prominent because of a large phase space, enhanced by having flat bands. (See also Haas *et al.*, 1995; Dahm *et al.*, 1996; Vilk, 1997; and Monthoux 1997.)

Salkola *et al.* (1996a,b) made use of the recent discovery of charged metallic stripes alternating with insulating antiferromagnetic stripes found for $La_{1.48}Nd_{0.4}Sr_{0.12}CuO_4$ (Tranquada *et al.*, 1996). They calculated the spectral density function for single-particle excitations of an ordered array of such stripes, and for a disordered array of them. The result for the disordered array was in reasonable agreement with the Fermi surface of Bi2212 as determined by Aebi *et al.* (1994), a region of flat bands near E_F and weaker shadow bands. (To date, stripes have not been found in all cuprates, and whether or not they will exist on the proper time scale, if at all, in Bi2212 at 300 K is not yet known.) Salkola *et al.* began with a one-particle Hamiltonian,

$$H = -t(c_{l\sigma}^\dagger c_{l\sigma} + \text{H.c.}) + V_\sigma(\mathbf{r})c_{l\sigma}^\dagger c_{l\sigma}.$$

The phenomenological effective potential, $V_\sigma(\mathbf{r})$, had two terms, one periodic in the spacing of the charged stripes and a spin-dependent part with the period of the antiferromagnetic domains, half that of the charged stripes. This potential has the effect of folding bands from higher Brillouin zones into the first, and producing splittings where the folding introduces band crossings. Parameters were chosen to give stripes of the width found by Tranquada *et al.* The calculated spectral density function integrated over an energy window of width $t/30$ about E_F is shown in Fig. 7.5(a). Without the potential simulating the effect of the stripes the Fermi surface would be a closed surface about the Γ point, one exhibiting four-fold rotational symmetry. With this potential, the rotational symmetry is two-fold and parts of the closed Fermi surface are gone. Also new features appear, including a weak structure identified with the shadow bands, but only

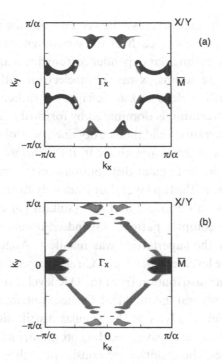

Fig. 7.5 Calculated spectral density function in the first Brillouin zone integrated over a window about E_Γ. (a) Result for an ordered array of vertical stripes. The density of shading indicates the relative magnitude of the spectral density function, $A(k_\Gamma, E_\Gamma)$. (b) Result for a disordered array of vertical stripes, and finite temperature, $k_B T = t/10$. (Salkola *et al.*, 1996a)

for an idealized single-domain crystal. They next assumed the stripes are disordered because the charged stripes are likely to be pinned at defects. This was simulated by using random stripe widths, and averaging the results of five calculations. The result, still for a single domain, neglecting the effect of other domains oriented at right angles, is shown in Fig. 7.5(b). Note that the closed Fermi surface about Γ reappears, resembling the measured Fermi surface of Bi2212. The dispersion relation for this band has a large flat region near the $\overline{M}$ point, again in agreement with the extended line of critical points found in angle-resolved photoemission (see below). The shadow band does not appear in Fig. 7.5(b), because it has been washed out in the averaging, but it appears if a logarithmic scale is used, again in qualitative agreement with the results of Aebi *et al.* Salkola *et al.* point out that the stripe domains running in two orthogonal directions break only the reflection symmetry through planes at 45° to the Cu–O bond, but not that for planes parallel or perpendicular to these bonds. The electric dipole selection rules are retained.

Osterwalder *et al.* (1995) later reported a much more extensive study by the same technique. In addition to the Fermi-surface study, they used the Mg-Kα

and Si-Kα lines to excite core levels so they could study the angular distribution
of electrons from core levels of each of the constituent atoms in Bi2212. The
core-level spectra show an angular dependence from diffraction of the photoelectrons from neighboring atoms, x-ray photoelectron diffraction, mentioned
briefly in Chapter 3. If the electron kinetic energy is rather high, above 800 eV
or so, the angular distribution is dominated by forward scattering from individual atoms. The stereogram should then be obtained by projection from the emitting atom of the atoms in the planes above it. If the surface left by cleavage is a
Bi–O plane, this effect in the angular distribution of electrons from Bi core levels
taken will not occur from the top layer, but from only the next-deepest Bi layer.
The latter has many atoms above it and the resultant pattern is expected to be
very complex. Such a complex pattern was indeed observed, and the two-fold
symmetry induced by the superlattice was manifest. Angular distributions of
electrons from the core levels of Ca, Sr, and Cu also showed a two-fold rotational
symmetry. The angular distribution from the O K level showed nearly four-fold
symmetry. There are several inequivalent O sites contributing incoherently to
this pattern. A complete analysis of the angular distributions was not carried
out. This would require an extensive modeling program, including multiple-scattering effects. However, the results were consistent with what is known about
the bulk crystal structure of Bi2212. The Fermi-surface map reported by
Osterwalder *et al.* (1995) is essentially that of Aebi *et al.* (1994b), Fig. 7.4. The
lines due to the 5 × 1 superlattice were added. In later work (Aebi *et al.*, 1995b;
Schwaller *et al.*, 1955) they extended their study to Pb-doped Bi2212. This sample had a Pb:Bi ratio of 0.24:1, a T_c of 83 K and, most importantly, no 5 × 1
superlattice. Both the Fermi surface and shadow bands were present, but the
structure previously assigned to the superlattice was not present, as expected. A
complicating factor was that the Pb-doped crystal exhibited its own superlattice,
weak diffraction spots with the same symmetry as those from ordered spins.
More work on the origin of the shadow bands is clearly of importance. Several
other interesting results were obtained by Schwaller *et al.* Bi $4f_{5/2}$ electrons
were used to produce stereographic projections of Bi2212 and the Pb-doped sample, and Pb $4f_{7/2}$ photelectrons used to produce stereograms of the latter crystal.
All three stereograms were nearly identical, reinforcing the conclusion from neutron scattering results that Pb substitutes for Bi. They also observed that the Pb-
doped samples did not cleave as well as the pure Bi2212 crystals.

The loss of the superlattice bands upon suitable replacement of some Bi by Pb
may not always occur. Clearly in the work of Aebi *et al.* it did occur. Chen *et al.*
(1992) made a study of the effects of annealing on the structure of several Pb-
doped Bi2212 crystals with a Pb:Bi ratio of 0.16. Annealing in oxygen usually
produced the 5 × 1 superlattice, ascribed to distortion in the Bi–O planes due to
interatomic-spacing mismatch with the other planes. Annealing in nitrogen

caused it to disappear in one crystal. Both samples had a weaker superlattice distortion along the b-axis, one with a period longer than $5b$, seen only in Pb-doped Bi2212. However, one sample, that with the least amount of oxygen, displayed neither superlattice. Chen et al. showed that although the structures changed with annealing and O doping, over the range of parameters varied T_c did not change appreciably. The lattice modulation by Pb appears to arise from compositional modulation.

Pb-doped Bi2212 was studied further by Ma et al. (1995c), using angle-resolved photoemission, as well as resistivity measurements and transmission electron microscopy. The Pb:Bi ratio was 1:4. Their oxygen-annealed single crystals showed a twinned orthorhombic structure and a Pb-type superlattice with an unusually long period, about $13b$, but no usual Bi-type superlattice. However, the symmetry of the orthorhombic Pb-doped Bi2212 (2mm) appeared to be lower than that of the orthorhombic undoped Bi2212 (mmm), but the actual crystal structure is not known. This different symmetry has not been reported by any other group working with Pb-doped crystals. The crystals also showed twinning. Annealing in He produced the same structures as annealing in oxygen. The Fermi-surface crossings were sought at 160 points in the first Brillouin zone, using 21 eV photons. The Fermi surface exhibited two-fold rotational symmetry indicative of an orthorhombic, not tetragonal, lattice. The Fermi surface tendency toward nesting was visible, but nesting was far from complete in the overdoped ($T_c = 75$ K) crystal. Parts of a hole pocket around the $\overline{M}$ point were found in the Pb-doped, O_2-annealed sample, but not in the undoped sample and not in a crystal annealed in He. This pocket was detected by Olson et al. (1990b), but not by Dessau et al. (1993a). The difference appears to be caused by different oxygen stoichiometries. They reported that there was no clear evidence of shadow bands crossing the Fermi level.

This work also examined the symmetry of some of the initial states as a function of doping, using the mirror-plane symmetry selection rules of Hermanson (1977). The results were unexpected and difficult to explain. Such measurements bear repeating. Angle-resolved EDCs were taken at a few points along the lines Γ–X, Γ–Y, and Γ–$\overline{M}$–Z, using radiation polarized parallel and perpendicular to $\mathbf{k}$. The initial states that were even or odd with respect to reflection in the mirror plane could then be identified. Along Γ–$\overline{M}$–Z both O_2-annealed and He-annealed samples gave the same result: the initial states just below E_F were even with respect to reflection in the mirror plane. Along Γ–X and Γ–Y the symmetries of the states just below E_F were not the same for the two types of samples. One conclusion reached was that along Γ–X for the O_2-annealed crystal, the initial state had either p_x or d_{xz} symmetry, not the anticipated symmetry of $d_{x^2-y^2}$ states. This unexpected conclusion had been reported earlier for Bi2212 without Pb by some of the same authors (Kelley et al., 1993b), but it has not been reported by any other group. This is a difficult region in which to work, for the band that does, or does not, cross the Fermi level disperses very little.

Ding *et al.* (1996a) took angle-resolved photoemission spectra after the work of Aebi *et al.* (1994) appeared, with the aim of clarifying which parts of the spectra were compatible with single-particle band theory. The Fermi-surface crossings were determined by the criterion that the integrated area of the background-subtracted EDC was one-half its maximum value at k_F (see next section). The Fermi-level crossings were mapped in the Brillouin zone. They fell into three categories: (i) those close to the LDA Fermi surface, (ii) those that could be obtained by translating that Fermi surface in one direction by the superlattice wave vector, and (iii) shadow-band Fermi-surface crossings which can be located by translating the LDA bands by a reciprocal lattice vector at 45° to the superlattice direction. These are shown in Fig. 7.6(a). The features arising from the superlattice could have two origins. They could be an intrinsic part of the

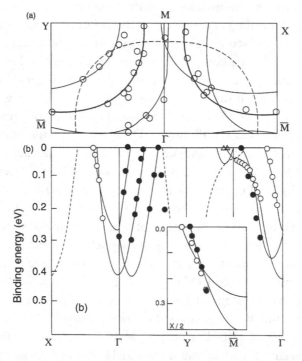

Fig. 7.6   (a) Fermi-surface crossing points in Bi2212 (circles) and (i) Fermi surface from a tight-binding fit to the dispersing bands (thick solid line); (ii) a Fermi surface produced by umklapp processes with a reciprocal lattice vector derived from the superlattice $(0.21\pi/a, 0.21\pi/a)$ (thin solid line); and (iii) a Fermi surface produced by umklapp with a half-Brillouin zone wave vector $(\pi/a,\pi/a)$ to produce shadow bands (dashed line). (b) Band dispersion below E_F for Bi2212. Filled circles: initial states odd with respect to the corresponding mirror plane; open circles: even states; and triangles: initial states with unknown parity due to the use of mixed polarization. (Ding *et al.*, 1996a)

bulk electronic structure or they could be the result of a surface umklapp process upon escape. Ding *et al.* favor the latter. Upon changing the photon energy from 19 to 22 eV, the shadow bands disappeared from the spectra, presumably from dipole matrix element effects, although they were observed by Aebi *et al.* with 21.2 eV photons, which, however, were not polarized. Figure 7.6(b) shows the dispersion of some of the bands below the Fermi level.

Several of the bands in Fig. 7.6(b) were determined by the use of mirror-plane symmetry selection rules. Note that the symmetry along Γ–Y is the not the same as that along Γ–X. The mirror planes are no longer the same when the superlattice plays a role. Ding *et al.* state that if the polarization is changed from **E** normal to the mirror plane, used for the spectra they reported, to **E** in the plane, the peaks vanish. The initial states are of odd symmetry with respect to reflection in the Γ–Y mirror plane. An orbital with $d_{x^2-y^2}$ symmetry at the Cu site has such mirror symmetry. Along Γ–X, however, changing the polarization does not cause the loss of the peak. The peak along Γ–X appears to have even polarization, while odd is expected from the LDA bands. Ding *et al.* attribute this peak to an even combination of states produced by the superlattice umklapp process. The Fermi surface is displaced along +**Q** and −**Q** by a surface umklapp process. **Q** is along Γ–Y, and its length is 21% of this distance. This produces two additional Fermi surfaces in each of the two quadrants, Γ–$\overline{M}$–Y–Γ and Γ–$\overline{M}$–X–Γ. Those in the former quadrant do not intersect, while those in the latter do. At the intersection of these two umklapp bands one can form states as the odd and even combinations of the two states, $\psi(\mathbf{k} + \mathbf{Q}) \pm \psi(\mathbf{k} - \mathbf{Q})$. Ding *et al.* found a peak due to the even state. They searched for the corresponding structure due to the odd combination but did not find it.

The effect of the superlattice was studied further by Kelley *et al.*, (1993a) and Yokoya *et al.* (1994). The former group studied bands near E_F, and the latter group, the entire valence band complex. Scanning along Γ–Y in the Brillouin zone is accomplished by scanning emitted electrons in the plane of the surface normal and the *b*-axis. Kelley *et al.* showed that a peak that apparently passes through the Fermi level disperses at least 300 meV when measured along Γ–X, but there is much less dispersion when it is measured along Γ–Y. For Γ–Y this peak is less pronounced and it remains in the 100–300 meV range of binding energy. However, the polarization was not reported, and the peaks would correspond to the same structure, because of dipole matrix anisotropy, only if the sample were rotated 90° about its normal between measurements to keep **E** ∥ **k** for both sets of spectra (or **E** ⊥ **k** for both) or if the peak intensity has been shown to be relatively isotropic. Yokoya *et al.* verified the results of Kelley *et al.*, attributing them to the loss of wave vector as a good quantum number when the superlattice causes the Brillouin zone to collapse to about 1/5 of its normal length in the Γ–Y direction. They also found new effects at greater binding energy.

Here, instead of losing dispersion of a peak along Γ–Y, they found dispersion comparable in size to that along Γ–X, but different from the dispersion along Γ–X. However, they did not state their polarization conditions, and given the complexity of bands in the 1–6 eV range of binding energies, it is not certain the same bands were being followed in the two directions. To separate the anisotropy brought about by the dipole matrix elements from that of the superlattice, such measurements should be made by rotating the sample 90° about its normal to keep the angle between **E** and **k** constant. Measurements on Pb-doped Bi2212 with no superlattice would also be helpful. Singh and Pickett (1995) have discussed the effect of a superlattice on the electronic structure of Bi2201, a Bi-based cuprate with just two CuO_2 planes per unit cell instead of two pairs of such planes, and a simpler superlattice. The effects are rather large, as can be seen in Fig. 2.16.

An important study by Randeria *et al.* (1995) was carried out on Bi2212 to show that angle-resolved photoemission with 19 eV photons is indeed described by the spectral density function $A(\mathbf{k}, E)$, i.e., the sudden approximation is valid at this photon energy. Because the states involved in the photoexcitation are derived from states based on the CuO_2 planes, this conclusion probably can be extended to other cuprates. They measured angle-resolved EDCs for a point on the Bi2212 Fermi surface at many temperatures between 13 and 95 K, as shown in Fig. 7.7. The spectral peaks obviously broaden with increasing temperature.

In the sudden approximation, the photocurrent at energy E in the direction of **k** is proportional to $I_0(\mathbf{k}) f_{FD}(E) A(\mathbf{k}, E)$. The first factor depends on **k** via the square of the dipole matrix element. Its dependence on photon energy and polarization are unimportant here, for these variables are held fixed. I_0 has little depen-

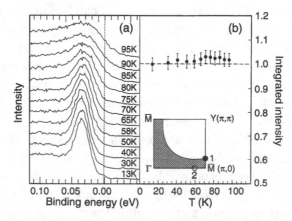

Fig. 7.7   (a) Angle-resolved EDCs for Bi2212 at a Fermi-surface crossing near the $\overline{M}$ point (inset in part (b)) at several temperatures. (b) The integrated areas under the spectra of part (a) after subtraction of a (temperature-independent) background. (Randeria *et al.*, 1995)

dence on temperature or final-state energy (related to E) over the limited range of the EDCs. The Fermi–Dirac function has a known temperature and energy dependence. The spectral function satisfies a few sum rules. As Randeria *et al.* point out, the most obvious one, $\int_{-\infty}^{\infty} A(\mathbf{k}, E)\mathrm{d}E = 1$, is not useful here, because it requires both angle-resolved photoemission and angle-resolved inverse photoemission spectra. Also, the sum or integral over $\mathbf{k}$, giving the single-particle density of states, is not useful with angle-resolved spectra. (Measuring over many values of $\mathbf{k}$ is not very promising, for the prefactor I_0 involves geometrical factors which are difficult to control.) The useful sum rule is the integral of $f_{\mathrm{FD}}A(\mathbf{k}, E)$ over energy, for the Fermi function vanishes where the part of A that requires inverse photoemission data differs from zero:

$$\int_{-\infty}^{\infty} f_{\mathrm{FD}}(E)A(\mathbf{k}, E)\mathrm{d}E = n(\mathbf{k}).$$

Here, $n(\mathbf{k})$ is the momentum distribution $\langle c_k^{\dagger} c_k \rangle$. If the right-hand side does not depend on temperature, then as long as I_0 does not depend on temperature, the integral over energy of the photocurrent at fixed $\mathbf{k}$ should not depend on temperature. For $\mathbf{k} = \mathbf{k}_{\mathrm{F}}$, $n(\mathbf{k}_{\mathrm{F}}) = \frac{1}{2}$ at all T for a free-electron gas and a BCS superconductor. Randeria *et al.* argue that this result is valid as long as there is particle–hole symmetry: $A(\mathbf{k}_{\mathrm{F}}, E) = A(\mathbf{k}_{\mathrm{F}}, -E)$. The problem then becomes one of separating the primary photocurrent from the rest of the spectrum.

Randeria *et al.* first subtracted a constant from each spectrum to bring each spectrum to zero several tens of meV above E_{F}. The background below the Fermi level is not understood but, except for very near E_{F}, it appears to be constant over a considerable range and the constant value does not depend on temperature, once the data have been suitably normalized for small changes in sample position from thermal expansion of the mount and from desorption of gases. This constant was extrapolated to E_{F} and subtracted from the spectrum. At low temperature, this subtraction led to negative values of the photocurrent around 60–70 meV binding energy. This is in the region of a "dip" to be discussed further. It was important to keep this part of the spectrum when carrying out the integration. The remaining photocurrent was then integrated over energy, with the results shown in Fig. 7.7; the integral varies only about 5% over the 13–95 K range.

The same procedure was carried out at a different point on the Fermi surface with similar results, but when repeated for a wave vector that was not on the Fermi surface, the background at large binding energy was temperature dependent. It was about 10% larger at lower temperatures, and this difference persisted to such large binding energies it could not be attributed to the opening of the superconducting gap. (This background was proportional to the current of electrons of higher kinetic energy, just as in one of the models used for corrections for the background of inelastically scattered electrons described in Chapter 3. This is a result worthy of further study.) Nonetheless, the integral was evaluated as before, and it varied 10% with temperature. However, for $\mathbf{k} \neq \mathbf{k}_F$, $n(\mathbf{k})$ need not be temperature independent.

It is clear from the above that although there is considerable agreement, there are some inconsistencies between data taken by several groups. To date, these inconsistencies are not all understood, and the effort to explain some of them may not be worth the gain. An additional difference between groups is the observation of, or failure to observe, the splitting of bands due to interaction between the two CuO_2 planes in the Bi2212 unit cell. These bands would be degenerate without such interaction and their splitting is a measure of such interaction. LDA band calculations yield a splitting of about 200 meV between Γ and X, at the point where the uppermost band crosses E_F near X. Ding et al. (1995, 1996a) reported only a single band crossing the Fermi level, while Shen et al. (1995) reported two. Near the Fermi level, band renormalization due to correlation is particularly strong. Liechtenstein et al. (1996) carried out a model calculation using a Hubbard model with band dispersion from a tight-binding model with parameters chosen to approximate the LDA bands of Bi2212 near E_F. The splitting due to interaction between the two CuO_2 planes turns out to be greatly reduced, to about 40 meV, and the location of the crossing of the uppermost band near X is unchanged. 40 meV is in principle resolvable, but in practice, it may prove rather difficult. Moreover, the bulk results may not be strictly applicable. In the bulk, the hybrid band states are the odd and even combinations of the individual plane states with equal weights. In the unit cell at the surface, the mirror plane between the two CuO_2 planes is no longer a plane of symmetry due to the surface and the weights from the two planes no longer must be equal. This could change the splitting as well as the dipole matrix elements which, in any case, need not be the same for the two split bands, making observation of the one with the smaller dipole matrix element more difficult. However, the outermost of the two CuO_2 planes in the surface unit cell may not "feel" the presence of the surface enough for this to be an observable effect. A thorough search using several photon energies and polarizations may be required for the observation of this splitting. This interaction between CuO_2 planes is also important for some models of superconductivity in cuprates (Chakaravarty et al., 1994).

A simple phenomenological model has been used for several years to fit the band that passes though the Fermi level in cuprates, a band based on the Cu $3d_{x^2-y^2}$ and O $2p_x$ and $2p_y$ hybrids for the atoms in the CuO_2 planes. It is a fit to a simple tight-binding model. A recent version is that of Norman et al. (1995a) for Bi2212 which also reproduces the Fermi surface and can be made to give the Fermi-surface crossing caused by the superlattice.

Resonant photoemission studies on Bi2212, CuO, and Cu_2O were carried out by Tjeng et al. (1992), exciting in the regions of the O 1s and Cu 2p core levels. These were undertaken to clarify a number of issues left over from previously observed resonances. The main resonance near the valence band from Cu $2p_{3/2}$ photoexcitation is a satellite between 12 and 15 eV binding energy, depending on which of the three materials is considered. There is only weak resonance enhancement at lower binding energies, i.e., in the valence band. The O 1s resonance is primarily of a satellite between 13 and 21 eV. Valence-band enhancement is also weak, especially in the first eV below E_F. The authors take this to mean that the "resonance" observed in this region from photoexcitation in the O 2s region (about 18 eV) (Takahashi et al., 1988, 1989; Manzke et al., 1989a,b,c; Wells et al., 1989) is not really a resonance as normally understood, but rather a peak in the photoexcitation cross-section or a final-state effect. They used a cluster model as discussed in Chapter 5 to interpret their enhancements. The initial states had principal components p^6d^{10}, p^6d^9, and p^5d^9 for Cu_2O, CuO, and doped Bi2212, respectively. The resonant satellites were assigned to final states primarily of p^6d^8s, p^6d^8, and p^5d^8 character for the Cu resonance and $p^4d^{10}s$, p^4d^{10}, and p^4d^9 for the O resonance. The latter assignment differs from the previous suggestion (Sarma et al., 1989) of a p^5d^8 final state, which would require an interatomic Auger transition, normally expected to have a small probability. Simmons et al. (1992) measured valence-band spectra in Bi2212 on and off the Cu 3d resonance and the O 2p "resonance." They had 50 meV energy resolution but $\pm14°$ angle resolution, giving spectra probably closer to angle integrated than angle resolved. The weak resonances they found indicated O 2p character around 180 meV and Cu 3d–3s character around 320 meV below E_F.

More resonance studies were carried out by Qvarford et al. (1995). Their samples were single crystals of Bi2212 and Bi2201. They excited photoemission from the valence band and Cu 3p levels with radiation around the Cu 2p photoexcitation threshold. For both crystals the satellite below the valence band, at about 13 eV binding energy, and the satellite pair at greater binding energy than the Cu 3p main line show very large enhancements at the Cu 2p excitation energy, while the valence band and Cu 3p main line do not. The final states are primarily Cu $3d^8$ for the valence-band satellite and Cu $3p^5 3d^9$ for the Cu 3p satellite. The line splittings and intensities were calculated for a Cu atom. When broadened, they fit the experimental spectra very well. The valence-band spectrum and that of its satellite were used to estimate $U - \Delta$ for a Cu-centered cluster, along the lines discussed in Chapter 5. This energy was slightly smaller, 8.2 eV, for Bi2201 than for Bi2212, 8.6 eV. The large value reconfirms the charge-transfer nature of these cuprates, U is considerably larger than Δ. The differences in the two cuprates can be assigned to different values of Δ, different because of the different oxygen coordination for the Cu atoms. More Cu 2p spectra were taken by this group (Chiaia *et al.*, 1995) and a fit to the cluster model described in Chapter 5 carried out with the new data. These were compared with similar data on CuO (Ghijsen *et al.*, 1988; Eskes, *et al.* 1990) and Y123 and La$_{1.8}$Sr$_{0.2}$CuO$_4$ (Shen *et al.*, 1987). Chiaia *et al.* discuss some of the trade-offs necessary in fitting data to such a model. The values of t, the Cu–O transfer integral, are nearly the same for all the materials. U, the Cu-site d–d Coulomb repulsion energy, is largest for CuO and smallest for the Bi-based cuprates; the range is 7.7 to 5.4 eV. Δ, the O–Cu transfer energy, varies the most on a relative scale, 0.3 to 1.75 eV. It is largest for CuO and smallest for LaCuO$_4$ and La$_{1.8}$Sr$_{0.2}$CuO$_4$. Δ is expected to depend on the number of oxygen nearest neighbors of the Cu atoms and on the nearest-neighbor distances. Before a reliable interpretation of these differences can be made, however, it might be necessary to carry out all the fits again with the same fitting criteria.

The valence-band satellite resonance from photoexcitation in the region of the Cu 3p binding energy was also studied by Flavell *et al.* (1995) as a function of Y concentration in polycrystalline samples of Bi$_2$Sr$_2$ Ca$_{1-x}$Y$_x$Cu$_2$O$_8$. The shape of the resonance in a CIS spectrum changed gradually (and slightly) as x varied from 0 to 1. Flavell *et al.* attributed this to a shift in the Cu 3p binding energy as the Cu valence changed with Y substitution.

Angle-resolved inverse photoemission spectra were taken by Watanabe *et al.*, (1991). They found some intensity at the Fermi edge for all values of wave vector along the Γ–X line. At two angles, corresponding to 20% and 60% of the distance from Γ to X, there were slight enhancements of this edge, each attributed to a band crossing through the Fermi level. Previous angle-resolved inverse photoemission spectra had seen one of these crossings (Claessen *et al.*, 1989) or the other (Wagener *et al.*, (1990b). At the time, this was an important issue, for there were discrepancies in the number of Fermi-surface crossings along this line in angle-resolved photoemission. Manzke *et al.* (1989a,b,c) and Olson *et al.* (1989, 1990b) had each reported one, but they were not necessarily the same one. Takahashi *et al.* (1989) had reported two, which matched up with the two inverse-photoemission points, within the rather large error bars. However, most recent work shows one crossing along Γ–X. The cause of the second band crossing the Fermi surface is not yet clear. Figures 7.2 and 7.5(b) show a single Fermi-surface crossing along Γ–X, but at a place where superlattice umklapp structures cross (but with negligible intensity) the LDA Fermi surface. Figure 7.6, however, clearly shows two Fermi-level crossings between Γ and X; the second one is from a shadow band.

7.1.2   Core levels

Several core-level studies have been repeated, but the resolution has not improved much over that of earlier studies. The studies were made on better-characterized samples, and sometimes on a series of crystals of varying composition. Nagoshi *et al.* (1993) made core-level XPS measurements on $Bi_2Sr_2Ca_{n-1}Cu_nO_{2n+2+\delta}$, with $n = $ one, two, and three, i.e., Bi2201, Bi2212, and Bi2223. These crystals have similar structures, but with 1, 2, and 3 CuO_2 planes per unit cell, respectively. They used polycrystalline samples and a resolution of 0.45 eV. The O 1s spectrum of each was a single asymmetric peak at about 528.5 eV, with the intensity on the low binding-energy side and in a high binding-energy tail increasing in the order Bi2201, Bi2212, Bi2223, i.e., with increasing number of CuO_2 planes per unit cell. (The peaks were normalized to give equal peak heights.) The spectrum for Bi2212 was close to that of Parmigiani *et al.* (1991a). Each asymmetric line was fitted with four overlapping Gaussian lines, peaking at 527.9, 528.7, 529.4, and 530.5 eV. The second peak was the largest. The peak at lowest binding energy was assigned to O in the CuO_2 planes because of its growth in relative intensity as more CuO_2 planes are added. This assignment disagreed with previous ones (Fujimori *et al.*, 1989; Parmigiani, 1991a), which, however, did not agree with each other. Nagoshi *et al.* assigned the main peak to O on the Bi–O planes and the next to O on the Sr–O planes. The high binding-energy part of the spectrum, represented by the fourth peak, was assigned to possible shake-up excitations. The assignments were made on the basis of the expected number of CuO_2 planes and the attenuation expected for electrons emitted from layers below those planes.

Sanada *et al.* (1994) studied doping in Bi2201 by varying the Bi:Sr ratio in $Bi_{2+x}Sr_{2-x}CuO_y$ polycrystalline samples. The values of x ranged from 0.1 to 0.4. The samples with $x = 0.10$ and 0.15 were metallic, with T_c presumably 10 K (a value from the literature) for the former. The others were semiconducting. The Bi 4f, Cu $2p_{3/2}$, and O 1s spectra showed little change with hole removal (increasing x), the only change being the narrowing of the O 1s line as x increased. Valence-band spectra taken with the He I line and inverse photoemission spectra taken at 9.6 eV showed a Fermi edge, but it was decidedly smaller for the samples with $x = 0.35$ and 0.40. The metal-to-semiconductor transition evidently occurs near $x = 0.35$. The core-level and valence-band spectra did not shift with x, leading to the conclusion that, unlike Bi2212, Bi2201 has a pinned Fermi level.

Qvarford *et al.* (1992) studied the electronic states of Ca in Bi2212 by resonance photoemission, as well as by x-ray absorption. The Ca 2p level was used for excitation and emission from the Ca 3s and 3p core levels (42 and 23 eV binding energy, respectively) was detected, along with the Ca $L_{2,3}M_{2,3}M_{2,3}$ Auger spectrum. The photoexcitation of the Ca 2p state is primarily into the Ca 3d state, which the authors concluded to be well localized around the 2p hole. They also concluded that the hole in the ionic Ca–O layer was also significantly screened by charges in nearby CuO_2 layers.

Leiro *et al.* (1995) worked with single crystals and polycrystals of Bi2212 and of Pb-doped Bi2212 (Pb to Bi ratio: 1:5.7). They measured the O 1s photoemission spectra in an attempt to clarify the interpretation of its components. The number of CuO_2 planes per unit cell was believed to be the same for all samples, so differences in the O 1s peak should not be caused by changes in the relative number of O atoms located in CuO_2 planes. They did not decompose the single peak into individual components. For the single crystals the effect of Pb addition was to cause an increase in a component on the low energy side of the main peak, which was at 529 eV. It caused about a 10% increase in the peak area. In view of the fact that the surface is expected to be a Bi–O plane, this new peak is attributed to O atoms neighboring Pb atoms in this plane, while the main peak is from the O atoms with Bi neighbors in this plane. The shift to lower binding energy of the O atoms neighboring Pb atoms is consistent with the 1s O binding energies in PbO, PbO_2, and Bi_2O_3.

The preceding work was done at 0.45 eV resolution. Qvarford *et al.* (1996) were able to work with 0.25 eV resolution. They also varied the emission angle detected to alter the relative surface sensitivity. They were able to decompose their spectra into three components (Doniach–Šunjić line shapes) which were used in a study of doping effects. As the emission angle changed, only the relative intensities of the components were allowed to change in the fitting. Their single-crystal samples were Bi2212, Bi2201, and $(Bi,Pb)_2Sr_2(Ca,Er)_1Cu_2O_8$. In addition to photoemission, they measured the O 1s absorption edge, since it can show changes with hole doping (Himpsel *et al.*, 1988). The analysis led to the assignment of the components to oxygen in CuO_2, Sr–O and Bi–O planes, in order of increasing binding energy. The relative intensity of the CuO_2 plane component varied as expected. This component was much weaker in Bi2201 than in Bi2212. Increasing the hole concentration causes this peak to shift to lower binding energy. In their Pb-doped crystal they did not find the increased intensity on the low binding-energy side of the spectrum reported by Leiro *et al.* (1995).

7.2 Bi2212 below T_c

We first discuss what might be expected in a photoelectron spectrum when a sample goes superconducting. As discussed in Chapter 3, for a normal metal (Fermi liquid), the photocurrent from valence electrons in a single crystal emitted into a small solid angle is a Lorentzian distribution in energy or, more generally, a function resembling a Lorentzian, with the resonant energy shifted by $\mathrm{Re}\widetilde{\Sigma}$ and a broadening parameter of $\mathrm{Im}\widetilde{\Sigma}$. Here, $\widetilde{\Sigma}$ is the self-energy, a function of energy and wave vector, but the dependence on **k** is usually neglected. If $\mathrm{Im}\widetilde{\Sigma}$ varies much over the line width, the shape will no longer resemble a Lorentzian.

Moreover, if $\mathrm{Im}\widetilde{\Sigma}$ is large, the line will be very broad, and the concept of a quasi-particle breaks down. Bi2212 is close to this situation, especially when under-doped (see below). This line shape due to the crystal potential and electron–electron interactions will be convolved with a Gaussian function representing phonon contributions to the width, sometimes not large enough to be considered. The energy dependence of this Lorentzian changes slowly with angle within the angular acceptance of the analyzer, the result of dispersion of the final-state and initial-state quasiparticles. Usually the observed spectrum is taken to be a convolution of the Lorentzian with an instrument function, assumed to be a function only of energy. This is not always the case. In the following we discuss this problem, then conclude with expectations for a metal in the superconducting state. An additional complication is the easily observed asymmetry of the peaks in the normal-state cuprates. They have rather long tails to larger binding energy. If a Lorentzian is used, a fit cannot be obtained with a constant broadening parameter. A broadening parameter that increases with binding energy over the width of the peak, usually linearly, produces a fit.

The emitted photocurrent will be proportional to

$$I(\mathbf{k}, E) = \frac{\Gamma f_{\mathrm{FD}}(E)}{[E - E(\mathbf{k})]^2 + \Gamma^2},$$

where $\mathbf{k}$ is used also to denote the angle of electron emission. The analyzer has an instrument function $G(\mathbf{k}, E)$. An approximation used in almost all analyses is that $G(\mathbf{k}, E)$ factors into $G_a(\mathbf{k})G_e(E)$. Then the observed signal is proportional to the convolution

$$S(\mathbf{k}, E) = \int I(\mathbf{k}', E')G_a(\mathbf{k} - \mathbf{k}')G_e(E - E')\mathrm{d}^2k'\mathrm{d}E'.$$

With the factoring of G, G_e can be determined by measuring a Fermi edge or an atomic photoelectron. It may be approximated by a Lorentzian or Gaussian. G_a is not well known, and it is more difficult to measure. Ray tracing probably is not accurate enough because of stray fields in the analyzer. A top-hat function or Gaussian may be used with a width given by the geometrical angular acceptance of the angle-limiting aperture in the analyzer, but the correct functional form is not known accurately. This will underestimate the angular broadening due to the neglect of stray fields. There is no known source of photoelectrons emitted as a near-delta function in angle, nor as a step function.

Liu *et al.* (1991) made an attempt to fit the spectra of Olson *et al.* (1989, 1990b) with a one-peak spectral density function. They used two forms for the self-energy: (i) that expected for a Fermi liquid, $\widetilde{\Sigma}(E) = \alpha E + i\beta E^2$, with E measured from E_F; and (ii) that expected for a marginal Fermi liquid (Varma *et al.*, 1989), $\widetilde{\Sigma}(E) = \gamma(E\ln(x/E_c) - i\pi x/2)$, with $x = \max(|E|, k_B T)$ and E_c a cutoff energy. The background subtraction they used, that of Shirley (1972), assumed the back-

ground arose from inelastically scattered primary electrons. They could fit the
spectral peaks equally well with either self-energy, but neither fit was in perfect
agreement with the measured spectral shape. Differentiating between the two
forms for $\widetilde{\Sigma}$ was best done at larger binding energies, but the uncertainty in the
subtraction of the large background precluded a choice.

Later, Claessen *et al.* (1992) obtained new angle-resolved photoemission data
on $TiTe_2$, e.g., see Fig. 7.8, and fitted them with a spectral function using the
Fermi-liquid self-energy. An excellent fit was obtained for peaks near E_F. As the
quasiparticle energy reached 20 meV or so below E_F, the fits became progressively poorer. This was ascribed to the expected failure of the simple Fermi liquid
self-energy, $\widetilde{\Sigma}(E) = \alpha E + i\beta E^2$, basically a Tayor-series expansion of $\widetilde{\Sigma}(E)$

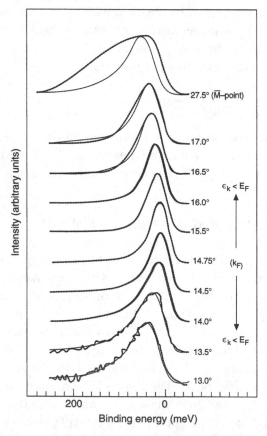

Fig. 7.8 Several angle-resolved EDCs (dots) near a Fermi-surface crossing in $TiTe_2$. The
crossing occurs near 15°. A small background due to inelastic scattering has been subtracted from only the lower four spectra. This background is much smaller than in similar
spectra for cuprates. The solid lines are fits to a Fermi-liquid model as discussed in the
text. The fit is very good for a band near the Fermi surface, but poor (top curve) for a
band 50–60 meV below E_F. (Claessen *et al.*, 1992)

about the Fermi energy, when the energy E became too large. These fits required no subtraction of a background, for it was very small. There were two effects of the self-energy on the spectra. First, it introduced a tail to higher binding energy, leading to a distinctly non-Lorentzian appearance. Second, the spectral density function was finite at and below E_F even though the quasiparticle band had dispersed through the Fermi energy. The spectral density function in a single-particle model is zero once the peak has passed above the tail of the Fermi function. Both effects were present in the experimental spectra. For a peak at the Fermi-level crossing, a fit using the self-energy for the marginal Fermi liquid, given above, was tried, but a fit as good as that achieved with the Fermi-liquid self-energy was not obtained. This work demonstrated that even for a metal of low electron density like $TiTe_2$, many-body effects are manifest in the line shape of a peak in an angle-resolved EDC. Moreover, the simple expression above for the Fermi-liquid self-energy may be used only for peaks close to E_F. Claessen *et al.* improved their fit by adding a correction term to $\widetilde{\Sigma}$, but this does not extend much the energy range of applicability of this form of $\widetilde{\Sigma}$.

Later, Harm *et al.* (1994) repeated the fits of Claessen *et al.* to a larger data set, with temperature as an additional variable. Improved fits were obtained, even at large angles, by adopting a more realistic self-energy. This self-energy, suggested in unpublished work by K. Matho, is more physical, but analytically more complicated. It can be handled numerically, however. It fitted the data quite well even out to $27°$, where the Taylor series expansion does not give a good fit (Fig. 7.8). At about the same time, Allen and Chetty (1994) analyzed the temperature dependence of the electrical resistivity of $TiTe_2$ and calculated the expected temperature dependence from the quasiparticle lifetime measured by Claessen *et al.* There was a large discrepancy, a predicted T^2 dependence vs. an observed T^1 dependence. Several possible reasons for such a discrepancy were suggested, but no resolution is yet at hand. Allen and Chetty advise caution in interpreting the lifetimes obtained from photoemission spectra, not only because they may be difficult to extract reliably from measured spectra, but also because "the theory of photoemission or layered metals (or both) is poorly understood."

Fehrenbacher (1996) has evaluated the above convolutions for a spectrum representing a simple spectral density function from BCS theory (Chang *et al.*, 1989a),

$$I(\mathbf{k}, E) = u_k^2 \delta[E - E(\mathbf{k})] + v_k^2 \delta[E + E(\mathbf{k})]$$

with a Gaussian for G_e and a top-hat function for G_a. The widths were $7\,\mathrm{meV}$ (σ) and $0.036\pi/a$ (radius), respectively. (Except for the choice for G_a similar calculations were done by Fehrenbacher and Norman, 1994 and Ding *et al.*, 1995.) The coherence factors are $u_k^2 = (1 + \epsilon_k/E_k)/2$ and $v_k^2 = (1 - \epsilon_k/E_k)/2$, with $E_k = (\epsilon_k^2 + \Delta_k^2)$, the energy in terms of the quasiparticle energy ϵ_k (measured from E_F) and order parameter Δ_k. The latter was taken to be approximately 15 meV $(\cos k_x a - \cos k_y a)$. The quasiparticle energies were taken from a tight-binding fit to angle-resolved photoemission data taken on Bi2212 in the normal state. Figure 7.9 shows the results of convolution with G_e only, with G_a only, and with both, at three different angles from the surface normal. Δ_k has a node at 45°, so there is only one peak in the original spectrum. The spectra shown are without the Fermi–Dirac factor so they represent both photoemission and inverse photoemission. The broadening due to the analyzer energy spread, the second row, is similar for all three spectra, but angular broadening, third row, is not. The 45° spectrum is broadened the most, for at this angle the quasiparticle band is very dispersive. The 0° spectrum is broadened the least because the quasiparticle band is very flat due to a saddle point. The bottom row shows the final spectra along with a Fermi function broadened by G_e. It is clear that the angular acceptance can contribute as much to broadening in some regions of reciprocal space as does the instrumental energy broadening, and possibly more.

Campuzano *et al.* (1996) recently published an esthetically satisfying study of the approach of the edge of the Bi2212 angle-resolved photoelectron spectrum

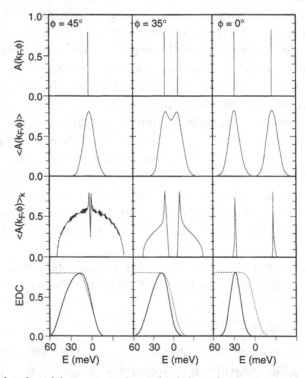

Fig. 7.9 Calculated model cuprate angle-resolved photoelectron spectra for three angles (*top row*), and the effect of instrumental energy resolution (*second row*), angle resolution (*third row*), and both (*bottom row*). The dashed curve is a Fermi–Dirac function convolved with the same energy function as the model spectra. (Fehrenbacher, 1996)

to E_F and its retreat from it. The BCS spectral density function was given in the preceding paragraph. The normal state spectrum results if $\Delta = 0$, and any spectral peak, followed as a function of **k**, should pass through E_F from below and disappear. Below T_c, the closest a peak in the spectral density function A can come to E_F is Δ. Scanning angle, starting from below E_F, the peak in A should follow the curve shown in Fig. 7.10, approaching E_F, but then retreating from it with decreasing strength, as the quasiparticle amplitude for these **k** values is predominantly above E_F. Such behavior was seen in the spectra of Campuzano *et al.* (Figs. 7.11 and 7.12). Fehrenbacher's analysis was directed at extracting the value of Δ from the spectrum with the closest approach to E_F. The curve above E_F in Fig. 7.10 contains a weak quasielectron component, so a weak contribution to the photoelectron spectrum could come from it. It is difficult to observe because of possible contributions from higher-order radiation from the diffraction grating.

One could obtain the emitted photoelectron spectrum from the measured spectrum by deconvolution, but the instrument function is not known sufficiently well. An alternate route is to simulate a spectrum from a microscopic model, convolve that with the instrument function, adjusting model parameters until a best fit is achieved. This is occasionally done, but again the instrument function is not well known. Moreover, such a simulation requires knowledge of the background, a large feature of cuprate ARUPS spectra which is not understood. The spectrum measured when a band has passed through the Fermi level has been used for this background, but the appropriateness of such a background is not known.

Instead, what has been done to date is to study peaks and edges, sometimes correcting the measured peak and edge energies for the effect of the spectrometer to give their true values. Foremost among these peaks and edges is the gap energy

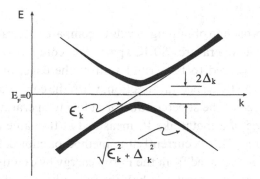

Fig. 7.10 Behavior of expected quasiparticle dispersion in the normal state (thin straight line) and in the BCS superconducting state. The line width of the lower curve for the superconducting state indicates the intensity expected in the photoelectron spectrum. (Campuzano *et al.*, 1996)

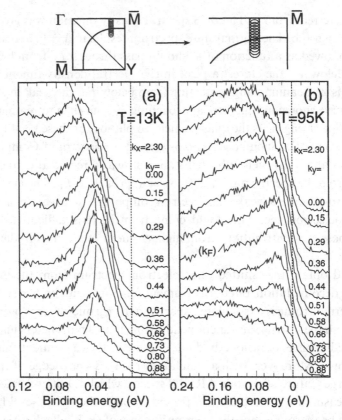

Fig. 7.11 Angle-resolved photoelectron spectra from Bi2212 above and below T_c. The solid curve is a guide to the eye. In the normal state (b), the curve approaches close to the Fermi level while in the superconducting state (a) it approaches, then retreats. The wave vectors are in units of (π/a). The scans in the Brillouin zone for each set of spectra are shown above the spectra. (Campuzano *et al.*, 1996)

Δ. The earliest approach to obtaining Δ was to compare spectra taken above and below T_c, often at temperatures 80 K apart. The comparison was made "by eye" when there appeared to be a rigid shift of the edge, or by fitting to an assumed function, the BCS spectral function, when the edge shapes were not the same. Later, the edge of the superconductor at low temperature was compared with the Fermi edge of a metal like Pt measured at the same temperature and nearly the same time. This is currently the method of choice. The midpoint of the rising edge is not displaced from the Fermi energy by exactly the gap energy, but by something less, of the order of half the full width at half maximum of the instrumental and intrinsic broadening functions. It is hoped that changes in the gap energy due to doping, and perhaps wave vector, will be tracked by shifts in the midpoint of the edge. Tracking shifts with temperature this way is more

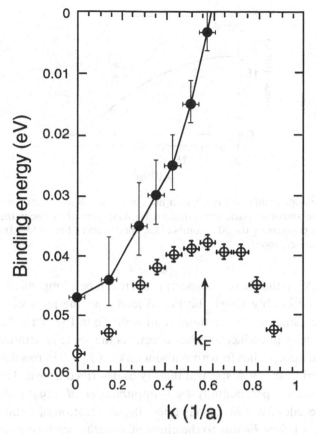

Fig. 7.12 Distribution of maxima in the spectra of Fig. 7.11 plotted as a dispersion curve. Solid points: normal state; open points: superconducting state. Compare with Fig. 7.11. (Campuzano *et al.*, 1996)

tricky. The rising edge is affected by the temperature-dependent width of the Fermi function.

Fehrenbacher (1996) studied this method of extracting Δ values by using synthetic spectra like those in the top row of Fig. 7.9, convolving a Fermi function with $G_e(E)$ as in the bottom panel. The spectrum of the superconductor was normalized to make its peaks equal to the limiting value of the Fermi function. Figure 7.13 shows the results, values of Δ obtained from comparing the broadened synthesized spectrum with a broadened Fermi edge, as a function of the angle of measurement. In this case the measured Δ is always significantly less than the input Δ. Near where Δ goes to zero the "measured" Δ has a different slope with respect to angle than the true one and, if experimental errors were to be considered, would show an extended range of angles for which Δ appears to be zero. Fehrenbacher then calculated the effect of scattering by non-magnetic impurities on the expected spectral function, including the effect of the spread in

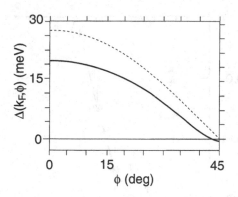

Fig. 7.13 Angle-dependent order parameter used in the calculated spectra of Fig. 7.9 (dashed curve) and the order parameters obtained from spectra like those in the bottom row of Fig. 7.9 by comparing the 50% points of the spectra and a Fermi edge (solid curve). (Fehrenbacher, 1996)

energy and angle on the measured spectra. There can be a long tail on the edge, a tail extending noticeably above the Fermi level, and the plots of Δ vs. angle tended to underestimate Δ even more than is shown in Fig. 7.13. The impurity scattering also may introduce another effect. As the angle is scanned, the peak in the spectrum approaches to within about $\Delta(\mathbf{k})$ of E_F, then recedes, in accord with the v_k^2 term in the BCS spectral density as described above. This behavior has been observed experimentally by Campuzano *et al.* (1996). As the peak weakens and recedes from E_F, it may allow the observation of a small, non-dispersive peak just below E_F due to the effect of impurity scattering on the quasiparticle self-energy.

The BCS spectral density function used above is not the only possibility considered for spectral line shapes from cuprates. Photoemission from a resonating valence bond (Anderson, 1987) system was considered by Huber (1988) and Chang *et al.* (1989b). The photohole of the Fermi liquid in the normal state of the cuprate is replaced by a spinon–holon pair, which may be assumed to be independent quasiparticles, the former, a fermion, the latter, a boson. The spinon spectrum extends over a range of the order of the exchange interaction, J. In the region of apparent binding energy between 0 and J, the phase space for spinon creation is reduced. Huber assumed a constant density of states for the spinons and a quadratic dispersion law for the holons, obtaining a photoelectron current rising linearly with binding energy from a small value at E_F, to a constant value for $E > J$. Well above T_c, the photocurrent at the Fermi edge is linearly proportional to the temperature. To date, such spectral shapes and temperature dependence have not been found in experiment.

After the first reports of the observation of a gap of about 20 meV in Bi2212, several groups made extensive measurements. The initial result, a gap that was isotropic in the plane of the sample, was superseded by several measurements that showed an angle dependence of the gap. The first hint of such anisotropy was reported by Dessau *et al.* (1991a). They measured EDCs of the Fermi-level crossing along $\Gamma-\overline{M}$ and $\Gamma-X$ at 10, 80 and 100 K on a Bi2212 crystal with $T_c = 91$ K. They observed the shift of the leading edge from the Fermi level at temperatures below T_c and reported this effect was larger for the Fermi-level crossing along $\Gamma-\overline{M}$ than for the one along $\Gamma-X$. Using a BCS line shape, as did Olson *et al.* (1989), they obtained 27 meV for the gap along $\Gamma-\overline{M}$ and 15 meV along $\Gamma-X$. The energy resolution used was 40 meV and the acceptance angle was $\pm 4°$. This spread in angle represents a spread in wave vector of $\pm 16\%$ of the distance from Γ to $\overline{M}$. Later work showed the gap along $\Gamma-X$ to be smaller than 15 meV. The large angle acceptance apparently caused the inclusion of points off the $\Gamma-X$ line where Δ is larger than on the line, thus increasing the measured value. There is a more extensive discussion of these data in a second publication (Dessau *et al.*, 1991b) which also reports a gap anisotropy in another Bi2212 crystal with a lower T_c (79 K). For this crystal the gap values were 14 and 6 meV for the $\Gamma-\overline{M}$ and $\Gamma-X$ directions, respectively.

Along the $\Gamma-\overline{M}$ line, where the gap is large, Dessau *et al.* found a dip in the spectrum at about 90 meV binding energy when T was below T_c. The dip increased in depth as the temperature was decreased further. This dip did not appear, or was extremely shallow, along the $\Gamma-X$ line. The dip was discussed as an anomalous transfer of spectral weight. In the BCS model for superconductivity, the single-particle states removed from the gap appear in the "pileup" region just below the gap, thereby conserving area in a single-particle density of states. This is true not only in the angle-integrated spectrum, but also in an angle-resolved EDC taken along one direction in the Brillouin zone because of the assumed spherical symmetry. In the angle-resolved spectra, the pileup for spectra along $\Gamma-X$ had a larger area than that lost at the gap, while along $\Gamma-\overline{M}$, the area in the pileup was less than was lost at the gap and in the dip. However, these data represent only a small fraction of the Brillouin zone volume with a non-spherical Fermi surface, so it is not clear that the BCS area conservation need occur. Moreover, the sum rule, integrating over energy at constant wave vector, requires an integration over empty states well above E_F in order to be satisfied, as discussed above.

Shen (1992) and Wells *et al.* (1992) reported improved measurements with 35 meV resolution, $\pm 1°$ angular acceptance and many more scans through the Brillouin zone. Several Fermi-surface crossing points were measured between the $\Gamma-\overline{M}$ and $\Gamma-X$ lines. The gap decreased gradually. This angular dependence was consistent with $d_{x^2-y^2}$ symmetry for the gap, but a detailed fit to a function of this symmetry was not carried out. The dip was studied further by Hwu *et al.* (1991) and Dessau *et al.* (1992). The two papers disagree on the wave-vector dependence of the dip. Later work is in agreement with Dessau *et al.*

Yokoya *et al.* (1993) demonstrated an anisotropic gap using only angle-integrated EDCs taken with 20 meV resolution. They fitted the spectrum above T_c to a simple density of states multiplied by a Fermi-Dirac function, then convolved with a Gaussian instrument function. Below T_c, the density of states was multiplied by a BCS function, but no single value of Δ could produce a fit of both the peak below E_F and the tailing of the EDC above E_F. The use of an angle-dependent function for Δ produced an excellent fit when a function with $d_{x^2-y^2}$ symmetry was used. This does not demonstrate that this is the correct symmetry, but it does show that an isotropic Δ does not fit the data.

The anisotropy of the gap was investigated further by Shen *et al.* (1993b). They used three samples, 30 meV overall resolution and (presumably) an angular acceptance of $\pm$ 1°. They comment on the difficulties of such measurements: finite energy and angle resolution; drift in the electronics leading to apparent Fermi level shifts (about 1 meV in their case); and the flatness of the surface, which was characterized by the cross-sectional shape of a reflected laser beam. They also found that after time, the apparent gap along Γ–$\overline{M}$ decreased from 19 to 14 meV. This decrease could be reversed by warming the samples to room temperature, then recooling. They concluded that adsorbed gases caused this effect. However, the lack of anisotropy of the gap found by Olson *et al.* (1990b) seems not to be due to adsorbed gases, for the isotropic gap they observed was about 20 meV, not diminished. Norman (1994) discussed the deterioration of the anisotropy of time as a possible effect from impurity scattering within the bulk, i.e., as an effect on the initial state, a Cooper pair. However, Dessau *et al.* (1994) replied that the scattering involved occurred in the photoemission process, e.g., upon escape from the surface in the three-step model.

Figure 7.14(a) shows the points in the Fermi surface at which Shen *et al.* determined the gap for the three samples, and Fig. 7.14(b) shows the values of the gap. In these plots, k_x and k_y are parallel to the Γ–$\overline{M}$ directions, i.e., parallel to the Cu–O bond directions. Gap values are displayed as a function of $|\cos k_x a - \cos k_y a|$. As discussed in the next paragraph, the simplest wave-vector dependence expected for a gap with $d_{x^2-y^2}$ symmetry is $\cos k_x a - \cos k_y a$, but photoelectron spectra cannot distinguish the sign of Δ or changes in sign. A straight line on these plots would indicate $d_{x^2-y^2}$ symmetry for the gap, and the fit is certainly better to this than to a constant value of Δ expected for a gap of simple s-wave symmetry. In fact, this fit was questioned by Mahan (1993) who showed that a fit to a gap of generalized s-symmetry gave a better fit. However, additional data taken subsequently by several groups have reinforced the fit first proposed by Shen *et al.* In particular, Dessau took considerably more data, some of it published in the response to Mahan's comments (Dessau *et al.*, 1993b) and some in the review by Shen and Dessau (1995), where they appear as Fig. 5.18. Also, Yokoya *et al.* (1995) studied the symmetry of the gap by angle-

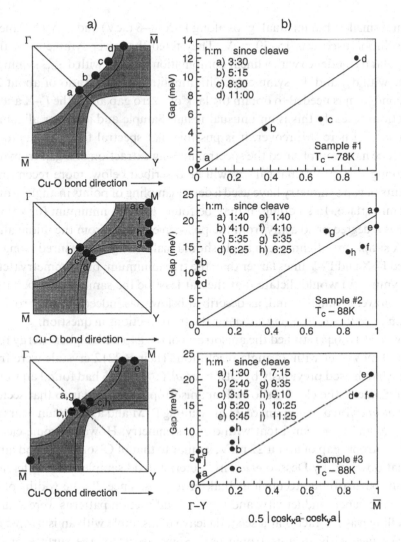

Fig. 7.14 (a) Brillouin zone points at which the gap was measured. The shaded regions indicate where a band was found close to E_F. Only the CuO_2 plane band Fermi surface is shown. (b) The measured gap vs. $0.5|\cos k_x a - \cos k_y a|$. A straight line is expected for a gap of $d_{x^2-y^2}$ symmetry. The spectra were taken at various times after the initial cleaves, as marked. (Shen *et al.*, 1993b)

resolved photoemission on single crystals. Their minimum gap was 0–4 meV, along both the Γ–X and Γ–Y directions, indicating no detectable influence of the superlattice along the **b** axis on the magnitude of the gap. The maximum gap, along Γ–$\overline{M}$, was 17–19 meV. They also observed a sample aging effect; the maximum gap was reduced with time and the minimum increased. Warming to room temperature could restore the original gaps. Kelley *et al.* (1994), using the shift of the rising edge with temperature, obtained gaps of 14–20 meV along

Γ–$\overline{\text{M}}$, and smaller, but unequal, gaps along Γ–X (4–8 meV) and Γ–Y (1–2 meV). The resolution used was 15–30 meV. They fitted the lower symmetry of their gap angular dependence with a linear combination of a gap with $d_{x^2-y^2}$ symmetry and one with d_{xz} and d_{yz} symmetry, with amplitudes in the ratio of about 2:1. The second term is needed to obtain the lack of a zero gap along the Γ–X and Γ–Y directions. Clearly this is an unusual result. Sample and analyzer alignment may be crucial here. Moreover, it is possible that spectral features due to the superlattice have complicated the spectra, making extraction of a gap from data taken along Γ–X problematical. As will be described below, more recent measurements of gap symmetry have used a dense sampling of points in angle around the Fermi surface. In so doing, it may be found that the minimum (or zero) in the gap is a degree or so away from the position expected from the initial alignment. A simple misalignment in angle should make the gap measured along the assumed Γ–X and Γ–Y lines larger than the true minimum, but symmetry (tetragonal symmetry) would dictate that they at least be the same. The superlattice causes a lower symmetry and, as described below, has indeed caused problems in obtaining values of Δ along and near the two directions in question.

Ding *et al.* (1994a) studied the anisotropy of the gap with higher energy resolution (10 meV) but with 2° angle resolution. Their Bi2212 crystals were from the same batch used previously by Olson *et al.* (1990) who had found an isotropic gap. Some of the cleaves showed an anisotropic gap similar to that seen by Dessau *et al.*, with a maximum of 22 meV along Γ–$\overline{\text{M}}$ and a minimum near zero along Γ–X and Γ–Y, consistent with $d_{x^2-y^2}$ symmetry. However, some cleaves gave an isotropic gap of about 22 meV, similar to that of Olson *et al.*, and larger than that observed by Dessau *et al.* in "deteriorated" samples. The differences could not be attributed to obvious bulk properties, since T_c, the widths of the transition to superconductivity, and the x-ray diffraction patterns were similar. Most telling was the fact that a second cleave of a sample with an isotropic gap gave a surface with an anisotropic gap. Some aspect of the surface quality appears responsible for the loss of anisotropy in the gap, but a microscopic explanation is not yet at hand. Ding *et al.* also reported the dip, at 65 meV in their case, that appeared only when the gap was anisotropic, and then only for the direction of largest gap.

Most recently Ding *et al.* (1996c) and Yokoya *et al.* (1996) re-examined the anisotropy of the gap. Ding *et al.* used 17 meV total energy resolution and an angle resolution corresponding to a circle in reciprocal space with a radius of $0.045\pi/a$. They took many spectra over a small range of angles near the directions of the minimum in Δ, spacing data points every $0.045\pi/a$ along the Fermi surface and $0.0225\pi/a$ normal to it. The peaks below T_c were fitted with a BCS spectral function plus a background to extract values of Δ. The resultant gaps as a function of angle around the Fermi surface are shown in Fig. 7.15. The

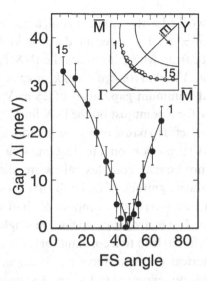

Fig. 7.15 Gap in superconducting Bi2212 vs. angle around the Fermi surface. The reciprocal-space locations of the points are shown in the inset. The solid curve is a fit with $d_{x^2-y^2}$ symmetry. (Ding et al., 1996c)

curve fitted to the data represents a gap with $d_{x^2-y^2}$ symmetry. The maximum gap, 33 meV, is larger than that usually reported by others, perhaps due to the method used to extract the gap from the spectra, although the authors also state that the use of the shifts of the leading edges gives gap estimates within 3 meV of those obtained from the fitting. (T_c was 87 K for the sample used for this figure.)

In earlier work this group had made measurements of the gap anisotropy in which minimum values of Δ were found not along Γ–X but about 10° off this line (Ding et al., 1995). This was later shown to be in error due to the near coincidence of the main Fermi surface along the Γ–X direction and a section of the Fermi surface due to an umklapp process with a wave vector of the superlattice, i.e., one of the 5 × 1 sections of the Fermi surface in Fig. 7.4 (Norman et al., 1995b). The small peak on the Δ value observed on the Γ–X line was not from the main CuO_2 Fermi surface itself, but from another segment elsewhere in the zone, translated to the Γ–X line by the umklapp process. The two contributions in this region of the zone have different polarization dependences, but in the original measurement, only one polarization was used. Ding et al. (1996c) later used the other polarization to study the Γ–X line, thereby increasing the ratio of the main band emission to that of the band produced by the superlattice. They obtained the expected zero or near-zero gap energy along the Γ–X line. Shen et al. (1993b) did not have this problem, for they did not observe the Fermi-surface crossing produced by the umklapp processes. They used a photon energy of 22.4 eV, 3.4 eV higher than that used by Ding et al. Again, dipole matrix elements probably account for the differences in the two spectra.

Yokoya *et al.* (1996) worked with a single crystal, a resolution of 23 meV, and an angular acceptance of $\pm 1°$. They measured above and below T_c in an arc around the Fermi surface from where it crosses the Γ–X line to about halfway to the $\overline{M}$–X line. A fit of Δ to $0.5|\cos k_x a - \cos k_y a|$ gave an excellent straight line, with maximum and minimum gap values of 25 meV and 0. The smallest measured gap was 3 meV for a point just off the Γ–X line.

At this point it may be useful to pause to discuss in more detail the gap symmetry. The gap function $\Delta(\mathbf{k})$ is proportional to the Ginzburg–Landau order parameter Ψ. Thus, Δ need not be real; complex values are allowed. Photoelectron spectroscopy gives only the magnitude of Δ. Until 1986 all known superconductors, except for several heavy-fermion compounds, had an order parameter ("gap") that was described as isotropic or an s-wave singlet. Measurements of Δ as a function of angle should lead to a constant for a material with a spherical Fermi surface. Non-spherical Fermi surfaces would impart an angular dependence to Δ, but one that is not expected to have nodes, hence sign changes. The cuprates long have been suspected of having non-s-wave coupling, coupling with $d_{x^2-y^2}$ symmetry, the symmetry of the Cu 3d hole orbitals in the CuO_2 planes. If we consider a spherical Fermi surface again, Δ will then have four angular nodes 90° apart in the CuO_2 plane. It will be positive in two quadrants and negative in the other two. The square-planar symmetry is compatible with the tetragonal lattice often assumed for Bi2212.

The simplest angular dependence for s-wave symmetry is a constant. Mahan (1993) showed that the more complicated form, $\Delta_0 + \Delta_4 \cos(4\phi)$, where ϕ is the angle in the plane from an axis often taken to be a Cu–O bond direction, also has s-symmetry, or more properly the symmetry of the Γ_1 representation of the D_{4h} point group. This represents the first two terms with this symmetry in a series expansion of Δ in angle. The first such term for $d_{x^2-y^2}$ symmetry is $\cos k_x a - \cos k_y a$, where $k_x = k_F \cos \phi$ and $k_y = k_F \sin \phi$. This plots as a function of angle into the familiar four-lobed pattern with the opposite signs on alternate lobes. These angular functions are plotted in Fig. 7.16. When k_x and k_y are small, $\cos k_x a - \cos k_y a = -(a^2/2)(k_x^2 - k_y^2) = -(a^2 k_F^2/2) \cos 2\phi$. When the Fermi surface is not a sphere, these patterns distort, but the angular nodes remain. Note that when Bi2212 is assumed to be tetragonal the Cu–O bond is parallel to the Γ–$\overline{M}$ direction. This means that k_x is directed at 45° to the Γ–X direction, not along it. When Bi2212 is treated as orthorhombic, the symmetry is lower (D_{2h} point group). The result is that both the s and $d_{x^2-y^2}$ symmetries for Δ of the tetragonal lattice now belong to the same representation, and they may occur simultaneously. Δ could have the symmetry of a linear combination of both. However, there is hope that the CuO_2 planes may be treated first as square, then a small distortion added to get the nearly tetragonal structure. This would make the symmetry of Δ predominantly s or $d_{x^2-y^2}$. The validity of this

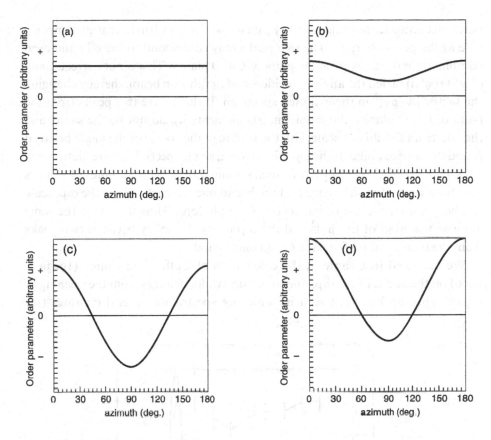

Fig. 7.16 Expected angle (or k-) dependence of the superconducting order parameter (assumed real) for several symmetries of the pairing interaction. Photoemission measures only the absolute value of the ordinate, shown dashed. (a) s-wave, with spherical Fermi surface. (b) s-wave with anisotropic Fermi surface (of $d_{x^2-y^2}$ symmetry), (c) d-wave ($d_{x^2-y^2}$) coupling, (d) s plus d coupling. A larger s-component could make part (d) resemble part (b). In part (c) the zeroes are at 45° and 135°, while in part (d), the zeroes have moved closer to 90°. The phases are arbitrary, but measurements on Bi2212 indicate that 0° corresponds to the Γ–X or Γ–Y direction.

may be questioned in view of the features with two-fold symmetry that appear in the Fermi surface in the measurements of Aebi *et al.* The combination s + $d_{x^2-y^2}$ with suitable coefficients will have the four lobes of the $d_{x^2-y^2}$ symmetry alone, but now the s-component may be large enough that there is no sign change from lobe to lobe and no angular nodes. Another possibility is that the coefficients could give angular nodes no longer 90° apart, but separated by angles alternating between larger and smaller than 90°. More extensive discussion may be found in Annett *et al.* (1990, 1996), Annett (1996), and Scalapino (1995).

Consider again the dip observed in the superconducting state for directions with larger values of the gap. A natural interpretation is that it is the valley between two adjacent peaks, a valley that grows deeper with decreasing temperature as the peaks sharpen. The two peaks may correspond to the odd and even combinations of states from the pairs of Cu–O planes. This is not correct. Ding *et al.* (1996a) varied the angle of incidence of the photon beam, thereby changing the factor $|\mathbf{A} \cdot \mathbf{p}_{if}|^2$ in the expected spectrum. If there were two peaks from the pairs of Cu–O planes, the dipole matrix elements would not be the same, and the spectrum should not scale with the square of the cosine of the angle between $\mathbf{A}$ and the surface plane. If the dip arises from a single spectral feature, designated a many-body effect by Ding *et al.*, such scaling should occur. Figure 7.17 shows that both peaks, the sharp one and the broad one, on each side of the dip, scale similarly with the angle of incidence of the photons. Thus they have the same angular variation of $|\mathbf{A} \cdot \mathbf{p}_{if}|^2$ and the dip is not the valley between two peaks from the pairs of states from the Cu–O plane bands.

We discussed in Chapter 3 the calculation of Coffey and Coffey (1993a,b, 1996) on the origin of the dip, due to lifetime width changes from the opening of a gap. This dip has been seen in tunneling spectra for several cuprates (see

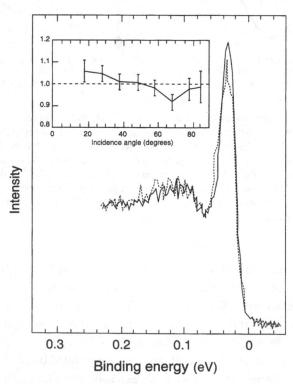

Fig. 7.17 EDCs at the $\overline{\mathrm{M}}$ point of Bi2212 at 13 K taken at two angles of incidence of the photon beam. Solid: 18°, dashed: 85° from the sample normal. The inset shows the height of the sharp peak normalized to the height of the broad peak as a function of angle of incidence of the photons. (Ding *et al.*, 1996a)

Chapter 12). Kouznetsov and Coffey (1996) re-examined tunneling calculations and added several effects: anisotropy; and the presence of a dead layer which leads to incoherent tunneling. This calculation led to the presence of dips in tunneling spectra and in angle-resolved EDCs. However, the location of the dip in an EDC is different, about Δ to 2Δ below E_F for assumed $d_{x^2-y^2}$ coupling. It depends on the magnetic fluctuation correlation length, for in the model used the dip arises from the superposition of two spectra, a broadly peaked one from inelastic processes (electron scattering by exciting spin fluctuations) and a narrow peak near E_F. The correlation length affects the shape of the former. However, it must be assumed that the background from these inelastic processes is the entire background in the photoelectron spectrum, in which case the shape is not right, or one must find a way to separate the measured background into at least two components, the one from the processes of Kouznetsov and Coffey and one from other processes. Xiao and Li (1994) carried out a similar calculation. The inelastic scattering processes were the excitation of antiferromagnetic fluctuations, the same interaction leading to superconductivity in this model. In the normal state, a highly correlated one, this interaction slightly weakened the quasiparticle peak. It strengthened this peak just below T_c when the peak was near enough E_F, i.e., it produced the "pileup." Neither of these effects involved the excitation of magnetic fluctuations. The inelastic excitation of antiferromagnetic fluctuations led to a large background at larger binding energies, extending without structure to at least 500 meV with the parameters used. Absorption of such excitations was also possible, leading to a tail of the background above E_F. Below T_c this background is reduced in a region below E_F a few Δ wide, leading to a dip in the spectrum around 3Δ below E_F. The inelastic background in this calculation does not satisfy a sum rule. Coffey et al. (1997) revisited this problem.

The dip appears at approximately 3Δ, becoming small, but not shifting in energy, for those directions for which Δ is small. Randeria et al. (1995) have shown that the angle-resolved photoemission spectrum is proportional to the spectral density function $A(E,k)$ at the photon energies used, i.e., about 20 eV. Then the sum rule described above makes $n(\mathbf{k}_F)$ a temperature-independent constant. Thus, the area of an angle-resolved EDC should be temperature independent, provided that it is taken at $\mathbf{k} = \mathbf{k}_F$ and that the background is subtracted. The peak or pileup seen upon going below T_c is then not just the pileup or peak in the BCS spectral density, but something more complicated, involving not just states removed from the region near E_F, but possibly states near the region of the dip, which also participate in the sum rule.

The temperature dependence of the gap has been measured several times, but rarely published. For example, in the early data of Olson et al. (1990b) in which an isotropic gap was reported, the temperature dependence was determined for comparison with the universal BCS plot of $\Delta(T)/\Delta(0)$ vs. T/T_c. It was not published because of the increasing error bars as T_c was approached from below. The data points all remained above the BCS curve at higher temperatures, characteristic of strong coupling, but the error bars increased rapidly, so that most of them reached the BCS curve. Similar results were obtained by other groups and not published. Ding *et al.* (1994b) reported the temperature dependence of the gap at three points on the Fermi surface, where the low-temperature gaps were 22, 6, and 5 meV. The temperature dependence of the gap at each point agreed with the BCS prediction, within estimated experimental errors. Ma *et al.* (1995a) described a very different result for the temperature dependence of the gap in Bi2212 for $\mathbf{k_F}$ along $\Gamma-\overline{M}$ and $\Gamma-X$. At low temperatures, the gaps reported for these two directions were 16 meV and about 10 meV, respectively. The gap for $\mathbf{k_F}$ along $\Gamma-X$ dropped to zero at a temperature around 0.8–0.9 T_c, while the gap along $\Gamma-\overline{M}$ did not begin to fall until about 0.85 T_c, above which it fell to zero at the expected temperature of T_c. These gaps were derived from the spectra by the BCS curve fitting used by Olson *et al.* (1989) and by the shift of the leading edge of the EDC. The differences in gap energy between the two methods were less than 2 meV. Most other studies of the gap anisotropy well below T_c found a larger gap along $\Gamma-\overline{M}$ and a smaller or zero gap along the $\Gamma-X$ line than those found by Ma *et al.*, except when the sample "ages." When ambient gas condenses on the surface, the gap along $\Gamma-\overline{M}$ is reduced, while that along $\Gamma-X$ increases, bringing the gap values in previous studies, e.g., that of Shen *et al.* (1993b), closer to those of Ma *et al.* However, even if the samples of Ma *et al.* were not the best, it seems worthwhile to remeasure the temperature dependence of the gap along $\Gamma-\overline{M}$ and, since the gap along $\Gamma-X$ is very small or zero, half-way between $\Gamma-\overline{M}$ and $\Gamma-X$ to see if both collapse to zero at T_c. One should also keep in mind that it is along the $\Gamma-X$ line that the superlattice bands complicate measurements near E_F (see Fig. 7.6). Thus these measurements may have been more readily interpreted had they been made along $\Gamma-Y$.

Another approach to the symmetry of the order parameter is to introduce suitable impurities or defects to convert the sample to a dirty superconductor. For an anisotropic s-wave superconductor this should cause, in effect, an average over a small range of angles in a plot such as Fig. 7.15. This will broaden the maxima and minima so the angles where Δ is zero become regions of finite angular extent with a small, but not zero, value for Δ (Borkowski and Hirschfeld, 1994; Fehrenbacher and Norman, 1994). For a d-wave superconductor the effects of impurity or defect scattering are more difficult to recognize in photoelectron spectra, but the zeroes remain, with broader minima about them. Large defect or impurity concentrations should cause Δ to go to zero for the d-wave, but not s-wave, case. Ideally these defects should be point scattering centers, but not dopants. Radiation damage can produce defects that satisfy this requirement (Giapintzakis *et al.*, 1994). Irradiation with 1 GeV electrons displaces oxygen atoms from the CuO_2 planes into interstitial positions. Ding *et al.* (1996d) carried out such an irradiation on a single crystals of Bi2212. The irradiation lowered T_c by 8 K, but the transition was not broadened. The leading edge of the EDC changed shape as a result of the irradiation; the peak became smaller and a region below the midpoint of the edge shifted toward E_F. The position of the midpoint of the edge as a function of angle around the Fermi surface was not changed by the irradiation. The apparent $d_{x^2-y^2}$ symmetry remained and the region around the zero in Δ vs. angle was not broadened. However, the region at the foot of the Fermi edge clearly indicated the presence of gapless superconductivity, compatible with a gap of $d_{x^2-y^2}$ symmetry.

A different type of photoelectron spectroscopy was carried out by Park *et al.* (1994). They scanned photon energies from below the threshold for photoemission ($h\upsilon = \Phi$) in Bi2212 to a bit above it, measuring the total yield of electrons. This was done between 30 and 110 K, below and above T_c. The threshold was 4.76 ± 0.05 eV. A small yield of photoelectrons was found below threshold, but it varied from sample to sample and was assigned to chemical inhomogeneities on their sample surfaces, which were cleaved at low temperature in ultrahigh vacuum. The yields were interpreted with a model in which the yield for $T > T_c$ was proportional to an integral over energy and angle of a density of states multiplied by a Fermi–Dirac function and the classical escape function discussed in Chapter 3. At threshold the photoelectrons escape into a very small cone around the sample normal. As the photon energy increases, the cone opens wider. Below T_c the integrand was multiplied by a BCS spectral density function. Yield vs. temperature spectra taken at threshold ($h\upsilon = 4.75$ eV) show a peak at about 80 K. This peak is attributed to the quasiparticles in the BCS spectral density function above E_F. At lower temperatures this contribution to the yield is cut off by the Fermi–Dirac function, leading to a maximum contribution at 80 K. The yields at fixed photon energies above threshold increase monotonically with increasing temperature, but for $h\upsilon = 4.95$ eV, 200 meV above threshold, there is a small depression around 95 K in the otherwise smooth curve. (Magnetic susceptibility measurements indicated $T_c = 86$ K.) They attribute the large (ca. 50%) drop in yield with cooling to the fact that the Cooper pairs have their momenta in or near the Cu–O planes, and the photon momentum cannot increase significantly the component normal to that plane. The photoexcited (broken) Cooper pairs fail to enter the narrow escape cone till the photon energy is some 500 meV above threshold. As the temperature decreases an increasing fraction of the electrons just below E_F are in Cooper pairs.

After devoting considerable attention to the main features of the angle-resolved photoemission spectra of Bi2212 as a function of temperature and doping, people began focusing on less prominent features. Shen and Schrieffer (1997) and Norman *et al.* (1997a) both re-examined angle-resolved photoelectron spectra in the first few hundred meV of binding energy at temperatures below and above T_c. In the spectra of Norman *et al.* for $T < T_c$ the sharp peak near E_F and the nearby dip are present over a range of wave vectors along the Γ–$\overline{M}$–Z line. This peak is present, but weak, over an even wider range of wave vectors along this scan, while for $T > T_c$, there is a peak near E_F for a more limited wave-vector range along this line. There is also a broad, dispersing peak in both phases. The broad peak dispersion flattens out in the superconducting phase when it approaches within about 100 meV of E_F, while it continues to move closer to E_F as the wave vector changes in the normal state. When it is near E_F it appears to lose spectral weight to the sharp peak nearer E_F. The positions of the peaks for T above and below T_c are plotted in Fig. 7.18. Norman *et al.* proposed a phenomenological model for the self-energy which mixes electronic states with a bosonic mode of unspecified nature. As a consequence, some of the structure in the spectral density function arises primarily from the excitation of this bosonic mode. They calculated a spectral density function, assuming a BCS-like density of states and a gap Δ of either s- or d-wave symmetry. The bosonic mode density of states was taken to be BCS-like, with its strength and energy as parameters to be varied. A mode energy of $1.3\Delta(\mathbf{k})$ gave the best fit. The model spectral density function peak energies are shown in Fig. 7.18, which also indicates, by the lengths of the plots, the loss in spectral weight found in both experiment and the model. The line shapes of the angle-resolved EDCs could also be fitted well with this model. The bosonic mode was associated with an excitation peaking at a wave vector of $(\pi/a,\pi/a)$ (Γ–Y). Such excitations had been observed in Y123 by neutron scattering, although there was not a single mode, but rather a distribution with a broad peak at $(\pi/a,\pi/a)$.

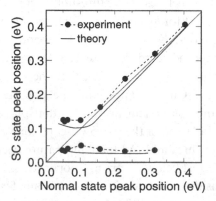

Fig. 7.18 Energies of the sharp and broad peaks for $T < T_c$ vs. energies of the same peaks above T_c (points connected by a dashed line). Solid lines are from a model calculation (see text) and the dotted line at 45° is the normal-state dispersion. (Norman *et al.*, 1997a)

Shen and Schrieffer focused on the broad peak around 100 meV binding energy that is present above T_c in underdoped samples near the $\overline{M}$ point, the peak just mentioned. This is the only spectral feature present in the first 500 meV of binding energy, even when spectra are taken at the anticipated Fermi surface. Below T_c, this feature remains, but the sharp peak and the dip appear near E_F. They assigned the broad peak to the excitation of a photohole accompanied by "sidebands" from the simultaneous excitation of collective bosonic modes. For some relative strengths of the photohole peak and the collective mode sidebands a dip can appear between them. In some regions of the Brillouin zone, coupling to this mode is strong and only this broad peak appears. In other regions, the coupling is much weaker and the sharper peak from the excitation of just the photohole dominates. The stronger coupling occurs for regions of large density of states (e.g., the flat bands near $\overline{M}$) separated from other regions of large density of states (e.g., near an equivalent $\overline{M}$ point) by a wave vector $(\pi/a,\pi/a)$. They suggest that the coupling that produces the broad peak may be the same coupling that causes pairing of electrons.

This coupling to a low-energy collective mode, or a spectrum of such modes, was foreshadowed by an analysis of angle-resolved photoelectron spectra for Bi2212 by Arnold et al. (1991). They used the Nambu–Eliashberg theory, often used to analyze tunneling data. They assumed a spherical Fermi surface. The result of the analysis is an average electron–phonon coupling parameter, λ, an average electron–electron coupling parameter, μ^*, and a spectral density function, $\alpha^2 F$, where α is a coupling parameter and F is the spectral density function for an unspecified boson to which the electrons couple. λ was large, 8.7, while μ^* was typical of many superconductors, 0.15. The spectrum of $\alpha^2 F$ was assumed to consist of three Lorentzian peaks whose parameters were adjusted to fit a spectrum based on the ratio of two EDCs, one for $T > T_c$ and one for $T < T_c$. The best fit had a sharp peak at 10 meV. The much weaker peaks at 30 and 50 meV are not significant. In the model of Norman et al. a boson energy that depended on wave vector ($d_{x^2-y^2}$ symmetry) was used. The fit gave a maximum value of 42 meV. It is not clear whether the difference in energies is meaningful because of differences in the modeling and analysis and in approximations used. Dispersion and the lifetime width of the boson mode were neglected. In the model of Shen and Schrieffer, the broad peak at about 100 meV occurs at an energy of several times the energy of the bosonic mode.

There are boson modes that may correspond to the mode used in the descriptions just given. Fong et al. (1995) used inelastic neutron scattering on a single crystal of $YBa_2Cu_3O_{6.95}$. An excitation at 41 meV peaking at a wave vector of $(\pi/a,\pi/a)$ was already known. Fong et al. showed that this mode originates from a phonon involving out-of-plane motion of the O atoms in a CuO_2 plane. (This phonon has an energy of 42.7 meV in Raman spectra, but due to the

resolution of the neutron spectrometer, it may appear in spectra taken at 41
meV.) For $T < T_c$ an additional contribution appears in the 41 meV spectrum.
It has the same wave vector, but is of magnetic origin, representing the excitation
of an antiferromagnetic fluctuation.

Norman *et al.* (1998) introduced a different way to display angle-resolved
EDCs near the Fermi level. Called symmetrization, it was designed to reduce
the apparent effect of the falloff in the Fermi–Dirac function on the appearance
of an edge or peak near E_F in an EDC. The spectra are proportional to
$I(E) = A(\mathbf{k}_F, E)f_{FD}(E)$. $I(-E)$ is generated by reflecting the data about E_F, the
zero of apparent binding energy. The symmetrized spectrum is $I(E) + I(-E)$.
If particle–hole symmetry is valid in a small energy range about E_F,
$A(\mathbf{k}_F, E) = A(\mathbf{k}_F, -E)$, and since $f_{FD}(-E) = 1 - f_{FD}(E)$, $I(E) + I(-E)$ is
$A(\mathbf{k}_F, E)$ itself (convolved with the instrument function). Errors of a few meV in
locating E_F can produce spurious peaks in the symmetrized spectrum, as can
departures in $\mathbf{k}$ from $\mathbf{k}_F$. The symmetrized spectrum differs from the raw spectrum, $I(E)$, only in a range $\pm 2.2 k_B T_{eff}$, where $4.4 k_B T_{eff}$ is the energy between
the 90%–10% width of the Fermi edge measured on a reference sample with no
structure near E_F, e.g., Pt. In this limited energy range no background spectrum
need be subtracted before data analysis. The values of gaps obtained this way
agree with those obtained by using the leading edge of the EDC. Figure 7.19
shows a symmetrized spectrum from a sample of overdoped Bi2212. The peak
appears about 25 meV below E_F in the raw spectrum, but this an artifact from
the falloff in f_{FD}. The symmetrized spectrum shows a peak at E_F.

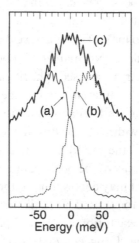

Fig. 7.19   (a) Measured angle-resolved EDC at $\mathbf{k}_F$ along $\overline{M}$–Y in Bi2212 above T_c. (b)
Spectrum of (a) reflected about E_F. (c) The sum of (a) and (b), the symmetrized spectrum.
(Norman *et al.*, 1988)

7.3   Doping studies

Shen *et al.* (1991b) varied the hole concentration in single crystals of Bi2212 by annealing in different atmospheres. From the Hall coefficient, the hole concentrations ranged from 4.3 to 5.2 in units of 10^{21} cm^{-3}; T_c varied from 89 to 77 K. These samples were overdoped. Angle-resolved photoemission showed that there was an increase in the relative size of the Fermi edge in the spectra with increasing hole concentrations. Angle-integrated photoemission spectra of both valence bands and shallow core levels gave spectra that were identical, except for a slight rigid shift attributed to a change in the Fermi level (chemical potential). (All spectra were referenced to the Fermi level of the spectrometer.) The edge in the sample with the higher hole density moves lower with respect to the bands and core levels by 0.15–0.2 eV, the direction expected. A shift of about 0.1 eV is expected from the density of states at E_F from the LDA band calculation and the change in hole density.

van Veenendaal *et al.* (1993a) carried out similar measurements, varying the hole density by varying the oxygen concentration and by replacing some Ca with Y. Over a range of hole densities from 0.01 to 0.23 per CuO$_2$ unit, a wider range than that of Shen *et al.*, the Fermi level shifted by up to 0.6 eV, as judged by the Fermi edge, and by up to 0.64 eV from the shift of several core lines. These two measurements of the shifts agree with each other when one considers the accuracy of the measurements of the shifts. The shifts were not linear, however; the rate of shift with hole concentration decreased as the hole concentration increased, i.e., as the samples went from insulating to metallic. The shifts were assigned to shifts in the Fermi level with doping. Core-level shifts due to other causes, e.g., a changing Madelung potential, were no larger than about 0.2 eV.

Ori *et al.* (1995) also compared Bi2212 as grown and after annealing in 12 atmospheres of O$_2$, using angle-integrated valence-band spectra and electron energy-loss spectra taken with 200 eV electrons in a reflection geometry. They also added a monolayer of Bi to the surface of the annealed sample and exposed the surface of the as-grown crystal to 100 L of O$_2$. As expected, the more metallic oxygen-annealed sample showed a larger step at the Fermi level in the photoelectron spectra. There were differences in the 0–7 eV range of binding energy as well. The annealed crystal had more obvious structures, and the largest structure (at 4.5 eV) was shifted 0.25 eV to lower binding energy in the annealed sample. Unfortunately, this large structure probably arises from the contributions of several bands, so the shift is not easily interpreted. The presence of the Bi overlayer caused the removal of the step at the Fermi level, an effect ascribed to the oxidation of the Bi by "disordered oxygen on the cleavage plane." Oxygen exposure of the surface of an as-grown sample caused the electron energy-loss spectrum to resemble that of the oxygen-annealed surface.

Rietveld *et al.* (1995) carried out their work with polycrystalline samples of $(Bi,Pb)_2 Sr_2Ca_2Cu_3O_{10+\delta}$. The addition of Pb was required to stabilize this phase. The samples were subjected to four surface treatments prior to XPS measurements, involving scraping or annealing in ozone or vacuum. Annealing in ozone affected only the surface regions, so the oxygen stoichiometry changes were monitored by a surface-sensitive technique, the relative areas of core-level peaks. The O 1s peaks had significant shoulders at 531 eV binding energy. The binding energies of all of the core levels measured, from Bi, Pb, Ca, Sr, Cu, and O, shifted with oxygen concentration. The shifts were not linear. They varied over 0.6 eV when δ varied from -1 to $+1$, in the direction expected if hole doping moved the Fermi level into the valence band. The shifts observed for this material were similar to those found by van Veenendaal *et al.* for Bi2212.

Kelley *et al.* (1996) varied the hole concentration by annealing in O_2 or in Ar, then measured the anisotropy of the gap as determined by its values along $\Gamma-\overline{M}$ and $\Gamma-X$. The gap was determined by the shift of the leading edge from the Fermi edge of an adjacent Au sample. An underdoped (Ar-annealed) crystal gave $\Delta_{\overline{M}} = 17$ meV, $\Delta_X = 0$, with an estimated error of ± 2 meV. An overdoped (O_2-annealed) sample gave 22 and 10 meV, respectively, for these gaps. Such results were obtained on four crystals of each type. The authors concluded that the symmetry of the order parameter changes with doping. For $d_{x^2-y^2}$ symmetry the gap along $\Gamma-X$ is zero, as found in the underdoped crystals. The non-zero gap in this direction, if not an artifact, precludes pure $d_{x^2-y^2}$ symmetry for the overdoped crystals. As mentioned above, the superlattice may have some influence on these results.

The substitution of trivalent Y for divalent Ca has been used to reduce the hole concentration. Bi $4f_{5/2}$, Sr $3d_{5/2}$, Ca $2p_{3/2}$, Y $3p_{3/2}$, and Cu $2p_{3/2}$ spectra of powdered samples of $Bi_2Sr_2Ca_{2-x}Y_xCu_2O_y$, with x between 0 and 1 were taken by Shichi *et al.* (1990). When x exceeds about 0.6 the samples are insulating. The Sr and Y level binding energies increased with x, while the Bi binding energy remained unchanged. The Sr $3d_{5/2}$ line shape changed progressively as x increased. There were two main components, one assigned to Sr atoms between the Bi–O and CuO_2 planes and the other to Sr between the pairs of CuO_2 planes. The latter weakened with respect to the former as x increased, indicating that the Y atoms substituted primarily for the Sr atoms between the CuO_2 planes. The Cu $2p_{3/2}$ satellite intensity weakened with respect to that of the main peak. This was taken to mean that the average valence of Cu decreased with increasing Y substitution. The core-level binding energy shifts were later correlated with bond-valence sums (Itti *et al.*, 1991).

The bond-valence sum originates with an ionic picture of binding (Brown and Altermatt, 1985). For a simple cubic ionic crystal AB, where the ions have charges $+Z$ and $-Z$, the crystal binding energy per ion pair, relative to separated free ions, is

$$U(R) = -\alpha_M (Ze)^2/R + Nb\exp(-R/\rho),$$

with α_M the Madelung constant, using the nearest-neighbor distance R as a basis, and N the number of nearest neighbor ions. b and ρ are parameters in the repulsive energy that are nearly constant over a wide variety of similar materials, e.g., oxides, where $\rho = 0.37$ Å. The equilibrium bond length d is given by setting $\partial U/\partial R = 0$ at $R = d$. The bond valence s is defined as $s = Z/N$, which becomes, after some simplification,

$$s = \exp[(R_0 - d)],$$

with

$$R_0/\rho = \ln(bd^2/\alpha_M \rho Ze^2).$$

For the simple AB crystal the bond-valence sum is the sum of N equal values of s, giving $+Z$ or $-Z$, depending on the ion. For more complex materials, there may be more than one contribution to the bond-valence sum. Values of s have been tabulated for many ion pairs (Brown and Altermatt, 1985).

If an element has more than one oxidation state (valence), e.g., Cu, the bond-valence sum becomes more complex. Brown (1989a) has tabulated values of 1.60, 1.68, and 1.73 for R_0 for Cu^+, Cu^{2+}, and Cu^{3+}, respectively, when bonded to O. Then one assumes a valence of $+2$, and evaluates $V(i) = \sum_j s_{ij}$ for Cu. If it is larger than 2 or smaller than 2, Brown (1989b) has proposed an interpolation scheme based on the bond-valence sum assuming Cu^{2+} and the bond sum for the next higher or lower oxidation state.

Itti et al. (1991) evaluated bond-valence sums for $Bi_2Sr_2CaCu_2O_y$ and $Bi_2Sr_2YCu_2O_y$. All sums except that for the Cu–O bond were larger in the latter material. The largest structural change upon the complete substitution of Ca by Y is a decrease in the unit cell length along c, by about 0.5 Å, and this plays a large role in the changes in the bond-valence sums. The larger bond-valence sums indicate a higher valence, which should cause an increased core-level binding energy. This correlation appears in the Sr, Ca, and Y core-level spectra measured on $Bi_2Sr_2Ca_{1-x}Y_xCu_2O_y$, but not the Bi 4f, which has the largest change in bond-valence sum. The Cu 2p core peaks shift with x, but the bond-valence sum does not change much. Clearly the bond-valence sums have to be used with some caution.

Similar samples of Y-doped Bi2212 were used in a study by inverse photoemission and by ultraviolet and x-ray-induced photoemission of the Cu 2p states and the valence-band region (van Veenedaal et al., 1994b). (Much of the original data appear in van Veenendaal et al., (1993a).) In addition to changes in the hole density by Y substitution, there were changes due to varying O stoichiometry in their seven samples. The hole concentrations from both causes were determined. Sample surfaces were prepared by scraping in vacuum. There was no high-energy structure in the O 1s spectra. They used a Cu_3O_{10} cluster with two or three holes per cluster to model the Cu 2p spectra. The Cu $2p_{3/2}$ main line, broad and asymmetric, narrows with increasing Y content. It arises from two final states, $\underline{c}3d^{10}$ at lower binding energy (the hole is on a neighboring Cu–O cell, forming a Zhang–Rice singlet) and $\underline{c}3d^{10}\underline{L}$ at higher. The satellite is primarily a $\underline{c}3d^9$ final state. Removing a hole in the cluster calculation results in a smaller splitting of the two components of the main line, in agreement with the measured spectra.

The valence-band spectra also change as the hole concentration varies. Starting with the fully Y-substituted sample, the insulating Bi2212 becomes metallic after enough holes are added by Y removal. The chemical potential starts in a gap, and either shifts into the valence band at the metal–insulator transition or remains in the gap while new states are transferred into the region of the chemical potential. A shift in the chemical potential should result in the same shift for all core-level binding energies. They do not all shift equally, but there are other effects. The Bi 4f level does not shift as much as Ca or Sr, but the Bi–O planes are the ones losing oxygen atoms when hole doping is accomplished by oxygen variation. Lattice parameter changes and ion valence changes can alter the Madelung potentials differently for different ions in the case of O or Y doping. Nonetheless, after an extensive discussion, van Veenendaal et al. conclude that the chemical potential shifts upon hole doping the insulator. It shifts rapidly, about 0.5 eV between $x = 1$ and $x = 0.6$, the latter at the metal–insulator transition, then more slowly as the metallic phase is hole doped more and the chemical potential enters the valence band. The valence-band photoemission and inverse photoemission spectra show a closing of the insulating gap upon hole doping and a slight shift in the position where the two spectra cross, along with an approach of the largest valence-band peak to the crossing point, i.e., to the chemical potential. This is consistent with the picture of a chemical potential that shifts with hole doping.

Small amounts of Cs have been used to "inject" electrons into the surface region of Bi2212, thereby reducing the hole concentration (Söderholm et al., 1996). The Cs is easily ionized, and the electron enters the surface layer(s). During the early stages of deposition of Cs the surface is not disrupted. The Cs is presumably in the form of atoms or clusters. Additional Cs deposition leads to the appearance of cesium oxides and reduced Ba, both judged by core-level spectroscopy. We concern ourselves here with the early stages of Cs deposition. The chemically reacted surfaces will be discussed in Chapter 10. With small amounts of Cs on the surface of Bi2212, several changes are observed in the photoelectron spectrum and in the soft x-ray absorption spectrum determined as partial and total yield spectra. Intensity is lost at and near the Fermi level in the valence band and the pre-edge structure of the O 1s absorption spectrum is diminished. Several core-level peaks (Ba 4f, O 1s) shift to higher binding energy by a few tenths of an eV, depending on the Cs coverage. This shift is with respect to the Fermi level, so it can be explained by an increase in the Fermi level due to the Cs coverage, which removes holes. The shift is comparable in magnitude, but not in direction, to that seen upon annealing in oxygen (Wells et $al.$, 1990b). Hole loss near E_F also explains the weakening of the pre-edge structure in the O 1s spectrum, for this structure is believed to arise from the excitation of O 1s electrons into empty O 2p states near E_F. The O 1s photoemission peak, a composite, also changes with Cs coverage. The low-binding energy component weakens. This component must be related to the loss of holes near E_F. The same group had already studied the components of this peak (Qvarford et $al.$, 1996), and the component in question assigned to oxygen in the CuO_2 planes in the bulk crystal. These are the same oxygen atoms responsible for the other spectral features that also weaken with Cs coverage. The electrons from the Cs atoms on the top Bi–O layer then are located in the CuO_2 plane(s) below. When the Cs coverage reaches the point at which all of the O 2p holes are filled, the disruption of the surface begins.

The underdoped region was explored early, e.g., Takahashi et $al.$ (1990), but it has taken on much more significance in recent years. It has proven difficult to carry out photoemission spectroscopy on the insulating parent compounds, then to dope them toward and into the metallic phase. Instead, underdoping metallic samples has been pursued, and after about eight years of photoelectron spectroscopy of Bi2212, some surprises were found.

Marshall et $al.$ (1996) varied the hole concentration in Bi2212 by substituting Dy for Ca in their epitaxial thin-film samples. This substitution reduces the hole concentration. With 1%, 10%, and 17.5% Dy, T_c had values of 85, 65, and 25 K, respectively. 50% Dy produced an insulating sample. Fresh surfaces of these films were produced by cleaving in vacuum at 110 K, the measurement temperature. Single crystals with no Dy had hole concentrations varied by annealing in O_2 or in Ar. The spectra (0–600 meV binding energy range, 35–45 meV resolution) for the sample with 1% Dy resembled those of optimally doped Bi2212, a single Fermi-surface crossing along each of the two directions scanned, Γ–X and $\overline{M}$–X. The sample with 50% Dy had very broad peaks on a large background, with no strength at E_F. Their dispersion resembled that of the previously measured insulator $Sr_2CuO_2Cl_2$ (Wells et $al.$, 1995). 10% Dy doping produced spectra of intermediate character. The Γ–X Fermi-surface crossing was present and the peaks were almost as narrow as those for the sample with 1% Dy, but the spectra along $\overline{M}$–X were broadened and displaced 20–30 meV to higher binding energy, as were the spectra of the insulating sample, and the Fermi-level crossing was lost. Very similar spectra could be produced by underdoping a single crystal by annealing in Ar. The effects appear to be due to the hole concentration, not to particular atom substitutions.

Spectral line-shape analysis for Bi2212 is not yet a science. In order to map bands, Marshall *et al.* subtracted a background in the usual way, one proportional to the current of elastically scattered electrons of higher kinetic energy, then fitted the peaks to the product of a Lorentzian and a Fermi–Dirac function. The parameters from such a fit may not have direct physical significance, but the changes with doping are qualitatively, perhaps quantitatively meaningful. Figure 7.20 shows the centroids of such fits as a function of wave vector. The resultant Fermi surface is shown in Fig. 7.21. Band theory would predict the underdoped sample to have a Fermi surface with approximately the same shape as that for the optimally doped or overdoped samples, just shrunk somewhat around the X point. However, part of it is missing, since it does not reach the $\overline{\text{M}}$–X line. One possible explanation is that a gap has opened up near the $\overline{\text{M}}$–X line. For example, quasiparticles could form pairs without the coherence between pairs required for superconductivity. A gap with $d_{x^2-y^2}$ symmetry would be consistent with the location of the missing Fermi-surface region. A fluctuating superlattice of suitable symmetry also could cause a gap in the proper part of the Brillouin zone. In addition to the Fermi surface changes, the region of the extended critical points near $\overline{\text{M}}$ moves quite far, about 200 meV, below E_F in the underdoped samples.

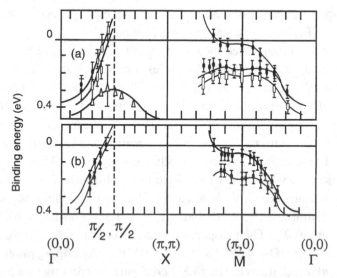

Fig. 7.20 Band maps for Bi2212 using the centroids of the fitted line shapes. In (a) the hole density was reduced by replacing Ca with Dy. In (b) the hole density was varied by annealing in oxygen or argon. In both panels, the highest curves have the largest hole density. (Marshall *et al.*, 1996)

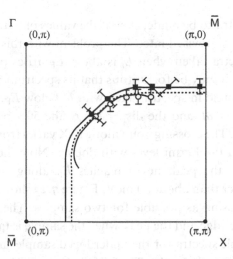

Fig. 7.21 Fermi surfaces for two samples from Fig. 7.20, one overdoped and one underdoped. The solid curves represent schematic Fermi surfaces based on the data for the two samples. The dashed line is the Fermi surface expected for the underdoped sample on the basis of band structure. (Marshall *et al.*, 1996)

Although there was a significant loss of Fermi surface, these samples exhibited a gap at temperatures far above T_c, the temperature increasing as the hole concentration fell. Loeser *et al.* (1996a,b) and Ding *et al.* (1996b) both report the occurrence in underdoped crystals of Bi2212 of a gap above T_c of the same magnitude and symmetry as the gap observed below T_c in superconducting samples. Such a gap is not present above T_c in overdoped crystals or crystals near optimal doping. This gap is widely called a pseudogap, a carry over from earlier measurements of several physical properties of Y123. We discuss this further below.

The samples of Loeser *et al.* were single crystals with oxygen concentrations varied on both sides of optimal doping by annealing. All had T_c in the range 70–95 K. The measurements were carried out at 100–110 K with 25–35 meV overall resolution. Scans were made along two symmetry lines, Γ–X and $\overline{\text{M}}$–X, and along several other lines crossing the Fermi surface. In all the spectra a peak was seen that dispersed toward the Fermi level. In the optimally doped and overdoped samples, this peak appeared to pass through the Fermi level and disappear at the expected point (or small area) in the Brillouin zone. The same thing happened for all samples for the scan along Γ–X. This is the direction along which the superconducting gap has its minimum value. For the underdoped sample in the scans along $\overline{\text{M}}$–X the peak approached E_F but did not cross it. At increasing wave vector, the peak disappeared without ever reaching E_F. Loeser *et al.* subtracted the background, using the extrapolation of the apparent photocurrent at negative binding energy, and used the point half-way down from the peak as a measure of the minimum energy of the photohole. It allows comparisons of spectra to be made, even if the values of parameters, e.g., a gap energy, may have to be revised later. The problem with this method is that the photocurrent in spectra taken when E_i is above E_F arises partly from second-order radiation, and it may be fortutitous that its spectrum has about the same shape as the background in spectra taken with E_i below E_F. Figure 7.22 shows the data of Loeser *et al.* and the dispersion of the 50% point on the leading edge of the spectrum. The crossing point along Γ–X varies from sample to sample due to the change in the Fermi level with doping. Note, however, that in the underdoped sample, the peak never reaches E_F along the $\overline{\mathrm{M}}$–X direction, approaching no closer than about 20 meV. Figure 7.23 shows spectra as near to a Fermi surface crossing as possible for two samples. The overdoped sample shows the shift of the edge and the peak when the sample is taken from above T_c to below. Comparable spectra for the underdoped sample show that the edge is shifted both below and above the T_c of this sample. When $T > T_c$, however, there is no peak in the spectral intensity at the top of the shifted edge, as there is in the superconducting state. For this sample, when **k** is moved away from $\mathbf{k}_F$ along the $\overline{\mathrm{M}}$–X line and the initial state has a binding energy of 0–50 meV, there

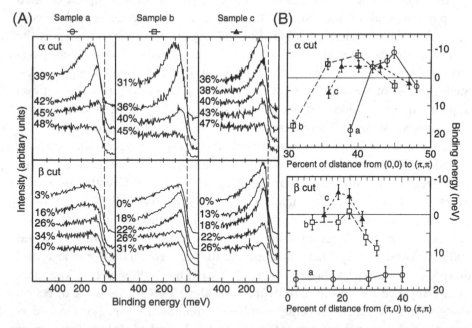

Fig. 7.22 (A) EDCs from three crystals of Bi2212 taken at 100 K. Crystal a is underdoped ($T_c = 84$ K) while b and c are overdoped ($T_c = 85$ and 80 K, respectively). The α cut is a scan from Γ to X and the β cut, from $\overline{\mathrm{M}}$ to X. The wave vector is indicated as a percent of the scan range. (B) Binding energies of the midpoints of the leading edges of the spectra in (A) vs. wave vector. In sample a the edge never reaches E_F in the scan from $\overline{\mathrm{M}}$ to X. (Loeser *et al.*, 1996a)

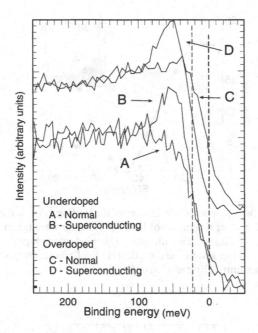

Fig. 7.23 EDCs of underdoped and overdoped Bi2212 taken below and above T_c. The spectra are from the scan along $\overline{M}$–X and are for the minimum binding energy (zero for the overdoped sample). The dashed lines are at binding energies of 0 and 23 meV. (Loeser *et al.*, 1996a)

is still a shift, but no peak. Similar spectra for the underdoped sample were taken for cuts through the Fermi surface at various angles. They show that the apparent gap at $T > T_c$ ranges from about 20–25 meV, as in Fig. 7.23, to zero, perhaps ± 5 meV, along the Γ–X direction. This is consistent with a gap of the same symmetry as that for superconductivity, but other symmetries may be possible.

Contemporaneously, Ding *et al.* (1996b) discovered the same gap above T_c in underdoped samples of Bi2212. One of their underdoped samples had $T_c = 10$ K. Another underdoped sample had $T_c = 83$ K. Another sample was electron irradiated till $T_c = 79$ K. Scans were made along $\overline{M}$–X. All these samples and an optimally doped one had similar spectra for $T < T_c$. Above T_c, however the spectral peaks near E_F were much broader in the underdoped samples than in samples near optimally doped, even at 14 K for the sample with a T_c of 10 K. Moreover, the spectra for the sample with $T_c = 83$ K resembled those of the irradiated sample. Thus, this broadening is not caused by disorder or phonons. A gap was observed at 14 K in the sample with $T_c = 83$ K and this gap persisted in spectra taken up to 170 K. The more underdoped $T_c = 10$ K sample had a gap that persisted all the way to 301 K. The angle dependence of this gap is shown in Fig. 7.24, and the temperature dependence of the maximum gap is in shown in Fig. 7.25. The angular symmetry and the magnitude of the gap above T_c is the

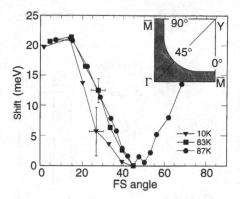

Fig. 7.24 Wave-vector dependence of the shift of the leading edge of the EDCs, taken as indicative of the gap, for several samples of Bi2212. The data were taken at 14 K, and the critical temperatures of the samples is indicated. The most underdoped sample is not super-conducting at 14 K. The inset shows the directions in the Brillouin zones of the angles of the minimum and maximum. (Ding *et al.*, 1996b)

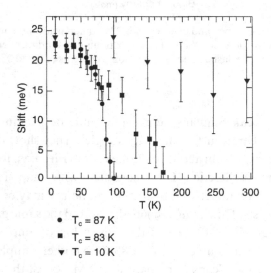

Fig. 7.25 Temperature dependence of the maximum gap for the three samples of Fig. 7.24. (Ding *et al.*, 1996b)

same for all three underdoped samples. It is also the same, within experimental uncertainty, as the magnitude and symmetry of the gap in superconducting samples, the samples used, and optimally doped samples. Figure 7.26 shows a phase diagram derived from these data.

There is not complete agreement on the interpretation of the data by the two groups, although the data generally agree. In later work Ding *et al.* (1997) reported that there were no missing parts of the Fermi surface for Bi2212 under-

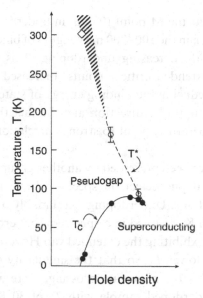

Fig. 7.26 Schematic phase diagram of Bi2212. The filled circles are from magnetic susceptibility measurements. The three open symbols are from Fig. 7.25. The open diamond is a lower bound. (Ding *et al.*, 1996b)

doped so that T_c was only 15 K. A complete quadrant of the Fermi surface for a slightly underdoped sample was measured at a temperature ·above that at which the pseudogap forms. They also measured below this temperature and found the minimum pseudogap tracked this Fermi surface. The more underdoped sample ($T_c = 15$ K) was measured in the pseudogap regime and the trace of the minimum gaps (as a function of wave vector along a cut) departed only slightly from those of a slightly underdoped crystal and an overdoped one. Measurements could not be taken above the temperature at which the pseudogap disappears, for it was too high. If the Fermi surface can be assumed to track the trace of the minimum gap in this greatly underdoped crystal, the Fermi surface was "large." These authors observed an interesting aging effect in underdoped samples, which appear not to have the stable cleavage surfaces exhibited by crystals doped nearer the optimal value. Below T_c the peak near the Fermi edge in an underdoped sample is not as sharp as that in a sample closer to optimal doping. However, after a day at low temperature in ultrahigh vacuum the peak becomes significantly sharper and the spectrum at higher binding energy shifts about 40 meV to lower binding energy. The authors suggest this arises as the surface becomes less underdoped with time. Their data on the most underdoped sample suggest that the Fermi surface of this sample expands slightly upon aging, an effect attributable to a decrease in the amount of underdoping.

The spectra taken near the $\overline{\text{M}}$ point $(0,\pi/a)$ in underdoped samples by both groups show a broad peak in the 100–300 meV region of binding energy. Its binding energy increases with increasing underdoping. This is the region in the Brillouin zone of the extended critical points. Increased hole doping should lower the Fermi level, reducing the binding energy of states around the critical points. The shift in the opposite direction and the fact that this peak is very broad, even at low temperature, point to strong correlation effects in the states involved.

Yet different data have been presented by another group. Grioni *et al.* (1996) compared angle-resolved photoelectron spectra of two underdoped samples with an optimally doped one. Underdoping, presumably by varying the oxygen content, to give T_c of 60 K, resulted in a spectrum that crossed the Fermi level along Γ–$\overline{\text{M}}$ rather than exhibiting the extended van Hove singularity just below E_F. (More holes could lower E_F so that the singularity occurred above E_F.) More importantly, at this Fermi-surface crossing, there was no evidence of a pseudogap. A more underdoped sample with T_c of 30 K was not measured along Γ–$\overline{\text{M}}$, but along Γ–Y the Fermi-surface crossing indicated a much smaller Fermi-surface than in the optimally doped sample.

Norman *et al.* (1997a) compared angle-resolved spectra taken near E_F for several directions in the Brillouin zone at several temperatures on two samples of Bi2212. The samples were overdoped ($T_c = 82$ K) and underdoped ($T_c = 83$ K). The differences were dramatic. At the Fermi-surface crossing along the Y–$\overline{\text{M}}$ direction, both samples showed a peak shifted below E_F by a gap, 50 meV in the underdoped case, 20 meV in the overdoped case, consistent with previous work (Harris *et al.*, 1996). The peak in the underdoped sample is much weaker and broader, interpreted as diminished quasiparticle weight in the spectrum. As the temperature increases, the gap in the overdoped sample decreases, closing at T_c. The peaks in the underdoped sample also disappear with increasing temperature, but the shift, or gap, remains well above T_c, vanishing only at about 200 K. At this point, the "peak" is so broad and weak that the concept of a quasiparticle may be inapplicable.

Similar data were taken on the underdoped sample for two other directions in the Brillouin zone, directions from Y about one-quarter- and one-half-way toward the Y–Γ line. The pseudogap, which closed at about 200 K along the Y–$\overline{\text{M}}$ direction, closed at 120 K and 95 K for the other two directions. There is no pseudogap along Y–Γ. This makes the Fermi surface for the underdoped Bi2212 exhibit interesting behavior as the temperature varies. Above 200 K, the Fermi surface is a complete arc in a quadrant of the projected Brillouin zone, a closed contour in the extended zone scheme, despite the apparent lack of well-defined quasiparticles. As the temperature is lowered, a pseudogap opens over a range of azimuths around the arc, leading to short arc segments for a Fermi surface. Then, just above T_c, it has been reduced to a very small arc segment along the Y–Γ line. Experimental errors preclude calling this a point.

These data were analyzed with symmetrized EDCs and by the more traditional shift of edges. The two methods gave comparable results. However, the symmetrized plots were useful to answer the question of whether the pseudogap vanishes with increasing temperature by closing or by filling in states in the gap. At the Y–$\overline{M}$ crossing, the latter seems to dominate, while at about half-way toward Y–Γ, the former occurs.

Norman *et al.* (1997b) continued work on the crystals just discussed. They wished to fit the line shape with the simplest self-energy possible, and to study the temperature dependence of its parameters. For the overdoped sample, with no pseudogap, they used $\Sigma(\mathbf{k}, E) = -i\Gamma_1 + \Delta^2/[(E + i0^+) + E(\mathbf{k})]$, for the limited range of energy and wave vector averaged in an angle-resolved EDC. Γ_1 is a single-particle scattering rate, assumed independent of E. The second term is from the BCS model of superconductivity, with Δ the gap and $E(\mathbf{k})$, the dispersion, with $E(\mathbf{k_F}) = 0$. This was inserted into $A(\mathbf{k}, E)$, convolved with the instrument function and fitted to the symmetrized data for the Fermi-surface crossing along the Y–$\overline{M}$ line over a range of 45 meV on each side of E_F. Γ_1 was constant above T_c, then fell by a factor of about 4.5, starting at T_c, as the sample cooled. This is consistent with expectations for electron–electron scattering. As T increased, Δ was constant, then dropped at T_c, but not quite to zero, remaining in the range of 5–10 meV above T_c to about 150 K, perhaps the result of a small pseudogap. The above self-energy could not fit the data for the underdoped sample. The denominator in the second term was replaced by $[E + E(\mathbf{k}) + i\Gamma_0]$, where Γ_0 represents the inverse pair lifetime. Γ_1 varied with temperature as in the overdoped case, although above T_c it was twice as large. Δ was temperature independent from about 180 K to the lowest temperatures used, about 10 K. The superconducting gap merged smoothly into the pseudogap in the plot. Γ_0 was zero below T_c, rising linearly with $(T - T_c)$ above T_c. Γ_0 reaches Δ at the temperature at which the pseudogap fills in. The linear temperature dependence of Γ_0 suggests that pairs are indeed present when the pseudogap is present.

The fits were different for the Fermi-surface crossing along a different direction in the Brillouin zone, roughly half-way in angle toward the Y–Γ line. As mentioned in previous work, the pseudogap closes at a lower temperature along this line than along the Y–$\overline{M}$. $\Delta(T)$ is constant only well below T_c. It falls sharply near T_c and reaches zero at about 120 K. Γ_0 is difficult to fit with the strong temperature dependence of Δ.

These are difficult studies. Improved resolution may help clarify some issues, but benefits may appear only in low-temperature data, not those at higher temperatures due to the broadening of the Fermi–Dirac function which dominates resolution effects in data taken well above 20 K. Higher resolution may not benefit spectra on strongly underdoped Bi2212. More information could be extracted reliably from the spectra by fitting to the correct model for the spectral shape. Unfortunately the final model does not yet exist. We note that the dip at about 70 meV should appear when the pseudogap is present if its origin is the increased photohole lifetime due to the decreased phase space for scattering when a gap or pseudogap is present.

The pseudogap has been detected in high-resolution core-level studies (Tjernberg *et al.*, 1997). They measured a number of core-level spectra in samples of $Bi_2Sr_2Ca_{1-x}Y_xO_{8+\delta}$ single crystals at room temperature. $x = 0$, 0.16, 0.55, with $T_c = 80$, 89, and < 5 K, respectively. The core-level shifts are shown in Fig. 7.27. The Ca, Sr, and Y core-level binding energies shift monotonically as holes are removed by Y addition. The Cu, Bi, and Cu levels shift similarly between $x = 0$ and $x = 0.16$, but they shift in the other direction upon going to $x = 0.55$, ending up at a lower binding energy. The O 1s spectra of all the samples, taken with 250 meV resolution, were decomposed into three peaks as discussed above. The same was also done for the O 1s spectrum of the sample with $x = 0$ taken below T_c. The three components of the spectra for $x = 0$ were assigned, in order of decreasing binding energy, to O in Bi–O planes, in Sr–O planes, and in CuO_2 planes. The latter component shifted monotonically to higher energy with x. The other two shifted to higher energy between $x = 0$ and 0.16, but shifted back for $x = 0.55$, perhaps interchanging their order in the binding-energy sequence. The authors conclude that the pseudogap expected at room temperature in the sample with $x = 0.55$ causes the anomalous binding-

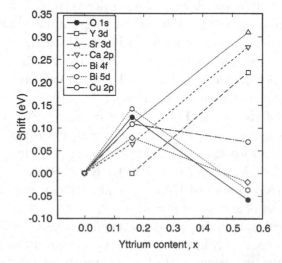

Fig. 7.27 Core-level binding-energy shifts in $Bi_2Sr_2Ca_{1-x}Y_xCu_2O_{8+\delta}$ as a function of Y content. (Tjernberg *et al.*, 1997)

energy shift observed for the Cu, O, and Bi core levels in that sample, and that the states involved at the pseudogap have some Bi character, as well as Cu and O. A comparison of these shifts with the shifts seen upon cooling the sample with $x = 0$ below T_c suggests that the effect of the pseudogap on the core-level binding energies is similar to the effect of the superconducting gap.

The occurrence of a gap in the normal phase of underdoped Bi2212 has several possible explanations. One of the simplest is that pairs form at a temperature above T_c, but they are ordered only at and below T_c. This was recently discussed by Emery and Kivelson (1995), but the idea has a long history, recounted by Gyorffy *et al.* (1991). Further study of underdoped samples has yielded more information. Harris *et al.* (1996) compared T_c with the measured values of the gap in Bi2212 underdoped by the substitution of Dy for Ca or by reducing the oxygen content. T_c fell by up to a factor of 2, but the gap did not fall. It may even have increased slightly. Above T_c the gap had either $d_{x^2-y^2}$ symmetry or something similar to it but with broadened angular regions of zero gap instead of gap zeroes at $\pm 45°$ and $\pm 135°$ (see Fig. 7.28). Thus the ratio of Δ to $k_B T_c$ increased by a factor of at least 2, a characteristic inconsistent with BCS-like theories. The gap varied continuously, within errors, as the temperature varied through T_c (see Fig. 7.29).

White *et al.* (1996) examined the overdoped region of the phase diagram. Here there is no pseudogap above T_c, but the ratio of Δ to $k_B T_c$, where Δ must be measured below T_c, is controversial; two determinations of Δ by Raman scattering disagree. Unlike the underdoped case, the superconducting gap size was found to decrease faster than T_c decreased as hole doping increased. Another difference was that in spectra taken along the Γ–$\overline{M}$ direction, where the superconducting gap is the largest, the pileup near the Fermi edge upon going below T_c is larger and sharper in the overdoped crystal, and the dip around 70–90 meV deeper than in the underdoped crystal. There seems to be a correlation between the shape and dispersion of the normal-state peak near $\overline{M}$ and the size of the gap. In an overdoped sample ($\Delta = 26$ meV), this peak is rather sharp and is at about 50 meV binding energy. In an underdoped sample ($\Delta = 15$–18 meV), it is much broader, is located around 15–200 meV, and it disperses more.

The gap Δ below T_c is still generally called a gap, while that above T_c, which may or may not be the same thing, is called a pseudogap. A pseudogap in a spectrum is a minimum with few states in it, but the spectrum does not vanish as it would for a true gap. There are a few states or excitations that have a zero value for their gap. (In this sense, Δ could qualify as a pseudogap if it has strict $d_{x^2-y^2}$ symmetry, for it then has a value of zero for only four directions of $\mathbf{k}$.) A well-known example of a pseudogap is found in amorphous semiconductors, in the energy region between the mobility edges.

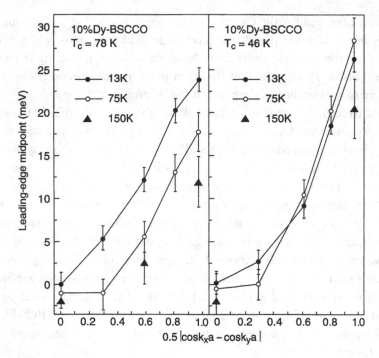

Fig. 7.28   Energy gap at several temperatures vs. $0.5|\cos k_x a - \cos k_y a|$ for two under-doped samples of Dy-doped Bi2212 (BSCCO). (Different oxygen content causes T_c to differ in the samples.) (Harris *et al.*, 1996)

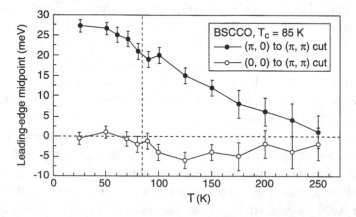

Fig. 7.29   "Fermi"-edge energies vs. temperature at two points in the Brillouin zone for an underdoped Bi2212 (BSCCO) crystal. The shift with temperature of the filled circles represents a shift of the gap with temperature, as the open circles are for points with a small or zero gap. (Harris *et al.*, 1996)

In Y123 there is a long history of measurements that seem to require for their interpretation the opening of a pseudogap in the electronic structure at some temperature $T^* > T_c$. They have been summarized by Egami and Billinge (1996). These include specific-heat measurements, transport measurements, infrared spectra, and spin–lattice relaxation rates in nuclear magnetic resonance. In these experiments a pseudogap, also called a spin gap, was invoked to explain an apparent reduction in phase space at energies near E_F into which electrons can scatter or be thermally excited. No suggestion of a such a pseudogap exits in the corresponding data for Bi2212, except for infrared spectra, and some recent transport measurements to be discussed below. However, as described in the next chapter, despite the considerable evidence for a pseudogap above T_c in Y123 from measurements other than photoemission, no pseudogap has been seen reliably in photoemission studies of Y123, and the gap itself has only recently been observed.

The Cooper-pair condensate in cuprates is qualitatively different from that in other superconductors. The coherence length is very short, 10 Å or so, compared to hundreds or thousands of Å in other superconductors. (An exception appears to be granular films of Al–Ga.) The density of Cooper pairs is not high so the overlap of Cooper pairs is poor, of the order of ten mutually overlapping pairs at a point in the CuO_2 plane of an optimally doped sample. Upon reducing the doping level the cuprates become insulating. This has led to a search for localization of the electronic states as the hole concentration diminishes, but before the insulating state is reached, and for superconductivity in any cuprates with demonstrably localized holes.

The first such study by photoemission was that of Quitmann et al. (1992). They grew crystals of Bi2212 with Y, Gd, or Nd substituting for some of the Ca atoms. For Y doping, which removes holes, the samples showed metallic electrical resistivity until the Ca density fell to 43% of the stoichiometric value. At this doping level, the resistivity was metallic at higher temperatures, but appeared to arise from hopping at lower temperatures. Such samples were superconducting, however, with T_c values up to 90 K. Further Y doping yielded insulating behavior. The valence-band EDCs, taken at room temperature apparently in an angle-integrated mode, showed that the step at the Fermi level grew weaker with increasing Y concentration. At the critical concentration for the metal–insulator transition, there was no step at E_F, just a gradual rise of intensity. Also a peak at about 3 eV, due primarily to Cu 3d states, appeared to shift to larger binding energy as the Y content increased. The reference for this shift is the Fermi level, so it is more likely that the Fermi level shifted upward as Y was added (holes removed), as noted by the authors. The authors discussed their data in terms of a band of extended states that narrows upon Y doping, both from disorder-induced localization and a decrease in the hole concentration. When a mobility edge reaches the Fermi level, the states at the Fermi level become localized instead of delocalized.

Quitmann *et al.* (1994) reported that a few (3–5%) of their as-grown Bi2212 crystals showed no quasiparticle peak in angle-resolved EDCs taken along Γ–X and Γ–Y in the normal state, but a retreat of the edge and a pileup of intensity just below the edge as though there were a gap. The other ca. 95% of their crystals appeared to be overdoped, for all were oxygen treated. They took these spectra as an indication of the loss of wave vector as a good quantum number in these few crystals, i.e., of localized electronic states rather than extended (Bloch) states. These states were stated to have formed an anisotropic gap below T_c (15 meV for the gap along Γ–Y and 6 meV for Γ–X). However, the Γ–$\overline{\text{M}}$ direction was not studied. This led to additional work. Ma *et al.* (1995d) reported more extensive measurements on the same samples, including x-ray, LEED, and electrical resistivity measurements and EDCs taken along the Γ–$\overline{\text{M}}$ direction. These gave a gap of 20 meV when no dispersing peak was found for the normal state, reinforcing the conclusions of Quitmann *et al.* Note, however, that no gap was reported for the data taken above T_c.

Alméras *et al.* (1994) produced an insulating state by doping Bi2212 with Co. 1.5% Co reduces T_c to 65 K. 1.6% Co produces a qualitative difference in several properties. In particular the slope of the resistivity vs. temperature changes sign, suggesting Anderson localization and conduction by hopping, rather than thermally activated carrier excitation across a gap. The resistivity extrapolated to 0 K was 60 and 600 $\mu\Omega$-cm for the 1.5% and 1.6% samples, respectively, and zero for undoped samples. However, the crystals are still superconducting, with $T_c = 57$ K. Angle-resolved photoelectron spectra resemble those reported in the preceding paragraph, evidence of a gap below T_c and no evidence of a dispersing band above for the sample with 1.6% Co, but a dispersing band in the sample doped with 1.5% Co when measured above T_c. A possible explanation, favored by the authors, is that the Co is situated randomly on the Cu lattice, and when a critical concentration of Co is reached the mobility edge reaches the Fermi level and the sample goes suddenly from metallic to insulating as the holes at the Fermi level localize. This explanation has a problem, noted by its authors, that for an insulating sample in the superconducting state a typical small-radius Cooper pair will not overlap a Co site. Most Cooper pairs are in a region of the CuO_2 planes undisturbed by the Co. Yet, T_c is lowered considerably by the Co. If it is so that the pairs are in undisturbed regions, the pairing mechanism appears to be of long range. Co is a unique dopant to date, the only one located on the CuO_2 planes.

Quitmann *et al.* (1995a) continued such studies. They used polycrystalline samples of Bi2212 with Pr substituted for Ca. The samples had either no Pr or 60% of the Ca sites occupied by Pr. The latter samples had a temperature-dependent electrical resistivity characteristic of an insulator. Angle-integrated photoelectron spectra were taken with about 250 meV resolution at 40 K, below T_c for the pure sample and above T_c for the sample with Pr, if that sample was superconducting. (It was insulating to about 5 K.) Even when 60% of the Ca sites were occupied by Pr there was little change in the EDC at the Fermi level. (This result conflicts with that of Mante et al. (1992) whose room-temperature study on Y doping in Bi2212 showed that both the step at the Fermi edge and a barely resolved peak at 1.5 eV both diminished continuously with increasing Y content.) Resonance photoemission showed the Pr 4f states peaked at a binding energy of 1.36 eV with no measurable intensity at the Fermi level. The Pr 4f asymmetric contribution of Pr to the EDC could be fitted with a Doniach–Šunjić line shape. The asymmetry parameter, α, was quite large, 0.28 ± 0.05, characteristic of a metal with good screening, i.e., no energy gap at E_F and perhaps a large density of states there and/or a large cross-section for the scattering of a Fermi-surface electron from the 4f photohole. This was believed to be compatible with a mobility gap at or near E_F. Chen and Kroha (1992) showed that localized states could give a non-zero value for α. Quitmann et al. argue that the large value of α is indicative of an interaction between the Pr 4f state and the states at E_F, unlike the conclusion of Ratner et al. (1993) for Pr-doped Bi2201, to be discussed below, who concluded there was no hybridization of Pr 4f states with states near E_F. A puzzling feature of the Pr resonance spectrum, noted by the authors, is the appearance of only one peak. Similar spectra from other compounds of Pr show two resonant peaks, typically at binding energies of 1–2 and 1–5 eV. Quitmann et al. also measured CIS spectra for various initial states. They found that for an initial state 0.25 eV below E_F, assumed to be characteristic of states at E_F, there was a resonance as the photon energy scanned through the Pr $4d \rightarrow 4f$ resonance. They took this as further evidence of coupling between the Pr 4f states and the states at or near E_F in this insulating sample.

Further measurements of the electrical resistivity and Hall coefficient on Bi2212 samples with a range of Pr doping were carried out by Beschoten et al. (1996). They analyzed the data with a model based on a single band of carriers and a single (temperature- and doping-dependent) scattering time. They concluded that increased doping with Pr leads to spatial localization of the carriers with electrical conduction occurring via thermally activated hopping. The critical composition occurs at $Ca_{0.52}Pr_{0.48}$. All samples that were superconducting at low temperatures had metallic resistivities above T_c. However, consideration of the temperature-dependent Hall effect data led Beschoten et al. to conclude, via the temperature dependence of the Hall mobility, that the superconducting samples with Pr content near the critical value had a mobility characteristic of hopping above T_c, although at yet higher temperatures, the mobility was characteristic of a metal. The minimum in the plot of mobility vs. temperature gave the temperature of the crossover between hopping and metallic conduction. This crossover temperature is plotted as a function of Ca concentration in Fig. 7.30. The plot is rather suggestive of the plot in Fig. 7.26.

Single crystals of Bi2212 doped with Co and Ni were studied by Quitmann *et al.* (1996). The Ni concentration was 3 at.%, and T_c was 77 K. The Co concentrations were 1.57 and 1.60 at.%, the same within the uncertainty of 0.3 at.%. However, they had very different electrical properties. The former sample was metallic with $T_c = 76$ K and the latter was semiconducting (negative temperature coefficient of resistivity) below about 190 K, with $T_c = 66$ K. The resistivities of the doped crystals were much higher than those of the undoped crystals in the linear region above T_c. Scans in the 0–1 eV range of binding energy were taken along the Γ–X and Γ–$\overline{\text{M}}$ directions at room temperature with a resolution of 120 meV. These were compared with spectra taken at 100 K and 35 meV resolution on an undoped reference Bi2212 crystal. The latter showed a broad dispersing peak that sharpened as it approached E_F. It crossed E_F along the Γ–X direction. Along Γ–$\overline{\text{M}}$, it remained just below E_F for a range of collection angles. This is the extended saddle-point region. The sample with 1.5% Co showed similar spectra, but the intensity of the peak was reduced. The dispersion of the peak was about the same and the Γ–X Fermi-level crossing was at about the same point in the Brillouin zone. The other Co-doped sample had qualitatively different angle-resolved EDCs. All the spectra are about the same, with no peak at all, but still with a significant intensity at the Fermi level. The authors ascribe this large effect to the localization of the single-particle states due to disorder. This removes wave vector as a good quantum number. The disorder may be

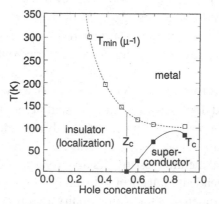

Fig. 7.30 Phase diagram for Bi2212 from electrical transport measurements. The phase boundary between insulating and metallic conduction, labeled T_{min}, was determined from the minimum in a plot of the reciprocal of the Hall mobility vs. temperature. (Beschoten *et al.*, 1996)

especially strong here, for Co doping puts impurities right in the CuO_2 planes, while most other forms of doping put the impurities or lattice defects elsewhere. Localization also accounts for the conduction by hopping believed to have been observed below 200 K in this crystal. The Ni-doped crystal had EDCs similar to those for the Co-doped crystal with the lower Co concentration. The authors conclude that they have observed the coexistence of superconductivity and electron localization, and that the delocalization was brought about by the disorder.

The localization just described results in a mobility gap. It is tempting to connect this gap with the pseudogap seen in underdoped Bi2212 above T_c, i.e., might the pseudogap be the result of carrier localization? It seems premature to make this connection yet, and further measurements should be undertaken.

Single crystals of Bi2212 intercalated with I_2 were studied by Fukuda *et al.* (1992). This intercalation lowers T_c, leaves the basal plane electrical resistivity essentially unchanged, but lowers the resistivity along the *c*-axis to a metallic value. The angle-integrated EDC of the valence band was identical to that of Bi2212, but there were changes in the XPS spectra. All core levels shifted by about 100 meV to lower binding energy. The Cu $2p_{3/2}$ spectrum had some overlap with the I $3p_{1/2}$ line, making its study difficult. By using the relative intensities of the I $2p_{1/2}$ and $2p_{3/2}$ lines in NaI and the intensity of the I $2p_{3/2}$ line in the cuprate, a correction was made. This led to a change in the intensity of the satellite relative to that of the Cu $2p_{3/2}$ main line. The ratio went from 0.35 to 0.46 upon intercalation with I_2. This would lead to an iodine-induced change in the parameters of a CuO_4 cluster if the cluster model were re-analyzed. The I $3d_{5/2}$ line had two components, the stronger attributed to I^- bound to Bi and the weaker to I^{7+} bound to O atoms as in an iodate.

7.4 Other Bi-cuprates

Bi2212 has two pairs of CuO_2 planes per unit cell. Bi2223 has two layers of three CuO_2 planes each. Bi2201 has just two separate CuO_2 planes. The transition temperature depends on the number of CuO_2 planes in a layer: 105 K for Bi2223, 90 K for Bi2212, and about 10 K for Bi2201. It is important to look for differences in the electronic structure of these three materials for clues to the differences in T_c.

Gu *et al.* (1995) measured a gap in polycrystalline samples of Pb-doped Bi2223. The samples of $Bi_{1.8}Pb_{0.4}Sr_2Ca_2Cu_3O_{10+\delta}$ were heat-treated rolled tapes. The 1–10 μm-sized grains had nearly parallel *c*-axes. They were cleaved at 13 K as though they were a single crystal. Because the *a*- and *b*- axes were not aligned in different grains the spectra represent an average of angle resolved EDCs over the basal plane. The valence-band EDC resembles those of other cuprates. Data taken at 13 K and 106 K show a retreat of the rising edge near E_F, a pileup of intensity below the edge, and a dip at 72 meV in the low-temperature spectra when compared to the data taken above T_c (104 K). The gap is estimated to be 29 meV, larger than the largest value in Bi2212 which was estimated to be 22 meV from data of the same authors. This is the maximum gap. Gap anisotropy is suggested by residual spectral weight near E_F for $T < T_c$ which leads to a broader edge than that found in angle-resolved EDCs of Bi2212.

Ratner *et al.* (1993) carried out angle-integrated and angle-resolved photoelectron spectroscopy on Bi2201, in the form of a single crystal of $Bi_2Sr_{1.9}Pr_{0.1}CuO_{6+\delta}$, the Pr being necessary to obtain a single crystal. T_c was about 10 K. A comparison of the angle-integrated valence-band EDCs of Bi2201 and Bi2212, which are quite similar, reveals three differences. There is a shift of about 250 meV to higher binding energy for Bi2201. Bi2201 has a smaller intensity at the Fermi level, and less intensity in the region of a sharp rise in the EDC at about 1 eV binding energy. Since Bi2201 has fewer CuO_2 planes, the latter two differences indicate that these parts of the spectra arise from states on the CuO_2 planes. The structure at 1 eV is common to many cuprates in which the only common structural elements are the CuO_2 planes. This feature will be discussed in the next chapter.

The angle-resolved spectra show one or two dispersing bands in the first few meV below E_F. Fermi-level crossings were sometimes difficult to identify. Comparison of the spectra with an LDA band structure was difficult, for at the time the only band calculation was a tight-binding calculation on the undoped material. The Fermi-level crossings observed were at different points along the symmetry lines from those in Bi2212, and not were expected from the tight-binding bands. The 4d–4f resonance of Pr was used to determine the energy range of the Pr-derived states in the valence band. They turned out to be fairly well localized in energy, peaking at about 1.3 eV binding energy, with no intensity at E_F. This indicates, as expected, they do not hybridize much with the CuO_2 plane-derived states responsible for the superconductivity. (Pr quenches superconductivity in Y123, and in Pr123 there is Pr 4f spectral intensity at E_F.)

King *et al.* (1994) mapped bands near E_F in another Bi2201 sample, a single crystal of $Bi_2Sr_{1.94}Pr_{0.6}CuO_{6+\delta}$. The extended saddle point that is present in Bi2212 was found in Bi2201 at about the same energy below E_F as in Bi2212, i.e., within about 30 meV of E_F, but it did not extend over as large a region of reciprocal space. This is shown in Fig. 7.31. This finding indicates that T_c, only 10 K in Bi2201, does not depend on how far below E_F the extended saddle point lies, but it could depend on the extent of the saddle point. Figure 7.31 also shows the same band in Bi2212, Y123 and Y124, and NCCO. (NCCO is an n-type conductor while the others are all p-type.) The latter compounds will be discussed in later chapters.

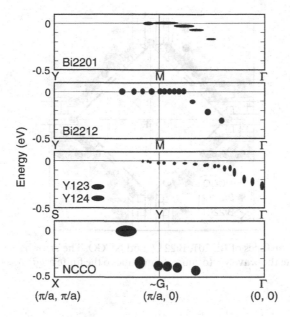

Fig. 7.31 Band dispersion parallel to the long axis of the extended saddle point in several cuprates. (King *et al.*, 1994)

A more detailed mapping of the Fermi surface than that of Ratner *et al.* (1993) was made. The Fermi surface is shown in Fig. 7.32, where the data, taken in two irreducible wedges of the (tetragonal) Brillouin zone, have been replotted throughout the zone using symmetry. This plot also includes the Fermi surfaces of Bi2212 and NCCO. The Fermi surfaces of Bi2201 and Bi2212 are similar, hole pockets centered at X and Y. The pocket is a little larger in "diameter" in Bi2201. The nesting wave vector is therefore a little smaller in Bi2201, $(0.8\pi/a)(1,1)$ vs.$(0.9\pi/a)(1,1)$ in Bi2212.

The anisotropy of the gap in Bi2201 was studied by Harris *et al.*, (1997). They used an optimally doped crystal of $Bi_2Sr_{1.65}La_{0.35}CuO_{6+\delta}$ with $T_c = 29$ K. The spectra taken below T_c showed a peak just below the Fermi level, but no dip and broad peak at higher binding energy. The gap, measured as the shift of the midpoint of the leading edge of the EDC, varied between 0 and 10 meV, with an estimated uncertainty of ±2 meV. The angular variation was consistent with $d_{x^2-y^2}$ symmetry for the gap. An overdoped crystal of Bi2201 showed a much sharper peak near the Fermi level than that in the optimally doped crystal, but no shift due to the opening of a gap could be detected, consistent with the much lower T_c of this sample, 8 K. (Spectra were taken between about 9 K and 100 K.) An underdoped crystal of Bi2201 showed a gap of 7 meV at 60 K, far above its T_c, which was below 4 K, presumably the pseudogap. Bi2201, with its

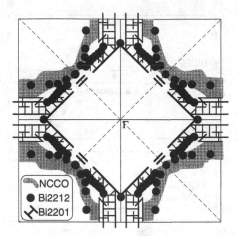

Fig. 7.32 Fermi surfaces of Bi2201, Bi2212, and NCCO. The error bars on the points for Bi2201 indicate the wave vector uncertainty due to the finite angle resolution. (King *et al.*, 1994)

single CuO_2 plane, has photoelectron spectra with many features in common with the spectra of Bi2212.

7.5   Summary

Photoelectron spectroscopy of Bi2212 and related compounds has led to a number of important conclusions about the electronic structure of this class of cuprates. Some of these conclusions have been corroborated by other types of experiment, while some are unique. Most importantly, the sudden approximation appears to be valid at the photon energies with which most angle-resolved photoelectron spectroscopy has been carried out. This makes comparisons with calculations valid, for it means the spectra should resemble spectral density functions, weighted by dipole matrix elements.

The electronic structure of the normal state in the first eV below the Fermi energy in samples of Bi2212 near optimal doping is described qualitatively by the LDA band structure in many, but not all, of the spectra reported. The number of bands, and their character, a mix of Cu $3d_{x^2-y^2}$ and O $2p_x$ and $2p_y$ states, was successfully predicted, but the measured effective masses are larger than the LDA masses. The splitting of the degenerate bands from the two equivalent CuO_2 layers has not yet been reliably resolved. As the doping is changed, the Fermi surface swells or shrinks in the direction predicted by the rigid-band model, but quantitative agreement cannot be checked due to large errors in determining k_F. The Fermi surface is, however, the large one, expected from

Luttinger's theorem. There is an extended van Hove singularity just below E_F with its long axis along the $\Gamma-\overline{M}$ direction. In compounds related to Bi2212, T_c does not seem to correlate with how far below E_F this line of critical points is located. If there is any correlation with T_c it is with the extent of the line of critical points. The direction of the axis of this line of critical points is parallel to the Cu–O bond axis in the CuO_2 planes, as it is for Y123, to be discussed in the next chapter. The degree of agreement with the LDA Fermi surface, however, falls far short of that for Cu. This is partly because of the small count rates from states near E_F in Bi2212, but the main reasons are the presence of a large background of unknown origin and the lack of an analytical line shape for fitting the data.

In addition to the bands predicted by the LDA, shadow bands have been found in angle-resolved photoelectron spectra. These bands are related to the LDA bands, but the route by which they appear in the spectra is not yet understood, although there are several potential explanations. These bands are difficult to find in the normal EDC scans; they are more readily seen in the angular scans, $E(\mathbf{k})$.

In the superconducting state, a gap is seen to open at the Fermi surface, and several groups now agree that it has angular anisotropy consistent with a gap with $d_{x^2-y^2}$ symmetry, although the negative values such symmetry requires are not observable in photoemission studies. The maximum gap occurs along the $\Gamma-\overline{M}$ direction, along the nearest-neighbor Cu–O bonds in the CuO_2 plane. Its value is about 20–25 meV, well above the BCS value. Not all groups who have measured this agree on the minimum value, but the most widely accepted value is 0 ± 5 meV. The temperature dependence of this gap appears to be more "square" than the universal BCS dependence, consistent with strong coupling, but not all groups agree that the temperature dependence is independent of angle in the CuO_2 plane. The approach of a spectral peak toward E_F and its retreat from E_F and loss of intensity at larger wave vector are in good agreement with expectations from the BCS picture.

Underdoping appears to cause a loss of part of the Fermi surface and the opening of a pseudogap. (This gap itself destroys a segment of the Fermi surface.) The gap appears to have the same size and symmetry as the gap in the superconducting state. It may represent the same physical process. There is some evidence that in underdoped samples there is also a mobility gap, the same size as the pseudogap, and with the same or similar temperature dependence. Overdoping causes T_c to drop, but the superconducting gap drops much less rapidly. Underdoping also causes T_c to drop, but Δ does not, so $\Delta/k_B T_c$ increases. Thus $\Delta/k_B T_c$ is a minimum at optimal doping, and this minimum value is about twice that of the BCS theory if the maximum Δ, along the $\Gamma-\overline{M}$ direction, is used.

Chapter 8

Y123 and related compounds

8.1   Introduction

Although Y123 and Bi2212 both contain CuO_2 planes and have similar values of T_c, there are notable differences which have affected the course of the study of each class of material. We have already mentioned two such differences, the reproducible cleavage of Bi2212 and the stability of the cleaved surface in ultrahigh vacuum. The cleavage plane(s) of Y123 is not really known and may vary from cleave to cleave. Moreover, a single plane is unlikely, and the atomic nature of the exposed surface may change at cleavage steps. Surface reconstruction, or at least relaxation, may be possible, even at low temperature. The surfaces of Y123 generally are not stable in ultrahigh vacuum except at temperatures below about 50 K. However, exceptions to this have been found by several groups, but the reasons for this stability are not yet known. T_c may be varied in Bi2212 by the addition or removal of oxygen and there is an optimum oxygen content at which T_c is a maximum. $YBa_2Cu_3O_x$ also has variable oxygen stoichiometry, but between $x = 6.8$ and $x = 7$, T_c remains very close to its maximum value. In Bi2212 the oxygen content changes occur on or adjacent to the Bi–O planes. In Y123 the changes occur in the Cu–O chains, although the holes provided by the oxygen atoms are believed to reside on the CuO_2 planes (Cava *et al.*, 1990). Starting from $x = 7$, removing oxygen atoms from the chains should lower the hole concentration. However, the O vacancies in the chains tend to order (Beyers and Shaw, 1989; de Fontaine *et al.*, 1990). Chain Cu atoms with only two nearest-neighbor (non-chain) oxygen atoms begin to appear, and these have a valence of +1. These then, by "capturing" an electron each, inject holes into the rest of the crystal, an effect opposite to that of O removal. As a result, at $x = 6.8$, T_c has not really fallen from its value at $x = 7$, and at $x = 6.5$, Y123 is still metallic with only one-half the hole concentration at $x = 7$. T_c has fallen from 92 K to about 50–60 K, however. (Conventional valence arguments lead to a half-filled band at $x = 6.5$, which should be a Hubbard insulator.) As a result of this behavior, it is possible to change the hole concentration without changing the oxygen stoichiometry by annealing, which changes the degree of vacancy ordering. This has not yet been applied in photoelectron spectroscopy.

Finally, we note that a superconducting gap had not been observed directly by photoemission in Y123 until very recently, with the exception of some reports by one group, in which a gap was seen on a few surfaces after "anomalous" cleaves. The recent work (Ding et al., 1996d; Schabel et al., 1997) demonstrated a gap and its symmetry. The reasons earlier spectra failed to display a gap are not clear. The bulk is known to be superconducting. The surface may not be superconducting, possibly because of the cleavage-induced disorder, although the surface is known to be metallic. The unit cell in the c-direction is about one-third as long as that in Bi2212. Valence-band photoelectron spectra are dominated by electrons in the topmost three unit cells. In Bi2212, the valence-band spectra near E_F arise from states derived from states on Cu and O atoms in the CuO_2 planes. In Y123 these states contribute, but there is an additional contribution from the Cu–O chain states. If there is no superconductivity on the chains, a gap opening in the spectrum of CuO_2 plane states might be masked by the spectrum at E_F from the chains. Such an effect, possibly present in angle-integrated spectra, should not occur in angle-resolved spectra from detwinned crystals. Despite the common failure to observe the gap, photoelectron spectroscopy has proven valuable in adding to our knowledge of the electronic structure of Y123, the cuprate most studied by other techniques. Photoelectron spectroscopic studies of Y123 and related cuprates has been reviewed by Brundle and Fowler (1993) and by Veal and Gu (1994).

8.2   Ba 4d spectra

The standard photoemission technique for studying surface composition is the spectroscopy of core levels. This has not completely clarified the uncertainty about surface planes after cleavage. Liu et al. (1989) resolved the Ba 4d spectrum into two spin–orbit split pairs. There is only one Ba bulk site, so one pair of peaks was assigned to Ba atoms in the surface plane. The spin–orbit pair at lower binding energy was assigned to the surface on the basis of its relative strength as the photon energy was varied, making use of the universal curve for the mean free path for inelastic scattering, Fig. 3.4. They used photon energies around 110–140 eV. (The Ba 4d binding energy is 90 eV, so kinetic energies were in the 20–50 eV range.) This assignment was the opposite of that in previous XPS studies. In those studies, the surface component was determined by its increasing relative contribution to the spectrum as the collection angle was increased away from the surface normal. Those studies were carried out at room temperature, and Liu *et al.* assumed the surfaces were not comparable. Moreover, the surface component in XPS is considerably smaller in comparison with the bulk, due to the longer mean free path for inelastic scattering. However, later XPS work on samples in which a Fermi edge was present in the XPS valence-band spectra confirmed the earlier XPS assignment. Additionally, List (1990) repeated spectra like those of Liu *et al.*, using a range of photon energies. He found that the relative intensities of the two spin–orbit split pairs changed with photon energy, but over a wider range of photon energies, the change was not monotonic in photon energy, as found by Liu *et al.* in their more limited range. This means that there are dipole matrix element effects even at these high photon energies. Final states 50 eV above E_F cannot be assumed to be in the XPS range, effectively a continuum. In addition, it is also possible that the inelastic mean free path for Y123 does not follow the universal curve. Measurements similar to those of Liu *et al.* subsequently were made as a function of angle (Veal and Gu, 1994) and the higher-energy pair of peaks (Fig. 8.1) assigned to the surface. Fowler *et al.* (1990) and Ziegler *et al.* (1990) also identified the higher-energy peaks as the surface peaks, using XPS on single crystals. An extensive account of O and Ba core-level studies on good and bad surfaces of Y123 is given in Brundle and Fowler (1993). Their best data were taken on single crystal surfaces at 300 K. They also report angle-integrated valence-band spectra, but the depth sampled is considerably larger than that sampled with the 15–50 eV photons usually used in valence-band studies.

Veal and Gu (1994) attempted to correlate the relative strength of the Ba 4d surface component with features in the valence-band spectra. The relative strength of the surface component varied from cleave to cleave, apparently from different relative areas of Ba–O planes on the surface. In Y123, even if all cleaves are the result of breaking the same bonds, there need not always be a distribution of surface planes with 50% of the area in Ba–O planes. However, the valence-band spectra were essentially the same for each cleave. These studies did not allow the identification of which bonds were broken in the cleaves, those between the Ba–O planes and the CuO_2 planes, those between the Ba–O planes and the planes of the Cu–O chains, or both.

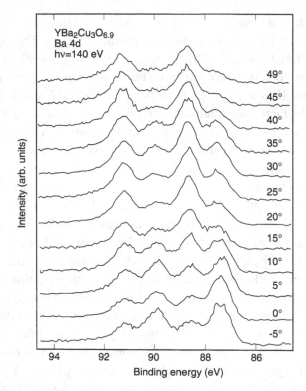

Fig. 8.1 Angle-resolved Ba 4d photoelectron spectra in $YBa_2Cu_3O_{6.9}$ for several collection angles from the surface normal. Spectra at larger angles are more surface sensitive. (Veal and Gu, 1994)

8.3   Valence bands and Fermi surface

The first determination of the Fermi surface of Y123, by Campuzano *et al.* (1990a,b) was described in Chapter 6, see Fig. 6.11. This work and the measurements to be described in this section were carried out on crystals doped to give as high a T_c as possible. Additional studies of the Fermi surface and energy dispersion were made on twinned single crystals by Mante *et al.* (1991), Campuzano *et al.* (1991), Liu *et al.* (1992a,b). In such samples, the measured Fermi surface will be the superposition of two actual Fermi surfaces, each with two-fold symmetry, rotated by about 90° with respect to each other to give a Fermi surface with four-fold rotational symmetry. Each segment due primarily to states on the Cu–O chains will appear twice, while the 90° rotation will not have such a dramatic effect on the segments from the CuO_2 plane states, for they have near four-fold symmetry. In fact, the rotation is not exactly 90°

(Islam and Baetzold, 1989), but with the angular resolution used, the small departure from 90° is not significant.

Figure 8.2 shows several EDCs in which peaks disperse through the Fermi level. These, and other EDCs, resulted in the experimental band structure shown in Fig. 8.3, which shows experimental bands plotted on top of calculated bands. Since the crystals were twinned, the Γ–X and Γ–Y lines are superimposed. A number of bands appear, many of them flatter than the calculated bands. Fermi-surface crossings appear at expected points, but not all calculated crossings were observed. In particular, a highly dispersive band due to Cu–O chain states was not observed and a Fermi-surface crossing near the S point, also due to chain states, was not observed. One early study (Campuzano *et al.*, 1990a) shows this crossing but later studies (Mante *et al.*, 1991; Liu *et al.*, 1992a; Tobin *et al.*, 1992) do not. The spectra in Fig. 8.3 were taken with a photon energy of 21.2 eV. The failure to see the Fermi-level crossing at S could be caused by a small dipole matrix element for this energy. Dipole matrix element effects are very large in cuprates, and changing the photon energy by only a few eV can

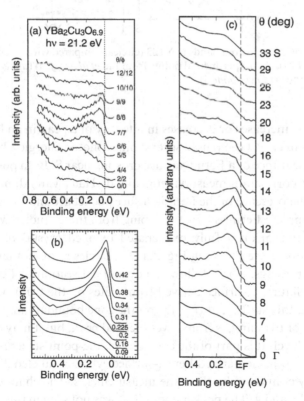

Fig. 8.2 Angle-resolved energy distribution curves taken along Γ–S for Y123. The data are from three different groups: (a) Liu *et al.* (1992a); (b) Campuzano *et al.* (1991); (c) Mante *et al.* (1991). (Veal and Gu, 1994)

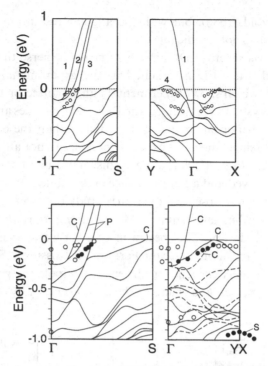

Fig. 8.3 Dispersion of measured bands in Y123 (circles) compared with calculated bands (Pickett *et al.*, 1990). Experimental points from Liu *et al.* (1992a) (*upper panels*) and Mante *et al.* (1991) (*lower panels*).

result in dramatic increases or decreases in intensity of a peak in an EDC. Liu *et al.* (1992b) and Gu *et al.* (1993) used several other photon energies, but these did not yield the appearance of a Fermi-surface crossing near S. Both positron-annihilation angular correlation measurements and de Haas–van Alphen measurements indicate the presence of the Fermi-surface pocket around the S point.

The most complete Fermi-surface mapping from these studies was reported by Liu *et al.* (1992b) on a fully oxygenated twinned crystal of Y123 with $x = 6.9$. Their results are shown in Fig. 8.4. The circles represent unambiguous Fermi-surface crossings. These follow the cylindrical section of Fermi surface labeled 3 in the figure, a surface derived from states on the CuO_2 planes. (The one circle not on this surface, at $\phi = 15°$ on the zone boundary and reflected to $\theta = 15°$ because of twinning, is from a well-defined peak, but it may not actually cross the Fermi level. It is part of the extended saddle-point structure described below.) The second plane-derived Fermi-surface cylinder, labeled 2, is also reasonably securely confirmed by these the measurements. The chain-derived sections are labeled 1 and 4. The pocket about S, 4, was not seen in this work nor in most other studies by angle-resolved photoelectron spectroscopy. Section 1 has three measured open circles in it, suggesting that it does exist.

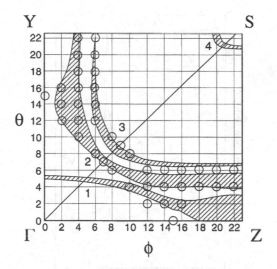

Fig. 8.4 Fermi surface of Y123 (circles) and projection of the calculated (Pickett *et al.*, 1990) Fermi surface. The points were measured (on a twinned crystal) on one side of the diagonal line, then reflected across that line. (Liu *et al.*, 1992b)

An extensive study of the Fermi surface of Y123 and of states in about the first eV of binding energy was carried out by Tobin *et al.* (1992) in detwinned single crystals. This allowed the separation of the contribution of the Cu–O chain states to the Fermi surface from those of the CuO_2 plane states. The dispersion of the most prominent feature of the valence-band photoelectron spectrum in Y123, a very narrow peak at about 1 eV, could also be followed. They used an energy resolution of 30 meV and angle resolution of 1°. This angular resolution translates to a spread in the parallel component of wave vector of about 10% of the length of an edge of the Brillouin zone for a photon energy of about 20 eV. The crystals of $YBa_2Cu_3O_{6.9}$ (fully oxygenated) were cleaved at 20 K and held there for almost the complete duration of the measurements. The samples were warmed only for attempts to detect the opening of a gap.

The most prominent feature in the valence-band spectra of Y123 is a sharp peak at about 1 eV. Tobin *et al.* followed this peak around the Brillouin zone and noted that it showed dispersion, a variable width and a variable intensity which also was polarization sensitive. Figure 8.5 shows this peak taken at several points in the Brillouin zone. The broad appearance of the peak taken at Γ is the result of the finite angular acceptance (only 1°) and the strong dispersion near the Γ point. At the S point, spectra were taken with the electric vector in the plane of incidence and normal to it. The strong anisotropy, evident in Fig. 8.5, is evidence that the initial state is even with respect to mirror symmetry about the Γ–S line. Tobin *et al.* argue that the "1-eV" peak is common to all cuprates. It arises from states on the CuO_2 planes. This is also supported by studies of the

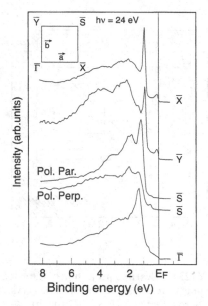

Fig. 8.5 Angle-resolved EDCs for Y123 measured on an untwinned crystal at various symmetry points in the Brillouin zone. (Tobin *et al.*, 1992)

Cu 3p resonance on several cuprates. Although the plane normal to the surface and passing through the Γ–S line is not a mirror plane in the orthorhombic crystal, the CuO_2 plane alone has four-fold rotational symmetry. Thus to the extent the states are localized on the CuO_2 planes, the selection rule, leading to even symmetry is valid. By scanning away from the S point, Tobin *et al.* showed that this sharp peak splits into two, one component dispersing very little, the other dispersing significantly. A similar behavior appears to occur near the X and Y points. The dispersion about the X point was measured using two different photon energies, 24 and 74 eV, with the same results. This indicates that the different values of $k_\perp$ played no role, i.e., the band was very close to two-dimensional.

This peak, first observed by Claessen *et al.* (1991), Mante *et al.* (1991), and Manzke *et al.* (1992), had been attributed to a surface state. They reported it to be highly sensitive to surface contamination. The difference between a surface state and a state localized on in internal CuO_2 plane is not great. The exponential tail occurs in vacuum for a surface state and in a Ba–O plane for the state on an internal plane. It now appears to be no more sensitive to the deterioration of the surface with time or to the adsorption of O_2 or Ar overlayers than the rest of the spectrum. (However, Ar does not chemisorb on cuprates, and O_2, although O is a constituent of the cuprates, may not dissociate on the surface.) So far, most cuprates have a similar structure near 1 eV (Olson *et al.*, 1995) but it is most prominent in Y123. Its occurrence in several cuprates suggests it does not depend on the plane terminating the surface, but it originates in the bulk, not the surface. Its prominence, however, appears to vary for a few "anomalous" cleaves, as we shall discuss shortly. At the present time, this peak appears to be from a bulk initial state, albeit one that is highly two-dimensional, with many of the attributes of a surface state, but the issue is not settled. Later work (Olson *et al.*, 1995) suggests that there are broader, dispersing transitions throughout the Brillouin zone for at least two bands. Near the Y point, these initial bands merge to give sharp intense peak.

EDCs were taken with a wide range of photon energies between 14 and 74 eV. The intensity of the 1-eV peak was largest around 24 and 74 eV, and smaller in between, an effect of the dipole matrix element. The small peak at the Fermi level also varied in intensity, but it was small at 24 and 74 eV and largest around 17 and 28 eV. This implies that the two peaks do not have a common origin, i.e., the 1-eV peak is not a satellite of the Fermi-level peak. (The data do not permit a conclusion on the question of whether or not the peaks vary out of phase when varying the photon energy or wave vector.) The same data also show a small resonance of the 1-eV peak at the X and Y points, but not elsewhere, when the photon energy scans through the Cu 3p resonance. The initial states must be largely of Cu 3d character at X or Y, but hybridize strongly away from X or Y.

Tobin *et al.* studied the small peak near E_F to map the Fermi level. Of particular importance was locating the parts of the Fermi surface based on states localized on the Cu–O chains, previously not separable from the plane states in twinned crystals. Scans were made along directions near Γ–X and Γ–Y, and also along several other directions. There were marked differences between Γ–X and Γ –Y. Along Γ–Y the peak at the Fermi level was quite intense and sharp, with widths of about 40 meV, most of which came from the instrumental resolution. Instead of dispersing with angle, the peak intensities varied at nearly constant binding energy. Along Γ–X, the peak near E_F was much broader and it dispersed with changing collection angle. This difference was not due to different polarizations, i.e., to the dipole matrix element anisotropy, for both spectra were measured with **E** nearly perpendicular to $\mathbf{k}_{\parallel}$. This peak at or near E_F near the Y point was assigned to a band derived from chain states. Its intensity varied from cleave to cleave, and was most intense when the cleave was believed, on the basis of core-level spectra, to have left a Cu–O chain layer at the surface. Later work (Liu *et al.*, 1992a) which showed its intensity to change when the oxygen concentration was changed, supports this assignment.

Many of the Fermi wave vectors determined by angle-resolved photoelectron spectra fall on segments of the calculated Fermi surface derived from states on the CuO_2 planes. The agreement with the Fermi surface from the LDA calculations was quite good in the vicinity of the X point, but not very good near Y. One or two points can be assigned to Cu–O chain-based segments near the Γ–X line. The sharp peak at the Fermi surface near the Y point, discussed in the preceding paragraph, is not present in the LDA Fermi surface. There was no evidence of a Fermi-surface crossing near the S corner of the surface Brillouin zone. The band calculations indicate a small hole pocket about the S point, although its shape and occurrence differ among different calculations. It has been detected in de Haas–van Alphen measurements (Mueller *et al.*, 1992; Kido *et al.*, 1991). Failure to detect this section of the Fermi surface is not proof that it does not exist. A thorough search would require measurements at many photon energies and with two polarizations, for dipole matrix element effects are very large in cuprates. To date, it has been seen by angle-resolved photoemission in only one set of measurements (Campuzano *et al.*, 1990a,b)

Extensive attempts were made to observe the opening of a gap in the spectrum at the Fermi level upon cooling below T_c. No clear-cut shift of the edge could be detected, leading to the conclusion that any gap, if present, was no larger than 5 meV. Possible reasons for this include oxygen loss upon cleaving and surface relaxation or reconstruction. The likelihood of the occurrence of these processes at 20 K may seem remote, but it is clear that these surfaces seemed not to be superconducting, while the bulk crystal was, as were surfaces of Bi2212.

Gofron *et al.* (1993) mapped the bands in both Y123 and Y124 along Γ–Y in the first 300 meV below E_F, taking a large number of EDCs and using a resolution of 10 meV, in some cases, and an angular acceptance of $1°$. The peak positions were mapped vs. wave vector, and a very flat band was revealed, one that did not cross the Fermi surface, as previously thought (see Fig. 8.6 for Y123). By scanning through the flat part of this band in a direction normal to the Γ–Y direction, the surface of constant initial energy revealed a saddle point. To the extent that the band is truly flat parallel to Γ–Y, this is not a single van Hove critical point, but rather a line of critical points, or an extended critical point. The distinction is important, for in two dimensions, a single saddle point at energy E_0 causes a singular structure in the density of states proportional to $\log(|E - E_0|)$. For an extended string of saddle points in two dimensions, the singularity becomes stronger, proportional to $|E - E_0|^{-1/2}$ (Abrikosov *et al.*, 1993), as for the extremum of a one-dimensional band. Dispersion in $k_\perp$ was sought by varying the photon energy, but there was little, suggesting the two-dimensional description of the line of saddle points is a good one. As can be seen in the figure, this line of saddle points lies close to E_F, approaching E_F to within the gap energy Δ, if we may assume the value of Δ for Bi2212 may be used for Y123 with a comparable value of T_c. Such a critical point near E_F can enhance any tendency toward an instability in the electronic structure, one of which could be the transition to a superconducting state. In this sense it plays a role similar to Fermi-surface nesting. Such an extended saddle point, if within $k_B T_c$ of E_F, can introduce the electronic energy scale, about 1 eV, (instead of the

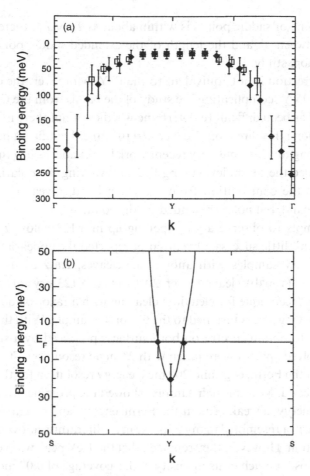

Fig. 8.6 Dispersion of bands for Y123 along (a) Γ–Y and (b) Y–S showing the extended saddle-point nature of a band just below E_F. The calculated band (solid curve in (b)) is from Massidda *et al.* (1991). (Gofron *et al.*, 1993)

much smaller phonon energy scale) into the weak-coupling BCS theory (Abrikosov *et al.*, 1993). In Y124 the corresponding structure lies only slightly further below E_F (Gofron *et al.*, 1994). Andersen *et al.* (1994) pointed out that the extended saddle points appear naturally in LDA calculations of the electronic structure when the CuO_2 planes in structures that have two of them per unit cell "dimple." The out-of-plane displacements lead to a stronger coupling between these planes and an extension in k-space of the critical points in unit cells with flat CuO_2 planes.

The extended line of critical points has been reported for several cuprates: Bi2201 and Bi2212 (Dessau *et al.*, 1993a), Y123 and Y124 (Gofron *et al.*, 1993, 1994), and NCCO (King *et al.*, 1993). There is no correlation between the energy of the critical point and T_c, but King *et al.*, (1994) suggest that for those cuprates for which the line of saddle points is within about $k_B T_c$ of E_F, there might be a correlation between T_c and the length of the extended saddle point, although this requires more study.

The Γ–X direction is not equivalent to the Γ–Y. Moreover, the chain bands run parallel to Γ–Y, complicating the study of the bands from the CuO_2 planes. As a result, it has been difficult to determine whether or not there is an extended saddle point along this direction. Veal et al. (1990b) did not find a peak near the Fermi level along this line, but very recent work by Schabel et al. (1997) found a peak that crossed the Fermi level along Γ–X. By varying the polarization, they could separate the contribution from the chain band, revealing a dispersing plane-derived band, but not an extended saddle point.

Many attempts to observe a gap opening up in Y123 below T_c have been reported, all with little success. One attempt reported the appearance of a gap, but only in a few samples with anomalous cleaves. Ratz et al. (1992) and Schroeder et al. (1993b) cleaved over 20 twinned Y123 crystals at 10–20 K using a different technique for cleaving: cleaving with a razor blade rather than breaking by moving a post fastened to the top of the sample. With this technique they were able to obtain spectra on both surfaces produced by a single cleave. The angle-resolved spectra were taken with 2° angle resolution, 70 meV energy resolution near the Fermi edge and 200 meV energy resolution for the remainder of the valence band. Most of their samples showed the prominent peak at about 1 eV binding energy, a weak peak at the Fermi energy, and no gap at the Fermi edge. These spectra resembled many of those in the literature, allowing for differences in resolution. However, they reported that the 1-eV peak was very sensitive to overlayers, losing much of its intensity with a coverage of 0.02 monolayers of H_2, while the Fermi-edge structure did not change (Schroeder et al., 1993c). This reinforces the view of the 1-eV peak as being of surface origin. Two crystals gave a different result. For them, the 1-eV peak was very weak or missing, but the Fermi edge was a little larger. Moreover, it shifted below T_c, corresponding to the opening of a gap of about 20 meV. (The temperature was kept at 10–20 K. The shift was with respect to the Fermi level of an adjacent Ag surface.) They suggest that these cleaves occur between the Y layer and the adjacent CuO_2 plane. This would leave all of the Y on one side or other of the cleave, or, as is more likely, some Y on each side. In fact, a 50–50 distribution optimizes charge balance. LEED patterns on these surfaces were different than those on the usual surfaces. They showed a c 2 × 2 superstructure, but the spots were rather diffuse, implying imperfect order.

Another interesting aspect of the anomalous surfaces is that a scan along the Γ–X(Y) direction reveals a Fermi-surface crossing near the X(Y) point. This region of the Fermi surface contains large contributions from the Cu–O chains. This is the best evidence to date for this part of the Fermi surface in Y123.

Unfortunately, since this work was published there have been no more reports of measurements on such cleaves. It would be especially interesting to repeat some of the measurements near the Fermi edge with higher resolution.

A recent study of Y123 (Schabel *et al.*, 1997) has revealed good evidence for a gap in Y123, but because of the instability of the surface, a comparison of EDCs taken above and below T_c could not be made. Schabel *et al.* used four surfaces on detwinned single crystals which were cleaved at or below 20 K and held there for all measurements. EDCs were taken with 28 eV photons polarized at 45° to the *b*-axis. The overall energy resolution was 50 meV and the angle resolution ±1°. The EDCs were all taken well below T_c. They show clear Fermi-surface crossings, similar to those in previous work, but no evidence of a pileup at or just below E_F. Fermi-surface crossings were identified two ways: the collection angle at which the midpoint of the leading edge had the least binding energy; and the angle at which the spectral peak area was reduced by half.

Figure 8.7 shows the (normalized) EDCs for four Fermi-surface crossings, along with the Fermi edge for Au in contact with the sample. It is clear that the leading-edge position depends on location in the Brillouin zone. The edge shifts monotonically upon moving from the Γ–S line toward the Γ–X line. Along the Γ–S line the edge appears to occur at negative binding energy. This can arise as an artifact of the instrument function. A sharp peak located just below E_F in the true spectrum can be shifted to above E_F by convolution of the true spectrum

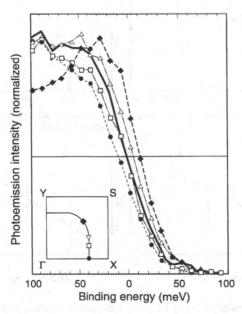

Fig. 8.7 EDCs for Y123 taken at Fermi-surface crossings at four points on the Fermi surface (*inset*) and the Fermi edge of Au (thick solid line). (Schabel *et al.*, 1997)

with the instrument function. The shift above E_F observed in Fig. 8.7 is consistent with the 50 meV resolution used. The gap for this spectrum must be zero or very small. The gap then grows roughly in accord with the shift of the leading edge as one goes around the Fermi surface.

The shifts of the leading edges are plotted in Fig. 8.8 as a function of $0.5|\cos k_x a - \cos k_y a|$, for which a gap of $d_{x^2-y^2}$ symmetry would give a straight line though the origin. For three of the cleaves, a straight line results. For the other cleave, there is an arc around the Fermi surface where the gap remains small, then it approaches a larger value. This could arise from impurities or imperfections which convert this sample to a dirty superconductor for which the minimum in $\Delta(\mathbf{k}_F)$ is broadened. These data strongly suggest a gap of $d_{x^2-y^2}$ symmetry, but the maximum value cannot be determined from the data without a model for the spectral shape. From Fig. 8.8 one sees that the intercepts range from 22 to 31 meV, so that fitting to a physically motivated spectral shape function would yield a comparable spread for the maximum gap.

Ding *et al.* (1996d) reported observation of the peak just below the edge by working with an untwinned single crystal of $YBa_2Cu_3O_{6.95}$, cleaved and measured at 13 K, then warmed to 95 K ($T_c = 90$ K) for a measurement, then cooled

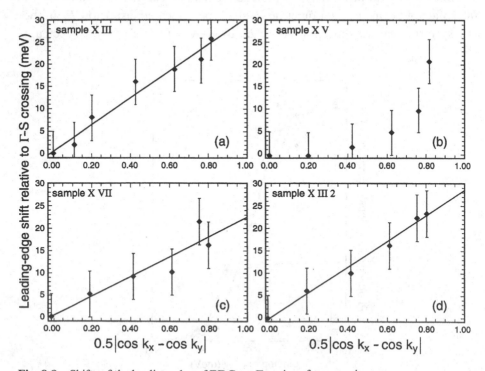

Fig. 8.8 Shifts of the leading edge of EDCs at Fermi-surface crossings vs. $0.5|\cos k_x - \cos k_y|$. A straight line through the origin would indicate a gap of $d_{x^2-y^2}$ symmetry except that the regions of negative gap can not be identified. (Schabel *et al.*, 1997)

back to 13 K for another measurement. Figure 8.9 shows all three EDCs taken with 18 meV overall resolution at the Fermi-surface crossing half-way between the Γ–X line and the Γ–S line, along with the Fermi edge of Pt taken at the same temperature as each Y123 spectrum. The gap between the rising edge of Y123 spectra at 13 K and the Pt edge is very clear. This gap is not present at 95 K. There is little difference between the two spectra taken at 13 K. The second one is slightly sharper, and the quasiparticle peak more prominent, an effect the authors attribute to the desorption of adsorbed gases upon heating to 95 K. Two of the spectra from Fig. 8.9 are replotted in Fig. 8.10, where one can see the pileup, and the opening of a gap. The areas of the two spectra are the same, within 0.4%, in accord with the momentum sum rule.

The 0–7-eV valence-band spectrum of the crystal used by Ding *et al.*, was different from the spectra reported in many previous publications by many authors. The sharp peak at 1 eV was somewhat less prominent, as it was in the anomalous cleaves of Schroeder *et al.* (1993b), again attesting to the variability in cleavage of Y123.

Because of the uncertainty in the identity of the surface plane(s) produced by cleaving, there has been considerable theoretical effort on the surface of Y123. This has followed two general approaches. One is the determination of the stability and structure of possible surfaces. The other has been to calculate the angle-resolved photoelectron spectra for all possible (unreconstructed, unrelaxed) surface planes for a comparison with experimental spectra.

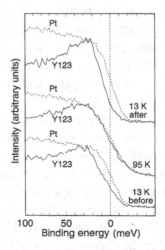

Fig. 8.9 Angle-resolved EDCs for Y123 (solid lines) taken at 13 K, then 95 K, then 13 K again and the EDC of Pt (dashed lines) taken at the same temperatures. The angle is such that the wave vector is at the Fermi surface, half-way between the Γ–X and Γ–S lines. (Ding *et al.*, 1996d)

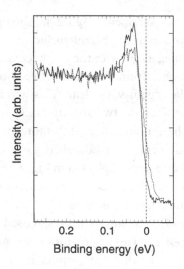

Intensity (arb. units)

0.2   0.1   0

Binding energy (eV)

Fig. 8.10 Angle-resolved EDCs from Fig. 8.9 replotted. Solid line: 13 K after recooling; dashed line: 95 K. (Ding *et al.*, 1996d)

Calandra *et al.* (1989, 1992) calculated the local density of states for the outer-most layers for a crystal of Y123 (fully oxygenated $YBa_2Cu_3O_7$) terminated by each of the six possible inequivalent layers. They used a tight-binding method with parameters derived from first-principles band calculations. They did not explicitly consider possible surface relaxation or reconstruction, but used the geometry of an ideally terminated surface, i.e., the atoms above the surface were simply removed, and the remaining atoms left in the locations they had in the bulk. Charge flow between atoms was permitted, so some surfaces may be left charged, leading to shifts in surface core-level binding energies. The band struc-ture of a slab four unit cells thick, with the *c*-axis parallel to the thickness, was calculated for each type of slab. The local density of states in the center of such a slab was the same as that of the bulk crystal, and interactions between states on the two surface planes were negligible.

Cleaving could break any one of three sets of bonds, each cleavage leaving two inequivalent surfaces. Calculations were done on three slabs, differing in the bonds broken to obtain the slab (Calandra *et al.*, 1992). The two surfaces of each slab were inequivalent, so all six possible terminations could be studied. The cleave between a Ba–O plane and a CuO_2 plane is believed to be the easiest. The Ba–O surface plane should be nearly electrically neutral, but the CuO_2 plane may have a net negative charge, compensated by the positive charge on the Y layer just below it. Cleaving between the CuO_2 and the Y planes leaves two charged surfaces which may be quite unstable. The other cleave leaves Ba–O and Cu–O chain surfaces, both possibly electrically neutral, but the chains may exhibit a tendency to reconstruct.

The calculated electronic structure for the Ba–O surface produced by cleaving between Ba–O and CuO_2 planes showed this surface to be very close to electrostatically neutral. Electronic charge flows from the bulk to the surface oxygens and to both Cu and O atoms in the planes below. Significant changes occur in the surface density of states, compared with that of the bulk. These changes were traced to changes in states on both the CuO_2 planes and the Cu–O chains in the layers nearest the surface. New peaks were produced and some structures were attenuated. The other surface produced by breaking the same bonds, a CuO_2 plane surface, loses electronic charge to the two layers below, and the chains below also lose electron density. The surface density of states also exhibits significant differences from that of the bulk.

Cleaving between the Ba–O plane and the Cu–O chain layer similarly produces surface densities of states that show several significant changes from that of the bulk. The CuO_2 plane layer below the Ba–O-terminated surface loses electrons to the surface oxygen atoms. The other surface, with Cu–O chains, turns out to be close to electrostatically neutral. Cleaving between the Y layer and a CuO_2 plane results in considerable charge flow. Charge flows from the nearest subsurface Cu–O chains, and even from the deeper CuO_2 plane, to the surface CuO_2 plane. This surface is expected to be unstable.

The two surfaces expected to be most stable, those produced by the cleave between the Ba–O and CuO_2 layers, were studied further by calculating the dispersion with $k_\parallel$ of the states localized at the surface, for comparison with the projected bulk density of states. The two surfaces showed five and seven, respectively, bands of surface states, many with binding energies greater than 1 eV. However, one surface state on the CuO_2 plane surface had a binding energy of about 1 eV, and it appeared over an extended range of $k_\parallel$ along Γ–Y and Γ–X. It was tentatively identified as the 1-eV peak believed at that time to be from a surface state (Claessen et al., 1991). The calculated character of this state was a hybrid of Cu $3d_{3z^2-r^2}$, O $2p_x$, and O $2p_z$ states based on atoms in the top CuO_2 plane. It appears as a dangling-bond state. The Ba–O surface also had surface states near 1 eV binding energy, but not along Γ–Y.

Calandra et al. also used the states of the surface layer (not just the surface states) to calculate the self-energy of a hole in the valence band. This leads via the spectral density function to the valence-band satellite spectrum, and how it might differ at the surface. They used the same formalism previously used (Calandra and Manghi, 1992; Calandra et al., 1991). They weighted the contributions to the spectral density function from each layer by an exponential factor to simulate the loss of peak intensity by inelastic scattering. There was not much difference between the surface and bulk contributions to the spectral shapes for the Ba–O surface, but for the CuO_2 surface there were some differences in the valence-band region of the angle-integrated spectral functions. (Dipole matrix elements were not calculated.) The calculated satellites were weaker from the surface regions for both terminations, more so for the Ba–O termination. They also simulated changes in oxygen stoichiometry on the angle-integrated valence-band photoelectron spectra, including satellites, for both the bulk and the Ba–O terminated surface. The satellites became weaker for the oxygen-poor surfaces, expected for they arise from two-hole final states on O atoms. In the oxygen-poor slab the valence-band spectrum was narrower than that from the bulk, and the step at the Fermi level considerably smaller.

Calandra and Manghi (1994) reviewed the foregoing and discussed comparisons with experimental attempts to determine the atomic nature and structure of surfaces produced by cleaving, by scraping, and of natural surfaces of films. The comparison is made difficult by the diverse nature of the experimental results.

Bansil *et al.* (1992) also assumed the six ideal terminations used by Calandra *et al.*, and calculated the angle-resolved valence-band photoelectron spectra along Γ–S for each, using the one-step formalism (Caroli *et al.*, 1973; Pendry, 1976; Hopkinson *et al.*, 1980). Their formalism is described in Lindroos and Bansil (1991). Muffin-tin potentials were used in the crystal. The surface was described empirically by a potential step, the photohole lifetime width, taken to vary from 0.1 eV at E_F to 1 eV at 3 eV below E_F, and the photoelectron lifetime width, given a constant 2 eV. Bulk energy bands calculated from the muffin-tin potentials agree with self-consistent LDA calculations as well as different LDA calculations agree among themselves. As with the work of Calandra *et al.*, fully oxygenated Y123 is assumed, and surface relaxation and reconstruction not considered.

The calculated angle-resolved EDCs are shown in Fig. 8.11, along with some experimental spectra. The spectra were calculated for unpolarized photons with 21.2 eV energy. The differences among the calculated spectra arise because of differences in multiple scattering and dipole matrix element effects. The nature of the layer below the top layer has a strong effect on the spectra as can be seen by comparing the spectra for two surfaces terminated by CuO_2 planes (T4 and T5) and for two Ba–O-terminated surfaces (T1 and T3). Because of these effects, none of the spectra can be expected to be close to what one would expect using bulk energy bands and bulk dipole matrix elements. Fermi-surface features in angle-resolved EDCs expected from bulk band structures may appear weaker, or even be missing, as a result of surface-specific effects.

Bansil *et al.* carried out more extensive calculations than shown in Fig. 8.11 (Bansil *et al.*, 1993). They calculated spectra to several eV binding energy with polarized radiation. As a result of comparing these, along with spectra in Fig. 8.11, with experiment, they found that termination T1 fits the spectra the best. T1 is a Ba–O surface with a CuO_2 plane just below. This is the surface produced

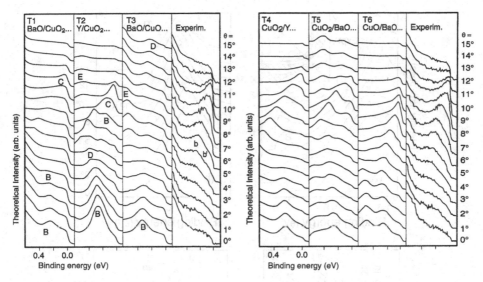

Fig. 8.11 Angle-resolved EDCs calculated for Y123 with six different terminations. (Bansil *et al.* 1993)

by breaking the shorter Cu–O bond, a cleave usually considered less likely than the cleave leading to the other Ba–O terminated surface. Thus the Ba–O surface producing the best agreement between the calculations of Bansil *et al.* and experiment is not the same Ba–O surface that Calandra *et al.* believe to be the most easily produced. Moreover the possibility of two different terminations produced by one cleave as a result of cleavage steps apparently was not considered, an effect complicating the choice of best fit.

In other publications (Lindroos *et al.*, 1993; Bansil *et al.*, 1993), more detailed polarization measurements are reported and compared with the predictions for the Ba–O terminated surface with CuO_2 planes just beneath it. An example is shown in Fig. 8.12, angle-resolved EDCs taken along the Γ–S line with polarization parallel and perpendicular to that line. The calculated peaks b and b' are from bands based on the CuO_2 planes. They have slightly different dipole matrix elements. They appear only for **E** parallel to $\mathbf{k}_{||}$, i.e., for **E** parallel to Γ–S. The calculations are for the Ba–O surface with a CuO_2 plane just below. The experimental counterpart in Fig. 8.12(a) is peak B. Peaks c appear for both polarizations. The authors demonstrate that the polarization dependence for the spectral features observed is different in the calculations for other surface terminations. They also indicate that it may not be necessary to break the short Cu–O bonds to obtain this surface. If the crystal separates on a Cu–O plane, leaving some of the chains on one side of the cleave and some on the other, both surfaces will have Ba–O layers on top with CuO_2 planes just below. Both will, however, have

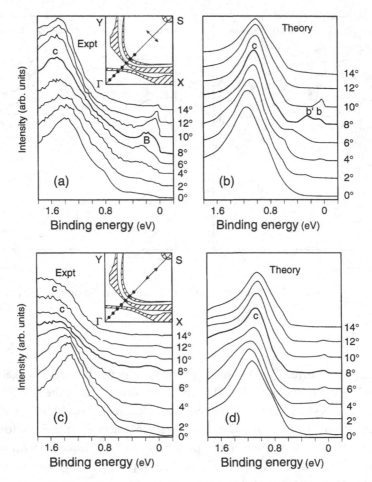

Fig. 8.12 (a) and (c) Experimental; and (b) and (d) calculated angle-resolved photo-electron spectra from Y123 taken along the Γ–S line. (a) and (b) are for the electric vector perpendicular to $\mathbf{k}_\parallel$, (c) and (d) are for $\mathbf{E}$ parallel to $\mathbf{k}_\parallel$. The angles from the surface normal determine the values of $\mathbf{k}_\parallel$, which are shown in the inset, along with the projection of the calculated Fermi surface. The thicker curves are those for which the bands due to the CuO_2 planes are just below E_F. (Bansil *et al.*, 1993)

Cu–O chains, perhaps randomly dispersed on top of the Ba–O layer. Such cleaved surfaces have been found in STM studies (Edwards *et al.*, 1992). This requires that the Cu–O chain fragments contribute little to the spectrum. As a matter of fact, the parts of the Fermi surface of Y123 derived from the Cu–O chain bands have yet to be reliably detected in angle-resolved photoemission spectra.

An important result of these studies (Campuzano *et al.*, 1991; Lindroos *et al.*, 1993; Bansil *et al.*, 1993) is that they observed the Fermi-surface crossing of two bands related to the CuO_2 planes. The splitting of the degeneracy of these two planes is much clearer in Y123 than in Bi2212, for it is uncertain whether or not it has really been observed in the latter.

Another approach to determining the surface layer was made by Frank *et al.* (1991), but unfortunately they used a different type of Y123 sample, epitaxial films with $T_c > 87$ K. They were studied after having been placed in the vacuum of their XPS spectrometer and after several subsequent annealing steps. The relative intensities of angle-resolved core-level spectra for the constituent atoms can be modeled, making use of the number of atoms of each type, their depth below the surface, the photoexcitation cross-section, the energy-dependent inelastic mean free path, and the angular distribution of photoelectrons after photoexcitation, all of which are known or can be calculated from free atoms. The usual model assumes the atoms are homogeneously distributed, but in Y123 the layers containing only a few of the constituent atom types are separated by distances of the order of the inelastic mean free path. Therefore Frank *et al.* treated the crystal as a series of discrete layers with the spacings given by the crystal structure. They modeled the relative intensity for all layers contributing to the spectrum, doing so for each of the six possible surface terminating planes of Y123 and compared the results with their measured spectra. They state the calculated relative intensities should be accurate to within about 15%. The spectra for a vacuum-annealed film (520 K), taken at $0°$ and $70°$ from the surface normal, were most consistent with the top layer being a CuO_2 plane with a Ba–O plane just below. After an anneal to 650 K in vacuum, the best fit suggested the surface had changed, with the outermost layer being Ba–O with a CuO_2 layer just below. Annealing in oxygen produced another change. The spectra implied a surface with Ba–O on top but with Cu–O chains just below. Their O 1s spectra had a significant component at 531–532 eV after each treatment.

The same approach to determining the surface layer was made by Terada *et al.* (1993), also on epitaxial films (*a–b* planes) produced by sputtering Y, Ba, and Cu in oxygen. The films thicker than 300 Å had values of T_c above 80 K. The samples were moved into an XPS spectrometer without breaking vacuum and XPS measurements carried out on the as-grown surfaces, not cleavage planes. Angle-resolved XPS measurements were made on the Y 3d, Ba 3d, and Cu 2p core levels as a function of collection angle. The relative intensities of these three lines were plotted as a function of angle. They did not vary as rapidly as might be expected from inelastic scattering only. The Cu peak fell with angle and the other two peaks increased slightly, all relative to the total intensity of the three peaks. They modeled the intensities as a function of angle with a single-scattering model, including inelastic scattering, for x-ray photoelectron diffraction, discussed briefly in Chapter 3. They assumed two possible terminations, CuO_2 and Y, Ba–O, and found only the former fitted the data.

8.4   Effects of stoichiometry

Liu *et al.* (1992a) studied the upper part of the valence bands in twinned crystals of $YBa_2Cu_3O_x$ as a function of oxygen stoichiometry, varying x from 6.35 to 6.9. Y123 is metallic and superconducting for x values of about 6.40 and above. The sample with $x = 6.35$ was insulating. Oxygen removal causes oxygen vacancies in the chains (Jorgensen *et al.*, 1990), and the vacancies tend to order (Reyes-Gasga et al, 1989; Hervieu *et al.*, 1989; de Fontaine *et al.*, 1990; Veal *et al.*, 1990a). Angle-integrated photoemission studies had showed a loss of intensity at the Fermi level as oxygen was removed (Veal *et al.* 1989). Scans along Γ–S and Γ–X(Y) were made. Along Γ–S two Fermi-surface crossings were seen. A crossing near Γ expected from the chain bands, according to the LDA calculations, was not seen with 21.2 eV photons. The region near S was not scanned. The Fermi-surface crossings occurred at the same values of **k** for $x = 6.7, 6.5$ and 6.4. In fact the spectral shapes were nearly the same for all three samples. For $x = 6.35$, the spectral peaks are much weaker and broader. For the sample with $x = 6.9$, the dispersion of the Γ–S spectrum below E_F is considerably less than that of the LDA bands. The scans along Γ–X(Y) showed just one Fermi level crossing, evidently the same as that seen by Tobin *et al.* (1992) near the Y point. Not all LDA calculations agree in this region of the Brillouin zone (Yu *et al.*, 1987; Freeman et al, 1987; Krakauer et al, 1988; Pickett *et al.*, 1990). Some describe this region as a chain/plane band. This peak decreases in intensity as oxygen vacancies are introduced. Such an effect is expected if the peak arises from oxygen states (at least partly) on the chains.

The foregoing work was extended by Liu *et al.* (1992b). They took many more scans on samples with $x = 6.9$, 6.5, and 6.3, covering a grid in the Brillouin zone. The grid spacing corresponded approximately to the wave-vector resolution determined by the angular resolution and the photon energy. The energy resolution was 55 meV, and 20–30 meV for some scans. For $x = 6.5$, the vacancies have ordered and the structure has been altered to a different orthorhombic structure and T_c has dropped from 92 K to around 50 K. For $x = 6.3$, the sample is insulating. In determining Fermi-surface crossings, data taken on a grid were especially helpful, for if a crossing were difficult to determine from a scan along a single line in **k**-space, often a parallel or perpendicular scan aided in the decision between a crossing or only a near crossing. The scans could be constructed from the data on the grid.

Their results for the Fermi surface of Y123 with $x = 6.9$ were shown earlier in Fig. 8.4. The data were taken only in the upper triangle, and reflected into the lower triangle by the symmetry operation expected for the twinned sample. The calculated Fermi surface is for an untwinned sample. The shaded regions are an indication of the dispersion with k_z of the Fermi surface. The experiment partially integrates over k_z for the resolution in $k_\perp$ corresponds to about one-third of the Brillouin zone height. There is good agreement with the Fermi-surface sections 2 and 3, both arising primarily from Cu 3d and O 2p states on the CuO$_2$ planes. Section 4, arising from Cu–O chain states, was not detected, despite using a range of photon energies and polarizations in an extensive attempt to see it. Section 1, based on Cu–O chain states, cannot be said to have been seen, for the points in the lower triangle are reflections of points in the upper triangle, points clearly due to section 2.

The sample with $x = 6.5$ had been heat treated to produce ordering of the oxygen vacancies. For $x = 6.5$ half the oxygen atoms in the chain basal plane are missing. The thermal annealing allows them to redistribute into alternating filled and empty chains, yielding the ortho II structure (Reyes-Gasga et al., 1989; Hervieu et al., 1989; de Fontaine et al., 1990). The scans along Γ–S were essentially identical to those for the sample with $x = 6.9$. The scans along Γ–X(Y) were similar, but the peak was weaker for $x = 6.5$, and its crossing of the Fermi level was not so obvious; it may not actually cross. Elsewhere in the Brillouin zone, the differences between the EDCs for the two samples are small and subtle.

Both samples showed the very sharp peak seen by Tobin et al. (1992) along the Γ–Y line, closer to Y than to Γ. Liu et al. altered their usual 21.2 eV photon energy to maximize this peak. It was somewhat weaker in the sample with $x = 6.5$, and it crossed the Fermi level in a slightly different place than for the sample with $x = 6.9$. This was the only difference in the Fermi surfaces of the two samples, and it suggests the Fermi-surface bulge on section 2 near the Γ–Y line is somewhat smaller for the sample with $x = 6.5$. One set of band calculations (Yu et al., 1987; Freeman et al., 1987) show a chain-related band crossing E_F in this region. For the sample with $x = 6.5$, this band would be lost near E_F, due to the decrease in chain oxygen atoms. The foregoing assumes the observed peak is near Y, not X. LDA calculations on Y123 with $x = 6.5$ (the ortho II structure) do not produce added insight (Yu, unpublished). The unit cell doubles in the a-direction and the side of the Brillouin zone along Γ–X is halved. Fermi-surface segments from the plane bands are folded back into the new zone, doubling their number, and the segments from the chain bands are not folded.

The samples with x in the range of 6.3 to 6.35 were semiconducting; the resistivities increased upon cooling below about 70 K. As in their previous work, these authors found no peak at E_F (near where such a peak was found in the samples with higher values of x) yet there is evidence of a broad dispersing structure in spectra taken along Γ–X(Y). They suggest that for such non-metallic crystals, a "small" Fermi surface may be present. However, such a Fermi surface should consist of small roughly elliptical closed surfaces centered on the lines from Γ to S. The EDCs showed only one Fermi-level crossing along this line, not the expected two, despite additional attempts to find the second crossing.

Liu *et al.* (1995) studied the effect of oxygen stoichiometry in Y123 once again, this time emphasizing the entire valence-band region in their angle-resolved EDCs. They used twinned single crystals, extensively annealed to stabilize the oxygen content. The resolution was between 20 and 100 meV, with a $\pm 1°$ angular acceptance. Various photon energies were used to bring out spectral features. The most prominent valence-band feature is the 1-eV peak discussed above. It is sharpest near the Y point (Tobin *et al.*, 1992). This peak appears with about the same intensity and dispersive characteristics in the crystals with $x = 6.9$, 6.5, and 6.4. The sample with $x = 6.3$, one that is semiconducting at low temperature, shows this peak, but it is much weaker, broader, and difficult to observe at some points in the Brillouin zone. The EDCs of an insulating sample with $x = 6.2$ do not show this peak at all. The 1-eV peak shifts slightly with decrease in oxygen content (see Fig. 8.13). The binding energy of this peak increases 0.20 eV as x goes from 6.9 to 6.3, using the Fermi edge of a Pt foil in electrical contact with the sample as a reference. This shift may be interpreted as an increase in the Fermi energy as the oxygen content drops. The direction of this shift is consistent with the picture of hole removal accompanying oxygen removal. The magnitude of the shift between $x = 6.9$ and 6.7 is close to the prediction of the rigid-band

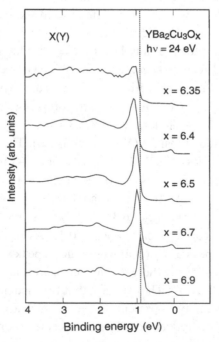

Fig. 8.13   Angle-resolved EDCs taken at the X(Y) point on several twinned crystals of Y123 with different oxygen stoichiometries. (Liu *et al.*, 1995)

model with the calculated LDA bands. This shift is also similar to that observed in Bi2212 as the hole concentration was varied (Shen *et al.*, 1991b).

The peak near the Fermi level was also re-examined. Along Γ–X(Y) a band disperses toward the Fermi level, but does not cross it. Instead, it remains just below E_F for a range of $\mathbf{k}_\parallel$ values, then disperses away. This is the extended saddle point along Γ–Y, and possibly along Γ–X. The EDCs showing this behavior are nearly the same in both line shape and intensity and in dispersion for the crystals with $x = 6.5$ and $x = 6.9$. There is not a shift with oxygen stoichiometry as might be expected from the rigid-band model and the shift in the 1-eV peak. This dispersing peak is not present in the corresponding EDCs for the crystal with $x = 6.3$.

The effect of substituting Zn and Co for Cu in Y123 was studied by Gu *et al.* (1993) using angle-resolved photoemission. Zn substitutes for Cu atoms primarily in the CuO_2 planes, while Co substitutes primarily for the Cu atoms in the Cu–O chains. Three fully oxygenated single crystals were used. One, $YBa_2Cu_{2.91}Zn_{0.19}O_{7-\delta}$, had $T_c = 58$ K and was twinned. The other two, $YBa_2Cu_{3-x}Co_xO_{7-\delta}$, were untwinned and had $x = 0.18$ and $T_c = 51$ K and $x = 0.54$ (not metallic). Along Γ–S the peaks near E_F due to CuO_2 plane states are very similar to those in fully oxygenated Y123 for the two superconducting samples, but they were not present in the non-superconducting crystal doped with Co. Zn, with its filled 3d shell with a binding energy of 9–9.5 eV, when substituted for CuO_2-plane Cu atoms is expected to diminish the number of states, based partly on Cu 3d states, in the plane bands. No such effect was seen. Since only about 3% of the total Cu content was replaced by Zn the expected effect may have been too small to have been seen. Co reduces the hole concentration, but the EDCs along Γ–S are hardly changed unless the sample becomes non-metallic.

Gu *et al.* also took angle-resolved EDCs on a fully oxygenated twinned crystal of $GdBa_2Cu_3O_{6.9}(T_c = 92.5$ K), as well as crystals of lower oxygen content, including one that was non-metallic. Gd is not expected to alter the electronic structure when it replaces Y. T_c does not change much. The spectra taken along Γ–X(Y) were compared with those for Y123. As expected, as the oxygen content decreases, a peak near E_F along Γ–X(Y) decreases in intensity without shifting in energy. It is absent in the non-metallic crystal. A close comparison with Y123 was somewhat limited by the rather low resolution, 140 meV, used in this study. Spectra were taken with 143 and 148 eV photons, just below and at the Gd 4d $\rightarrow$ 4f resonance. The difference spectra revealed the Gd 4f states occurred at 3–12 eV binding energies; there was little 4f intensity within 3 eV of E_F. The strong Gd 4f contribution to the spectrum below 3 eV was considerably broader than expected. The additional width was tentatively assigned to Gd at or near the surface.

8.5   Core-level spectra

The most interesting core-level spectra, the Cu 2p, were measured quite early and there have been no new developments. Core-level spectra from other elements have been studied as a function of hole concentration by Nagoshi *et al.* (1995), not only for Y123, but also for Bi2212 and $Pb_2Sr_2YCu_3O_y$. All cuprates studied have two CuO_2 planes per unit cell. The core levels used were Sr $3d_{5/2}$, Ba $3d_{5/2}$, Ca $2p_{3/2}$, and Y $3d_{5/2}$. The resolution was 0.46 eV. The Ca and Sr peaks are composites; the peak center was used as a measure of binding energy. The peak energies were plotted as a function of the measured T_c, which is a monotonic function of hole density, although not necessarily a linear one (see their reference 31). Their data, and those of others, are shown in Fig. 8.14. It is clear that there is a continuous distribution of binding energies, but with the exception of a few points in the plot for the Ca $2p_{3/2}$ panel, Fig. 8.14 (c), all the superconducting samples have smaller binding energies than all of the non-superconducting samples. The hole concentrations are expected to change primarily on the CuO_2 planes, but the core levels measured are on atoms in different planes. As discussed in Chapter 6, there are several contributions to the core-level binding-energy shifts with changes in chemical composition. The changes observed in the chemical potential (Fermi level) with hole doping are considerably smaller than the

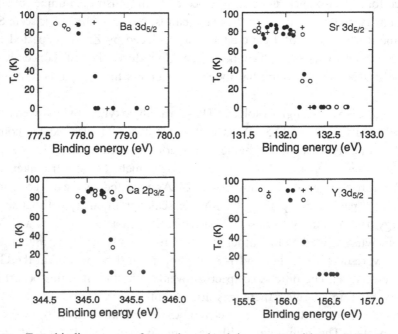

Fig. 8.14 T_c vs. binding energy of several core levels in cuprates with two CuO_2 planes per unit cell. (Nagoshi *et al.*, 1995)

core-level shifts. Nagoshi *et al.* suggest the major contribution to the shifts with doping comes from the slight reduction in the *a* and *b* lattice parameters with hole doping. This shortens the Cu–O bonds in the CuO_2 planes, and it should shorten the bond lengths between alkaline earth (and Y) atoms and their neighboring oxygens. This, in turn, increases the Madelung potential at the alkaline-earth site, which decreases the binding energy. However, no estimates of the magnitude of this effect were made. (There may also be changes in covalency with bond-length changes, making the estimate more difficult.)

Nagoshi *et al.* (1996) augmented the above study, reporting also the Bi $4f_{7/2}$ peak as a function of T_c as the hole concentration was changed, but only in Bi2212. In this case, there was a difference between changing the hole concentration by substituting Y for Ca and by varying the oxygen stoichiometry. The two plots were not the same; Y substitution did not produce much of a binding-energy shift while oxygen loss did. However the Sr $3d_{5/2}$ and Ca $2p_{3/2}$ lines gave "universal" plots for both types of change in composition. Nagoshi *et al.* attribute this to the fact that the oxygen loss occurred on or near the Bi–O planes, while Y replacing Ca and the concomitant loss of holes occurred at some distance from the Bi–O planes.

8.6   Related cuprates

$YBa_2Cu_4O_8$, Y124, is similar in structure to Y123 except that it has a second set of chains between the Ba–O planes. Both cuprates are orthorhombic. The calculated LDA bands are very similar, except for the number and location of Cu–O chain bands near the Fermi level. Campuzano *et al.* (1992) made angle-resolved photoemission measurements on Y124. As with Y123, there is a dispersing 1-eV peak and a peak dispersing through the Fermi level. Fermi-surface crossings along Γ–S and Γ–Y were in accord with the LDA Fermi surface (Yu *et al.*, 1991). The crossing along Γ–S had odd mirror-plane symmetry, consistent with the calculated Cu $d_{x^2-y^2}$ nature of the CuO_2 plane states involved. The Γ–Y crossing is by a band based on Cu–O chain states. One discrepancy with the LDA Fermi surface is the occurrence of a large peak right at E_F near the Y point at a photon energy of 28 eV. This peak is not present at all when 21.2 eV photons are used. It is not known whether this is an intrinsic bulk state or a surface state.

Aside from assigning the "1eV" peak to a transition originating on the CuO_2 plane, its nature remains problematical. Recently Pothuizen *et al.* (1997) carried out angle-resolved photoemission studies on $Sr_2CuO_2Cl_2$. This insulator is isostructural with La_2CuO_4 and has exhibited photoelectron spectra widely believed to be related to the spectra of the cuprates. In particular, a highly dispersive peak at lowest binding energy, around 1 eV, is believed to be the Zhang–

Rice singlet produced by photoexcitation. This state can be described roughly as an O 2p hole distributed on four O sites in a square, hybridized with a Cu $3d_{x^2-y^2}$ state on the central Cu atom. It appears with very low intensity in the spectrum because strong electron–electron correlation causes most of the weight in the spectral density function to appear in the structureless, incoherent part of the spectrum. Pothuizen *et al.* sought other O 2p states that did not hybridize much with the Cu 3d states.

The most intense peak in the spectrum dispersed between 2.5 and 3 eV, but it was not present in all parts of the Brillouin zone scanned. A second, almost dispersionless, peak appeared at about 4 eV in those parts of the Brillouin zone where the 2.5–3 eV peak was absent. Pothuizen *et al.* believe both these peaks arise from O 2p orbitals with little hybridization from Cu 3d states, the lower-energy peak primarily from p_x and p_y orbitals and the 4-eV peak primarily from p_z orbitals, based on the angle-dependence of their photoexcitation and on a comparison with a tight-binding model calculation. Little hybridization with Cu should lead to smaller correlation effects, hence a larger fraction of the spectral weight in the coherent peak. They identify these peaks with the "1 eV" peak found in several cuprates (Olson *et al.*, 1995). In regions of the Brillouin zone where there is some hybridization of the O 2p states with the Cu 3d, the intensity of these peaks should be smaller as correlation effects increase and place more of the spectral weight in the incoherent part of the spectrum. They also conclude that a self-energy dependent only on energy and not on wave vector will be insufficient to give a good account of the electronic structure.

A different oxychloride, $Ba_2Cu_3O_4Cl_2$, also an insulator, was studied by Golden *et al.* (1997) with a similar view. This material has a few properties different from many cuprates. There are CuO_2 planes like those in the cuprates, in which the Cu atoms order antiferromagnetically at 330 K, but there are other Cu atoms, also in the CuO_2 planes, in the centers of squares of the other Cu atoms, which order magnetically only at 31 K. Angle-resolved EDCs show three dispersing peaks, at roughly 1, 2, and 3 eV binding energy. The peak near 1 eV disperses, with a minimum binding energy, 800 meV, at the $(\pi/a, 0)$ point in the Brillouin zone. In $Sr_2Cu_2O_2Cl_2$ the Zhang–Rice singlet band also disperses. Its minimum binding energy occurs at $(\pi/2a, \pi/2a)$, a point equivalent to the $(\pi/a, 0)$ point in the Brillouin zone of $Ba_2Cu_3O_4Cl_2$. However, in $Sr_2Cu_2O_2Cl_2$ the intensity of this peak is large enough to detect only in a limited area of the Brillouin zone, while in $Ba_2Cu_3O_4Cl_2$, it appears throughout the sampled regions of the zone. This was ascribed to the occurrence of two Zhang–Rice singlet bands, a view reinforced by the line shape in some parts of the zone. The second component is at slightly larger binding energy, about 1 eV. The two spectral components were assigned to Zhang–Rice singlets centered on the two types of Cu atoms in the plane. The peak at lower energy, assigned to the

Zhang–Rice singlet like that in the cuprates, also exhibits the mirror-plane symmetry expected for the "common" Zhang–Rice singlet when measured at $(\pi/a, 0)$. They modeled these bands with tight-binding bands fitted to other calculated bands and a t–J model for hole motion on each of the two Cu sublattice. A reasonable fit to both bands was achieved. An unusual feature is that the usual Zhang–Rice singlet moved on an antiferromagnetic lattice but the other one was centered on the Cu atoms on a paramagnetic sublattice.

8.7 Summary

Far less can be said with certainty about the electronic structure of Y123 and related compounds than for Bi2212. The cleaved surfaces of Y123 are not as simple as those of Bi2212. The surfaces of Y123 are less stable, and several types of surface are exposed in a single cleave. The twinning of as-grown single crystals also hinders studies on Y123 because of the high failure rate of the detwinning procedure. Nonetheless, the Fermi surface has been mapped, and it agrees qualitatively with the LDA Fermi surface. A small pocket expected about the S point has not been reliably observed. As in Bi2212 this is a large Fermi surface. Doping causes changes in the direction predicted by the rigid-band model. Effective masses are larger than LDA band masses. There is an extended line of saddle-point singularities with its axis along the nearest-neighbor Cu O bonds in the CuO_2 planes along the Γ–Y direction. This line of singularities lies just below E_F, within an energy of a few times $k_B T_c$ of E_F. The states at the Fermi level are a combination of Cu $3d_{x^2-y^2}$ and O $2p_x$ and $2p_y$ states. It has not been observed along the (non-equivalent) Γ–X direction, but the chain bands complicate the search.

The origin of the most dominant feature of the photoelectron spectrum, a narrow dispersing peak at about 1 eV, has not been securely identified.

In the superconducting state there is a shift of the edge at E_F to greater binding energy in a number of spectra but determining a gap energy and symmetry has proven elusive. There are reports now from three groups which are in agreement on a number of features. The maximum gap appears to be around 20–30 meV, and it is more evident in just a few of many cleaves. The symmetry of the gap appears to be consistent with $d_{x^2-y^2}$ symmetry.

Chapter 9

NCCO and other cuprates

9.1 NCCO

$Nd_{2-x}Ce_xCuO_{4-y}$ (NCCO) was the first cuprate superconductor to exhibit n-type conduction. The substitution of Ce^{4+} for Nd^{3+} introduces electrons which go to the CuO_2 planes. (There are no Cu–O chains.) T_c peaks at 24 K for $x = 0.15$. The importance of NCCO is that is can be used to address the question of electron–hole symmetry. The electron doping swells slightly the large Fermi surface, while hole doping in other cuprates shrinks it slightly. In the localized picture with its small Fermi surface, how do the electron and hole Fermi surfaces compare? NCCO has a simpler LDA band structure near E_F than many other cuprates because there is only one CuO_2 plane per unit cell. Fewer bands cross E_F. What are the new states in NCCO the doping introduces along with the new electrons, or do those electrons occupy existing states as in the independent-electron picture? Other trivalent rare earths may be substituted for Nd. Why do the magnetic moments of the rare-earth ions not destroy superconductivity as they do in many other superconductors? A brief review of photoemission studies of NCCO was written by Sakisaka (1994).

Most of the studies of NCCO have been carried out on polycrystalline, i.e., sintered, samples. The earliest study on a single crystal (an epitaxial film), that of Sakisaka et al. (1990a,b), was mentioned in Chapter 6. Since then, and since the review of Sakisaka was written, there have been several more studies on single crystals.

King et al. (1993) and Anderson et al. (1993) made angle-resolved photoemission measurements on single crystals of $Nd_{2-x}Ce_xCuO_{4-y}$, mapping the Fermi

surface. Both groups cleaved the crystals at about 20 K and made measurements
at this temperature. King *et al.* used two crystals with $x = 0.15$ and 0.22. The latter was metallic but not superconducting. Its Hall coefficient was positive, unlike
that of the sample with $x = 0.15$. They used the relatively high photon energy of
70 eV, which caused their resolution to be about 140 meV, with an angular acceptance of $\pm 1°$. They measured at temperatures of 20, 35, and 80 K, but found the
sample surface deteriorated rather rapidly at 80 K; the 9-eV peak grew faster at
this temperature. Figure 9.1 shows some of their scans, while Fig. 9.2 shows the
measured Fermi surface of both samples and the Fermi surface from LDA calculations. They are hole surfaces, not electron surfaces. The agreement is very
good. Figure 9.2(c) shows the experimental Fermi surfaces superimposed. That
for $x = 0.22$ is slightly smaller than that for $x = 0.15$, although the displacement
is just larger than the combined errors reported. This is the direction expected
from the LDA calculations if the Fermi level rises with respect to the bands
upon adding electrons. The small change in Fermi surface is not easy to reconcile

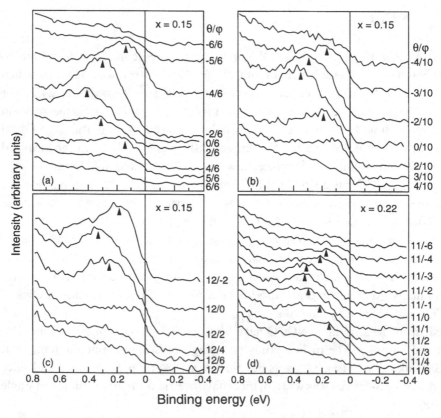

Fig. 9.1   Angle-resolved energy distribution curves for $Nd_{2-x}Ce_xCuO_{4-y}$.
For (a)(b)(c) $x = 0.15$, and for (d) $x = 0.22$. (King *et al.*, 1993)

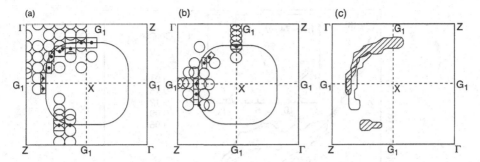

Fig. 9.2 Experimental Fermi surfaces (filled dots) and the Fermi surface of NCCO from LDA calculations (Massidda *et al.*, 1989) (solid lines) for (a) $x = 0.15$ and (b) $x = 0.22$. Open circles represent wave vectors at which spectra were taken but no Fermi-surface crossing was found. The circle size indicates the region sampled due to the finite angle acceptance. (c) Comparison of the two measured Fermi surfaces. Shaded region: $x = 0.15$; open region: $x = 0.22$. (King *et al.*, 1993)

with the significant changes in transport properties between samples with these two Ce concentrations. As with other cuprates, the bands below E_F were found to be flatter than the LDA bands. They were narrower by about a factor of 2. The 1-eV peak, so prominent in Y123, was not reported. The peak in the EDC that crosses the Fermi level as it disperses appeared much broader than the corresponding peaks in Bi2212 and Y123. It was considerably broader, its width not resolution limited. Finally, a flat band some 300 meV below E_F is part of an extended saddle point found much closer to E_F in other cuprates.

Anderson *et al.* (1993) worked with two single crystals of NCCO, $x = 0$, 0.10, and 0.15, and eight polycrystalline samples with x between 0 and 0.22. The former were cleaved and measured at 20 K. The latter were broken at 100 K. The angle-resolved EDCs were taken with 150 meV and $\pm 1°$ resolution, for the peaks were broad, and the ceramic samples were studied with 170–230 meV resolution. The Fermi surface of the crystal with $x = 0.15$ was mapped, and was in agreement with the LDA Fermi surface. Figure 9.3 shows how the lowest 0.5 eV of the gap for the sample with $x = 0$ fills in as the Ce concentration increases. Even at $x = 0.06$, the smallest non-zero value, the entire region from E_F to a binding energy of 0.5 eV starts to fill in. The Fermi energy of the metallic samples is known from the Fermi edge of an adjacent Pt sample. For the insulating $x = 0$ sample, the large band peaking around 4 eV was aligned with that of the other samples. Both King *et al.* and Anderson *et al.* used mirror-plane photoemission to establish that the band crossing E_F was consistent with $d_{x^2-y^2}$ symmetry, in agreement with the symmetry of the LDA band.

A complication to the interpretation of these data arose when Lindroos and Bansil (1995) published a one-step calculation of the angle-resolved photoelectron spectra expected from NCCO. They considered all four possible inequiva-

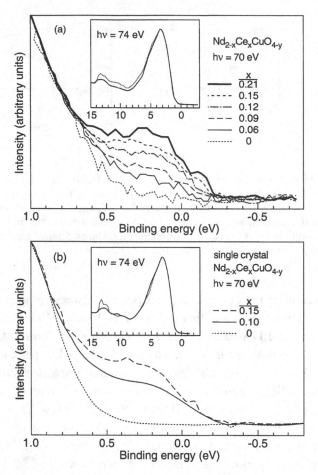

Fig. 9.3 Angle-integrated energy distribution curves of $Nd_{2-x}Ce_xCuO_{4-y}$ for several values of x. (a) Polycrystalline samples; (b) single crystals. The insets show several of the spectra over a wider range of binding energy. The peak in the insert was used to locate the Fermi energy for the insulating sample with $x = 0$. (Anderson *et al.*, 1993)

lent terminations for a crystal cleaved leaving an a–b plane, and did not consider lattice relaxation or reconstruction in their LDA calculation. Of these four, two, both terminated by a layer of Nd, gave rise to surface states which dispersed and crossed the Fermi level. They differed in the ordering of the layers below the surface: CuO_2, Nd, O_2 and O_2, Nd, CuO_2. Over part of the range of $\mathbf{k}_\parallel$ studied, one of the two surface states was above E_F, and the other below. The calculated bulk and surface bands are shown in Fig. 9.4. Along Γ–X where bulk band B crosses the Fermi surface, surface band SS1 is rather close, remaining below E_F. The broad asymmetric peak observed by King *et al.* and Anderson *et al.* could be a composite of the bulk band and this surface band. Lindroos and Bansil calculated the polarization dependence of the surface-band peaks in the EDCs and

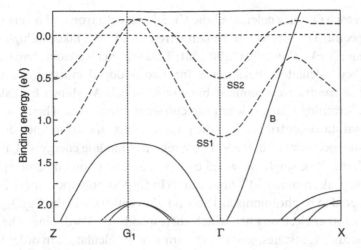

Fig. 9.4 Calculated LDA bands for NCCO with $x = 0.15$ and a surface terminated by a Nd plane. B denotes a bulk band that dispersed through the Fermi level. SS1 and SS2 are surface bands. (Lindroos and Bansil, 1995)

the energy dependence of their photoexcitation cross-sections. Measurements of these quantities could distinguish between the two surface states and the bulk. The only data to date are the relative photoexcitation cross-section measurements of the peak crossing the Fermi level made by Sakisaka *et al.* (1990b), which support the identification of this peak as arising from SS1, the peak expected from Fig. 9.4. Liu and Klemm (1994, 1995) also considered the role of surface states in generic cuprates with one and two superconducting layers per unit cell. In the latter case they found that the superconducting order parameter at the surface could be significantly larger than that in the bulk, especially when surface states were present.

The apparent filling in of the gap in NCCO when Ce substituted for Nd, i.e., upon electron doping, did not agree with much theoretical work at the time. However, it was in accord with photoemission and inverse photoemission studies on hole doping of insulating cuprates. There, too, the gap filled in uniformly with new states. Many cluster calculations (Dagotto, 1994) found that electron or hole doping led to a chemical potential near the conduction or valence-band edge, respectively, with retention of a gap. It is possible that our understanding of doping or of the photoemission process is inadequate. This was addressed by Tanaka and Jo (1996), who pointed out that theoretical studies of doping using a CuO_4 cluster, or even a Cu_2O_7 cluster, might be misleading because of the small size of the cluster. They used clusters with 2, 3, and 4 Cu atoms, and a larger basis set. The calculation followed that of Eskes and Sawatzky (1991). They used the same Hamiltonian, which includes multiplet structure, but they used a larger basis set and cluster size. They used 10 3d orbitals on each Cu atom and 6 2p orbitals on each O. They calculated the Cu 3d photoelectron and inverse photo-electron spectra. The former can be compared with EDCs taken at high photon energies, e.g., 70 eV, where the Cu 3d contribution is much greater than the O 2p.

First, they calculated the spectra for two undoped clusters, Cu_2O_7 and Cu_3O_{10}. The spectra were similar, but not identical. As shown by Eskes and Sawatzky, screening changes when the cluster size increases. There were three kinds of final states contributing to the photoemission spectrum, each dominant in part of the spectrum: d^8 in the 9–17 eV region of binding energy, $d^9\underline{L}$ (2–5 eV) and the Zhang–Rice singlet (0.9–1.5 eV). The inverse photoemission spectrum had a single peak from the $3d^{10}$ final state. (In Cu_3O_{10} only the central Cu atom was considered to be photoionized.) Second, the central Cu in the Cu_3O_{10} cluster was given another electron (3d), which corresponds to 33% doping. The energy eigenvalues and eigenstates, and the spectra were recalculated. In order to com-pare the spectrum of the doped cluster with that of the undoped clusters, the peaks due to the d^8 final state of each were aligned. There were several changes in the photoemission spectrum. The satellites at 10–15 eV became weaker. More importantly, a new type of final state contributes to the spectrum, one with a d^9 configuration on each Cu atom. This state contributes to the spectrum at lower binding energies than the other three. The gap between it and the lowest inverse photoemission peak is about 0.7 eV, instead of the value of about 2 eV for the insulating gap of the undoped cluster. Thus electron doping has produced a filling-in of the gap in the spectrum, as seen by Anderson *et al.* (1993), but it appears only in the final states near the gap, which are not the same as those in the undoped cluster. They also comment that hole doping produces similar changes in the lowest conduction band, filling in the gap in photoemission spec-trum of the insulator from above.

The calculation was repeated for larger clusters with different geometries, Cu_4O_n, with $n = 8, 10, 12,$ and 13. These had different photoelectron spectra, but some trends could be identified. The Cu atom which received the extra elec-tron had 1, 2, 3, and 4 nearest Cu neighbors in these clusters. The d^9 peak increased in intensity as the number of Cu neighbors increased. This arises because of charge transfer between Cu sites, an effect not possible in CuO_4 clusters.

The superconducting gap has not been detected by photoemission, and it seems likely not to be in the near future. Raman scattering measurements (Stadlober *et al.*, 1995) showed a gap appearing below T_c that was smaller than expected on the basis of the gap in other cuprates, and less anisotropic. The max-imum and minimum values of Δ obtained on a crystal with $x = 0.16$ ($T_c = 20.5$ K) were 4.1 and 3.5 meV. The ratio $2\Delta : k_B T_c$ was 4.1–4.9, less than the 6–8 obtained for other cuprates using the maximum values of the anisotropic order parameter.

Core-level studies on NCCO carried out more recently than those reported in Chapter 6 have been puzzling. First of all, not all the data agree. Then the results have been interpreted primarily in terms of an ionic model, the bond-valence sum, leaving little room for screening or final-state effects, except insofar as they are partially accounted for in a parameter-fit to experiment.

Itti *et al.* (1993) measured the O 1s peak on $La_{1.85}Sr_{0.15}CuO_4$ and $Nd_{1.85}Ce_{0.15}CuO_4$, cuprates with similar structures, but p-type and n-type respectively, using both single crystals and polycrystalline samples. For a superconducting sample of each the O 1s peaks differed by 0.4 eV. Each cuprate has oxygen atoms in CuO_2 planes and an equal concentration in out-of-plane sites. The spectra were decomposed into two peaks of approximately equal areas. The identification of the peaks was made with the help of the bond-valence sums for the two inequivalent oxygen sites. For $La_{1.85}Sr_{0.15}CuO_4$ the bond-valence sums were -2.26 for the in-plane O and -1.63 for the out-of-plane O. For $Nd_{1.85}Ce_{0.15}CuO_4$, these sums were -1.79 and -2.18, reversed in relative magnitude. Itti *et al.* suggest that the higher the effective valence, measured by the bond-valence sum, the higher the O 1s binding energy. Thus they assigned the component at lower binding energy to out-of-plane O in $La_{1.85}Sr_{0.15}CuO_4$ and to O in the CuO_2 planes for $Nd_{1.85}Ce_{0.15}CuO_4$. The peaks for the in-plane O were separated by 0.4 eV for the two cuprates. They also increased the doping of each material and found the peaks assigned to in-plane oxygen shifted in opposite directions, as expected.

The bond-valence sum is an empirical parameter, and its interpretation is not always straightforward. Itti *et al.* (1990b) applied the method to $Bi_2Sr_2Ca_{1-x}Y_xCu_2O_y$ with $x = 0$ and 1. The core-level binding energies for most of the elements (except O, which is complicated by having inequivalent sites) shifted about 0.4 to 0.8 eV to higher binding energy as x varied from 0 to 1. The bond-valence sums also increased, but the magnitudes of the increases did not correlate with the observed shifts. The Bi 4f sum shifted the most, but the Bi 4f binding energy shift was the smallest observed.

XPS spectra were taken by Cummins and Egdell (1993) on polycrystalline samples of $Nd_{2-x}Ce_xCuO_4$ for $0 < x < 0.20$. Sample preparation was by annealing at 823 K in good vacuum. A Fermi edge was observed in the metallic samples ($x > 0.1$). The O 1s spectrum was a single peak at 528.8 eV for all samples. The Cu 2p spectrum did not vary with x. The constant binding energy of the O 1s state was taken as evidence that the Fermi level did not move into the conduction band as x increased. The Cu 2p spectra did not change with x. A fit to a cluster model gave a negative value for the charge-transfer energy, -0.9 eV, which implies that NCCO may not be a charge-transfer insulator.

Wang *et al.* (1995) measured the $Cu\ 2p_{3/2}$ spectrum of NCCO with $x = 0.15$ before and after annealing in a reducing atmosphere. $Cu\ 2p_{3/2}$ core spectra had been measured previously as a function of Ce concentration, but different groups had obtained different results. The polycrystalline samples were measured as prepared, and after annealing for 7 and 42 hours in N_2. All three samples exhibited semiconducting behavior, but the two annealed samples showed a sharp drop in resistivity at 15 and 20 K, respectively, indicative of superconductivity. The O 1s spectra showed a significant shoulder at about 532 eV. The Cu $2p_{3/2}$ spectra showed the strong peak and satellite, but the satellite intensity was low, much lower than in the p-type cuprates. The anneals did not change the intensity of the satellite with respect to the main peak. The ratio of satellite to main peak intensities and the energy between these two peaks are used frequently to determine parameters in the model for cuprates based on a Cu–O cluster, discussed in Chapter 5 for CuO. In that model the charge-transfer energy, Δ, the Cu 3d–O 2p hopping integral, t, and the repulsive energy of the Cu 2p core hole and the Cu 3d hole, U_{cd}, are interrelated by these two parameters. In one previous analysis of the Cu $2p_{3/2}$ spectrum of NCCO, the small relative intensity of the satellite line led to a negative value for Δ. Wang et al. evaluated bond-valence sums for all the ions in NCCO, obtaining 3.07 for Nd, 1.70 for Cu, -1.75 for the in-plane O and -2.18 for the out-of-plane O. The principal difference between these values and those for Y123 is that in the latter, the bond-valence sum for Cu is 2.3. The assumption that all the Cu in NCCO is divalent is probably incorrect, and this assumption was used in fitting the data to the cluster model. NCCO, especially when doped with electrons, contains some monovalent Cu. The Cu $2p_{3/2}$ spectrum from divalent Cu is a single peak with a binding energy near that of the main peak of Cu^{2+}. No such peak was resolved in the spectra of Wang et al., taken with an unmonochromated Mg-Kα source. Wang et al., went on to relate changes in Δ to changes in the difference in Madelung energy between the Cu and in-plane O sites. This way, they obtained a positive value of Δ, but one smaller than in other cuprates.

Nd_2CuO_4 can be electron doped in yet another way, by replacing some oxygen atoms by fluorine. Sugiyama et al. (1992) made XPS and UPS measurements on several polycrystalline samples of $Nd_2CuO_{4-x}F_x$ with $x = 0$, 0.1, 0.16, 0.2, and 0.3. The samples with $x = 0.2$ and 0.3 were superconducting at 20 K and 19 K, respectively. The F 1s core spectrum was a single peak that did not shift in position with x, but with a strength that grew with increasing x. Comparison to its binding energy from that of other F-containing compounds suggested that there were Cu–F bonds, i.e., that F replaced in-plane O atoms. (This was not known from neutron diffraction studies because of the similar neutron scattering cross-sections of O and F.) Nd core-level spectra and the O 1s spectra were independent of x. The Cu $2p_{3/2}$ spectrum showed a slight growth in the main peak with respect to the satellite as x increased. This was taken to mean that F-doping led to the conversion of some Cu^{2+} ions to Cu^+. The Cu $2p_{3/2}$ spectrum of the latter added to the intensity of the main line from the Cu^{2+}, which it overlapped, but not to the satellite. The UPS spectra showed a slight increase in the intensity at E_F as x increased. However, for the insulating sample with $x = 0$ there was no gap region in the first 0.5 eV of the spectrum as observed by Anderson *et al* . (1993).

Klauda *et al.* (1993) studied NCCO and two related compounds. NCCO is an n-type conductor. Sr can be added along with the Ce to make $Nd_{1.4}Ce_{0.2}Sr_{0.4}CuO_{4-y}$, which is an n-type conductor. Unfortunately, the structures of these two materials are only similar, not identical. The structure of NCCO is the so-called T′ structure (Fig. 2.6), while that of the Sr-doped material is the so-called T* structure. In the T′ structure each Cu has four nearest-neighbor coplanar oxygen atoms. In the T* structure each Cu atom has the same four plus a fifth O atom above it. Klauda *et al.* also studied samples of $Sr_{0.85}Nd_{0.15}CuO_{4-y}$, a material with a negative Seebeck coefficient but positive Hall coefficient. It has yet a different structure, but the Cu atoms have four coplanar oxygen neighbors at about the same distances as in NCCO.

Polycrystalline samples were made of each compound. T_c values were measured as 24 K for NCCO, 23 K for $Nd_{1.4}Ce_{0.2}Sr_{0.4}CuO_{4-y}$, and 31 K for $Sr_{0.85}Nd_{0.15}CuO_{2-y}$. Photoelectron spectra were obtained on surfaces scraped *in situ* using Al-Kα, He I, He II, and Ne I radiation. Additional measurements were made with synchrotron radiation. The O 1s spectrum showed no "impurity" peak at 532 eV for the NCCO samples, but one was always present in the spectra of $Nd_{1.4}Ce_{0.2}Sr_{0.4}CuO_{4-y}$, possibly from the sample holder. These samples also had a peak at 9.5 eV, below the valence-band spectrum, a peak which grew noticeably over an hour or so in ultrahigh vacuum, accompanied by changes in the valence-band spectrum. No intensity was observed at the Fermi edge in any of the samples, which may therefore have lost enough surface oxygen to render the surface regions non-metallic. The Fermi-edge measurements were made with He or Ne sources, and the sampling depth was less than that for the core levels studied with Al-Kα radiation.

The Cu $2p_{3/2}$ spectra were studied extensively. Adding Ce to NCCO causes the ratio of the intensity of the main peak to that of the satellite to increase, but with no shape changes or shifts in binding energies. This had been attributed to an increase in Cu^+ brought about by electron doping, for the $2p_{3/2}$ spectrum of Cu^+ should be one peak, overlapping the main peak of Cu^{2+}, but no satellite. Cu $2p_{3/2}$ spectra for each compound were fitted to the CuO_4 cluster model (Fujimori and Minami, 1984; Zaanen *et al.*, 1986), frequently used for such spectra. (A CuO_5 cluster was needed for $Nd_{1.4}Ce_{0.2}Sr_{0.4}CuO_{4-y}$.) The new input parameters were the splitting between the main line and satellite, which ranged from 8.6 to 8.7 eV for the three compounds, and the intensity ratio of main line to satellite, ranging from 2.2 to 3.1. The valence-band spectra of the T′ and T* samples were slightly different only in the rise toward the main peak, i.e., in the 1–3 eV region of binding energy. These spectra were fitted to a cluster model like the one used for CuO by Eskes *et al.* (1990).

The conclusion of their analysis was that the 3d–3d correlation energy was larger for the two n-type cuprates studied than that for the p-type studied (and other cuprates, all p-type), although this parameter is not a direct result of the fitting. As noted by the authors, these conclusions rest both on experimental data, some of which were valence-band spectra taken on surfaces that showed no Fermi edge, and on fits to a model known at that time to lack some of the features found in more sophisticated (and complicated) models, e.g., those based on larger clusters. See also the more recent work of Tanaka and Jo (1996), discussed above. Klauda *et al.* also obtained spectra for Nd. In the case of $Sr_{0.85}Nd_{0.15}CuO_{2-y}$ they showed that the valence of the Nd was indeed 3+, so that electron doping is expected, despite the ambiguous sign of the carrier charge obtained from transport measurements.

9.2   Other cuprates

$La_{2-x}M_xCuO_4$ with M = Ba or Sr was the first of the cuprate superconductors to be discovered. When x is near 0.125 and M is Ba, there is a structural transition from orthorhombic to tetragonal at about 60 K and T_c is suppressed. If M is La, this does not occur. Maruyama *et al.* (1994) took angle-integrated EDCs on single crystals of both structural phases of $La_{1.875}Ba_{0.125}CuO_4$, and found no difference in the first eV of binding energy. The Fermi edge was very small. Comparison with EDCs from a single crystal of $La_{1.85}Sr_{0.15}CuO_4$ showed the latter to have a smaller EDC and Fermi step than the Ba-doped crystal, despite having a higher T_c. For this system T_c does not scale with the density of states near E_F. These data also suggest that the structural phase transition is not the cause of the lower T_c.

Soda *et al.* (1993) took angle-integrated photoelectron spectra on single crystals of $(La_{1-x}Sr_x)_2CuO_{4-y}$ for several values of x. Data were taken at 300 K, 80 K, and 30 K. At each temperature the sample surface was prepared by fracturing or scraping. Data taken at 80 K indicated that many scrapes, several hundred, were needed to obtain a spectrum believed to be free of surface contaminants. That surface was then stable in ultrahigh vacuum for at least 10 hours. The spectra consisted of a single broad feature between 1 and 7 eV, with its peak at 3.5 eV. There was little or no strength at the Fermi level for x as large as 0.08. They compared their spectra with those taken previously at room temperature by others by fracturing and scraping, but different photon energies were used, which could contribute to differences. Room-temperature spectra were observed to change noticeably within half-an-hour after scraping, with changes continuing for at least 8 hours. Two peaks grew with time and a region of the EDC was reduced in intensity. The origin of the changes was suggested to be the adsorption of water molecules which react to yield surface hydroxides. The changes were partly reversible by exposure to ultraviolet radiation. The instability of the surface of La_2CuO_4 and related cuprates is probably caused by the lack of a stable terminating plane upon cleavage or fracture, much the same as in Y123. Figure 2.6 shows one of the structures for this type of cuprate. See also Fig. 2.2(a). There are planes of Cu and O atoms. Between each of them are two "planes" with the La and remaining O atoms, but these are strongly buckled. Cleavage between La–O and Cu–O planes would leave a surface terminated with either type of plane, and in practice, both would occur. The La–O-terminated surface would be likely to reconstruct. Cleavage between La–O planes is more symmetric, leaving La–O surfaces on each side of the cleave. Each seems likely to reconstruct.

Law *et al.* (1996) studied resonance photoemission in La_2CuO_{4+x} using the La 5p core level with a binding energy of about 16 eV. The samples, single crystals with x between 0.005 and 0.03, were cleaved in vacuum at a temperature of 40 K or below. Electrons were collected at normal emission. The valence-band spectrum covered the range 0–7 eV in binding energy. By measuring the EDC with photon energies covering the 15–19 eV range in steps, a resonance was found at about 6 eV binding energy. If the angle of incidence of the photons was increased from 20° to 70°, thereby increasing the component of the electric field along the c-axis, a second resonance peak appeared at about 5.5 eV binding energy, along with the 6-eV peak. The 6- and 5.5-eV peaks had maximum intensity at photon energies of 17 and 16.5 eV, respectively. Proof that these were La resonances was obtained by repeating the measurements on Nd_2CuO_4. Excitation with photons in the region of the Nd 5p core-level excitation energy produced no resonance enhancements. Spectra were taken on La, partially oxidized La, and fully oxidized La, presumably La_3O_3. Only partially oxidized La showed a resonance similar to that found in La_2CuO_4.

Rietveld *et al.* (1993) studied the shift of the chemical potential in $La_{2-x}Sr_xCuO_4$ using polycrystalline pellets and thin films. x varied from 0 to 0.25 in steps of 0.05. The pellet samples were scraped in vacuum. A great deal of material, up to several hundred μm, had to be removed before an XPS spectrum with a minimum in the strength of the C 1s line could be obtained. The thin-film samples were prepared by laser ablation and transferred in vacuum to the XPS chamber, where no further surface treatment was carried out. The O 1s "contaminant" peak at 231 eV was well resolved, but as the XPS measurements progressed, it weakened, apparently from loss of surface contaminants. XPS spectra from a core level of each element and the valence band were taken. The shift in binding energy with x was linear, within estimated errors, about 0.2 eV for x varying from 0 to 0.25. There was no change in slope at the metal–insulator boundary, unlike the increased rate of shift found for Bi2212 by van Veenendaal *et al.* (1993a). The Fermi level did not appear to jump across the insulating gap when the sample was doped to become a metal. However, the shift with x of the chemical potential for $La_{2-x}Sr_xCuO_4$ was similar to that for Bi2212 in the region of metallic conductivity of the latter. These data also differed from the data of Allen *et al.* (1990) on NCCO where no shift was found upon electron doping. The authors suggested that data describe the drift with doping of the Fermi level through a rigid narrow band lying near the center of the insulating gap.

Recently, Ino *et al.* (1997) reinvestigated the chemical potential shift in $La_{2-x}Sr_xCuO_4$ with XPS and BIS spectra taken on single crystals with values of x of 0, 0.1, 0.15, 0.2, and 0.3. There was no detectable shoulder on the high-energy side of the O 1s peak. The O 1s and La 3d and 4f peaks shifted rigidly with changes in x, with no change in shape. The Cu 2p peaks changed shape as x changed, and the shift observed was due, at least in part, to changes in relative strength of unresolved components. For x between 0 and 0.15, there was little or no shift of the O and La core-level binding energies. For larger x, the overdoped region, the binding energies decreased with increasing hole concentration. The authors ruled out major contributions from doping-induced changes in the Madelung potential, ionic charge, and extra-atomic relaxation energy. Within experimental errors, the chemical potential shift, to smaller binding energy with increasing hole concentration as though the Fermi level moved toward or into the valence band with increasing hole doping, is quadratic in the hole concentration. The magnitude of the shift, about -1.5 eV/hole, is similar to that found in Y-doped Bi2212 by van Veenendaal *et al.* (1993a), about 1 eV/hole.

We have discussed the well-known 4d $\rightarrow$ 4f resonance in rare earths. The 4f states lie near the Fermi level and a 4d electron is excited into one of them. The 5p–5d resonance is somewhat different, for the La 5d level in La_2CuO_4 is found at least 3 eV above E_F in inverse photoelectron spectra. Thus the 5p–5d photoexcitation threshold is above the energy measured for the resonance, about 1 eV above the binding energy of the 5p electron. This is explained qualitatively by the effect on the 5d state on the 5p hole; it lowers the system energy by about 2 eV. An atomic calculation to test this explanation may not be relevant, for the authors note that these La resonances apparently occur only with an oxygen atom nearby, or the non-integral valence that the oxygen may cause, and on the density of states in the first few eV above E_F.

Several cuprates contain Hg and Tl oxides. These include one with the highest value of T_c reported for any material. However, because of the toxicity of Hg and Tl, and their high vapor pressures, efforts to grow single crystals of them have been far less extensive than for many other cuprates. Consequently no single crystals of adequate quality for angle-resolved photoemission have yet been grown. More correctly, no attempt to obtain angle-resolved photoelectron spectra on these cuprates has been published. The only studies to date are a few XPS studies of core levels and angle-integrated spectra of the valence-band region, often on sample surfaces that are demonstrably inferior to those in studies of other cuprates. We describe a few of these below, the most recent measurements. References to earlier work, and comments on that work may be found in the references we cite.

Vasquez et al. (1995) worked with epitaxial films of $HgBa_2CaCu_2O_{6+\delta}$, Hg1212. The films had values of T_c of 120 K and 123 K. The samples were handled in a nitrogen atmosphere, then etched with 0.2% Br_2 in absolute ethanol in a dry box. The surfaces are metallic (they show no charging effects) and they appear to have lost some Hg possibly due to preferential etching or the high vapor pressure of Hg. The O 1s spectrum has quite a large peak at 531.5 eV which was even larger before etching. The remainder of the spectrum, a peak at 528 eV was decomposed into two peaks, one, at 527.8 eV, assigned to CuO_2 planes and the other, at 528.8 eV, to O bonded to Hg. The Hg 4f spectrum was a simple spin–orbit split pair, indicating a single site and valence for Hg atoms in Hg1212. The Ba 4d spectrum was composed of two sets of spin–orbit pairs. The one at larger binding energy increased in relative intensity at larger collection angles, so it was assigned to a surface Ba, as in, e.g., Y123. The Ca 2p spectrum was also a pair of spin-orbit pairs, separated by 1.1 eV. The peak at lower binding energy is the less surface sensitive when data are collected away from the normal. It was assigned to Ca in the layer between the CuO_2 layers. The Ba and Ca peaks are both at lower binding energies than in the elemental metals, an effect not fully understood, but presumably due to Madelung potentials (Vasquez, 1994b).

The Cu $2p_{3/2}$ spectrum consisted of the familiar main peak and satellite. The ratio of the intensities of the satellite to that of the main peak was 0.36, smaller than the value for Tl2212. It may be premature to go to the effort of fitting a CuO_4 cluster model to these data. The valence-band spectrum showed a broad peak from 0 to 8 eV, with a width of the order of that expected from the Cu 3d states in the LDA band calculations (Novikov and Freeman, 1993). There was a small step at the Fermi edge. No satellite or 9.5-eV peak could be sought at higher energy because of the large peak from the Hg 5d levels, which produced a sharp rise beginning just below 8 eV.

Gopinath et al. (1995) worked with a polycrystalline sample of $HgBa_2Ca_{0.86}Sr_{0.14}Cu_3O_{8+\delta}$, Hg1223, with a T_c of 132 K. The only core-level spectrum they reported is the Hg 4f, which consists of two spin–orbit split pairs, unlike the case of Hg1212 reported by Vasquez et al. (1995). The difference may be intrinsic or arise from the difference in sample and in surface preparation. Vasquez et al. cite earlier work where scraping has produced changes in spectra. Gopinath et al. attribute the pair of peaks to two oxidation states for Hg, Hg^{2+} and Hg^{3+}. As supporting evidence, the bond-valence sum of Hg was reported to be 2.22. However, Kim et al. (1995), working with surfaces prepared by breaking polycrystalline samples, found no evidence of Hg^{3+} in their spectra. They also found that the Hg 4f spectra changed with time, changes being observed in ten minutes, and the vacuum deteriorated with time. They found scraped surfaces did not yield reproducible spectra.

Vasquez and Olson (1991) took XPS spectra on epitaxial films of $Tl_2Ba_2CaCu_2O_{8+x}$, Tl2212, that had been etched in 0.1% Br_2 in absolute ethanol. The O 1s spectrum had a significant peak at 531 eV, although, before the etch, this peak was larger than the intrinsic peak around 528 eV. (Exposure of the etched surface to air for one hour resulted in the same O 1s spectrum as before the etch. Only one minute in air was enough to degrade the O 1s spectrum.) Nonetheless, the etched sample showed a Fermi edge in the valence-band spectrum. The Cu $2p_{3/2}$ peak and its satellite appeared much like the corresponding peaks in other cuprates. The ratio of satellite intensity to that of the main peak was 0.45. The Ba and Ca core-level spectra showed two inequivalent Ca sites but only one Ba site.

Vasquez et al. (1996) carried out similar measurements on Br_2-etched epitaxial films of $Tl_2Ba_2CuO_{6+\delta}$, Tl2201, and compared them with the corresponding spectra for Tl2212. T_c values ranged from 10 K to 80 K, depending on annealing conditions, which alter δ. The component at 531 eV in the O 1s spectrum was comparable to that in the spectrum of Tl2212. The reminder of the O 1s spectrum consisted of two peaks, assigned to O in CuO_2 planes and to O in Tl–O bonds. The Tl 4f spectrum was a single spin–orbit split pair, and the Ba 4d was a pair of such doublets, at binding energies comparable to those in other cuprates. The Cu $2p_{3/2}$ spectrum was nearly identical to that of Tl2212. The valence-band spectrum was similar to that of Tl2212, but the main peak occurred about 1 eV closer to E_F and the intensity at E_F was higher. All core-level binding energies became slightly larger upon oxygen removal by annealing in Ar. This shift is in the direction expected if the chemical potential increases as oxygen is removed, i.e., if holes are removed.

$Tl_2Ba_2Ca_2Cu_3O_{10-\delta}$, Tl2223, was studied by Chen et al. (1996) using x-ray absorption (by total photoelectron yield) and total x-ray fluorescent yield spectra on powder samples. The electron-yield technique samples a depth of about 100 Å, while the fluorescence yield samples a greater depth, about 1000–5000 Å. The two spectra did not agree near the O 1s edge. Chen et al. attribute this to the loss of oxygen from the region sampled by the total electron yield, a region deeper than that sampled by photoelectron spectroscopy. Suzuki et al. (1989)

had already compared the Tl $4f_{7/2}$ spectrum of this cuprate, also taken on poly-crystalline samples, with the corresponding spectra of Tl_2O_3 and Tl_2O. Interpolation led them to conclude the valence of Tl in this cuprate was between 1+ and 3+. However, this could have been affected by the oxygen loss found by Chen et al.

XPS spectra from polycrystalline samples of $(Pb_{0.5}Cd_{0.5})(Sr_{2-x}Ba_x)(Ca_{0.5}Y_{0.5})Cu_2O_7$ were obtained by Yu et al. (1996). This cuprate has a resemblance to Y123, but there are no Cu–O chains. Instead, the charge reservoirs are (Pb,Cd)O rocksalt-structured layers. A series of samples was prepared with x between 0 and 0.5. For $x = 0.4$ and 0.5, the samples were superconducting (T_c was about 35 and 50 K, respectively). Samples with lower values of x were not metallic at low temperatures, although all spectra were taken at 300 K. The O 1s spectrum had a peak at 530.0 eV and a shoulder at 527.7 eV which grew as x increased from 0 to 0.5. The Pb $4f_{7/2}$ peak shifted to higher binding energy as x increased, a sign of increasing oxidation, evident by comparing the spectra with those of PbO, Pb_3O_4, and PbO_2. Finally, the Cu $2p_{3/2}$ peak showed a loss of satellite intensity as the Ba-level increased, and the main line developed an asymmetric tail to higher binding energies. The authors interpreted these changes with the standard cluster model; Ba doping decreases the Cu–O charge-transfer energy Δ. The growing asymmetry of the main line was attributed to increasing hybridization between Cu $3d_{3z^2-r^2}$ and O $2p_z$ states, the latter on out-of-plane oxygen atoms. However, the valence-band spectra, measured with 1245-eV photons, did not show a Fermi edge for the two superconducting samples. This calls into question the homogeneity of the samples within the depth being sampled.

Chapter 10

Surface chemistry

Photoelectron spectroscopy, especially XPS, can be used to study some aspects of the surface chemistry of cuprates. Examples of this are the changes seen in the photoelectron spectra of Y123 as a freshly prepared surface deteriorates with time at 300 K in ultrahigh vacuum. In this, the evidence that the changes are due to loss of oxygen is not direct. A better example, to be discussed below, is the study of a metal overlayer on a cuprate. The overlayer may just sit there, making it a candidate for an ohmic contact, or it may disrupt the surface, replacing some of the atoms in the cuprate, which then appear in the surface layer. The appearance or disappearance of core-level peaks can be used to track the depths of constituents, and binding-energy shifts may give clues about oxidation states and location of atoms, i.e., surface or bulk.

Most of the studies of interface reactions have been carried out on the same type of surfaces used in the study of the cuprates themselves, cleaved single crystals or scraped or fractured polycrystalline samples. These cannot be considered technologically important surfaces unless processing in the future is carried out under ultrahigh vacuum conditions. Some of the studies reported below may need to be repeated on air-exposed surfaces, or surfaces handled in inert atmospheres or poorer vacua.

10.1   Oxygen removal and replacement, water adsorption

A number of early photoemission studies made attempts to vary the surface stoichiometry by annealing the sample *in situ* in O_2 at various pressures and temperatures. One of the reasons for so doing was the failure to see a photoelectron signal from the Fermi level when the bulk sample was known to be metallic. These attempts failed probably because the O_2 was pumped out at room temperature and, before the vacuum was adequate for photoelectron spectroscopy, some surface oxygen atoms were able to leave the surface. Photoelectron spectroscopy with ultraviolet photons samples a surface layer no more than a few unit cells deep. The difference between metallic $YBa_2Cu_3O_7$ and insulating $YBa_2Cu_3O_{6.5}$ is the loss of only one oxygen atom out of every 14.

Oxygen can be removed from the surface by exposure to vacuum-ultraviolet or soft x-ray photons. Normally, the beam of monochromatic photons used for photoelectron spectroscopy is not adequate for producing a noticeable change in a spectrum after exposures of many hours. However, exposure to the zero-order beam from a monochromator on a source of synchrotron radiation, or from the raw beam emitted from an uninstrumented port on such a source, produces a flux of ionizing radiation orders of magnitude greater, and effects on the photoelectron spectra have been observed. Rosenberg and Wen (1998a) exposed pressed tablets of polycrystalline samples of Y123 at about 300 K to the raw beam and to the zero-order spectrum from a monochromator, both from an 800 MeV storage ring. Fresh surfaces were produced by scraping. Desorbed O_2 was detected in a mass spectrometer. There was a burst of O_2 upon scraping the surface, assigned to the release of O_2 trapped in the interstices between grains. Exposure to the zero-order beam in the monochromator produced a large release of O_2 which decayed non-exponentially with time to a steady rate of release about one-tenth of the initial rate. Periods of darkness lasting up to 15 hours initiated after the initial irradiation, followed by a second irradiation, did not produce an initial spike in O_2 release rate, but rather a continuation of the steady-state rate. A few runs taken at 110 and 50 K gave the same results, eliminating diffusion as a rate-limiting step in desorption. Rosenberg and Wen were able to fit their rate data to a sum of three decaying exponentials. The shortest lifetime was inversely proportional to the photon flux, while the other two lifetimes, 25 and 550 s, were independent of flux. Rosenberg and Wen (1988b) also found desorption of O_2 induced by 60–140 eV photons, those absorbed by Cu 3p and Ba 4d core electrons. A monochromator with a wide spectral bandpass, about 2 eV, was scanned for these measurements. The spectrum for O_2 desorption followed the total electron yield spectrum (proportional to the absorption coefficient). The ratio of the two spectra is an indication of the efficiency of desorption by a photon. It showed broad peaks at 90 and 110 eV, near the Ba 4d and Cu 3p absorption peaks. These peaks did not occur at the energies of the absorption peaks, however. These peak displacements were explained by assigning the desorption to only specific transitions in the core-level excitation spectra, to excited, but not ionized, final states of the Ba 4d initial state and to localized excited states well above the Cu 3p threshold in the region of the satellites in XPS.

Other simple attempts to increase the oxygen concentration at the surface have failed. Condensing O_2 on cuprate surfaces at low temperature results in spectral contributions characteristic of solid O_2. Warming such an overlayer until it evaporates results in spectra like those before the condensation of O_2. The O_2 bond does not break on the cuprate surface.

Terada et al. (1993, 1994) were able to insert oxygen into the surface layers of Y123 epitaxial films, converting them from apparently insulating to metallic. The films had been grown by magnetron sputtering. After several days in air photoelectron spectra showed no intensity at the Fermi level. The films were then placed in a vacuum system and annealed for 30 minutes at 573–873 K. During this time the sample was exposed to a beam of atomic oxygen produced in a microwave discharge. The flux incident on the sample was 2×10^{15} atoms $cm^{-2} s^{-1}$. It was then transferred to a photoelectron spectrometer without breaking vacuum. Angle-integrated EDCs of the valence-band region taken with a He I source showed a distinct Fermi edge. The 9-eV peak, indicative of surface deterioration, was absent. These studies were carried out on films oriented with the c-axis normal to the plane of the films and with the c-axis in the plane. In the former case, the restored Fermi edge was smaller than in the latter, possibly a result of the anisotropy of the diffusion of oxygen.

Aprelev et al. (1994) were able to restore oxygen to Bi2212 and Bi2201 in a different way. Both types of crystals were cleaved at 80 K in ultrahigh vacuum, and angle-integrated valence-band spectra taken. Upon heating in vacuum to 300 K, intensity was lost in the first 0.5 eV below E_F, as well as at larger binding energies. The Bi2212 crystal was then subjected to optical radiation while at 300 K. The maximum photon energy was 5.7 eV and the power density did not exceed 50 mW cm^{-2}. This caused an additional loss in intensity in the same parts of the EDC, but this loss was restored when the illumination was stopped. Illumination for 30 minutes under 0.2 Torr of O_2 restored much of the original, as-cleaved spectrum.

(Another way of introducing atomic oxygen to the surface of a cuprate is to adsorb N_2O at low temperature, then photodissociate it. It has not been reported as successful, although it may not have been tried.)

Flavell et al. (1990) cleaved a single crystal of Bi2212 at room temperature in ultrahigh vacuum, then used valence-band EDCs to follow changes induced by the adsorption of water vapor. The sample was cooled to 90 K for the water adsorption. Exposures of 1 and 2 L were used. ($1L = 10^{-6}$ Torr s.) The changes consisted of additional peaks in the EDCs around 6.5, 8.7, and 12.4 cV, readily interpreted as the spectrum from physisorbed water molecules which do not dissociate. In a later study, Flavell *et al.* (1993) used XPS and valence-band photoelectron spectra to study the effects of air, presumably those of water vapor and CO_2, on Y123 and Bi2212. The surface region sampled by XPS with polycrystalline Y123 deteriorates under air exposure to the point where the 531-eV contamination peak is the only component in the O 1s spectrum. The effects of the air exposure can be reversed nearly completely by annealing in O_2. Infrared reflectance spectra on the deteriorated samples revealed the presence of $BaCO_3$, not present in the starting material. Similar exposures of polycrystalline Bi2212 to air led to much less degradation.

10.2   Metal overlayers on cuprates

XPS has been used frequently to study overlayers on surfaces. In the simplest case of epitaxial growth, the overlayer grows one layer at a time. There is no intermixing of the two materials. In such a case, the Frank–Van der Merwe growth mode (Zangwill, 1988), the intensities of the XPS peaks from the substrate should decrease linearly with coverage as the first overlayer grows, then linearly but with a larger slope as the second layer grows, and so on. The intensity of a peak of unit area from a substrate core level is reduced to an intensity of $\exp(-d/\Lambda)$ when the first overlayer of thickness d is complete, then to $\exp(-2d/\Lambda)$ when the second layer is complete, and so on. Different peaks from the same element will not lose intensity at the same rate with coverage because the mean free paths for inelastic scattering, Λ, depend on the kinetic energy of the electrons. The peaks from the core levels of the overlayer atoms should increase in strength linearly with increasing coverage till the first monolayer is complete then continue to increase linearly, but with a reduction in slope as each succeeding layer is completed. This is because each new layer adds new atoms to contribute to the growing peak intensity, but it also contributes to inelastic scattering of the primary photoelectrons from the layers below, removing a fraction of them from the peak. Similar behavior occurs for peaks in Auger spectra, which are frequently used for the study of overlayers. Failure to see such thickness dependence of intensities indicates a more complicated growth morphology. The "opposite" growth mode has the impinging atoms form isolated clusters which grow till they merge. This is the Volmer–Weber growth mode. An intermediate mode, Stranski–Krastanov growth, has a few monolayers form, after which clusters grow on top of them. For both the latter modes, the expected behavior of the XPS spectra is different from that found for layer-by-layer growth, although for Stranski–Krastanov growth, it follows the same behavior as for Frank–Van der Merwe growth for the first few monolayers. For Volmer–Weber growth, the loss of substrate peak intensity with coverage will follow that for growth of a single layer for a small coverage, less than a monolayer, then the loss will be more gradual as clusters grow, leaving some areas of substrate uncovered by the additional deposition. Polycrystalline over-layers generally begin as clusters, but sometimes there is a thin overlayer between the clusters as in Stranski–Krastanov growth.

When the thickness of the overlayer is large enough, for example when there are no XPS peaks from those substrate atom core levels that do not intermix, valence-band spectra from the overlayer may be useful.

The impinging overlayer atoms may disrupt the surface, causing intermixing. The energy for bond breaking may be supplied by (i) the heat of condensation of the added atoms, (ii) their kinetic energy, typically $3k_B T/2$ per atom, where T is the temperature of the evaporation source, and (iii) thermal energy in the sub-strate. (We do not consider more energetic ways to produce adsorbed atoms such as sputtering.) Once the intermixing begins, the impinging atoms arrive at a surface of changing composition. This can be a complex process, even for monatomic substrates and adatoms. The complexity is exacerbated for cuprates, for the composition of the outermost layer will depend on how the surface was produced, by cleaving, by etching, or by use of the natural growth surface. It may also depend on the history of that surface, how long it is held at what tem-perature and pressure. For Y123, we do not expect a single surface plane to be exposed by cleaving. A rough idea of what to expect is that an overlayer on a cup-rate of a metal whose oxide is more stable than CuO is expected to remove oxy-gen from the cuprate to form an oxide of the overlayer metal. Oxide stability may be estimated from tables of the heats of formation of oxides, although there may be differences for the heats on surfaces. Different oxides may form first. The reaction rate may also be slow. Considerable work on the chemistry of metal overlayers on cuprates has been done. It has been reviewed by Meyer and Weaver (1990) and by Weaver (1994). A table on page 415 of Henrich and Cox (1994) lists 29 entries for studies of metal (and Si) overlayers on various cuprates.

10.2.1 Metals on CuO

Weaver et al. (1987) deposited Ti on CuO at room temperature. TiO_2 is more stable than CuO, so it is the overlayer expected at low coverages. The CuO was a pressed polycrystalline bar. Fresh surfaces were prepared by fracturing in ultra-high vacuum. Figure 10.1 shows the development of the Ti $2p_{3/2}$, $2p_{1/2}$, and Cu $2p_{3/2}$ core-level spectra and the valence-band spectrum with increasing Ti cover-age. The clean CuO spectrum has the broad main peak and satellite discussed

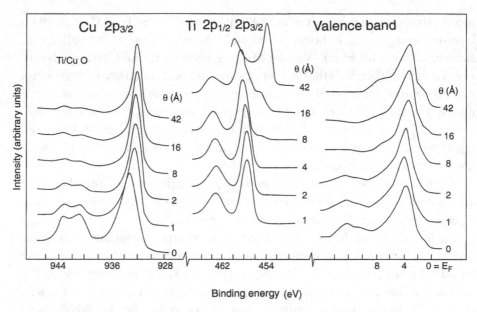

Fig. 10.1   Valence-band, Ti $2p_{1/2,3/2}$, and Cu $2p_{3/2}$ spectra (*right to left*) of CuO with various coverages, θ (Å), of Ti. (Weaver *et al.*, 1987)

frequently in this book. As Ti is added, the satellite almost disappears and the main peak shifts and sharpens. At the highest coverage shown, there is still considerable Cu contributing to the spectrum, but very little of it is CuO. The pair of peaks in the Ti 2p spectrum appears at low coverage at a binding energy characteristic of that of TiO_2. At higher coverages another pair of peaks appears superimposed on the TiO_2 peaks, and at the highest coverage only one pair is left, at the binding energy of metallic Ti. Note that at 2 Å coverage of Ti, the Cu 2p satellite has lost about 80% of its intensity with respect to the main peak. The main peak has nearly completely shifted to its new position, indicative of Cu^+. The O 1s peak is present even at 42 Å coverage of Ti. The presence of the Cu $2p_{3/2}$ satellite and the O 1s peak at this coverage need not mean the Cu and O are present near the surface, for some primary photoelectrons can escape through several tens of Å.

The Ti 2p spectra were analyzed into up to three sets of spin–orbit split pairs, corresponding to Ti^0, Ti^{2+} and Ti^{3+} together, and Ti^{4+}. (Four sets of spin–orbit split pairs had been used in XPS studies of Ti on TiO_2 (Rocker and Göpel, 1987).) Figure 10.2 shows the integrated intensity of the individual component Ti spectral peaks, normalized to the total Ti peak intensity at the highest coverage and plotted on a logarithmic scale as a function of Ti coverage. For the case of the total Ti 2p spectrum (the sum of the components) $\ln[1 - I(\theta)/I(\infty)]$ is plotted, where θ is the coverage. (These data were obtained from the data leading

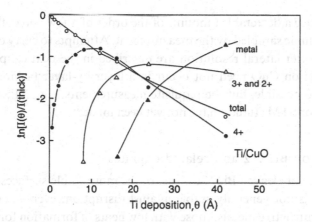

Fig. 10.2   Intensities of Ti 2p components as a function of coverage on CuO. (Weaver *et al.*, 1987)

to Fig. 10.1, but before normalization.) The drop in the line for the Ti total spectrum with increasing coverage represents a growth in coverage. The Ti^{4+} strength rises upon initial Ti coverage, peaks at about 10 Å total coverage, then falls. This component is from Ti in Ti–O bonds similar to those in TiO_2, although the atomic geometry is not known. Each of the initial Ti atoms has adequate O with which to react, and the highest oxide is formed. Lower oxides (Ti^{3+} and Ti^{2+}) begin to appear at about 8 Å coverage, peak in concentration at about 20 Å, then remain at a concentration not much reduced from this maximum. The metallic Ti component, Ti^0, appears at around 15 Å coverage and grows monotonically. The overall picture is one of a rather thick, spatially inhomogeneous interface between CuO below the original surface and metallic Ti at the top. Some O has been removed from the original surface layers to form TiO_2 at the stage at which there was much oxygen and little Ti. As more Ti is added, and perhaps less O is available at the interface, lower oxides of Ti appear in the first few layers of added material. Finally, the last added Ti remains unreacted as metallic Ti. The depth of each region presumably is limited by the diffusion rate of O atoms to the region of the changing interface.

The valence-band spectra in Fig. 10.1 are more surface sensitive than the core-level spectra. As a result, the satellites at 10–12 eV disappear at a lower Ti coverage than in the Cu $2p_{3/2}$ core-level spectra. (The conversion of Cu^{2+} to Cu^+ at the interface weakens the satellites in both spectra.) The Fermi edge appears in the spectrum at 16 Å Ti coverage, the coverage at which metallic Ti first appears.

It must be remembered that there could be considerable lateral inhomogeneity in these interface reactions. The spectra shown sample an area a few hundred μm in diameter. The reactions may take place initially at special sites, e.g., steps, and their effects not be detected in XPS spectra till they have proceeded far enough to produce a detectable amount, of the order of a few percent, of a new species in the volume sampled by the measurement. Attempts to carry out photo-emission with better lateral resolution are described in the next chapter. Also, the growth of Ti on CuO need not occur in a layer-by-layer fashion. Cluster formation is also possible, but the requisite measurements, e.g., XPS at several different angles or STM studies, have not yet been made.

10.2.2   Metals on Bi2212 and related cuprates

The Bi–O surface of cleaved Bi2212 is one of the more stable surfaces found on cuprates, but it is not generally stable against disruption, even when covered with relatively unreactive metals, those with low heats of formation for their oxides. In the following, all studies were carried out at room temperature, unless noted otherwise. Many overlayers have been studied on Bi2212 for various reasons. Cu, Ag, Au, Al, and In were potential ohmic contacts, and Ag is used as an encapsulating material for wires and tapes. Rb and Cs are electron donors to the superconductor, Bi is a constituent, and Pb can substitute for it. Several transition metals have been studied. Although they are metallic and could form contacts, it was expected that they would disrupt the surface to form their own oxides.

Lindberg *et al.* (1989d) evaporated overlayers of Rb on a cleaved single-crystal surface of Bi2212. Rb deposition leaves the Cu $2p_{3/2}$ and Sr 3p spectra unchanged, but the O 1s and Bi 4f peaks shift about 0.5 eV toward higher binding energy, indicative of a possible transfer of electrons from the Bi–O layer to the Rb. The O 1s spectrum becomes broader, suggesting a second O site. The angle-integrated valence-band spectrum changes in several ways. A shoulder at 2 eV becomes more prominent, and peaks grow at about 8.5, 9.5, and 11.5 eV. Two of these peaks are assigned to Rb–O states, since upon subsequent exposure to 1000 L of O_2, two of them become more prominent. The 8.5-eV peak disappears, and a new peak grows at 5 eV. An additional component appears at higher binding energy in the O 1s spectrum after the oxygen exposure (with the Rb in place). The 5-eV valence-band peak and this new component in the O 1s spectrum appear to grow together. Lindberg *et al.* suggest this is a peroxide ion, O_2^{2-}. The Rb disrupts the Bi–O surface, removing O, but leaving the CuO_2 planes intact.

Lindberg *et al.* (1990b) evaporated overlayers of Rb and Al on cleaved surfaces of Bi2212. The effects of the two overlayers were quite different. Al overlayers produced a more dramatic effect than Rb, discussed above. The O 1s spectrum shape changed and the peak of the composite spectrum shifted almost 3 eV to higher binding energy. This appeared to occur by a loss of intensity in the original peak and the growth of a new peak at high binding energy, where the peak from Al_2O_3 occurs. The Bi 4f spectrum shifts to lower binding energy, but not as a rigid shift. Instead a spin–orbit split pair of peaks grows with Al coverage at the expense of the original pair. The Sr 3p spectrum shifts to higher binding energy. Also the Cu $2p_{3/2}$ spectrum loses the satellite and has the main peak sharpen as Al is added. The overall intensity of the O 1s peak does not drop as fast with Al coverage as that from Cu, Sr, and Bi. It is clear that Al greatly disrupts the surface and pulls O atoms from the CuO_2 layers, leaving Cu^+. The Sr–O layer is also affected. It may lose O or gain Al or both. The average valence of Bi appears also to be reduced as O is removed from the Bi–O layer to enter the overlayer. Diffusion of O into the Al is rather extensive, keeping the O 1s peak strong as Al is added. (The maximum coverage in these studies was 17 Å.)

Bi was evaporated on fractured surfaces of Bi2212 by Meyer et al. (1988c). For thicknesses up to 8 Å, there were slight changes in the XPS spectra indicating some reduction of Cu^{2+} to Cu^+. All core-level spectra shifted 0.2 eV to higher binding energy, an indication of charge transfer to the Bi overlayer. New Bi peaks appeared, characteristic of "metallic" Bi. Adding Bi beyond 8 Å equivalent coverage just increased the Bi^0 component of the spectrum. No other changes occurred. This early study had the problem, noted by the authors, that parts of the fractured surfaces shadowed other parts of the surface from the incoming Bi atoms. The shadowed parts contributed to the spectrum although they had no Bi coverage. Pb behaves similarly on Bi2212, except that changes occur up to an overlayer depth of 40 Å (Kulkarni et al., 1990).

Balzarotti et al. (1992) carried out an extensive study with overlayers of Ti, Fe, Cu, Cr, Pd, and Au on single crystals of Bi2212 and Pb-doped Bi2212. All except Cu and Au are expected to disrupt the surface, removing O from the outer Bi–O layers and possibly the nearest CuO_2 layers. Balzarotti et al. found that Ti, Cr, and Fe were the most reactive, Cu, Pd, and Au less so, with Au perhaps not reacting at all. The cleavage surfaces showed cleavage steps with flat terraces between them about 200 μm wide. O 1s, Cu $2p_{3/2}$, Bi 4f, Ca 2p, Sr 3d, Pb 4f (when applicable, but these peaks changed only in intensity with adsorbate coverage) and valence-band spectra were taken as a function of metal coverage for coverages up to several tens of Å. Ti, Cr, and Fe 2p spectra, the Pd 3d spectrum, and the Au 4f spectrum were taken when these elements were used as overlayers. Figure 10.3 presents a series of spectra for various coverages of Fe. Small amounts of Fe apparently disrupt the Bi–O bonds in the surface layer. This produced a weaker Bi 4f spin-orbit split pair and the growth of another at lower binding energy, characteristic of Bi metal or of a lower oxide. The initial pair shifted 0.2 eV to higher binding energy upon Fe deposition. The Fe 2p spectrum shifted to lower energies from its initial position, characteristic of a change from the FeO which formed upon initial deposition to metallic Fe at greater coverage, beginning at the equivalent of 1 Å. The Fe spectrum ceased changing after about 40 Å Fe coverage, but the metallic Bi 4f intensity remained detectable at a cover-

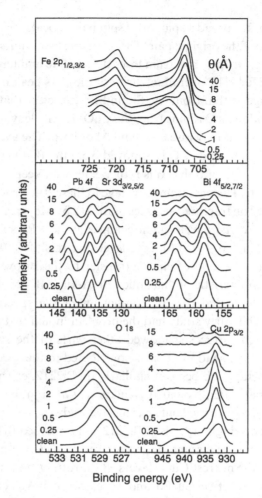

Fig. 10.3 Core-level spectra for $Bi_{1.7}Pb_{0.3}Sr_2CaCu_2O_8$ with varying coverages, $\theta(\text{Å})$, of Fe. The Fe spectra have been normalized to give comparable intensity at each coverage. (Balzarotti *et al.*, 1992)

age of 110 Å of Fe. Evidently Bi migrates through the Fe to the surface. The O 1s spectrum shifts to higher binding energy because of the rise of new components and the loss of others as the Fe deposit builds up. The Cu $2p_{3/2}$ spectrum loses the satellite while the main peak sharpens. By 2 Å Fe coverage, the satellite is gone, and the Cu^{2+} ions in the CuO_2 plane have become effectively Cu^+. The Cu signal is gone after about 15 Å coverage, indicating no migration of Cu into the Fe. Finally, the Sr 3d peaks broaden, indicative of the appearance of inequivalent sites; the Sr–O plane becomes disrupted.

Ti overlayers behave similarly with respect to the Bi2212 constituents. The Ti 3p spectrum indicates that TiO_2 forms immediately after Ti deposition, and Ti metal appears in the final states, but intermediate oxides of Ti appear at intermediate stages, much as in the case of Ti on CuO, discussed above. Cr behaves like Fe. Cr_2O_3 forms immediately and metallic Cr at later stages. Pd also reacts with the Bi–O surface, but to a lesser extent. Its 4d band is full, while Ti, Cr, and Fe have partially filled 3d bands. The Cu and Bi core-level spectra do not change much till a coverage of about 4 Å is reached. Metallic Pd contributes to the valence-band spectrum starting at 2 Å coverage. Cu overlayers are more puzzling. Bi–O is more stable than Cu_2O and CuO. Cu deposition quickly causes the loss of the Cu $2p_{3/2}$ satellites. They are gone by 0.5 Å coverage, and the main line has sharpened by then. This would seem to indicate that all the Cu^{2+} in the sampling depth had been converted to Cu^+. However, the O 1s and Bi 4f spectra do not change until a coverage of 2 Å is reached. The valence-band spectra show "metallic" Bi contributions starting at 2 Å coverage. Clearly there is some surface disruption, but a picture of what happens in the early stages of Cu coverage is not yet at hand. The disruption of the neighboring Bi–O surface may contribute to the instability of the CuO_2 plane when Cu is adsorbed on the surface.

Au simply covers the surface. The core-level spectra of the Bi2212 constituents lose intensity as Au coverage increases. There are no shifts or shape changes. However, this offers an opportunity to study the growth mode. For layer-by-layer growth, the core-line areas should decrease linearly during the deposition of each layer, but with changing slopes. The intensities at the end of each successive layer should fall exponentially with layer number. For Au on Bi2212 this is not what happens. The falloff with coverage is more gradual, without straight-line segments. Balzarotti et al. were able to fit the data Bi, Sr, and Cu core levels to an expression derived for a modified Volmer–Weber (cluster) growth. They also state that a fit to a Stranski–Krastanov mode does not fit well at large coverages.

Cu, Ag, and Au on cleaved surfaces Bi2212 were studied by Bernhoff et al. (1991, 1993). They used synchrotron radiation with photon energies up to 1000 eV. The Cu $2p_{3/2}$ spectra then arise from a region closer to the surface than when Al- or Mg-Kα lines are used for photoexcitation. Cu clearly disrupts the Bi–O layer. At 1 Å coverage a new pair of Bi 5d peaks has appeared. At 24 Å coverage, the new pair, due to "metallic" Bi is the only pair in the spectrum. Spectra taken from the surface normal and at 60° to the normal indicate the Bi is on the surface of the Cu. Bi has a lower surface energy than Cu, so this is the expected configuration. Deposition at 100 K instead of 300 K also causes the same Bi–O plane disruption, but at a reduced rate. The Sr–O layer appears not to be disrupted. The Cu $2p_{3/2}$ spectra were not taken in photoemission. Instead, yield spectra gave the absorption spectra at and above the 2p absorption edge. Reference spectra were synthesized from spectra of Cu and CuO. Metallic Cu appears in the spectrum at a coverage of about a monolayer. The Fermi edge decreases somewhat upon Cu deposition, then grows as metallic Cu appears. The decrease is probably due to the disruption of the Bi–O layer and its influence on the CuO_2 layer below.

Au is far less reactive than Cu, and Ag reacted less than Au. The Bi 5d spectra at 24 Å of coverage showed a small pair of peaks due to "metallic" Bi superimposed on a much larger pair of peaks from the original Bi–O plane, in the case of Au. For 24 Å of Ag, there was only a slight indication of the metallic Bi peaks in the spectrum. Changes in the O 1s spectra were more subtle, better seen after decomposing the broad peak into components. They were largest for Cu. Coverage by each of the three metals initially reduces the size of the Fermi edge in the valence-band spectra. This suggests some disruption of the Bi–O surface even by Ag. For Cu coverage, the intensities of the Sr 3d and O 1s peaks decay exponentially with coverage, indicating layer-by-layer growth of Cu. This was not the case for Ag and Au, for which cluster growth probably occurs.

Qvarford *et al.* (1996) augmented these studies by measuring absorption spectra above the O 1s and Cu $2p_{3/2}$ edges via Auger yield spectra and total yield spectra, the former being more surface sensitive. The final states are the empty states of O 2p and Cu 3d character, respectively, just above E_F. The O 1s spectrum of the clean surface has a peak at 528.3 eV which is rapidly reduced in intensity with Cu coverage, disappearing at 6 Å in the spectra taken by Auger yield. The empty O 2p states just above E_F are removed by the Cu overlayer. The total yield spectrum shows a small peak at the same coverage, indicating that it originates near the original surface of the Bi2212. The same coverages of Ag and Au alter the O 1s spectrum only by reducing the area of this peak by about 25% (Au) and 20% (Ag) after 24 Å of coverage. Most of the O 2p states are undisturbed. The Cu $2p_{3/2}$ absorption spectra do not change upon deposition of 6 Å of Ag or Au. (Cu coverage does alter this spectrum as discussed in the preceding paragraph.) Thus, although Ag and Au may disrupt the Bi–O surface layer to some extent, the CuO_2 layer just below appears unaffected. The occupied states, studied by photoemission, appear to be more strongly affected by the overlayers than the empty states studied by x-ray absorption, although this could arise, wholly or in part, from differences in sampling depths. Qvarford *et al.* discuss the possible effects on the outer planes of Bi2212 of the overlayers of these three metals in terms of charge transfer to Bi2212 and a shift in the Fermi level, largest for Cu, in addition to the disruption of the Bi–O and, for Cu, CuO_2 planes.

Earlier, Wagener *et al.*(1990c) used inverse photoemission to study overlayers of Ag and Au on Bi2212 polycrystalline samples. The spectra for a clean surface show a peak just above E_F for emission of 18 eV photons which is more prominent than that found with 16 eV photons. This is a resonance, the O 2s–2p resonance. It indicates a peak in the density of empty O 2p states in the first 1 eV above E_F. Evaporation of 2 Å of Cu eliminates this resonance, i.e., destroys the

O 2p empty states in this region. An evaporation of Au also eliminates this resonance, but it takes more than 4 Å coverage before it disappears.

Balzarotti *et al.* (1994) studied further the Fe–Bi2212 interface by XPS and STM. They also made extensive STM studies of the Fe overlayers. The cleaved surfaces were imaged with atomic resolution, and the superlattice was obvious. There were cleavage steps with flat terraces between them averaging 200 μm in width. The Fe deposits began as clusters which grew with increasing coverage (Volmer–Weber growth). After a deposition that would be 5 Å thick if deposition were layer-by-layer, they estimated that about half the surface was covered by clusters. The clusters formed upon low deposition are rather large, 10–20 Å in diameter and about 6 Å high at a coverage equivalent to 0.07 Å of Fe. The volume of the clusters is estimated to be larger than that of the deposited Fe, by about a factor of 1.6 for coverages up to 1 Å. This is due to the uptake of O by Fe in the clusters. At this stage, the superlattice is still visible in the STM images. This O comes from directly below the cluster, for there is no evidence of surface disruption on the areas not covered by a cluster.

Balzarotti *et al.* developed a phenomenological microscopic model for the growth of a cluster of nominal FeO on the surface. The model assumes Bi^{3+} below the surface is converted to Bi^0 from the loss of O to the Fe, the Bi^0 diffuses to the surface of the cluster, and the cluster grows with Fe deposition according to Volmer–Weber kinetics. From these assumptions they derive expressions for the size and composition distribution of the cluster as a function of Fe coverage. From these, they then derive expressions for the relative intensities of the core-level spectra for Bi^{3+}, Bi^0, and Fe (as FeO) as a function of Fe coverage, using mean free paths for inelastic scattering from the universal curve. There were five fitting parameters, and three curves, two of which exhibit a maximum, to use in the fitting. Agreement with their data is quite reasonable.

Balzarotti *et al.* (1995) used the same model to fit similar data for Cr and Ge on Bi2212. Cr grows much like Fe, but the Ge growth is layer by layer. A comparison of the fitting parameters for Fe and Cr shows clearly that Cr is more reactive than Fe on Bi2212. The model for cluster growth could not be fit to the data for all the core levels measured for Ge on Bi2212 The model was modified to try to describe layer-by-layer growth, and a fit was obtained that accounted for the intensity dependence of all peaks as the coverage increased. A thin GeO_2 layer forms at the interface.

Cs is an unlikely candidate for an electrode on Bi2212, but it offers the interesting possibility of doping the surface layers. A simplified ideal monolayer of Cs should consist of Cs^+ ions, the electrons having been donated to the underlying metal, Bi2212 in this case. Söderholm *et al.* (1996) studied this system for low coverages, where electron donation indeed dominates, and for higher coverages, which disrupt the surface of Bi2212. They took valence-band and core-level spectra and core-level absorption spectra on cleaved crystals of Bi2212 with various amounts of Cs on the surface. The Cs thickness was not measured, but the linear growth of the Cs 5p photoelectron peak with time implies that the deposition rate was constant. The rate of the Bi $5d_{3/2}$ peak intensity reduction with coverage changed after 2 min of deposition, implying a change in the Cs growth mode at this coverage. The first spectra with Cs were taken after 1 min of deposition. The Bi 5d pair of peaks was shifted about 0.3 eV to higher binding energy and the Bi 4f pair by 0.9 eV. At higher coverages, each of these spectra developed two new pairs of peaks at lower binding energy, one in the region of the "metallic" Bi (Bi^0) peaks, and one where Bi_2O_3 may be expected to contribute. The valence-band spectrum shifted to higher binding energy at low coverage, and there were some shape changes. At higher coverage peaks assigned to an oxide of Cs emerged. The Fermi edge was lost before the deposition lasted 1 min, and it did not return with additional coverage. Most of the loss occurred after a 10-s deposition. The Sr 3d peak shifted to higher binding energy with Cs coverage, saturating at a shift of 1.3 eV. The O 1s peak also shifted to higher binding energy, saturating at 0.6 eV, and at high coverages, new low binding-energy components appeared. These were tentatively assigned to Cs_2O. A peak at 9 eV in the valence band grew along with this component.

Cs deposition causes the disappearance of the O 1s absorption peak at very low coverages, after 10 s. At the highest coverage, a new structure develops simultaneously with the peak at 531 eV in the O 1s photoelectron spectrum. The Cu 2p absorption spectrum changes with Cs coverage is a way explicable by the loss of some Cu^{2+} and the gain of some Cu^+ as the coverage increases beyond 2 min.

The spectral changes found in the early stages of Cs deposition can be ascribed to electron injection into the Bi2212, filling primarily O 2p states just above E_F. This raises the Fermi level, causing the core-level shifts. No Cs oxide has appeared at this stage. There appears to be no disruption of the surface. Additional coverage leads to a disruption of the Bi–O surface layer and the appearance of Cs_2O and metallic Bi. This seems to occur when all the O 2p holes have been annihilated. The two new pairs of Bi core-level peaks imply there are two Bi sites in this late stage, Bi metal clusters and clusters of a Cs–Bi alloy. When the Bi–O layer is disrupted, there is some reduction of Cu and Ca. Söderholm *et al.* discuss extensively their data and other spectroscopic studies of doping, using several available models for the electronic structure, finding no completely consistent picture.

Kimachi *et al.* (1991) evaporated Ti, Cr, and Cu on single-crystal surfaces of Bi2212 held at 20 K. Ti overlayers cause effects very similar to those that occur for deposition at 300 K, except the changes in the spectra are not as great. One difference is that no Bi atoms reach the surface of the growing Ti layer, although some metallic Bi appears in the Ti at an earlier stage of growth. Warming the sample to 300 K after deposition at 20 K, produces little further reaction. The changes observed were attributed to a roughening of the surface when the Ti atoms became more mobile. Low-temperature deposition of Cr produced a broader and shifted Cr $2p_{3/2}$ spectrum than found for deposition at 300 K. This was attributed to less well-defined oxides of Cr because of low O-atom mobility at 20 K. The Cu and O core-level spectra resembled those obtained for deposition at 300 K, but there was less reaction at low temperature. Warming to 300 K produced no further reaction at the interface, but the surface became rougher. The Ti- and Cr-oxides at the interface produced at 20 K evidently protect the interface from further reaction upon warming, probably because of low diffusion coefficients in the oxides. Low-temperature deposition of Cu resulted in less Bi–O and Bi surface segregation than at 300 K. There was a conversion of Cu^{2+} to Cu^+, probably due to the loss of O from the CuO_2 plane as some of the deposited Cu was oxidized. The main effect of warming to 300 K was to allow some of the Bi in the Cu to migrate to the surface.

Luo *et al.* (1992) carried out an STM study of the deposition of Ag, Au, and Cr on single-crystal surfaces of Bi2212 at 300 K. The clean surfaces were imaged with atomic resolution, and bands due to the superlattice were manifest. A small coverage of Ag produced clusters 16–25 Å in diameter and about 6 Å high. The volume of a cluster was about 9 times larger than expected from the volume of Ag deposited. The clusters tended to be located on the bands. At 0.2 Å coverage there were more clusters, but the size distribution remained the same, and the alignment on stripes continued. At 0.6 Å coverage, the cluster diameter distribution narrowed to the 20–25 Å range, the heights were still 6 Å and the total cluster volume was about 5 times that of the Ag deposited. The clusters coalesce at 1.6 Å coverage. Luo *et al.* coupled these observations with the results of photoelectron spectroscopy to construct a microscopic picture of the interface reactions. Upon deposition of Ag, O left the Bi–O and CuO_2 planes, but AgO was not found. AgO is far less stable than, e.g., TiO_2, and its formation cannot be the driving force for surface disruption. They suggest that the Bi–O plane, stable in the bulk, is not so stable at the surface. They propose that the surface Bi–O plane converts to clusters of Bi_2O_3, which is much more stable than AgO. This requires additional oxygen, which could come from the conversion of some of the Bi–O plane to Bi and O, or from the removal of O from the CuO_2 plane, or both. Core-level spectra indicating Bi^0 and Cu^+ support the presence of both sources of oxygen. This surface disruption may be favored thermodynamically, but it does not occur spontaneously because of an activation barrier. They propose the energy to surmount the barrier is released as Ag atoms form clusters. The barrier may be lower along the stripes, where the clusters are formed. Au deposition follows the same pattern of clusters forming along the stripes.

Photoelectron diffraction from Ag 4d photoelectrons was used by Schwaller *et al.* (1993) to study surfaces similar to those studied by Luo *et al.* At low Ag coverage, below 7 Å, there was little order around the Ag sites. At larger coverages order appeared that was attributed to epitaxial Ag (110) layers occurring in domains with two orientations. In both, a diagonal of the surface unit cell was aligned along the Bi2212 *a*-axis. A single-scattering calculation of the expected angular pattern agreed with this picture. These ordered domains were believed to form in the regions between the Ag clusters that were formed at low coverages. Schwaller *et al.* estimated that at an average coverage of 84 Å about half of the deposited Ag is in the regions of epitaxial layers, the other half in the clusters which formed upon initial deposition and which grew as deposition continued.

Cr deposition was different. At low initial coverage the clusters were fewer but larger, and they did not occur on the superlattice-induced stripes. The cluster volume is about 10 times the volume of Cr deposited. Luo *et al.* proposed that the topmost Bi–O layer is completely converted to Bi_2O_3 and Bi, while some of the Cr has oxidized to Cr_2O_3. The oxidation of an impinging Cr atom releases enough energy to destabilize the Bi–O plane, creating Bi and Bi_2O_3. In this picture, the large clusters are not clusters of Cr, but of Bi_2O_3. At higher coverage, the Bi_2O_3 clusters remain about the same, but smaller Cr_2O_3 clusters grow larger, and eventually begin to reduce the Bi in the Bi_2O_3 clusters. At still larger coverages, when all the available surface oxygen is tied up, Cr metal appears on the clusters.

Ohno *et al.* (1991) were able to produce metal clusters on Bi2212 by a novel technique, assembling the metal clusters before placing them on the surface, thereby making the heat of cluster formation unavailable at the interface. This technique had been used successfully on semiconductors (Waddill *et al.*, 1990). The crystal was cleaved at 20 K. Then a layer of solid Xe about 50 Å thick was condensed on it at 20 K, and the metal then condensed on the Xe. The adatoms are mobile enough to form clusters at this temperature. The clusters were metallic, for the valence-band XPS spectrum was that of the metal. Then the sample was warmed up to about 100 K, the Xe atoms evaporated, and the clusters reached the surface. XPS at 20 K was used to monitor the oxidation states of the elements in the interface region. Finally the sample was warmed to 300 K and the spectra retaken. At 20 K, Cr and Cu clusters produced this way react only slightly with the cleaved Bi2212 surface. Similar clusters of Au and Ge appear not to react at all. Upon warming to 300 K, the Au clusters still did not react, Ge reacted some, Cr more, and Cu the most. There were changes in the cluster morphology upon warming. Simply depositing Cr on Bi2212 at 20 K also produces less disruption of the surface. The Bi–O planes are disrupted, but not the CuO_2 planes.

10.2.3 Metals on Y123 and related cuprates

Meyer *et al.* (1988a) deposited Ti and Cu on fractured polycrystalline samples of Y123. Not all exposed planes need be the same for this compound. The effects of the deposition were similar to those found on Bi2212, but the surface disruption was greater and it occurred sooner. Y123 has a more reactive surface. The Ti $2p_{3/2}$ peak shifts to higher energy as the Ti coverage increases to about 4 Å. At this point a second peak due to a lower oxide of Ti begins to grow on the low binding-energy side. This grows and shifts with coverage until at about 16 Å coverage it reaches its minimum binding energy. At this point the Ti is metallic, although oxide peaks can still be seen, persisting to at least 48 Å coverage. The O 1s peak shifts to higher binding energy at 0.5 Å coverage as O bonds with Ti. At a coverage of 8 Å the component of the O 1s spectrum from the substrate is reduced to 10% of its original value, indicating considerable disruption of the original surface layers. At larger coverages, the O 1s peak shifts even further to higher binding energy. This peak is due to O dissolved in the metallic Ti. The Ba $3d_{3/2}$ peak shifts steadily to higher binding energy with increasing coverage to about 8 Å coverage, after which the energy and shape no longer change. The Ba atoms should remain in their original locations because of their size, so the spectral shift arises from the altered environment of the Ba atoms in the topmost layers of the Y123. The Cu $2p_{3/2}$ spectra change very quickly with coverage. The satellite vanishes between 2 Å and 4 Å coverage of Ti. The sampling depth is of the order of 30 Å so the loss of Cu^{2+} occurs in more than the CuO_2 plane nearest the surface. This cannot be due to the loss of such a large amount of oxygen, for 1 Å of Ti corresponds to about 0.28 monolayers. The Ba spectra confirm this, for they show considerable shifts for most of the Ba atoms in the sampling depth, not just those in the layer closest to the surface. These effects are assigned to structural changes well below the surface, perhaps as far as 30 Å below. Similar effects occur with Cu coverage, but, except for the Cu $2p_{3/2}$ spectrum, at a reduced rate. The Cu $2p_{3/2}$ satellite disappears quickly with Cu coverage, vanishing between 4 and 7 Å coverage. Additional data for Ti and Cu on Y123 have been reported by Meyer *et al.* (1990).

The work described in the foregoing paragraph, and the work on metal overlayers on Bi2212 allows us to dispense with a detailed description of other metal overlayers on Y123. A given metal on Bi2212 produces a similar effect on Y123, but it occurs sooner and to a greater depth on Y123 than on Bi2212. Metals studied include Bi (Meyer *et al.*, 1988c), Cu and Au (Wagener *et al.*, 1990c), and Ti, Cr, and Cu (Kimachi *et al.*, 1991), both at 20 K and 300 K.

10.2.4 Metals on other cuprates.

$La_{1.85}Sr_{0.15}CuO_4$ reacts with Ti very much like Y123 (Meyer *et al.*, 1987b). TiO_2 forms quickly and Cu^{2+} within the sampling depth is reduced to Cu^+ by about

2 Å coverage. At this coverage metallic Ti begins to appear on the surface. Fe-covered surfaces behave similarly (Weaver *et al.*, 1987; Hill *et al.*, 1987). Au-covered $La_{1.85}Sr_{0.15}CuO_4$, however, behaves differently (Meyer *et al.*, 1987a). There appears to be no disruption of the surface. There is complete Au coverage at the early stages of Au deposition but clustering occurs at later stages. By an equivalent uniform coverage of 100 Å the photoelectron spectrum from the $La_{1.85}Sr_{0.15}CuO_4$ substrate has been extinguished.

Chapter 11

New techniques in photoelectron spectroscopy

We have already mentioned some incremental improvements in photoelectron spectroscopy, increased energy- and angle resolution through improvements in instrumentation. Another is the angle-scanned photoelectron spectroscopy pioneered by Aebi *et al.* (1994a,b), described in Chapter 5. Lindroos and Bansil (1996) recently pointed out additional advantages of this technique when coupled with suitable calculations. They found not only the usual features produced by direct transitions from initial states on the Fermi surface, but additional features from indirect transitions whose intensities are proportional to the one-dimensional densities of initial states for transitions at a fixed $k_\parallel$. They illustrated these aspects with calculations for several surfaces of Cu. A re-examination of such spectra for Bi2212 is likely to occur soon. There are several other variations of photoelectron spectroscopy already tested at some level, some of which may play a role in future work on cuprates.

11.1 Photoelectron microscopy

The photoemission studies normally carried out collect electrons from an illuminated area usually no smaller than about 100 μm × several hundred μm. There may be lateral inhomogeneities on a scale smaller than this, so photoelectrons from special sites may be averaged with those from the rest of the area sampled. This may be particularly misleading in studies of reactions with overlayers which may take place preferentially at steps. Photoelectron microscopy has been developed over the past fifteen years or so, along with x-ray microscopy, which uses absorption differences for contrast. At present, the best lateral resolution achieved is about 0.1 μm, but instruments with improved resolution are being built and tested. They are sure to be used on cuprate surfaces. Photoelectron spectromicroscopy allows not only the identification of atoms within a small area, but also bonding states via core-level shifts. A recent (1996) special issue of the *Journal of Electron Spectroscopy and Related Phenomena*, Volume 84, is devoted entirely to spectromicroscopy.

There are two general types of photoelectron microscopes, focusing and imaging (Margaritondo and Cerrina, 1990; De Stasio and Margaritondo, 1994). In the former, photoelectrons come from a small spot on the sample surface. In the latter, the sample surface is imaged by the photoelectrons.

In focusing photoelectron microscopy, photoelectrons are collected from a small area. If there is no subsequent energy analysis, the contrast arises from total yield differences. If the incident photons are near the threshold for photoemission, the contrast arises largely from work-function differences which control the yields, and from surface topography. For higher photon energies, the yield is proportional to the absorption coefficient. The energy distribution of the photoelectrons provides more useful information. Usually soft x-rays are used so the electron kinetic-energy spectrum contains peaks from shallow core levels. The images then are maps of the distribution of the elements in the sample. With adequate energy resolution much more is available, for then one can resolve different peaks from one core level or decompose a composite peak into components. One then has chemical binding information along with the maps of the distribution of the elements.

In focusing photoelectron microscopy, as small a spot of soft x-ray radiation as possible is produced. The photoelectrons come only from the small illuminated area, so the electron optics of the electron energy analyzer need not collect only from a small area. Imaging photoelectron microscopy does not require such tight focusing of the soft x-ray beam, for the electron optics collects photoelectrons only from a very small spot. The illuminated area can be larger, since photoelectrons from the outer regions of the illuminated spot are not collected. We discuss these in turn.

A small spot size for a beam of monochromatic soft x-rays has been achieved in two general ways. The first is to use a Fresnel zone plate, a series of rings of varying radius which produce a focus by Fresnel diffraction. The focal length is strongly wavelength dependent, so if different wavelengths are needed, the zone plate or the sample must by moved along the optical axis. (We assume here that monochromatic radiation is incident on the zone plate, but that the wavelength may need to be changed.) Monochromators based on the chromatic focusing of zone plates have been constructed. The alternating rings of the zone plate are "open" and absorbing. "Open" may mean truly open if the plate is an unsupported thin film, in which case the open rings have to be bridged by narrow bars to hold the device together. Alternatively, "open" may mean less absorbing than the closed part if both are mounted on a thin supporting substrate, which will have some absorption. It is also possible to make the absorbing part only slightly absorbing, but make its thickness such that there is a phase shift of $\pi/2$ for a wavelength of particular interest. This enhances the intensity from interference at that wavelength by a factor of 4, less some loss for absorption, but at other wavelengths the intensity is reduced. The radii of the rings in a Fresnel zone plate for soft x-ray use are very small and the plates are made by the lithographic techniques used in the semiconductor industry. The pattern is produced in a photoresist film by interference of a spherical beam with a plane beam, both derived from the same laser. A typical zone plate may have several hundred zones, and have an outer diameter less than 100 μm. A zone-plate x-ray microscope operating in the sample-transmission mode has been described by Rarback et al. (1988).

The second method uses all-reflecting microscope objective mirrors, usually two. Since the reflectance of all materials near normal incidence is so low for short wavelengths the mirrors must be used at grazing incidence or be made of multiple layers to enhance reflectance by interference effects. Both approaches are used. In the first approach, two mirrors are used, each near grazing incidence, in the so-called Wolter I configuration (Wolter, 1952). One mirror is an ellipsoidal mirror, the other, a hyperboloid. In the second approach, the mirrors are used near normal incidence in the Schwarzschild arrangement. Multiple-layer coatings are used for the mirrors, typically alternating layers of a high- and low-atomic number material, e.g., W and C, producing an interference enhancement of the reflectance (Spiller, 1983). Another approach has used a single ellipsoidal mirror at near grazing incidence (Voss et al., 1992).

A number of such microscopes have been built and several more are under construction. The earlier ones were placed on bending magnets or undulators, but in the latter case the beam line was not designed as a whole; the undulator often was one already in existence. Beam lines more recently built for photoelectron microscopy have matched the undulator to the rest of the beam line. An early version of such a microscope had a resolution of about 0.5 μm (Capasso et al, 1991). This was later improved to just under 0.1 μm by improving most of the components (Capasso et al. 1993). For example, at Sincrotrone Trieste there are two beam lines for microscopy. The ESCA microscopy beam line provides a beam of soft x-rays focused to a spot about 0.15 μm in diameter. The monochromator provides a resolving power of 3000 at a photon energy of 300 eV. The scanning range is 200–1200 eV. The spectromicrosopy beam line focuses to a spot size of about 500 Å. The photon energy range is 20–300 eV with a resolving power of 3000 or more. The overall energy resolution to be achieved with such beam lines depends on the sample under study. For example, the yield of photoelectrons from the first eV below E_F in a cuprate is quite small, and resolution may have to be sacrificed for better signal-to-noise ratio. Core-level spectra of major components present fewer problems, and spectra can be obtained in a matter of seconds.

Iketaki *et al.* (1996) have made a reflecting Schwarzschild objective for use at 141 Å (88 eV) which produced a measured focal spot only 450 Å in diameter. The two mirrors were each coated with 41 layers of Mo alternating with 41 layers of Si. The overall transmission of the mirror pair, including geometrical obstruction, was 9%. A zone plate has been used at a photon energy of 680 eV. It produced a focal spot about 0.5 μm in diameter, limited by vibrations, and had a transmission of 8% (Ade *et al.*, 1990).

Imaging photoelectron microscopes have been used less frequently. Here the emphasis is on the electron optics. Tonner and Harp (1988, 1989) built an electrostatic immersion lens microscope which imaged the sample in total yield onto a channel plate multiplier, which placed the image on a phosphor screen. The photon energy could be scanned and a total yield (absorption) spectrum obtained from a spot about 10 μm in diameter. Tonner and Dunham (1994) describe an improved version. One limit to the spatial resolution, the energy spread of the electrons and the chromatic aberration of the lens, was overcome by the use of energy filtering (Tonner, 1990), resulting in an improved lens (Dunham *et al.*, 1994) now installed on a beam line for photoelectron microscopy at the Advanced Light Source at Berkeley (Tonner *et al.*, 1996).

Both types of microscopies image a "point." To produce a map or image, the sample usually is scanned, rather than scanning the x-ray beam or using deflection plates in front of the analyzing electron optics.

Another type of photoelectron microscope, a magnetic projection microscope, has been used. In this, the diverging magnetic field from a solenoid produces the magnification (Beamson *et al.*, 1980; King *et al.*, 1990; Waddill *et al.*, 1991). Such instruments are commercially available. The sample is placed near the open end of a superconducting solenoid to give a high flux density B_s. Photoelectrons produced at a point spiral around the flux line passing through that point. Because the flux lines diverge, the photoelectrons from adjacent points on the sample spiral about diverging flux lines, becoming more widely separated. The areal magnification at a point well outside the end of the solenoid, where the field is B, is B_s/B, which can be of the order of 10^4. The electrons then pass through a retarding-field analyzer to a position-sensitive detector, the output of which displays a map of the sample with photoyield as the cause of the contrast. This detector can be moved out of the way to allow electrons from one part of the sample to be passed through an deflection energy analyzer to obtain a spectrum from a small area of the sample. The lateral resolution is limited by the diameters of the electron orbits, $[(8mE)^{1/2}/eB_s]\sin\theta$, where E is the electron kinetic energy and θ the angle of emission with respect to $\mathbf{B}_s$. The spatial resolution achievable may be estimated by assuming an isotropic distribution of photoelectrons, which leads to 0.6 μm $[E(\text{eV})]^{1/2}$ for a value of B_s of 7 T (Waddill *et al.*, 1991). The retarding-field analyzer does not have as good a signal-to-noise ratio as deflection analyzers (Chapter 4). It has been replaced by an imaging bandpass analyzer (Kim *et al.*, 1995).

Komeda *et al.* (1991) used a magnetic projection microscope to image with photoelectrons the surface of Bi2212 cleaved at 300 K. The resolution was 10 μm. They also could select spots about 20 μm in diameter and take photoelectron spectra. In the former mode, secondary electrons dominate the current to the detector, so the contrast is caused by differences in total yield from region to region. Small changes in local work function can produce large changes in photoelectron yield. Shadowing and other topographic effects may have played a larger role on the actual surfaces studied. (The beam of 70 eV photons was incident at an angle to the surface normal.) Bi 5d spectra (EDCs) taken on selected small spots showed differences in line shapes attributed to varying peak energies and strengths of spin–orbit split pair components, representing a clean Bi–O plane surface or one that was oxidized. The spectra showing evidence of further oxidation of the Bi were taken from regions at or step boundaries between single-crystal stacks in their sample, which was an ensemble of thin crystallites stacked with a common c-axis but with varying directions for the a- and b-axes. Cleaving produced flat areas about 1 mm × 1 mm, with steps at the edges to adjacent flat areas.

Hwu *et al.* (1996) used a similar system (Hwu *et al.*, 1995) on cleaved surfaces of Bi2212, as cleaved, and with small coverages of Cu. The performance was improved over that of previous work because of the higher brightness of the source, a third-generation storage ring. The spectral resolution was 100 meV and the lateral resolution was estimated to be 0.5 μm. Images of areas on the as-cleaved surface using contrast due to yield differences showed large areas with no chemical inhomogeneities, only small topographic structure. The two could be distinguished by using different photon energies. Occasional inhomogeneities with a size of 10–50 $(\text{μm})^2$ were found. These were assigned to areas with lower Sr concentrations. The Ca 2p yield spectra, also taken, showed a 200 meV binding-energy shift just in these regions. Upon Cu deposition, only the Sr-deficient areas reacted with the Cu, judged by a 400-meV shift in the binding energy of the Ca 2p spectrum in those regions, but not elsewhere.

11.2   Angle-resolved resonance photoemission

Resonance photoemission was described in Chapter 3 and some examples of its use appeared in Chapter 5. In those descriptions, the valence electron states used as intermediate and final states were treated as atomic states; angular momentum was used as a quantum number. This was justified in part by simplicity, atoms and solids could be treated alike, and by the fact that the only part of the valence-electron Bloch function that overlapped the core state was the part near the nucleus. The rest of the Bloch function played no role. Agreement with experiment, usually taken in the angle-integrated mode, was acceptable in most cases. Recently, resonance spectra have been taken with angle resolution. In most such measurements, the effect of varying the collection angle was a change in relative intensity of parts of the spectra (López *et al.*, 1995). This could be understood as coming from the angular part of the matrix elements involved. More recently, in one system, LaSb, an angle-resolved study of the 4d–4f resonance-enhanced valence-band spectrum showed changes in the energies of the peaks with collection angle, i.e., dispersion (Olson *et al.*, 1996). This cannot be explained with atomic wave functions expressed in spherical coordinates. The wave vector of the valence electrons played a role in the resonance spectrum.

Molodtsov *et al.* (1997) then used this effect as a way of characterizing the degree of spatial localization of the valence electrons around the site of the core hole. Their example was an epitaxial film of metallic La on W. They measured the valence-band spectrum with photons on and off the 4d–4f resonance at several collection angles. The peak energies as a function of wave vector were fitted to a band calculation in a specific way. The band structure was calculated with Bloch wave functions composed of a sum of atomic orbitals $|n\ell m\rangle$ on sites i:

$$|\mathbf{k}\rangle = \sum_{n\ell mi} C(k)_{nm\ell i} \exp(i\mathbf{k} \cdot \mathbf{R}_i)|n\ell mi\rangle.$$

The C coefficients are known from the band calculation. In practice, for La they used only three values of n and ℓ, corresponding to 6s, 6p, and 5d final states. The quantity $N_{n\ell i}(\mathbf{k}) = \sum_m |C_{n\ell mi}(\mathbf{k})|^2$ was used as a measure of the net $n\ell$-character about atom i of the state $|\mathbf{k}\rangle$. After determining which initial-state calculated band was to be identified with each of the three measured dispersing bands, $N_{5d}(\mathbf{k})$ was calculated as a function of $\mathbf{k}$ along the measured dispersion line. ($k_\perp$ was not measured. Several values were assumed, and the one giving the best fit was chosen.) The magnitude of the resonance enhancement of each of these three bands was different, and one of them had an angular dependence that was different from that of the other two. The (normalized) $N_{5d}(\mathbf{k})$ tracked rather closely the measured on-resonance intensity as a function of $\mathbf{k}$ for each band, both in relative magnitude of the enhancement and in angle dependence.

The angle dependence of the resonance enhancement was traced to the angle dependence of the Auger (interelectron Coulomb) matrix element which is involved in the interference effect leading to the Fano line shape. The wave functions of the 5d states overlap those of the 4d states better than do those of the 6s and 6p states. Their Auger matrix elements are therefore much larger and dominate the angle dependence of the resonance enhancement.

It is not clear how easily this technique can be applied to cuprates. Their spectra are more complicated, and the strong lanthanide 4d → 4f resonance will not tell much about the states of most interest, those based on O and Cu. The Cu 3p → 3d resonance is much weaker and less well understood. It may not be adequate for application of the analysis of Molodtsov *et al.* In any case, application of this technique to a few more systems simpler than cuprates, e.g., Ni or Cu, should be made in order to test its suitability for determining the amount of localized nature in the Cu 3d wave functions of cuprates.

11.3 Two-photon photoemission and related techniques

Pulsed lasers offer several possibilities for photoemission spectroscopy. Non-linear effects arise from the high power density delivered in a pulse (Bloembergen, 1965; Harper and Wherrett, 1977; Butcher and Cotter, 1990). The short pulse length also offers the possibility of measuring relaxation times of excited states, which may be carried out in the linear regime. Two photon absorption is a non-linear process, one requiring large optical electric fields at the sample. Usually such fields are attainable only with pulsed lasers. However, it is possible that only one of the two photons absorbed comes from a pulsed laser. The other could come from a continuum source and a monochromator (Hopfield *et al.*, 1963). Photoexcitation of electrons across a band gap is possible with two photons, neither of which has an energy as large as the gap energy, but whose sum equals or exceeds the gap. The transition may be viewed as occurring via a virtual intermediate state which need not lie in the gap. One useful application of two-photon absorption is that it reaches final states that are not the same as those reached in one-photon absorption. The dipole operator is applied twice, once between the ground state and the intermediate state, and again between the intermediate state and the final state. Thus for an atom, $\Delta\ell$ is 0 or ±2 in two-photon absorption instead of the ±1 in one-photon absorption. Similarly in a crystal, different final states are reached in two-photon absorption than in linear (one-photon) absorption. If these states lie above E_F by at least the work-function energy, photoelectrons are emitted which may be energy analyzed.

If the laser photon energy matches the energy between the ground and an excited state of the sample, there will be one-photon absorption. If the final state does not produce any photoelectrons, a second pulse, usually derived from the same laser as the exciting pulse but delayed in time by an additional path length, may photoionize some of the previously excited electrons. By delaying the second pulse for different amounts of time, the relaxation of the excited state can be studied.

Non-linear two-photon photoemission and pump-probe photoemission have been used on a number of prototypical materials. Steinmann (1989) has reviewed the work of his group on the two-photon photoemission from image-potential states on noble metals and Ni. (We previously mentioned these states only briefly, for they have not been found yet on cuprates.) Image-potential state lifetimes were measured subsequently by Schuppler *et al.* (1992) using a pump-probe technique, in which the pump pulse excited the electrons by two-photon absorption. A model theory for describing these processes was presented by Ueba (1995). Bokor *et al.* (1985) studied the dispersion of a normally empty surface state on the (110) surface of InP, carrying out angle-resolved photoemission with a second pulse after filling the surface state with a first pulse. (It was filled indirectly, from the plasma produced by the laser pulse.) They also followed the temporal decay of its occupancy. They carried out similar experiments on Si (111) (Bokor *et al.*, 1986). The thermalization of hot electrons in Au was followed by this technique (Fann *et al.*, 1992). Hertel *et al.* (1996) worked on a 10 fs time scale in studying the initial relaxation of excited electrons in Cu. Xu *et al.* (1996) carried out similar work on graphite, with the result that the excited electron lifetime was inversely proportional to the energy of the electron above the Fermi energy. It is tempting to suggest this may be in accord with the similar relationship obtained from photoelectron line shapes for Bi2212 (Olson *et al.*, 1990a), since both may be considered as composed of two-dimensional conducting sheets. However, as mentioned previously, the analysis of photoelectron line shapes in cuprates is not yet reliable. None of these techniques has yet been applied to cuprates, probably because interest in the cuprates has focused largely on the states within a few $k_B T_c$ of the Fermi level and on the bands that produce these states. Recent reviews of relaxation of hot electrons in metals (Bokor and Fann, 1996), of electron dynamics at surfaces (Haight, 1995), and of two-photon photoelectron spectroscopy of metals (Steinmann and Fauster, 1996) have appeared.

11.4   Coincidence techniques

Auger spectroscopy has been used for the study of the electronic structure of solids. The final state contains two holes, giving an opportunity to study hole–hole interactions. Auger spectra often are complex, composed of overlapping components. This overlap may arise from the proximity of two initial excitations, e.g., L_2 and L_3, which may be separated by less than the width of the Auger spectrum of either. Even if these are sufficiently widely separated to avoid overlap, there may be a different threshold for core-hole production at the surface than in the bulk. Thresholds for Auger emission by exciting a "main line" and a satellite will also differ. These overlapping Auger spectra may be simplified by a coincidence technique (Haak *et al.*, 1978; Sawatzky, 1988). If the core hole is excited by a photon and a photoelectron emerges, the "subsequent" Auger electron may be detected in coincidence with the photoelectron. (The Auger decay rate is very rapid compared with typical windows in coincidence counting, of the order of a few ns.) This offers the possibility of separating overlapping Auger spectra.

It is not always possible to separate photoexcitation and Auger electron emission into two temporally sequential events. The initial N-electron system plus a photon is transformed to an $(N - 2)$-electron system plus two detected electons. Various intermediate states may interfere to give the final spectrum. An advantage to Auger-photoelectron coincidence spectroscopy is that the lifetime width of the photoexcited (intermediate) state does not contribute to the width of spectral features. This can be seen by invoking energy conservation. The difference between the energies of the initial and final states of the sample is $E_{N-2} - E_N$. This is equal to the energy of the photon minus the kinetic energies of the two detected electrons. Intermediate-state energies do not appear. Lifetime broadening of the intermediate state does not appear in the coincidence spectrum, although it does occur in the photoelectron spectrum alone.

Haak *et al.* (1978) demonstrated this separation of overlapping Auger spectra by coincidence techniques. The Cu $L_2M_{45}M_{45}$ and $L_3M_{45}M_{45}$ Auger spectra overlap (Fig. 11.1), but in the coincidence spectra they can be separated. Coincidence experiments have low count rates, but this negative aspect is offset by a low background. What is not shown in Fig. 11.1 is that each point in the coincidence spectra took several hours to obtain, while the entire Auger spectrum took only a minute or so. Thurgate *et al.* (1994) carried out similar coincidence measurements on the $L_{23}VV$ Auger spectra of Ni and $Ni_{0.5}Fe_{0.5}$. Bartynsky *et al.* (1992) used this technique to separate a surface Ta N_7VV Auger spectrum from the bulk Ta Auger spectrum in TaC. Thurgate (1996) recently presented a detailed review of Auger-photoelectron coincidence spectroscopy, including prospects for its improvement and application. Sawatzky (1988) has provided an excellent introduction to the technique.

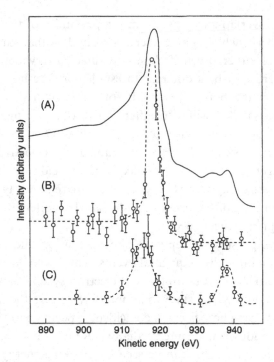

Fig. 11.1 (A) $L_{23}M_{45}M_{45}$ Auger spectrum for Cu. (B) Auger spectrum taken in coincidence with $2p_{3/2}$ photoelectrons. (C) Auger spectrum taken in coincidence with $2p_{1/2}$ photoelectrons. (Haak *et al.*, 1978)

11.5   Spin-resolved photoemission

The spin of photoelectrons may be analyzed. This was first done by accelerating the electrons, then scattering them from a large-Z target, usually Au. A left–right asymmetry in the scattering occurs via the spin–orbit interaction if the electrons are partly polarized (Mott scattering). The spin analyzer was large, and required voltages of the order of 100 keV. The efficiency of spin analysis was rather low, of the order of 10^{-4}. Smaller analyzers have been developed in the past decade. These operate at lower voltages, e.g., 150 eV, and they have larger efficiencies, of the order of 1%. These have led to the application of spin-polarized photoelectron spectroscopy by a larger number of groups. The samples generally, but not exclusively, have been ferromagnetic. Special precautions had to be taken to reduce the distortions of the electron paths caused by the magnetic field of the sample. The early work on spin-polarized photoelectron spectroscopy was done in the angle-integrated mode, but more recent work has been done with electrons that have been angle- and energy resolved. This is the most "differential" photoelectron spectroscopy to date, capable of determining E, **k**, spin, and perhaps some symmetry information. It has led to determinations of the wave-vector dependence of the exchange splitting of bands, and to studies of magnetic fluctuations above the Curie temperature (Smith, 1994b). Spin-resolved core-level spectroscopy is also a burgeoning field, for from the spectra one can separate the spin and orbital magnetic moments from each other (Stöhr and Wu, 1994). Here the spectra typically are absorption spectra, but they may be taken by measuring the photoelectron yield spectrum. Spin-polarized photo-electron spectra may be taken with linearly polarized radiation or with circularly (or elliptically) polarized radiation. The latter is obtained at a storage ring by taking synchrotron radiation from above and below the orbital plane, or by using special insertion devices. It can also be produced by retarders using only reflections.

There is a large literature on spin-polarized photoemission from ferromagnetically ordered materials. The parent compounds of the cuprates are antiferromagnetically ordered, and only recently has spin-polarized photoelectron spectroscopy been carried out on a system with such ordering. Interestingly, that system was CuO. Tjeng et al. (1997) used circularly polarized photons with an energy near the Cu $2p_{3/2}$ absorption peak to excite, resonantly, valence-band electrons which were spin analyzed. The resonance is required not just to obtain large photocurrents, but because a large spin–orbit splitting of the initial state is needed to yield much spin polarization from circularly polarized radiation. Direct valence-band photoemission will not provide a large effect. The sample was an oxidized polycrystalline film. The measurements were carried out at 300 K, above the Néel temperature of 230 K. It is not necessary to have aligned magnetic moments or a single crystal in this experiment because one is photoexciting core electrons. The spectrum obtained by taking the difference between a spectrum with electrons with spins parallel to the photon spin and a spectrum with antiparallel electron spin did not have same shape as the sum spectrum or the spectra in the literature taken without spin analysis. For an atomic singlet final state, the expected degree of polarization (difference spectrum divided by sum spectrum) is $+42\%$, while that for a triplet state is -14%, both neglecting spin–orbit effects. The peaks at 12.5 and 16.2 eV, previously identified as Cu $3d^8$ 1G and ^{1}S, respectively, had polarizations of $+41\%$ and $+35\%$, respectively. The valence-band region had a polarization varying between $+7\%$ and $+22\%$, except just below E_F where it rose to $+30\%$. The occurrence of triplet and singlet final states explains the lower polarization in the valence band, while in the first eV below E_F the final states are nearly pure singlets, which the authors suggest are the Zhang–Rice singlets (Zhang and Rice, 1988). Cluster calculations (Eskes et al., 1990) had shown that the final states near E_F and the final state for the 16.2 eV peak had the same symmetry. From the new type of information gained in this work on CuO, it is obvious that similar work on cuprates will be forthcoming.

Chapter 12

Results from selected other techniques

Several techniques not closely related to photoelectron spectroscopy give results similar to some of those obtained from photoelectron spectroscopy. They may also give other types of information as well. In the following we describe a few of them, but only with respect to data that can be compared with photoelectron spectroscopy. Other useful results from these techniques are not mentioned. We finish this chapter with some results from the spectroscopies closely related to photoelectron spectroscopy: electron energy-loss spectroscopy, soft x-ray absorption, and soft x-ray emission.

12.1 Infrared spectroscopy

Infrared spectroscopy probes the sample with photons of energies below about 1 eV, down to perhaps 1 meV. The interaction Hamiltonian is the same as for photoexcitation, proportional to $\mathbf{A} \cdot \mathbf{p} + \mathbf{p} \cdot \mathbf{A}$, and similar selection rules apply. The electronic excitations usually are divided into intraband excitations, often called the Drude contribution, and interband excitations. Because infrared photons have such small wave vectors, an additional scattering mechanism, phonons or impurities, is required in the Drude contribution, leading to a second-order process, and often to a strong temperature dependence. However, this can still lead to intense absorption for systems with metallic electron densities. The Drude term is the high-frequency analog of the electrical conductivity. In a superconductor, one then expects a delta function in the optical conductivity at zero energy and, perhaps, zero absorption until an energy equal to the gap energy

2Δ, is reached. In addition to electronic excitations, some phonons may also be excited directly, leading to absorption of photons. These phonons come from optical branches, and because the wave vector of infrared photons is so small, they are always phonons with wave vectors near the center of the Brillouin zone. Only phonons which have a value of the transverse electric dipole moment per unit cell that changes with vibrational amplitude contribute to infrared absorption.

Because the absorption is so large, most infrared measurements on cuprates are made by reflectance techniques. The optical conductivity or dielectric function may be recovered from the data by a Kramers–Kronig treatment of the data to recover the unmeasured phase spectrum. Since the cuprates are optically anisotropic, the conductivity tensor has two (for tetragonal lattices) or three (for orthorhombic lattices) independent complex components, so two or three complete reflectance spectra must be made for complete optical characterization. Ideally, one would like one of them to be made on a face terminated by an a–c or b–c plane, but the crystal habits usually preclude this. Instead, a measurement usually has to be made at a large angle of incidence with p-polarized radiation. The optical penetration depth is of the order of 1000 Å, considerably larger than that for photoelectron spectroscopy, so infrared absorption or reflection measurements are far less surface sensitive than photoelectron spectroscopy.

Timusk and Tanner (1989) and Tanner and Timusk (1992) reviewed measurement techniques and interpretation of infrared studies on cuprates. We describe here only those results that can be compared with photoemission data.

The Drude contribution to one component of the optical conductivity cannot be fitted by a single expression for a free-electron gas, whose two parameters are a constant ratio of carrier density to carrier mass, N/m^* (or the plasma frequency) and a constant scattering time. One option is to use two such expressions, but the more common approach is to allow these parameters to be energy dependent. If so, they must be constrained to keep the real and imaginary parts of the complex conductivity related by a dispersion (Kramers–Kronig) integral. The additional parameters are usually taken to be $\tau(E)$ and $\Sigma_1(E)$, where Σ_1 is the real part of the self-energy of the electron gas, and the dependence on wave vector often is neglected. (τ^{-1} is proportional to the imaginary part of the self-energy.) τ is usually found to vary with temperature as T^{-1}, thus making the optical resistivity vary as T^{+1}, as the d.c. resistivity of a cuprate often does. The energy dependence of τ^{-1}, the scattering rate, is linear also. Since this scattering rate is for electrons within a photon energy of the Fermi level, at the longer infrared wavelengths, these are the same electrons that contribute to the superconductivity. This linear energy dependence of τ^{-1} is consistent with the linear dependence of the photohole lifetime found for Bi2212 by photoelectron spectroscopy. Upon cooling below T_c, τ gets very much larger over an energy region of

100 meV or so. This increase actually begins at temperatures above T_c and may be taken as evidence of a pseudogap. τ increases because of a loss of phase space for scattering when a gap is present.

Below T_c one might expect a gap to appear in the conductivity over the first 2Δ (4Δ in some descriptions) of the spectrum. If the gap has nodes, part of such an anticipated region of zero conductivity will be filled in, and the structure will be more difficult to observe. Such a region of reduced conductivity does indeed appear, but relating it to a gap has not been easy. BCS calculations of the optical conductivity for a superconductor give results that differ for the clean and dirty limits (Bickers *et al.*; 1990, Nicol *et al.*, 1991). In the clean limit, the mean free path for carrier scattering is much larger than the coherence length. In the dirty limit, the reverse is the case. In the clean limit, believed to be the more appropriate for cuprates because of their small coherence lengths, the region of reduced conductivity is not very prominent, although it still should be measurable. Tanner and Timusk (1992) nonetheless conclude that no reliable determination of the gap had been made by infrared measurements up till the time of writing their review.

12.2 Raman spectroscopy

Raman scattering, the inelastic scattering of photons, is usually associated with the spectroscopy of phonons. The energy- and momentum differences between the incoming and scattered photon are related to the energy and momentum of a small wave-vector phonon in one of the optical branches. The study of phonons in cuprates is an important field, but here we are interested only in information from Raman spectroscopy directly comparable with results from photoemission investigations. The phonons do, however, interact with electrons, and from the line shapes of the phonon spectral peaks information on the electrons may be obtained. Moreover, inelastic scattering of photons by electrons also contributes directly to the spectrum of inelastically scattered photons. However, it is a rather complex many-body process, for many electrons screen the one-electron excitation. Finally, Raman scattering by the excitation of magnons is of interest in the study of the antiferromagnetic insulating parent compounds of the superconducting cuprates. There are several general references on Raman scattering (Hayes and Loudon, 1978; Cardona and Güntherodt, 1975–1991). Recent brief reviews on electronic Raman scattering in cuprates are those of Cardona *et al.* (1996) and Hackl *et al.* (1996).

The interaction Hamiltonian is again proportional to $\mathbf{A} \cdot \mathbf{p} + \mathbf{p} \cdot \mathbf{A}$, but it now is used in second order. In such a case, the term in the Hamiltonian proportional to A^2, negligible when the terms linear in $\mathbf{A}$ are used in first order as in infrared absorption, now has to be considered. Its effect in first order can be comparable to the second-order effect of the other two terms. Such is the case for inelastic scattering due to the excitation of conduction electrons. In all cases selection rules may be worked out. The Raman-active phonons may be the same as the infrared-active phonons, completely different, or some may be infrared-active and some not. Because of the tensorial nature of the crystal's response, for an anisotropic crystal more measurements are needed to characterize completely the Raman tensor. For example, if the incoming radiation is propagating along the a-axis, it may be polarized along the b- or c-axis. In each case, the scattered radiation need not have the same polarization as the incident radiation. Both polarizations should be measured. However, a complete set of measurements allows the assignment of symmetry to many of the phonon modes in the spectrum. The cuprates are opaque in the visible, so measurements are made in reflection geometry. The optical penetration depth is of the order of 100 Å, so the measurements are less surface sensitive than photoemission measurements. However, the volume of the sample contributing to the scattering is much smaller than that of crystals transparent to laser light, so the scattered photon flux is very small. The measurements are therefore rather difficult.

An example of the electron–phonon interaction is provided by early studies of Raman scattering in Y123 by Cooper et $al.$ (1988) and Thomsen et $al.$ (1988). With incident light propagating along the c-axis polarized along an a-axis, excitations of A_g symmetry ($d_{3z^2-r^2}$ symmetry about a Cu site) were observed if inelastically scattered reflected light polarized along the a-axis was measured. At about T_c the spectrum consisted of several peaks due to phonons and a continuum due to electronic excitation. The phonon peaks were asymmetric, exhibiting Fano line shapes, discussed in Chapter 3. The line shapes arise from the interaction of each phonon with the continuum of electronic excitations of the same symmetry. Upon cooling to 3 K, the electronic continuum changed shape, exhibiting the effects of the opening of a gap, with a pileup of states above the gap. Values of the gap could not be extracted without better knowledge of the expected spectral shape, but it was clear that there was not a single value of the gap. Rather, there appeared to be some electronic excitations at energies below the maximum gap energy. Also, a phonon line at an energy below the maximum gap value had its width decrease upon cooling, while the width of another phonon line whose peak is near the maximum in the low-temperature continuum had its width increase. Both width changes are consistent with the temperature-induced changes in the electronic background. One phonon had a lower density of excitations with which to interact, while the other had a higher density. Nearly simultaneously Hackl et $al.$ (1988) varied the polarization and showed from the same Raman continuum that the gap was anisotropic. In their fit, a peak value of 21 meV was found for Δ at 4.2 K.

Suitable doping can cause the energy of the Raman-active B_{1g} phonon ($d_{x^2-y^2}$ symmetry about a Cu site), about 42 meV, to vary from above 2Δ to below it. Freidl *et al.* (1990) substituted several rare earths for Y in Y123, and ^{16}O by ^{18}O to vary the phonon frequency. The Fano line shapes of the Raman spectrum for this mode over a range of temperatures were fitted to extract broadening parameters. Below T_c the mode softens considerably and its width almost doubles in the Er-substituted crystal, for which this mode has the highest energy. This mode can break a Cooper pair upon de-excitation. No such effect occurs for a Eu-substituted crystal which has the lowest energy for this mode, too low to break a Cooper pair. This, and data from the other crystals, when fitted to a many-body model for the electron-phonon contribution to the width, allow 2Δ to be bracketed between 38.3 and 40.0 meV. This supports strong coupling superconductivity in Y123, for it yields $2\Delta/k_B T_c = 4.95 \pm 0.10$.

One can vary 2Δ instead, keeping the phonon energy more nearly constant. This was done by Altendorf *et al.* (1992). They used five crystals of varying oxygen content and measured the temperature dependence of the line shape of a phonon of B_{1g} symmetry at an energy of about 42 meV. Fitting the Fano line shapes gave a peak position and a width parameter for each that varied linearly with temperature above T_c. Below T_c the widths for three samples decreased, while that for the other two increased and the peak positions all shifted downward by various amounts. As mentioned above, this mode has an energy just above or below 2Δ, and since 2Δ may be different for each sample, different behavior is expected. Upon fitting to a many-body calculation poor correlation between Δ and T_c was found. Such lack of correlation has been found for variously doped Bi2212 samples in photoemission studies.

Chen *et al.* (1993) took polarized Raman spectra at several temperatures on three samples of YBa$_2$Cu$_3$O$_y$ with y values of 7.00, 6.99, and 6.93. The broad peak in the continuum with B_{1g} symmetry shifted with oxygen doping. These shifts correlated well with the shifts obtained from fitting a phonon peak of the same symmetry in these samples with a model for the electron–phonon coupling. In a later study (Chen *et al.*, 1994b) they showed that the broad peak in the Raman continuum with B_{1g} symmetry shifted its position upon changing the oxygen concentration, while that with A_{1g} symmetry did not shift. They modeled this continuum using the theory of Klein and Dierker (1984) for Raman scattering by electronic excitations and a gap of $d_{x^2-y^2}$ symmetry. Both calculated spectra peaked at about $3.3\Delta_0$, where Δ_0 is the maximum gap value (Fig. 12.1(b)). However the peak in the experimental spectrum of B_{1g} symmetry was at a higher energy than that in the A_{1g} spectrum. They suggested that the B_{1g} continuum arises instead from the excitation of electrons across a pseudogap produced by spin fluctuations. More evidence for this is the fact that the broad peak in the

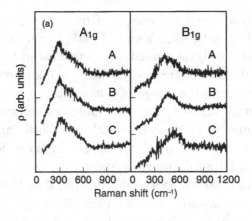

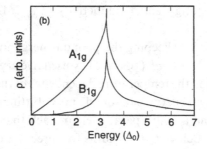

Fig. 12.1 Electronic Raman spectra of Y123. (a) Measured spectra for three different samples, A, B, and C, in each of two polarization arrrangments to isolate excitations of A_{1g} and B_{1g} symmetry. (b) Calculated spectra assuming a gap with d-wave symmetry. The energy scale is in units of the maximum gap energy. (Chen *et al.*, 1994b)

B_{1g} continuum occurs at temperatures above T_c in Y123 an underdoped sample with a value of y of about 6.7 (Slakey *et al.*, 1990).

Among the Raman scattering studies on Bi2212, we summarize only two. Staufer *et al.* (1992) carried out polarized measurements on three samples, doped differently by annealing in Ar and O_2. The electronic continuum below T_c showed an incomplete gap region which had a finite or zero intercept at zero energy, depending on polarization. It also was different for the one underdoped sample of the three. The position of the broad peak also depended on polarization. The peak positions, normalized to the values at $T = 0$, when plotted as a function of temperature, remained above the BCS plot of normalized gap vs. reduced T_c, much as the plots of $\Delta(T)/\Delta(0)$ from photoemission data. This is indicative of strong coupling. Upon extracting values of $\Delta(T)$ from the spectra using a BCS-like model, a BCS-like temperature dependence was found for the underdoped sample. The other two gave plots that remained above the BCS plot. Later Devereaux *et al.* (1994) made new measurements and carried out an extensive calculation of the electronic continuum, assuming s-like or $d_{x^2-y^2}$ symmetry for the order parameter. The most prominent structure in the calculated spectra for $T = 0$ was a peak near 2Δ in the spectrum for B_{1g} symmetry. The spectrum for each of the other symmetries had a weaker, broader peak at lower energy. The extrapolation of the spectrum within the gap to zero energy was calculated as a function of temperature, and this, too, showed an anisotropy consistent with measured spectra.

Chen et al. (1994a) studied the angle dependence of the electronic Raman scattering of $La_{2-x}Sr_xCuO_4$ above and below T_c (37 K). The three scattering geometries they used allowed spectra for excitations of three symmetries, $B_{1g}(x^2-y^2)$, $B_{2g}(xy)$, and $A_{1g}(s)$. Each spectrum showed a continuum which, upon cooling below T_c, showed an incomplete gap-like region opening up at low energies and a broad maximum. The energy of the maxima varied with the symmetry. The spectra were modeled by using a tight-binding band passing through the Fermi level and assumed $d_{x^2-y^2}$ and anisotropic s-wave gap functions. The Raman scattering cross-section is proportional to the product of the absolute square of the gap function, thereby insensitive to sign changes, and an angle- and polarization-dependent coupling parameter which is proportional to an average over the Fermi surface of components of the inverse effective mass tensor (Klein and Dierker, 1984). The result of the model was that the gap appeared to have symmetry, with a maximum value of approximately $2|\Delta| = 24.8$ meV for the directions along the CuO_2 plane bond directions. The gap appeared to be nearly zero for directions in the plane at $45°$ to the bond axes.

Devereaux and Einzel (1995) have carried out extensive calculations of the electronic Raman scattering in superconductors and compared them with measured spectra for Y123, Bi2212, and Tl2201. Data taken with several polarizations are necessary for such a comparison. For all three cuprates they find the data are consistent with a gap of $d_{x^2-y^2}$ symmetry. The best fit for Bi2212 was obtained with a value of 36 meV for Δ along the Cu–O bond axis. The maximum gap from the fitting was 26 meV for Y123 and 29 meV for Tl2210. One of the features of the fit is the different location of the broad peak for different symmetry of excitation. Also important is the region of energies less than $2|\Delta|$ in which the non-zero intensity is different for excitations of different symmetry.

Extensive polarized Raman spectra on NCCO have shown a peak in the electronic Raman continuum that is only slightly anisotropic (Stadlober et al., 1995). The gap values range from 6.9 to 8.2 meV as a function of angle in the basal plane. The value of $2\Delta/k_B T_c$, 4.1 to 4.9, is smaller than in other cuprates. These results suggest that superconductivity in n-type NCCO is not like the superconductivity in several p-type cuprates; s-wave coupling appears to occur.

12.3   Tunneling spectroscopy

Tunneling of electrons to or from a cuprate sample should be able to yield a spectrum of excitations in the meV range from the current–voltage curve (Wolf, 1985). Tunneling has provided values of superconducting gaps and phonon spectra, multiplied by coupling coefficients, for a number of conventional superconductors. The tunneling may be carried out using a point contact or by using a large-area electrode, e.g., a thin film condensed on the sample. The literature on tunneling in the cuprates is quite large. Tunneling measurements on cuprates were carried out very early in attempts to obtain values of the gap. Few of these early measurements gave a value for the gap and the spectra had unusual shapes that were not understood. The uncertain nature of the surfaces and their instability, as well as the large anisotropy and short coherence length contributed greatly to the problem. Reviews have been written by Hasegawa *et al.* (1992) and van Bentum and van Kampen (1992). Kitazawa (1966) discussed briefly some of the problems that have plagued tunneling measurements, and some of the promise for tunneling spectroscopy with atomic-scale lateral resolution.

Usually one measures the tunneling conductance of the sample, dI/dV. For a sample consisting of a normal metal–insulator–superconductor junction, the conductance shows a jump at an applied voltage of $\pm \Delta/e$. If the insulator separates two identical superconductors, the jump is at $\pm 2\Delta/e$. In between the two peaks the conductance is zero for a BCS superconductor with s-wave coupling. At voltages larger than those of the peaks, the conductance is constant. The insulator must be very thin, and it need not be a solid; vacuum suffices. Many of the early tunneling studies on cuprates gave quite different conductance spectra. There may have been more than one pair of peaks. The conductance at voltages between the pair of peaks with the lowest separation was not zero, and at voltages above those at the conductance peaks, the conductance often increased steadily with voltage.

Over the years sample preparation has improved greatly, leading to tunneling spectra that are more reproducible than the early spectra. Some, but not all, of the unusual features, are no longer present. Those remaining are now believed to be intrinsic to tunneling in the cuprates. The sample preparation techniques currently used often are similar to those used for angle-resolved photoelectron spectroscopy: low temperature cleavage in ultrahigh vacuum. We report below only a few recent results.

Renner and Fischer (1995) carried out an extensive study of tunneling from an STM tip into Bi2212 at 4.8 K, taking great care about reproducibility. The results are believed to be characteristic of tunneling in the a–b plane of Bi2212, i.e., the current is averaged over direction in this plane. The cleaved sample surfaces are believed to be Bi–O planes, for they can be imaged with atomic resolution and the superlattice is evident. Surfaces with unusual tunneling characteristics (semi-conductor-like) apparently have the same Bi–O surface, but the CuO_2 plane below may be deficient in holes. Good surfaces give similar tunneling spectra at many points over the surface, implying that special tunneling sites play no role. Their spectra (Fig. 12.2(b)) exhibit the following features. There are two nearly identical sharp peaks, symmetrically disposed about zero voltage. If the maximum gap corresponds to the point at which the extrapolated high-voltage background intersects the peak structure, $\Delta = 29.5 \pm 0.4$ meV. A more precise value for Δ requires fitting to a particular model for the superconductivity in Bi2212. The BCS description does not fit the shape of the spectra. d-wave or extended s-wave models give a gap that is 5–10 meV larger than the value given above, essentially half the peak-to-peak distance. The region between the peaks has a non-zero minimum, implying a spectrum of gaps smaller than 29.5 meV, possibly going to zero. For negative polarity, where the spectrum arises from structure in the filled density of states there is a dip in the spectrum at about 70 meV, followed by a weak peak. Both are reminiscent of the structures seen in angle-resolved photoemission spectra below T_c. Finally, the conductance at larger voltages was closer to constant than in many previous studies; it tended to decrease with

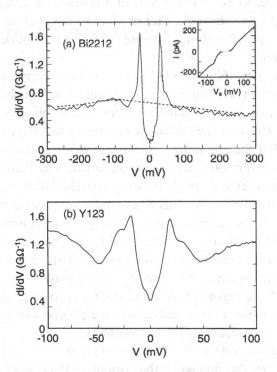

Fig. 12.2 Metal–insulator–superconductor differential tunneling conductances for (a) Bi2212 and (b) Y123 taken at 4.2 K. The inset to (a) is an I–V curve. (Renner and Fischer, 1995; and Maggio-Aprile *et al.*, 1995)

increasing voltage of either polarity. There was a slight asymmetry about zero voltage. Additional modeling led to the authors to conclude the tunneling spectra were consistent with a gap of d$_{x^2-y^2}$ or extended s-wave symmetry, although tunneling, like photoemission, can give only the magnitude of the gap, not its phase. Moreover, there were residual discrepancies between the measured and the fitted spectra.

Bok and Bouvier (1997) have modeled the data of Renner and Fischer. They used a tight-binding band with its logarithmic van Hove critical point to represent the quasiparticles in the normal state. They first located the Fermi level at the energy of this critical point. They used a simplified angle-dependent gap function, $\Delta(\mathbf{k}) = \Delta_M - \Delta_2 \sin 2k_x a$. The minimum gap is $\Delta_m = \Delta_M - \Delta_2$. A phenomenological broadening parameter was also introduced. Such a model gave a best fit to the tunneling conductance spectrum of Bi2212 for $\Delta_M = 27$ meV, $\Delta_m = 11$ meV, but the calculated conductance was symmetric about zero voltage. They then allowed the Fermi level to move above the energy of the singularity and found that a dip developed on the conductance for negative applied voltages. The dip was due to a singularity in the quasiparticle spectrum located at a negative energy of magnitude $(\Delta^2 + E_s^2)^{1/2}$, where E_s is the energy of the singularity with respect to E_F. A reasonable fit was obtained with $\Delta_M = 22$ meV, $\Delta_m = 6$ meV, and $E_s = -60$ meV. This produced a dip at about -40 meV, followed by a peak around -60 meV whose height and width depended on the damping parameter.

Kane and Ng (1996), using an STM with an unusual sample geometry, measured tunneling current as a function of direction in the a–b plane of Bi2212 at low temperature and found significant anisotropy, compatible with a gap of d$_{x^2-y^2}$ symmetry. However the minimum gap was about 20 meV and the maximum 40 meV. That the minimum gap was not closer to zero probably arose from the finite angular resolution of the arrangement. The direction of the minimum gap was along the Cu–O bond, 45° away from the direction of the minimum gap obtained by photoemission. This was found in other angle-resolved tunneling measurements (Tanaka et al., 1995). The origin of this discrepancy with the reuslts from photoemission is not clear, and further analysis of both type of experiments seems required. Miyake and Narikiyo (1995) discussed the possibility that these tunneling data, and some on Y123 (Tanaka et al., 1994) that also showed that a "45° discrepancy," could be due to a gap of d$_{xy}$ symmetry. They showed how this might be compatible with the photoemission data, but this seems not to have been discussed further by others.

Y123 has proven more difficult. The review of Hasegawa et al. (1992) reports a range of gap energies determined by the position of the peaks ranging from 18 to 100 meV from tunneling studies made before early or mid 1992. A recent study (Maggio-Aprile et al., 1995) is typical of the best recent work. They used an Ir tip in an STM to tunnel across a vacuum gap into as-grown surfaces of single crystals at 4.2 K. A well-defined pair of peaks in the tunneling conductance gives a value of 20 ± 2 meV for the gap (Fig. 12.2(b)). But there is a second set of conductance peaks, indicating a value for Δ of 25–30 meV, and still a third, less reproducible pair, corresponding to $\Delta = 4$–10 meV. The gap region has a rather high value of conductance, and the conductance at higher voltages is not constant, but increases with voltage. Zasadzinski et al. (1996) recently reviewed some aspects of tunneling in many cuprates. One of the more interesting results is a plot of tunneling conductance vs. applied voltage with the latter expressed in units of Δ for each of four cuprates, Bi2212, Bi2201, NCCO, and $Tl_2Ba_2CaCu_2O_8$. The curves were all quite similar in shape and each had a dip at about 3Δ on one side only. (These data were for tunneling between two superconductors. If one of the two metals had not been superconducting the dip would have appeared at 2Δ.) Note that of these, NCCO is n-type. One interpretation of the dip, as due to a change in quasiparticle lifetimes due to the presence of a gap, predicts the gap to be at 2Δ for tunneling from a normal metal into a d-wave superconductor and at 3Δ if the superconductor is an s-wave material. This interpretation would then fit with all four of the above cuprates if they had d-wave superconductivity, but NCCO is believed to be an s-wave material from the broad minimum in the tunneling conductance around zero applied potential difference.

12.4 de Haas–van Alphen measurements

de Haas–van Alphen oscillations occur in plots of the low-temperature magnetic susceptibility of a metal as a function of the inverse of the applied magnetic field. These oscillations arise from the passage of individual Landau levels outward through the Fermi surface as the magnetic field increases. The oscillation periods are proportional to extremal cross-sectional areas of the Fermi surface normal to the applied field. However, to observe the oscillations, the electrons at the Fermi surface must make several complete orbits around the Fermi surface before scattering. This condition is usually stated as $\omega_c \tau \gg 1$, where ω_c, the cyclotron frequency, is eB/m^*c. This is met in pure metals by working at low temperature with very pure single-crystal samples. Under such conditions the scattering rate by impurities and phonons is low and τ is long. Cuprates have rather high resistivities and the values of τ are too short for de Haas–van Alphen studies to be carried out with ordinary laboratory magnetic fields, even those from superconducting solenoids. However some pulsed fields can make ω_c large enough to overcome the small values of τ, and such measurements can be carried out. Ordinary pulsed fields from electromagnets are not adequate, but by imploding a current-carrying solenoid with an explosive charge, fields of 100 T can be produced. The implosion destroys the magnet, the sample, and more, but the experiments were considered important enough to be carried out nonetheless (Fowler *et al.*, 1992; Mazin *et al.*, 1992; Mueller *et al.*, 1992). de Haas–van Alphen oscillations were found in Y123. Fowler *et al.* used fine-grained powder samples in order to obtain good flux penetration into the volume during the short pulse duration. They found three fundamental periods and the harmonics of each in their spectrum. These were consistent with extremal areas from the Fermi surfaces from LDA band calculations. Kido *et al.*(1991) used a static field of 27 T for a similar measurement, also on finely divided powdered Y123, and found one de Haas–van Alphen period, which they assigned to a piece of the Fermi surface derived from Cu–O chain states.

12.5   Angular correlation of positron–electron annihilation gamma rays

Measurements of the angular correlation of gamma rays from electron–positron annihilation in the sample provides another measurement of the Fermi surface. Such measurements require single crystals, but often they need not be of such high purity and they can be carried out at room temperature. However, cuprates necessarily have a large number of charged defects which may trap positrons, and temperatures above room temperature may have to be used to reduce such trapping.

Annihilation of a stationary electron and a stationary positron produces two γ rays propagating in opposite directions. The sum of their momenta is zero. Call these directions $\pm x$. If the electron is moving with a z-component of momentum, p_z, the two γ rays are no longer quite $180°$ apart, and their momenta add to that of the original electron. Each has deviated from the $180°$ axis by p_z/mc, a small angle. The experiment consists of counting γ rays in coincidence, one in each of two counters, nominally $180°$ apart, as a function of the symmetric deviation of the counters from $180°$, i.e., the detectors are scanned in the z direction or the sample is scanned in the $-z$ direction. When the angle corresponding to the Fermi momentum is reached, the coincidence rate should drop precipitously. The shape of the drop-off depends on the shape of the Fermi surface. In practice, measurements are carried out in more than one Brillouin zone and folded back into the first zone. The independent-particle model gives (De Benedetti *et al.*, 1950) the coincidence rate as proportional to the projection on the x–z plane of the two-photon electron–positron momentum density $\rho_{2\gamma}(\mathbf{p})$.

$$\rho_{2\gamma}(\mathbf{p}) = \sum |\int_{-\infty}^{\infty} e^{i\mathbf{p}\cdot\mathbf{r}}\psi_{n\mathbf{k}}(\mathbf{r})\psi_{+}(\mathbf{r})d^3r|^2,$$

where $\psi_{n\mathbf{k}}$ is the electron wave function and ψ_+ that of the positron. The sum runs over occupied electron states. The projection is

$$N(p_z, p_y) = \int \rho_{2\gamma}(\mathbf{p})dp_x.$$

This is integrated over p_y in the case of long narrow slits parallel to y to get the coincidence rate as a function of p_z. These functions can be calculated from the band structure for comparison with the measurements. The crystal may be rotated so several projections may be measured. The positron is assumed to be thermalized. It is repelled by the nuclei and attracted to regions of high electron density. The positron wave function may be found from the same potential, with the sign reversed, as used for the LDA band calculation, except that the exchange potential is removed. Such calculations for Y123 have shown that ψ_+ is delocalized in regions between the CuO_2 planes, causing only small contributions from the CuO_2 plane states to appear in the final spectrum (von Stetten *et al.*, 1988). Experimentally it has been found from positron lifetime measurements that at low temperatures, including room temperatures, the positrons may be trapped at oxygen vacancies, and ψ_+ changes. In NCCO, ψ_+ is much more spread out in the unit cell (Blandin *et al.*, 1992).

Several cuprate Fermi surfaces have been mapped by angular correlation of positron annihilation γ rays. These have been summarized by Manuel *et al.* (1995). For Y123 the parts of the Fermi surface due to the Cu–O chains have been found to fit the coincidence spectra. These are the strip parallel to the X–Γ edge and the small arc around S in Fig. 2.13 (Shukla *et al.*, 1995). The other two sheets of the Fermi surface, composed primarily of states based on the CuO_2 planes, are not seen for reasons given above. The structure around S disappears in $PrBa_2Cu_3O_{7-\delta}$, in agreement with LDA calculations. The Fermi surface of NCCO was only recently detected by positron annihilation studies (Manuel *et al.*, 1995) and it, too, fits well with LDA calculations. Positron annihilation studies on Bi2212 have an additional complication, the effects of the superlattice. This makes a direct confrontation of measurements with theory not possible and more analysis must be carried out. To date, the data seem to be what is expected from the CuO_2 and Bi–O plane contributions to the Fermi surface (Chan *et al.*, 1991).

12.6 Electron energy-loss spectroscopy, soft x-ray absorption, and soft x-ray emission

Spectroscopic techniques related to photoelectron spectroscopy were applied early in the study of cuprates, for they are less surface sensitive than photoelectron spectroscopy. Two such are x-ray absorption and x-ray fluorescence. They may be applied equally well to metallic and insulating samples, for surface charging plays no role. The resultant spectra are functions of energy. No wave-vector information is available. Instead, atom- and angular-momentum-specific information may be obtained, and with single crystals and polarized radiation, one can get information about orbital orientation. For each of these techniques a core electron is either the initial or final state. The dipole matrix element has contributions from only those valence states that have significant overlap with the core region of the atom being studied. The valence states can be expanded in spherical harmonics about that atom and only those components with an angular momentum different by $\hbar$ from that of the core state contribute to the matrix element. A closely related technique is electron energy-loss spectroscopy. It can provide spectra of $\tilde{\epsilon}(E, \mathbf{k})$, and, under suitable conditions, dipole selection rules are obeyed. Both core and valence electron contributions to the spectra may be studied.

Fink *et al.* (1994) reviewed the electron energy-loss spectroscopy and x-ray absorption spectroscopy from O 1s and Cu 2p excitations in cuprates. It is straightforward to treat these together, since in the dipole and independent-electron approximations each spectrum is proportional to

$$\sum_{i,f} H'_{if} \delta(E_f - E_i - \hbar\omega),$$

where for electron-energy loss $H' \propto \mathbf{q} \cdot \mathbf{p}_{if}$ and for x-ray absorption $H' \propto \mathbf{e} \cdot \mathbf{p}_{if}$, where $\mathbf{q}$ is the scattering wave vector and $\mathbf{e}$ the x-ray polarization unit vector. $\mathbf{q}$ (or $\mathbf{e}$) can be aligned parallel and perpendicular to the CuO_2 planes. Near threshold, the independent-electron approximation breaks down as the excited electron and hole interact, along with screening by the other electrons. However, these effects are less prominent than in core-level photoemission well above threshold, for in photoexcitation near threshold the crystal is perturbed not by a charged hole but by a hole and an electron, i.e., a (dynamic) dipole, a weaker perturbation. This phenomenon, the x-ray edge problem, was described in Chapter 3. In the independent-electron approximation the final states are those just above E_F. Absorption spectra can discern the atomic states from which they are derived. Spectra on a series of variously doped samples may validate or invalidate some models for the nature of the states involved in superconductivity. These techniques have been especially valuable in the study of doping. Single crystals may be used in transmission electron energy-loss spectroscopy if the incident electron energy is high enough. X-ray absorption may be carried out on single crystals in the region of the O 1s and Cu 2p edge not by transmission, but by detection of total electron yield or of fluorescence yield. Measurements by total yield are less surface sensitive than photoelectron spectroscopy, and fluorescence detection is even less surface sensitive. Typical resolution is 100 meV in energy-loss spectroscopy and 260 and 600 meV in x-ray absorption spectroscopy at the O 1s and Cu 2p edges, respectively.

Fink *et al.* describe work on several cuprates. We restrict ourselves here to Bi2212, Y123 and NCCO, the cuprates most studied by photoelectron spectroscopy. In Bi2212 the O 1s absorption edge, measured by either technique, shows an anisotropic pre-edge peak that depends on hole concentration. Replacing Ca by Y reduces the hole concentration and the main change in the O 1s absorption spectrum is the loss of this peak with increasing Y concentration. Thus, it is natural to assign the peak to transitions into the empty valence band just above E_F in the metallic samples, and to assign significant O 2p character to these states. The pre-edge peak is present for $\mathbf{q}$ in the a–b plane and missing for $\mathbf{q}$ parallel to the c-axis, leading to the conclusion that the final states are O $2p_x$ and $2p_y$ states with little $2p_z$ character. This is illustrated in Fig. 6.9. The pre-edge peak is missing in insulating samples, but there is a polarization-dependent shoulder on the edge, attributed to final states of O $2p_x$ and $2p_y$ character in the lowest empty band, a band based on the states from the Bi–O planes. The Cu 2p absorption spectrum consists of a single peak that is much stronger with $\mathbf{q}$ in the basal plane than for $\mathbf{q}$ parallel to the c-axis. This peak, a white line, with only a weak continuum above it, results from the strong Coulomb interaction between the excited electron and the core hole, an exciton effect. The O 1s core hole is a much weaker perturbation so the O 1s spectrum resembles a density of states, not an exciton. The Cu 2p absorption spectrum can be modeled quite well with the cluster calculations described for CuO in Chapter 5, for the Cu 2p absorption spectra of CuO and the cuprates are much alike. (See also Eskes and Sawatzky, 1991 and Hybertsen *et al.*, 1992.) One can conclude that the final states have considerable Cu $3d_{x^2-y^2}$ character with some admixture of Cu $3d_{3z^2-r^2}$. The latter contributes to the spectrum in both polarizations, while the former is absent in the spectra for $\mathbf{e}$ parallel to the c-axis. Use of the theoretical angle dependence of the two components leads to an admixture of the $3d_{3z^2-r^2}$ state of about 10%. Final states with Cu 4s character are neglected because their contribution to the absorption is smaller by a factor of about 30 than that of the 3d states, partly from the smaller multiplicity and partly from a smaller dipole matrix element.

Y123 has a similar anisotropic pre-edge peak in the O 1s spectrum, but for $\mathbf{q}$ parallel to c there is also a pre-edge peak, weaker, and at lower energy. As the hole concentration decreases the pre-edge peak for $\mathbf{q}$ in the basal plane decreases in intensity and a second peak, about 1.6 eV higher in energy, increases in intensity. The polarization dependence has been used to determine the hole distribution between CuO_2 planes and Cu–O chains. This required taking spectra with $\mathbf{e}$ parallel to the a-, the b-, and the c-axis separately (Nücker *et al.*, 1995). Hole distributions among the four inequivalent O sites were determined. The Cu 2p absorption is an anisotropic peak as in Bi2212, much larger for $\mathbf{q}$ in the a–b

plane. The line shapes differ for **e** parallel to a and for **e** parallel to b. With decreasing doping a second peak appears at higher energy in all three polarizations. This peak is assigned to final states that are Cu $3d_{3z^2-r^2}$–O $2p_z$ hybrids centered on Cu^+ ions.

NCCO has a similar pre-edge peak for **e** in the a–b plane, but not for **e** parallel to the c-axis. This pre-edge peak is weaker in n-type cuprates than in p-type cuprates. Doping with Ce to produce an n-type superconductor has almost no effect on the peak for **e** perpendicular to c. However, for **e** parallel to c, a peak, apparently the same peak, grows with Ce doping. The pre-edge peak for **e** in the a–b plane should not have the same origin as in p-type cuprates, for the final-state band should be full in doped NCCO. It has been ascribed to O $2p_x$ and $2p_y$ final states in the upper Hubbard band, a band based predominantly on Cu 3d states, as is the lower, completely filled Hubbard band, but with some hybridization with O $2p_x$ and $2p_y$ states (Alexander *et al.*, 1991; Pellegrin *et al.*, 1993). The 3d states are predominantly $3d_{x^2-y^2}$ states. It should be noted that upon electron doping, no new absorption occurs below the edge. This indicates that if such doping creates new states in the insulating gap, as photoemission studies seem to show (Allen *et al.*, 1990; Namatame *et al.*, 1990; Anderson *et al.*, 1993), they are all filled or have little O 2p character. The Cu 2p peak is anisotropic, as with the other cuprates. It is also narrower and more symmetric than in p-type cuprates. This has been interpreted as due to the lack of holes in the n-type doped material. The peak weakens with n-type doping, which fills the lowest-lying 3d final states for the Cu 2p absorption.

These techniques have been applied to many other cuprates, including those not yet amenable to study by valence-band angle-resolved photoelectron spectroscopy. Comparison of the spectra for the cuprates as a function of doping has shown that p-type doping leads to empty states just above E_F that are primarily of O $2p_x$ and $2p_y$ character, while n-type doping leads to empty states just above E_F that are primarily of Cu $3d_{x^2-y^2}$ character. This is in accord with the charge-transfer picture for doped cuprates. States from both types of atoms are mixed in each doping regime, however, for the spectra of both Cu- and O-core levels show absorption peaks at threshold for each type of doping.

An example of x-ray fluorescence is the study by Butorin *et al.* (1995) of O 1s fluorescence in Bi2212. They used a single crystal and excited with polarized monochromatic radiation with energies at and a few eV above the O 1s absorption threshold. The emission spectrum was a double-peaked structure about 8 eV wide from states with O 2p character in the filled part of the valence band. The shape of the spectrum varied with the energy of the exciting photon, an effect attributed to the excitation of holes on inequivalent O sites with photons of different energy. The spectra could be fitted reasonably well with the site-specific (O1 and O3) O2p density of states from an LDA calculation. The pre-peak on the O

1s absorption spectrum of Bi2212 was assigned to absorption primarily on O1 sites.

Another example is the study of Ba 4d fluorescence in Y123 and Pr123, the former sample a superconductor, the latter an insulator. Mueller *et al.* (1994) showed that both had similar Ba fluorescence spectra from states near the Fermi level. The close resemblance of the two spectra is not in accord with a previous explanation for the insulating nature of Pr123: significant hybridization of Pr and Ba states near E_F.

These techniques have been very important in studies of doping. There is another type of electron energy-loss spectroscopy which we only mention here. By using very low electron beam energies, e.g., 5 eV, and a narrow energy spread by using an electron energy analyzer as a monochromator, as well as a second analyzer to determine the energy-loss spectrum, one can obtain very high resolution, e.g., 5–10 meV. The low electron energy leads to extreme surface sensitivity, and the technique is used to study surface phonons and the vibrations of adsorbates. It has been applied to cuprate superconductors, and there is some agreement among groups about the surface phonons. Upon going below T_c some groups have seen an energy-loss peak assigned to the superconducting gap, but others have not. For a discussion, see Phelps *et al.* (1994).

References

Abbati, I., Braicovich, L., Rossi, G., Lindau., I., del Pennino, U., and Nannarone, S. (1983) *Phys. Rev. Lett.* **50**, 1799.

Abrikosov, A. A., Campuzano, J. C., and Gofron, K. (1993) *Physica C* **214**, 73.

Adawi, I. (1964) *Phys. Rev.* **134**, A788.

Ade, H., Kirz, J., Hulbert, S., Johnson, E., Anderson, E., and Kern, D. (1990) *Nucl. Instr. and Meth.* **A291**, 126.

Aebi, P., Kreutz, T. J., Osterwalder, J., Fasel, R., Schwaller, P., and Schlapbach, L. (1996) *Phys. Rev. Lett.* **76**, 1150.

Aebi, P., Osterwalder, J., Fasel, R., Naumović, D., and Schlapbach, L. (1994a) *Surf. Sci.* **307–309**, 917.

Aebi, P., Osterwalder, J., Naumović, D., Fasel, R., and Schlapbach, L. (1993) *Solid State Commun.* **88**, 19.

Aebi, P., Osterwalder, J., Schwaller, P., Schlapbach, L., Shimoda, M., Mochiku, T., and Kadowaki, K. (1994b) *Phys. Rev. Lett.* **72**, 2757.

Aebi, P., Osterwalder, J., Schwaller, P., Schlapbach, L., Shimoda, M., Mochiku, T., and Kadowaki, K. (1995a) *Phys. Rev. Lett.* **74**, 1886.

Aebi, P., Osterwalder, J., Schwaller, P., Berger, H., Beeli, C., and Schlapbach, L. (1995b) *J. Phys. Chem Solids* **56**, 1845.

Al Shamma, F. and Fuggle, J. C. (1990) *Physica C* **169**, 325.

Alders, D., Voogt, F. C., Hibma, T., and Sawatzky, G. A. (1996) *Phys. Rev.* **B54**, 7716.

Alexander, M., Romberg, H., Nücker, N., Adelmann, P., Fink, J., Markert, J. T., Maple, M. B., Uchida, S., Takagi, H., Tokura, Y., James, A. C. W. P., and Murphy, D. W. (1991) *Phys. Rev.* **B43**, 333.

Allen, J. W. (1992) in *Synchrotron Radiation Research–Advances in Surface and Interface Science*, Vol. 1, ed. R. Z. Bachrach (Plenum Press, New York) p. 253.

Allen, J. W., Olson, C. G., Maple, M. B., Kang, J.-S., Liu, L. Z., Park, J.-H., Anderson, R. O., Ellis, W. P., Markert, J. T., Dalichaouch, Y., and Liu, R. (1990) *Phys. Rev. Lett.* **64**, 595.

Allen, P. B. and Chetty, N. (1994) *Phys. Rev.* **B50**, 14855.

Almbladh, C.-O., and Hedin, L. (1983) in *Handbook of Synchrotron Radiation*, Vol. 1b, eds. D. E. Eastman, Y. Farge and E.-E. Koch, (North-Holland Publishing Co., Amsterdam) p. 607.

Alméras, P., Berger, H., Margaritondo, G., Ma, J., Quitmann, C., Kelley, R. J., and Onellion, M. (1994) *Solid State Commun.* **91**, 535.

Altendorf, E., Irwin, J. C., Liang, R., and Hardy, W. N. (1992) *Phys. Rev.* **B45**, 7551.

Altmann, W., Dose, V., Goldmann, A., Kolac, U., and Rogozik, J. (1984) *Phys. Rev.* **B29**, 3015.

Andersen, O. K., Jepsen, O., Liechtenstein, A. I., and Mazin, I. I. (1994) *Phys. Rev.* **B49**, 4145.

Anderson, P. W. (1961) *Phys. Rev.* **124**, 41.

Anderson, P. W. (1987) *Science* **235**, 1196.

Anderson, R. O., Claessen, R., Allen, J. W., Olson, C. G., Janowitz, C., Liu, L. Z., Park, J.-H., Maple, M. B., Dalichaouch, Y., de Andrade, M. C., Jardim, R. F., Early, E. A., Oh, S.-J., and Ellis, W. P. (1993) *Phys. Rev. Lett.* **70**, 3163.

405

Andrews, P. T., Collins, I. R., and Inglesfield, J. E. (1992) in *Unoccupied Electronic States*, eds. J. C. Fuggle and J. E. Inglesfield (Springer-Verlag, Berlin) p. 243.

Anisimov, V. I. (1995) in *Spectroscopy of Mott Insulators and Correlated Metals*, eds. A. Fujimori and Y. Tokura (Springer-Verlag, Berlin) p. 106.

Anisimov, V. I., Korotin, M. A., Zaanen, J., and Andersen, O. K. (1992) *Phys. Rev. Lett.* **68**, 345.

Anisimov, V. I., Kuiper, P., and Nordgren, J. (1994) *Phys. Rev.* **B50**, 8257.

Anisimov, V. I., Solovyev, I. V., Korotin, M. A., Czyzyk, M. T., and Sawatzky, G. A. (1993) *Phys. Rev.* **B48**, 16929.

Anisimov, V. I., Zaanen, J., and Andersen, O. K. (1991) *Phys. Rev.* **B44**, 943.

Annett, J. F. (1996) *Contemp. Phys.* **36**, 423.

Annett, J. F., Goldenfeld, N., and Leggett, A. J. (1990) in *Physical Properties of High Temperature Superconductors*, Vol. 2, ed. D. M. Ginsberg (World Scientific, Singapore) p. 571.

Annett, J. F., Goldenfeld, N., and Renn, S. R. (1996) in *Physical Properties of High Temperature Superconductors*, Vol. 5, ed. D. M. Ginsberg (World Scientific, Singapore) p. 375.

Aprelev, A. M., Grazhulis, V. A., Ionov, A. M., and Lisachenko, A. A. (1994) *Physica C* **235–240**, 1015.

Arko, A. J., List, R. S., Bartlett, R. J., Cheong, S.-W., Fisk, Z., Thompson, J. D., Olson, C. G., Yang, A.-B., Liu, R., Gu, C., Veal, B. W., Liu, J. Z., Paulikas, A. P. Vandervoort, K., Claus, H., Campuzano, J. C., Schirber, J. E., and Shinn, N. D. (1989) *Phys. Rev.* **B40**, 2268.

Arnold, G. B., Mueller, F. M., and Swihart, J. C. (1991) *Phys. Rev. Lett.* **67**, 2569.

Aryasetiawan, F. and Gunnarsson, O. (1995) *Phys. Rev. Lett.* **74**, 3221.

Aryasetiawan, F. and Karlsson, K. (1996) *Phys. Rev.* **B54**, 5353.

Asada, S. and Sugano, S. (1976) *J. Phys. Soc. Jpn.* **41**, 1291.

Aspnes, D. E. (1982) *J. Opt. Soc. Amer.* **72**, 1056.

Baalmann, A., Neumann, M., Braun, W., and Radlik, W. (1985) *Solid State Commun.* **54**, 583.

Baalmann, A., Przybylski, H., Neumann, M., and Borstel, G. (1983) *Physica Scripta* **T4**, 148.

Babbe, N., Drube, W., Schäfer, I., and Skibowski, M. (1985) *J. Phys. E* **18**, 158.

Bachrach, R. Z. (ed.) (1992) *Synchrotron Radiation Research, Advances in Surface and Interface Science*, Vol. 1 (Plenum Press, New York).

Bachrach, R. Z., Brown, F. C., and Hagström, S. M. B. (1975) *J. Vac. Sci. Technol.* **12**, 309.

Baer, Y. (1984) in *Handbook on the Physics and Chemistry of the Actinides*, Vol. I, eds. A. J. Freeman and G. H. Lander (North-Holland, Amsterdam) p. 271.

Baer, Y. and Schneider, W.-D. (1987) in *Handbook on the Physics and Chemistry of Rare Earths*, Vol. 10, eds. K. A. Gschneidner, L. Eyring, and S. Hüfner (North-Holland, Amsterdam) p. 1.

Bala, J., Oleś, A. M., and Zaanen, J. (1994) *Phys. Rev. Lett.* **72**, 2600.

Baltzer, P., Wannberg, B., and Carlsson Göthe, M. (1991) *Rev. Sci. Instr.* **62**, 643.

Balzarotti, A., De Crescenzi, M., Motta, N., Patella, F., Sgarlata, A., Paroli, P., Balestrino, G., and Marinelli, M. (1991) *Phys. Rev.* **B43**, 11500.

Balzarotti, A., Fanfoni, M., Patella, F., Sgarlata, A., and Sperduti, R. (1994) *Phys. Rev.* **B49**, 9103.

Balzarotti, A., Fanfoni, M., Patella, F., Sperduti, R., and Licci, F. (1995) *Physica C* **250**, 382.

Balzarotti, A., Patella, F., Arciprete, F., Motta, N., and De Crecenzi, M. (1992) *Physica C* **196**, 79.

Bandarage, G. and Lucchese, R. R. (1993) *Phys. Rev.* **A47**, 1989.

Bansil, A., Lindroos, M., Gofron, K., and Campuzano, J. C. (1993) *J. Phys. Chem. Solids* **54**, 1185.

Bansil, A., Lindroos, M., Gofron, K., Campuzano, J. C., Ding, H., Liu, R., and Veal, B. W. (1992) *J. Phys. Chem. Solids* **53**, 1541.

Baraldi, A. and Dhanak, V. R. (1994) *J. Electron Spectrosc. Relat. Phenom.* **67**, 211.

Barr, T. L. and Brundle, C. R. (1992) *Phys. Rev.* **B46**, 9199.

Bart, F., Guittet, M. J., Henriot, M., Thromat, N., Gautier, M., and Duraud, J. P. (1994) *J. Electron Spectrosc. Relat. Phenom.* **69**, 245.

Bartynsky R. A., Yang, S., Hulbert, S. L., Kao, C.-C., Weinert, M., and Zehner, D. M. (1992) *Phys. Rev. Lett.* **68**, 2247.

Bassani, F. and Pastori Parravicini, G. (1975) *Electronic States and Optical Transitions in Solids* (Pergamon Press, Oxford)

Bassani, F. and Altarelli, M. (1983) in *Handbook on Synchrotron Radiation* Vol. 1a, ed. E.-E. Koch (North-Holland, Amsterdam) p. 463.

Bassani, G. F. (1966) in *The Optical Properties of Solids, Proc. of the International School of Physics "E. Fermi," Course 34*, ed. J. Tauc (Academic Press, New York) p. 33.

Beamson, G., Porter, H. Q., and Turner, D. W. (1980) *J. Phys. E: Sci. Instr.* **13**, 64.

Becker, H., Dietz, E., Gerhardt, U., and Angermüller, H. (1975) *Phys. Rev.* **B12**, 2084.

Behner, H., Rauch, W., and Gornik, E. (1992) *Appl. Phys. Lett.* **61**, 1465.

Benbow, R. L. (1980) *Phys. Rev.* **B22**, 3775.

Berglund, C. N. and Spicer, W. E. (1964a) *Phys. Rev.* **136**, A1030.

Berglund, C. N. and Spicer, W. E. (1964b) *Phys. Rev.* **136**, A1044.

Bernhoff, H. J., Tsushima, K., and Nicholls, J. M. (1990) *Europhys. Lett.* **13**, 537.

Bernhoff, H., Qvarford, M., Söderholm, S., Flodström, A. S., Andersen, J. N., Nyholm, R., Karlsson, U. O., and Lindau, I. (1991) *Physica C* **180**, 120.

Bernhoff, H., Qvarford, M., Söderholm, S., Nyholm, R., Karlsson, U. O., Lindau, I., and Flodström, A. S. (1993) *Physica C* **218**, 103.

Beschoten, B., Sadewasser, S., Güntherodt, G., and Quitmann, C. (1996) *Phys. Rev. Lett.* **77**, 1837.

Beyers, R. and Shaw, T. M. (1989) in *Solid State Physics* **42**, eds. H. Ehrenreich and D. Turnbull, p. 135.

Bickers, N. E., Scalapino, D. J., Collins, R. T., and Schlesinger, Z. (1990) *Phys. Rev.* **B42**, 67.

Björneholm, O., Andersen, J. N., Wigren, C., Nilsson, A, Nyholm R., and Mårtensson, N. (1990) *Phys. Rev.* **B41**, 10408.

Blandin, P., Massidda, S., Barbiellini, B., Jarlborg, T., Lerch, P., Manuel, A. A., Hoffmann, L., Gauthier, M., Sadowski, W., Walker, E., Peter, M., Yu, J., and Freeman, A. J. (1992) *Phys. Rev.* **B46**, 390.

Bloembergen, N. (1965) *Nonlinear Optics* (W. A. Benjamin, New York).

Bocquet, A. E. and Fujimori, A. (1996) *J. Electron Spectrosc. Relat. Phenom.* **82**, 87.

Bok, J. and Bouvier, J. (1997) *Physica C* **274**, 1.

Bokor, J. and Fann, W. S. (1996) in *Laser Spectroscopy and Photochemistry on Metal Surfaces, Part I*, eds. H.-L. Dai. and W. Ho (World Scientific, Singapore) p. 327.

Bokor, J., Haight, R., Storz, R. H., Stark, J., Freeman, R. R., and Bucksbaum, P. H. (1985) *Phys. Rev.* **B32**, 3669.

Bokor, J., Storz, R. H., Freeman, R. R., and Bucksbaum, P. H. (1986) *Phys. Rev. Lett.* **57**, 881.

Borkowski, L. S. and Hirschfeld, P. J. (1994) *Phys. Rev.* **B49**, 15404.

Borstel, G. and Thörner, G. (1987) *Surf. Sci. Rep.* **8**, 1.

Borstel, G., Neumann, M., and Wöhlecke, M. (1981) *Phys. Rev.* **B23**, 3121.

Borstel, G., Przybylsky, H., Neumann, M., and Wöhlecke, M. (1982) *Phys. Rev.* **B25**, 2006.

Böttner, R., Schroeder, N., Dietz, E., Gerhardt, U., Assmus, W., and Kowalewski, J. (1990) *Phys. Rev.* **B41**, 8679.

Braun, J. (1996) *Rep. Prog. Phys.* **59**, 1267.

Brenig, W. (1995) *Phys. Rep.* **251**, 153.

Brown, G. and Lavender, W. (1991) in *Handbook on Synchrotron Radiation*, Vol. 3, eds. G. S. Brown and D. E. Moncton (North-Holland, Amsterdam) p. 37.

Brown, I. D. (1989a) *J. Solid State Chem.* **82**, 122.

Brown, I. D. (1989b) *J. Solid State Chem.* **90**, 155.

Brown, I. D. and Altermatt, D. (1985) *Acta Cryst. B* **41**, 244.

Brundle, C. R. and Baker, A. D. (eds.) (1977) *Electron Spectroscopy: Theory, Techniques, and Applications*, Vol. 1 (Academic Press, New York).

Brundle, C. R. and Baker, A. D. (eds.) (1978) *Electron Spectroscopy: Theory, Techniques, and Applications*, Vol. 2 (Academic Press, New York).

Brundle, C. R., and Fowler, D. E. (1993) *Surf. Sci. Rep.* **19**, 143.

Burns, G. (1992) *High-temperature Superconductivity* (Academic Press, New York).

Butcher, P. N. (1951) *Proc. Roy. Soc.* (London) A**64**, 50.

Butcher, P. N. and Cotter D. (1990) *The Elements of Nonlinear Optics* (Cambridge University Press, Cambridge).

Butorin, S. M., Guo, J.-H., Wassdahl, N., Skytt, P., Nordgren, J., Ma, Y., Ström, C., Johansson, L.-G., and Qvarford, M. (1995) *Phys. Rev.* **B51**, 11915.

Calandra, C. and Manghi, F. (1992) *Phys. Rev.* **B45**, 5819.

Calandra, C. and Manghi, F. (1994) *J. Electron Spectrosc. Relat. Phenom.* **66**, 453.

Calandra, C., Goldoni, G., Manghi, F., and Magri, R. (1989) *Surf. Sci.* **211/212**, 1127.

Calandra, C., Manghi, F., and Minerva, T. (1991) *Phys. Rev.* **B43**, 3671.

Calandra, C., Manghi, F., and Minerva, T. (1992) *Phys. Rev.* **B46**, 3600.

Campagna, M., Wertheim, G. K., and Baer, Y. (1979) in *Photoemission in Solids II*, eds. L. Ley and M. Cardona (Springer-Verlag, Berlin) p. 217.

Campuzano, J. C., Ding, H., Norman, M. R., Randeria, M., Bellman, A. F., Yokoya. T., Takahashi, Y., Katayama-Yoshida, H., Mochiku, T., and Kadowaki, K. (1996) *Phys. Rev.* **B53**, R14737.

Campuzano, J. C., Gofron, K., Liu, R., Ding, H., Veal, B. W., and Dabrowski, B. (1992) *J. Phys. Chem. Solids* **53**, 1577.

Campuzano, J. C., Jennings, G., Faiz, M., Beaulaigue, L., Liu, J. Z., Veal, B. W., Paulikas, A. P., Vandervoort, K., Claus, H., List, R. S., Arko, A. J., Bartlett, R. J., Olson, C. G., Yang,

A.-B., Liu, R., and Gu, C. (1990b) *J. Electron Spectrosc. Relat. Phenom.* **52**, 363.

Campuzano, J. C., Jennings, G., Faiz, M., Beaulaigue, L., Veal, B. W., Liu, J. Z., Paulikas, A. P., Vandervoort, K., and Claus, H. (1990a) *Phys. Rev. Lett.* **64**, 2308.

Campuzano, J. C., Smedskjaer, L. C., Benedek, R., Jennings, G., and Bansil, A. (1991) *Phys. Rev.* **B43**, 2788.

Capasso, C., Ng, W., Ray-Chaudhuri, A. K., Liang, S. H., Cole, R. K., Guo, Z. Y., Wallace, J., Cerrina, F., Underwood, J., Perera, R., Kortright, J., De Stasio, G., and Margaritondo, G. (1993) *Surf. Sci.* **287/288**, 1046.

Capasso, C., Ray-Chaudhuri, A. K., Ng, W., Liang, S., Cole, R. K., Wallace, J., Cerrina, F., Margaritondo, G., Underwood, J. H., Kortright, J. B., and Perera, R. C. C. (1991) *J. Vac. Sci. Technol. A* **9**, 1248.

Cardona, M. and Güntherodt, G. (1975–1991) *Light Scattering in Solids*, Vols. 1–6 (Springer-Verlag, Heidelberg).

Cardona, M. and Ley, L. (eds.) (1978) *Photoemission in Solids I* (Springer-Verlag, Berlin).

Cardona, M., Strohm, T., and Kirchner, J. (1996) in *Spectroscopic Studies of Superconductor,* eds. I. Bozovic and D. van der Marel (SPIE, Bellingham) p. 182.

Carlson, T. A. (1975) *Photoelectron and Auger Spectroscopy* (Plenum Press, New York).

Carlson, T. A. and Krause, M. O. (1965) *Phys. Rev.* **140**, A1057.

Caroli, C., Lederer-Rozenblatt, D., Roulet, B., and Saint-James, D. (1973) *Phys. Rev.* **B8** 4552.

Cava, R. J., Hewatt, A. W., Hewatt, E. A., Batlogg, B., Marezio, M., Rabe, K. M., Krajewski, J., J., Peck, W. F., and Rupp, L. W. (1990) *Physica C* **165**, 419.

Cerrina, F. and Lai, B. (1986) *Nucl. Instr. and Meth.* **A246**, 337.

Cerrina, F., Myron, J. R., and Lapeyre, G. J. (1984) *Phys. Rev.* **B29**, 1798.

Chakravarty, S. (1995) *Phys. Rev. Lett.* **74**, 1885.

Chakravarty, S., Sudbø, A., Anderson, P. W., and Strong, S. (1994) *Science* **261**, 337.

Chan, L. P., Harshman, D. R., Lynn, K. G., Massidda, S., and Mitzi, D. B. (1991) *Phys. Rev. Lett.* **67**, 1350.

Chang, Y., Tang, M., Hwu, Y., Onellion, M., Huber, D. L., Margaritondo, G., Morris, P. A., Bonner, W. A., Tarascon, J. M., and Stoffel, N. G. (1989b) *Phys. Rev.* **B39**, 7313.

Chang, Y., Tang, M., Zanoni, R., Joynt, R., Huber, D. L., Margaritondo, G., Morris, P. A.,

Bonner, W. A., Tarascon J. M., and Stoffel, N. G. (1989a) *Phys. Rev.* **B39**, 4740.

Chang, Y., Tang, M., Zanoni, R., Joynt, R., Huber, D. L., Margaritondo, G., Morris, P. A., Bonner, W. A., Tarascon J. M., and Stoffel, N. G. (1989c) *Phys. Rev. Lett.* **63**, 101.

Chauvet, G. and Baptist, R. (1981) *J. Electron Spectrosc. Relat. Phenom.* **24**, 255.

Chen, C. (1990a) *Phys. Rev. Lett.* **64**, 2176.

Chen, C. (1992a) *J. Phys. Condens. Matter* **4**, 9855.

Chen, C. (1992b) *Phys. Rev.* **B45**, 13811.

Chen, C. (1993) *Phys. Rev.* **B48**, 1318.

Chen, C. and Falicov, L. M. (1989) *Phys. Rev.* **B40**, 3560.

Chen, C. H. (1990b) in *Physical Properties of High-Temperature Superconductors*, Vol. 2, ed. D. M. Ginsberg (World Scientific, Singapore) p. 199.

Chen, C. T. and Smith, N. V. (1987) *Phys. Rev.* **B35**, 5407.

Chen, J. M., Chung, S. C., and Liu, R. S. (1996) *Solid State Commun.* **99**, 493.

Chen, X., Qian, Y., Chen, Z., Lin, C., Yang, L., Mao, Z., and Zhang, Y. (1992) *Phys. Rev.* **B46**, 9181.

Chen, X. K., Altendorf, E., Irwin, J. C., Liang, R., and Hardy, W. N. (1993) *Phys. Rev.* **B48**, 10530.

Chen, X. K., Liang, R., Hardy, W. N., and Irwin, J. C. (1994b) *J. Supercond.* **7**, 435.

Chen, X. K., Irvin, J. C., Trohdahl, H. J., Kimura, T., and Kishio, K. (1994a) *Phys. Rev. Lett.* **73**, 3290.

Chen, Y., and Kroha, J. (1992) *Phys. Rev.* **B46**, 1332.

Chiaia, G., Qvarford, M., Lindau, I., Söderholm, S., Karlsson, U. O., Flodström, S. A., Leonyuk, L., Nilsson, A., and Mårtensson, N. (1995) *Phys. Rev.* **B51**, 1213.

Childs, T. T., Royer, W. A., and Smith, N. V. (1984) *Rev. Sci. Instr.* **55**, 1613.

Ching, W. Y., Xu, Y.-N., and Wong, K. W. (1989) *Phys. Rev.* **B40**, 7684.

Cini, M. (1976) *Solid State Commun.* **20**, 605.

Cini, M. (1977) *Solid State Commun.* **24**, 681.

Cini, M. (1978) *Phys. Rev.* **B17**, 2788.

Citrin, P. H., Wertheim, G. K., and Baer, Y. (1983) *Phys. Rev.* **B27**, 3160.

Claessen, R., Anderson, R. O., Allen, J. W., Olson, C. G., Janowitz, C., Ellis, W. P. Harm, S., Kalning, M., Manzke, R., and Skibowski, M. (1992) *Phys. Rev. Lett.* **69**, 808.

Claessen, R., Mante, G., Huss, A., Manzke, R., Skibowski, M., Wolf, Th., and Fink, J, (1991) *Phys. Rev.* **B44**, 2399.

Claessen, R., Manzke, R., Carstensen, H., Burandt, B., Buslaps, T., Skibowski, M., and Fink, J. (1989) *Phys. Rev.* **B39**, 7316.

Coffey, D. and Coffey, L. (1993a) *Phys. Rev. Lett.* **70**, 1529.

Coffey, D. and Coffey, L. (1996) *Phys. Rev.* **B53**, 15292.

Coffey, L. and Coffey, D. (1993b) *Phys. Rev.* **B48**, 4184.

Coffey, L., Lacy, D., Kouznetsov, K., and Erner, A. (1997) *Phys. Rev.* **B56**, 5590.

Collins, I. R., Law, A. R., and Andrews, P. T. (1988) *J. Phys. C* **21**, L655.

Coluzza, C., Almeida, J., dell'Orto, T., Gozzo, F., Alméras, P., Berger H., Bouvet, D., Dutoit, M., Contarini, S., and Margaritondo, G. (1994) *J. Appl. Phys.* **76**, 3710.

Con Foo, J. A., Stampfl, A. J. P., Ziegler, A., Mattern, B., Hollering, M., Denecke, R., Ley, L., Riley, J. D., and Leckey, R. C. G. (1996) *Phys. Rev.* **B53**, 9649.

Connerade, J. P., Esteva, J. M., and Karnatak, R. C. (1987) *Giant Resonances in Atoms, Molecules, and Solids, NATO ASI Series, Series B: Physics*, Vol. 151 (Plenum Press, New York).

Cooper, J. W. (1962) *Phys. Rev.* **128**, 681.

Cooper, J. W. (1975) *Photoionization of Inner-Shell Electrons* in *Atomic Inner-shell Processes*, Vol. I, ed. B. Crasemann (Academic Press, New York), Ch. 3, p. 159.

Cooper, S. L., Klein, M. V., Pazol, B. G., Rice, J. P., and Ginsberg, D. M. (1988) *Phys. Rev.* **B37**, 5920.

Cornwell, J. F.(1966) *Phys. Kondens. Materie* **4**, 327.

Courths, R. (1981) *Solid State Commun.* **40**, 529.

Courths R., Bachelier, B., Cord, V., and Hüfner, S. (1981) *Solid State Commun.* **40**, 1059.

Courths, R., Cord, V. Wern, H., and Hüfner, S. (1983) *Physica Scripta* **T4**, 144.

Courths, R. and Hüfner, S. (1984) *Physics Reports* **112**, 53.

Courths, R., Wern H., Leschik, G., and Hüfner, S. (1989) *Z. Phys. B* **74**, 233.

Cummins, T. R. and Egdell, R. G. (1993) *Phys. Rev.* **B48**, 6556.

Cyrot, M., and Pavuna, D. (1992) *Introduction to Superconductivity and High-T_c Materials* (World Scientific, Singapore).

Dagotto, E. (1994) *Rev. Mod. Phys.* **66**, 763.

Dahm, T., Manske, D., and Tewordt, L. (1996) *Phys. Rev.* **B54**, 602.

Daniels, J., von Festenberg, C., Raether, H., and Zeppenfeld, K. (1970) *Springer Tracts in Modern Physics* **54**, 77.

Das, P. and Kar, N. (1995) *Mod. Phys. Lett.* **B15**, 947.

Davis, L. C. and Feldkamp, L. A. (1979a) *Phys. Rev. Lett.* **43**, 151.

Davis, L. C. and Feldkamp, L. A. (1979b) *J. Appl. Phys.* **50**, 1944.

Davis, L. C. and Feldkamp, L. A. (1980) *Solid State Commun.* **34**, 141.

Davis, L. C. and Feldkamp, L. A. (1981) *Phys. Rev.* **B23**, 6239.

Davison, S. G. and Levine, J. D. (1970) in *Solid State Physics* **25**, eds. F. Seitz and D. Turnbull, p.1.

Davison, S. G., and Stęślicka, M. (1992) *Basic Theory of Surface States* (Clarendon Press, Oxford).

De Benedetti, S., Cowan, C. E., Konneker, W. R., and Primakoff, H. (1950) *Phys. Rev.* **77**, 205.

de Fontaine, D., Ceder, G., and Asta, M. (1990) *Nature* (London) **343**, 544.

de Gennes, P. G. (1966) *Superconductivity of Metals and Alloys* (W. A. Benjamin, New York).

Dehmer, J. L. and Starace, J. F. (1972) *Phys. Rev.* **B5**, 1792.

Dehmer, J. L., Starace, A. F., Fano, U., Sugar, J., and Cooper, J. W. (1971) *Phys. Rev. Lett.* **26**, 1521.

Delwiche, J., Hubin-Franskin, M.-J., Furlan, M., Collin, J. E., Roy, D., Leclerc, B., Roy, P., Poulin, A., and Carette, J.-D. (1982) *J. Electron Spectrosc. Relat. Phenom.* **28**, 123.

Desjonquères, M. C. and Spanjaard, D. (1993) *Concepts in Surface Physics, Springer Series in Surface Science*, Vol. 30 (Springer-Verlag, Berlin).

Dessau, D. S., Shen, Z.-X., and Marshall, D. M. (1993b) *Phys. Rev. Lett.* **71**, 4278.

Dessau, D. S., Shen, Z.-X., and Wells, B. O. (1994) *Phys. Rev. Lett.* **73**, 3045.

Dessau, D. S., Shen, Z.-X., King, D. M., Marshall, D. S., Lombardo, L. W., Dickinson, P. H., Loeser, A. G., DiCarlo, J., Park, C.-H., Kapitulnik, A., and Spicer, W. E. (1993a) *Phys. Rev. Lett.* **71**, 2781.

Dessau, D. S., Shen, Z.-X., Wells. B. O., King, D. M., Spicer, W. E., Arko, A. J., Lombardo L. W., Mitzi, D. B., and Kapitulnik, A. (1992) *Phys. Rev.* **B45**, 5095.

Dessau, D. S., Shen, Z.-X., Wells. B. O., Spicer, W. E., and Arko, A. J. (1991b) *J. Phys. Chem. Solids* **52**, 1401.

Dessau, D. S., Shen, Z.-X., Wells. B. O., Spicer, W. E., List, R. S., Arko, A. J., Bartlett, R. J., Fisk, Z., Cheong, S.-W., Mitzi, D. B., Kapitulnik, A., and Schirber, J. E. (1990) *Appl. Phys. Lett.* **57**, 307.

Dessau, D. S., Wells, B. O., Shen, Z.-X., Spicer, W. E., Arko, A. J., List, R. S., Mitzi, D. B., and Kapitulnik, A. (1991a) *Phys. Rev. Lett.* **66**, 2160.

De Stasio, G. and Margaritondo, G. (1994) in *New Directions in Research with Third-Generation Soft X-Ray Synchrotron Radiation Sources*, ed. A. S. Schlachter and F. J. Wuilleumier (Kluwer Academic Publishers, Dordrecht) p. 299.

Devereaux, T. P. and Einzel, D. (1995) *Phys. Rev.* **B51**, 16336.

Devereaux, T. P., Einzel, D., Stadlober, B., Hackl, R., Leach, D. H., and Neumeier, J. J. (1994) *Phys. Rev. Lett.* **72**, 396.

Dietz, E. and Eastman, D. E. (1978) *Phys. Rev. Lett.* **41**, 1674.

Dietz, E. and Himpsel, F. J. (1979) *Solid State Commun.* **30**, 325.

Ding, H., Bellman, A. F., Campuzano, J. C., Randeria, M., Norman, M. R., Yokoya, T., Takahashi, T., Katayama-Yoshida, H., Mochiku, T., Kadowaki, K., Jennings, G., and Brivio, G. P. (1996a) *Phys. Rev. Lett.* **76**, 1533.

Ding, H., Campuzano, J. C., Bellman, A. F., Yokoya, T., Norman, M. R., Randeria, M., Takahashi, T., Katayama-Yoshida, H., Mochiku, T., Kadowaki, K., and Jennings, G. (1995) *Phys. Rev. Lett.* **74**, 2784 and **75**, 1425 (E).

Ding, H., Campuzano, J. C., Bellman, A., Yokoya, T., Norman, M. R., Takahashi, T., Katayama-Yoshida, H., Mochiku, T, and Kadowaki, K. (1994b) *Abstracts of the 1994 Synchrotron Radiation Center User's Group Meeting*, p. 123 (unpublished).

Ding, H., Campuzano, J. C., Gofron, K., Gu, C., Liu, R., Veal, B. W., Jennings, G. (1994a) *Phys. Rev.* **B50**, 1333.

Ding, H., Norman, M. R., Campuzano, J. C., Randeria, M., Bellman, A. F., Yokoya, T., Takahashi, T., Mochiku, T., and Kadowaki, K. (1996c) *Phys. Rev.* **B54**, R9678.

Ding, H., Norman, M. R., Giapintzakis, J., Campuzano, J. C., Claus, H., Wühl, H., Randeria, M., Bellman, A., Yokoya, T., Takahashi, T., Mochiku, T., Kadowaki, K., and Ginsberg, D. M. (1996d) in *Spectroscopic Studies of Superconductors*, eds. I. Bozovic and D. van der Marel (SPIE, Bellingham) p. 497.

Ding, H., Norman, M. R., Yokoya, T., Takeuchi, T., Randeria, M., Campuzano, J. C., Takahashi, T., Mochiku, T., and Kadowaki, K. (1997) *Phys. Rev. Lett.* **78**, 2628.

Ding, H., Yokoya, T., Campuzano, J. C., Takahashi, T., Randiera, M., Norman, M. R., Mochiku, T., Kadowaki, K., and Giapintzakis, J. (1996b) *Nature* (London) **382**, 51.

Doniach, S. and Šunjić, M. (1970) *J. Phys. C* **3**, 285.

Dose, V. (1983) *Prog. Surf. Sci.* **13**, 225.

Dose, V., Fauster, T., and Scheidt, H. (1981) *J. Phys. F* **11**, 1801.

Drube, R., Himpsel, F. J., Chandrashekhar, and Shafer, M. W. (1989) *Phys. Rev.* **B39**, 7328.

Drube, R., Noffke, J., Schneider R., Rogizik, J., and Dose, V. (1992) *Phys. Rev.* **B45**, 4390.

Dunham, D., Desloge, D. M., Rempfer, G. F., Skoczylas, W. P., and Tonner, B. P. (1994) *Nucl. Instr. and Meth.* **A347**, 441.

Eastman, D. E. (1971) *J. Phys.* (Paris) **C1**, 293.

Eastman, D. E. (1972) in *Techniques of Metal Research*, Vol. VI, part 1, ed. E. Passaglia, series ed. R. P. Bunshah (Interscience Publishers, New York) p. 411.

Eastman, D. E. and Cashion, J. K. (1970) *Phys. Rev. Lett.* **24**, 310.

Eastman, D. E. and Freeouf, J. L. (1975) *Phys. Rev. Lett.* **34**, 395.

Eastman, D. E., Donelon, J. J., Hien, N. C., and Himpsel, F. J. (1980) *Nucl. Instr. and Meth.* **172**, 327.

Eastman, D. E., Knapp, J. A., and Himpsel, F. J. (1978) *Phys. Rev. Lett.* **41**, 825.

Eberhardt, W. and Himpsel, F. J. (1980) *Phys. Rev.* **B21**, 5572; (E), ibid, **B22**, 5014.

Eberhardt, W. and Plummer, E. W. (1980) *Phys. Rev.* **B21**, 3245.

Eckardt, H., Fritsche, L., and Noffke, J. (1984) *J. Phys. F: Met. Phys.* **14**, 97.

Edwards, H. L., Markert, J., T., and de Lozanne, A. L. (1992) *Phys. Rev. Lett.* **69**, 2967.

Egami, T. and Billinge, S. J. L. (1994) *Progress in Materials Science* **38**, 359.

Egami, T. and Billinge, S. J. L. (1996) in *Physical Properties of High-Temperature Superconductors*, Vol. 5, ed. D. M. Ginsberg (World Scientific, Singapore) p. 265.

Egelhoff, W. F. (1984) *Phys. Rev.* **B29**, 4769.

Egelhoff, W. F. (1987) *Surf. Sci. Rep.* **6**, 253.

Eisaki, H., Takagi,H., Uchida, S., Matsubara, H.,Suga, S., Nakamura, M., Yamaguchi, K., Misu, A., Namatame, H., and Fujimori, A. (1990) *Phys. Rev.* **B41**, 7188.

Emery, V. J. and Kivelson, S. A. (1993) *Physica C* **209**, 597.

Emery, V. J. and Kivelson, S. A. (1995) *Nature* (London) **374**, 434.

Emery, V. J., Kivelson, S. A., and Lin, H. Q. (1990) *Phys. Rev. Lett.* **64**, 475.

Engelhardt, H. A., Bäck, W., and Menzel, D. (1981) *Rev. Sci. Instr.* **52**, 835.

Erdman, P. W. and Zipf, E. C. (1982) *Rev. Sci. Instr.* **53**, 225.

Erskine, J. L. (1995) in *Atomic, Molecular, and Optical Physics: Charged Particles*, eds. F. B.

Dunning and R. G. Hulet (Academic Press, San Diego) p. 209.

Ertl, G. and Küppers, J. (1985) *Low-Energy Electrons and Surface Chemistry*, 2nd edn. (VCH Verlagsgesellschaft, Weinheim).

Eskes H. and Sawatzky, G. A. (1991) *Phys. Rev.* **B43**, 119.

Eskes, H., Meinders, M. B. J., and Sawatzky, G. A. (1991) *Phys. Rev. Lett.* **67**, 1035.

Eskes, H., Tjeng, L. H., and Sawatzky, G. A. (1990) *Phys. Rev.* **B41**, 288.

Fadley, C. S. (1984) *Prog. Surf. Sci.* **16**, 275.

Fadley, C. S. (1992) in *Synchrotron Radiation Research – Advances in Surface and Interface Science*, Vol. 1, ed. R. Z. Bachrach (Plenum Press, New York) p. 422.

Fann, W. S., Storz, R., Tom, H. W. K., and Bokor, J. (1992) *Phys. Rev.* **B46**, 13592.

Fano, U. (1961) *Phys. Rev.* **124**, 1866.

Fano, U. and Cooper, J. W. (1965) *Phys. Rev.* **137**, A1364.

Fano, U. and Cooper, J. W. (1968) *Rev. Mod. Phys.* **40**, 441 (addendum in *Rev. Mod. Phys.* **41**, 724 (1969)).

Fauster, T., Himpsel, F. J., Donelon, J. J., and Marx, A. (1993) *Rev. Sci. Instr.* **54**, 62.

Fauster, T., Schneider, R., Dürr, H., Engelmann, G., and Taglauer, E. (1987) *Surf. Sci.* **189/190**, 610.

Fauster, T., Straub, D., Donelon, J. J., Grimm, D., Marx, A., and Himpsel, F. J. (1985) *Rev. Sci. Instr.* **56**, 1212.

Fehrenbacher, R. and Norman, M. R. (1994) *Phys. Rev.* **B50**, 3495.

Fehrenbacher, R. (1996) *Phys. Rev.* **B54**, 6632.

Feibelman, P. J. and Eastman, D. E. (1974) *Phys. Rev.* **B10**, 4932.

Feuerbacher, B. and Willis, R. F. (1976) *J. Phys. C* **9**, 169

Feuerbacher, B., Fitton, B. and Willis, R. F. (eds.) (1978) *Photoemission and the Electronic Properties of Surfaces* (John Wiley and Sons, New York).

Fiermans, L., Hoogewijs, R., and Vennik, J. (1975) *Surf. Sci.* **47**, 1.

Fink, J., Nücker, N., Pellegrin, E., Romberg, H., Alexander, M., and Knupfer, M. (1994) *J. Electron Spectrosc. Relat. Phenom.* **66**, 395.

Flavell, W. R., Hoad, D. R. C., Roberts, A.J., Egdell, R. G., Fletcher, I. W., and Beamson, G. (1993) *J. Alloys and Compounds* **195**, 535.

Flavell, W. R., Hollingworth, J., Howlett, J. F., Thomas, A. G., Sarker, M. M., Squire, S., Hashim, Z., Mian, M., Wincott, P. L., Teehan, D., Downes, S., and Hancock, F. E. (1995) *J. Synchrotron Rad.* **2**, 264.

Flavell, W. R., Laverty, J. H., Law, D. S.-L., Lindsay, R., Muryn, C. A., Flipse, C. F. J., Raiker, G. N., Wincott, P. L., and Thornton, G. (1990) *Phys. Rev.* **B41**, 11623.

Flodström, A., Nyholm, R., and Johansson, B. (1992) in *Synchrotron Radiation Research, Advances in Surface and Interface Science*, Vol. 1, ed. R. Z. Bachrach (Plenum Press, New York) p. 199.

Fong, H. F., Keimer, B., Anderson, P. W., Reznik, D., Doğan, F., and Aksay, I. A. (1995) *Phys. Rev. Lett.* **75**, 316.

Forstmann, F. (1978) in *Photoemission and the Electronic Properties of Surfaces*, eds. B. Feuerbacher, B. Fitton, and R. F. Willis (John Wiley and Sons, New York), p. 45.

Fowler, C. M., Freeman, B. L., Hults, W. L., King, J. C., Mueller, F. M., and Smith, J. L. (1992) *Phys. Rev. Lett.* **68**, 534.

Fowler, D. E., Brundle, C. R., Lerczak, J., and Holtzberg, F. (1990) *J. Electron Spectrosc. Relat. Phenom.* **52**, 323.

Frank, G., Ziegler, Ch., and Göpel, W. (1991) *Phys. Rev.* **B43**, 2828.

Fraxedas, J., Trodahl, J., Gopalan, S., Ley, L., and Cardona, M. (1990) *Phys. Rev.* **B41**, 10068.

Freeman, A. J., Min, B. I., and Norman, M. R. (1987) in *Handbook on the Physics and Chemistry of Rare Earths*, Vol. 10, eds. K. A. Gschneidner, L. Eyring, and S. Hüfner (North-Holland, Amsterdam) p. 165.

Freeman, A. J., Yu, J., Massidda, S., and Koelling, D. D. (1989) *Physica B* **148**, 212.

Freeouf, J., Erbudak, M., and Eastman, D. E. (1973) *Solid State Commun.* **13**, 771.

Freidl, B., Thomsen, C., and Cardona, M. (1990) *Phys. Rev. Lett.* **65**, 915.

Fuggle, J. C. (1981) in *Electron Spectroscopy: Theory, Techniques, and Applications*, Vol. 4, eds. C. R. Brundle and A. D. Baker (Academic Press, New York) p. 85.

Fuggle, J. C. (1992) in *Unoccupied Electronic State*, eds. J. C. Fuggle and J. E. Inglesfield (Springer-Verlag, Berlin) p. 307.

Fuggle, J. C., Fink, J., and Nücker, N. (1988) *Intl. J. Mod. Phys.* **1**, 1185.

Fujimori, A. Uchida, S., Eisaki, H., Takagi, H., Uchida, S., and Sato, M. (1989) *Phys. Rev.* **B40**, 7303.

Fujimori, A. and Minami, F. (1984) *Phys. Rev.* **B30**, 957.

Fujimori, A., Takayama-Muromachi, E., Uchida, Y., and Okai, B. (1987) *Phys. Rev.* **B35**, 8814.

Fujimori, A., Tokura, Y., Eisaki, H., Takagi, H., Uchida, S., and Takayama-Muromachi, E. (1990) *Phys. Rev.* **B42**, 325.

Fukuda, Y., Nagoshi, M., Sanada, N., Pooke, D., Kishio, K., Kitazawa, K., Syono, Y., and Tachiki. M. (1992) *J. Phys. Chem. Solids* **53**, 1589.

Gadzuk, J. W. (1974) *Phys. Rev.* **B10**, 5030.

Gadzuk, J. W. and Šunjić, M. (1975) *Phys. Rev.* **B12**, 524.

Garcia-Moliner, F. and Flores, F. (1979) *Introduction to the Theory of Solid Surfaces* (Cambridge University Press, Cambridge).

Ghijsen, J., Tjeng, L. H., van Elp, J., Eskes, H., Westerink, J., Sawatzky, G. A., and Czyżyk, M. T. (1988) *Phys. Rev.* **B38**, 11322.

Ghosh, P. K. (1983) *Introduction to Photoelectron Spectroscopy*, (John Wiley and Sons, New York).

Giapintzakis, J., Ginsberg, D. M., Kirk, M. A., and Ockers, S. (1994) *Phys. Rev.* **B50**, 15967.

Ginsberg, D. M. (ed.) (1989, 1990, 1992, 1994, 1996) *Physical Properties of High-Temperature Superconductors,* Vols. 1–5 (World Scientific, Singapore).

Gobeli, G. W., Allen, F. G., and Kane, E. O, (1964) *Phys. Rev. Lett.* **12**, 94.

Godby, R. W. (1992) in *Unoccupied Electronic States*, eds. J. C. Fuggle and J. E. Inglesfield (Springer-Verlag, Berlin) p. 51.

Gofron, K., Campuzano, J. C., Abrikosov, A. A., Lindroos, M., Bansil. A., Ding, H., Koelling, D., and Dabrowski, B. (1994) *Phys. Rev. Lett.* **73**, 3302.

Gofron, K., Campuzano, J. C., Ding, H., Gu, C., Liu, R., Dabrowski, B., Veal, B. W., Cramer, W., and Jennings, G. (1993) *J. Phys. Chem. Solids* **54**, 1193.

Golden, M. S., Schmelz, H. C., Knupfer, M., Haffner, S., Krabbes, G., Fink, J., Yushanakhai, V. Y., Rosner, H., Hayn, R., Müller, A., and Reichardt, G. (1997) *Phys. Rev. Lett.* **78**, 4107.

Goldmann, A. and Matzdorf, R. (1993) *Prog. in Surf. Sci.* **42**, 331.

Goldmann, A., Altmann, W., and Dose, V. (1991) *Solid State Commun.* **79**, 511.

Goldmann, A., Westphal, D., and Courths, R. (1982) *Phys. Rev.* **B25**, 2000.

Goodman, K. W. and Henrich, V. E. (1994) *Phys. Rev.* **B49**, 4827.

Gopinath, C. S., Hurm N. H., and Subramanian, S. (1995) *Phys. Rev.* **B52**, R9879.

Gourieux, T., Krill, G., Maurer, M., Ravet, M. F., Menny, A., Tolentino, H., and Fontaine, A. (1988) *Phys. Rev.* **B37**, 7516.

Granneman, E. H. A. and van de Wiel, E. J. (1983) in *Handbook on Synchrotron Radiation,* Vol. 1a,

ed. E.-E. Koch (North-Holland, Amsterdam) p. 367.

Grass, M., Braun, J., and Borstel, G. (1993) *Phys. Rev.* **B47**, 15487.

Grass, M., Braun, J., and Borstel, G. (1994) *Phys. Rev.* **B50**, 14827.

Grassmann, A., Strobel, J., Saemann-Ischenko, G., and Johnson, R. L. (1990) *Physica B* **163**, 261.

Grepstad, J. K. and Slagsvold, B. J. (1980) *Solid State Commun.* **34**, 821.

Grioni, M., Berger, H., La Rosa, S., Vebornik, I., Zwick, F., Margaritondo, G., Kelley, R., Ma, J., and Onellion, M. (1996) preprint

Grioni, M., Malterre, D., Dardel, B., Imer, J.-M., Baer, Y., Muller, J., Jorda, J. L., and Pétroff, Y. (1991) *Phys. Rev.* **B43**, 1216.

Gschneidner, K. A., Eyring, L., and Hüfner, S. (eds.) (1987) *Handbook on the Physics and Chemistry of Rare Earths*, Vol. 10, (North-Holland, Amsterdam).

Gu, C., Veal, B. W., Liu, R., Ding, H., Paulikas, A. P., Campuzano, J. C., Kostic, P., Wheeler, R. W., Zhang, H., Olson, C. G., Wu, X., and Lynch, D. W. (1993) *J. Phys. Chem. Solids* **54**, 1177.

Gu, C., Veal, B. W., Liu, R., Paulikas, A. P., Kostic, P., Ding, H., Campuzano, J. C., Andrews, B. A., Blyth, R. I. R., Arko, A. J., Manuel, P., Kaufman, D. Y., and Lanagan, M. T. (1995) *Phys. Rev.* **B51**, 1397.

Gudat, W. and Kunz, C. (1972) *Phys. Rev. Lett.* **29**, 169.

Gudat, W. and Kunz, C. (1979) in *Synchrotron Radiation – Techniques and Applications,* ed. C. Kunz (Springer-Verlag, Berlin) p. 55.

Guillot, C., Ballu, Y., Paigné, J., Lecante, J., Jain, K. P., Thiry, P., Pinchaux, R., Pétroff, Y., and Falicov, L. M. (1977) *Phys. Rev. Lett.* **39**, 1632.

Gunnarsson, O., Allen, J. W., Jepsen, O., Fujiwara, T., Andersen, O. K., Olson, C. G., Maple, M. B., Kang, J.-S., Liu, L. Z., Park, J.-H., Anderson, R. O., Ellis, W. P., Liu, R., Markert, J. T., Dalichaouch, Y., Shen, Z.-X., Lindberg, P. A. P., Wells, B. O., Dessau, D. S., Borg, A., Lindau, I., and Spicer, W. E. (1990a) *Phys. Rev.* **B41**, 4811.

Gunnarsson, O., Andersen, O. K., Jepsen, O., and Zaanen, J. (1989) *Phys. Rev.* **B39**, 1708.

Gunnarsson, O., Jepsen, O., and Shen, Z.-X. (1990b) *Phys. Rev.* **B42**, 8707.

Gurzhi, R. N., Kalinenko, K. N., and Kopeliovich, A. I. (1995) *Phys. Rev.* **B52**, 4744.

Gurzhi, R. N., Kopeliovich, A. I., and Rutkevich, S. B. (1987) *Adv. in Physics* **36**, 221.

Gyorffy, B. L., Staunton, J. B., and Stocks, G. M. (1991) *Phys. Rev.* **B44**, 5190.

Haak, H. W., Sawatzky, G. A., and Thomas, T. D. (1978) *Phys. Rev. Lett.* **41**, 1826.

Haas, S., Moreo, A., and Dagotto, E. (1995) *Phys. Rev. Lett .* **74**, 4281.

Hackl, R., Gläser, W., Müller, P., Einzel, D., and Andres, K. (1988) *Phys. Rev.* **B38**, 7133.

Hackl, R., Krug, G., Nemetschek, R., Opel, M., and Stadlober, B. (1996) in *Spectroscopic Studies of Superconductors,* eds. I. Bozovic and D. van der Marel (SPIE, Bellingham) p. 194.

Hadjarab, F. and Erskine, J. L. (1985) *J. Electron Spectrosc. Relat. Phenom.* **36**, 227.

Haight, R. (1995) *Surf. Sci. Rep.* **21**, 275.

Hansen, E. D., Miller, T., and Chiang, T.-C. (1997) *Phys. Rev.* **B55**, 1871.

Hansson, G. V., Goldberg, B., and Bachrach, R. Z. (1981) *Rev. Sci. Instr.* **52**, 517.

Harada, T., and Kita T. (1980) *Appl. Opt.* **19**, 3987.

Harm, S., Dürig, R., Manzke, R., Skibowski, M., Claessen, R., and Allen, J. W. (1994) *J. Electron Spectrosc. Relat. Phenom.* **68**, 111.

Harper, P. G. and Wherrett, B. S. (1977) *Nonlinear Optics* (Academic Press, London)

Harris, J. M., Shen, Z.-X., White, P. J., Marshall, D. S., Schabel, M. C., Eckstein, J. N., and Bozovic, I. (1996) *Phys. Rev.* **B54**, R15665.

Harris, J. M., White, P. J., Shen, Z., X., Ikeda, H., Yoshizaki, R., Eisaki, H., Uchida, S., Si, W. D., Xiong, J. W., Zhao, Z. X., and Dessau, D. S. (1997) *Phys. Rev. Lett.* **79**, 143.

Harting, E. and Read, F. H. (1976) *Electrostatic Lenses* (Elsevier, New York).

Hasegawa, H., Ikuta, H., and Kitazawa, K. (1992) in *Physical Properties of High-Temperature Superconductors,* Vol. 3, ed. D. M. Ginsberg (World Scientific, Singapore) p. 525.

Hass, K. C. (1989) in *Solid State Physics* **42**, eds. H. Ehrenreich and D. Turnbull, p. 213.

Hatterer, C. J., Mairet, V., Beuran, F. C., Xie, X. M., Xu, X. Z., Laguës, M., Coussot, C., Eustache, B., Adrian, H., Deville Cavelin, C., Indlekofer, G., and Taleb, A. (1996) *Physica C* **261**, 153.

Hayes, W. and Loudon, R. (1978) *Scattering of Light by Crystals* (John Wiley and Sons, New York).

Hazen, R. M. (1990) in *Physical Properties of High-Temperature Superconductors,* Vol. 2, ed. D. M. Ginsberg (World Scientific, Singapore) p. 121.

Hedin, L., and Lundqvist, S. (1969) in *Solid State Physics* **23**, eds. F. Seitz, D. Turnbull and H. Ehrenreich, p. 1.

Heimann, F., Himpsel, F. J. and Eastman, D. E. (1981) *Solid State Commun.* **39**, 219.

Heimann, P., Miosga, H., and Neddermeyer, H. (1979) *Solid State Commun.* **29**, 463.

Heitler, W. (1954) *The Quantum Theory of Radiation,* 3rd edn. (Clarendon Press, Oxford) p. 231.

Henrich, V. E. and Cox, P. A. (1994) *The Surface Science of Metal Oxides* (Cambridge University Press, Cambridge).

Hermanson, J. (1977) *Solid State Commun.* **22**, 9.

Hertel, T., Knoesel, E., Wolf, M., and Ertl, G. (1996) *Phys. Rev. Lett.* **76**, 535.

Hertz, H. (1887) *Wiedemann's Ann. d. Phys.* **31**, 383.

Hervieu, M., Domenges, B., Raveau, B., Post, M., McKinnon, W. R., and Tarascon, J. M. (1989) *Mater. Lett.* **8**, 73.

Hicks, P. J., Daviel, S., Wallbank, B., and Comer, J. (1980) *J. Phys. E: Sci. Instr.* **13**, 713.

Hill, D. M., Meyer, H. M., Weaver, J. H., Flandermeyer, B., and Capone, D. W. (1987) *Phys. Rev.* **B36**, 3979.

Himpsel, F. J. (1983) *Adv. in Physics* **32**, 1.

Himpsel, F. J. (1990) *Surf. Sci. Rep.* **12**, 1.

Himpsel, F. J. and Eastman, D. E. (1978) *Phys. Rev.* **B18**, 5236.

Himpsel, F. J. and Eberhardt, W. (1979) *Solid State Commun .* **29**, 747.

Himpsel, F. J., Chandrashekhar, G. V., McLean, A. B., and Shafer, M. W. (1988) *Phys. Rev.* **B38**, 11946.

Himpsel, F. J., Knapp, J. A., and Eastman, D. E. (1979) *Phys. Rev.* **B19**, 2920.

Höchst, H., Hüfner, S., and Goldmann, A. (1977) *Z. Physik B* **26**, 133.

Hohenberg, P. and Kohn, W. (1964) *Phys. Rev.* **136**, B864.

Hopfield, J. J., Worlock, J. M., and Park, K. (1963) *Phys. Rev. Lett.* **11**, 414.

Hopkinson, J. F. L., Pendry, J. B., and Titterington, D. J. (1980) *Computer Physics Commun.* **19**, 69.

Howells, M. R. (1994) in *New Directions in Research with Third-Generation Soft X-Ray Synchrotron Radiation Sources,* ed. A. S. Schlachter and F. J. Wuilleumier (Kluwer Academic Publishers, Dordrecht), p. 359.

Howells, M. R. and Kincaid, B. M. (1994) in *New Directions in Research with Third-Generation Soft X-Ray Synchrotron Radiation Sources,* eds. A. S. Schlachter and F. J. Wuilleumier (Kluwer Academic Publishers, Dordrecht) p. 315.

Hu, C.-R. (1994) *Phys. Rev. Lett.* **72**, 1526.

Hubbard, J. (1955) *Proc. Phys. Soc.* (London) **68A**, 976.

Hubbard, J. (1963) *Proc. Roy. Soc.* (London) *A* **276**, 238.

Huber, D. L. (1988) *Solid State Commun.* **68**, 459.

Hüfner, S. (1979) in *Photoemission in Solids II*, eds. L. Ley and M. Cardona (Springer-Verlag, Berlin) p. 173.

Hüfner, S. (1994) *Adv. in Physics* **43**, 183.

Hüfner, S. (1995) *Photoelectron Spectroscopy* (Springer-Verlag, Berlin).

Hüfner, S. and Wertheim, G. K. (1972) *Phys. Rev.* **B7**, 2333.

Hüfner, S. and Wertheim, G. K. (1975a) *Phys. Lett.* **51A**, 299.

Hüfner, S. and Wertheim, G. K. (1975b) *Phys. Lett.* **51A**, 301.

Hüfner, S. and Wertheim, G. K. (1975c) *Phys. Rev.* **B11**, 5197.

Hüfner, S., Steiner, P., Reinert, F., and Schmitt, H. (1992) *Z. Phys.* **86**, 27.

Hüfner, S., Wertheim, G. K., and Wernick, J. H. (1973) *Phys. Rev.* **B8**, 4511.

Hüfner, S., Wertheim, G. K., Smith, N. V., and Traum, M. M. (1972) *Solid State Commun.* **11**, 323.

Hulbert, S. L., Johnson, P. D., Stoffel, N. G., Royer, W. A., and Smith, N. V. (1985) *Phys. Rev.* **B31**, 6815.

Hwu, Y., Cheng, N.-F., Lee, S.-D., Tung, C.-Y., Alméras, P., and Berger, H. (1996) *Appl. Phys. Lett.* **69**, 2924.

Hwu, Y., Lozzi, L., La Rosa, S., Onellion, M., Alméras, P., Gozzo, F., Lévy, F., Berger, H., and Margaritondo, G. (1993) *Phys. Rev.* **B48**, 624.

Hwu, Y., Lozzi, L., Marsi, M., La Rosa, S., Winokur, M., Davis. P., Onellion, M., Berger, H., Gozzo, F., Lévy, F., and Margaritondo, G. (1991) *Phys. Rev. Lett.* **67**, 2573.

Hwu, Y., Marsi, M., Terrasi, A., Rioux, D., Chang, Y., McKinley, J. T. Onellion, M., Margaritondo, G., Capozi, M., Quaresima, C., Campo, A., Ottaviani, C., Perfetti, P., Stoffel, N. G., and Wang, E. (1990) *Phys. Rev.* **B43**, 3678.

Hwu, Y., Tung, C. Y., Pieh, J. Y., Lee, S. D., Alméras, P., Gozzo, F., Berger, H., Margaritondo, G., De Stasio, G., Mercanti, D., and Ciotti, M. T. (1995) *Nucl. Instr. and Meth.* **A361**, 349.

Hybertsen, M. S., Steckel, E. B., Foulkes, W. M. C., and Schlüter, M. (1992) *Phys. Rev.* **B45**, 10032.

Ibach, H. (1993) *J. Electron Spectrosc. Relat. Phenom.* **64/65**, 819.

Iketaki, H., Horikawa, Y., Mochimaru, S., Nagai, K., Kiyokura, T., Oshima, M., and Yagishita, A. (1996) *J. Electron Spectrosc. Relat. Phenom.* **80**, 353.

Imer, J.-M., Patthey, F., Dardel, B., Schneider, W.-D., Baer, Y., Pétroff, Y., and Zettl, A. (1989a) *Phys. Rev. Lett.* **62**, 336.

Imer, J.-M., Patthey, F., Dardel, B., Schneider, W.-D., Baer, Y., Pétroff, Y., and Zettl, A. (1989b) *Phys. Rev. Lett.* **63**, 102.

Inglesfield, J. E. and Holland, B. W. (1981) in *The Chemical Physics of Solid Surfaces and Heterogeneous Catalysis*, Vol. 1, eds. D. A. King and D. P. Woodruff (Elsevier Scientific Publishing Co., Amsterdam) Ch. 3, p. 183.

Ino, A., Mizokawa, T., Fujimori, A., Tamasaku, K., Eisaki, H., Uchida, S., Kimura, T., Sasagawa, T., and Kishio, K. (1997) *Phys. Rev. Lett.* **79**, 2101.

Islam, M. S. and Baetzold, R. C. (1989) *Phys. Rev.* **B40**, 10926.

Itchkawitz, B. S., Lyo, I.-W., and Plummer, E. W. (1990) *Phys. Rev.* **B41**, 8075.

Itti, R., Inoue, Y., Munakata, F., and Yamanaka, M. (1990b) *Phys. Rev.* **B42**, 939.

Itti, R., Isawa, K., Sugiyama, J., Ikeda, K., Yamauchi, H., Koshizuka, N., and Tanaka, S. (1993) *J. Phys. Chem. Solids* **54**, 1199.

Itti, R., Miyatake, T., Ikeda, K., Yamaguchi, K., Koshizuka, N., and Tanaka, S. (1990a) *Phys. Rev.* **B41**, 9559.

Itti, R., Munakata, F., Ikeda, K., Yamauchi, H., Koshizuka, N., and Tanaka, S. (1991) *Phys. Rev.* **B43**, 6249.

Iwan, M. and Koch, E.-E. (1979) *Solid State Commun.* **31**, 261.

Iwan, M., Himpsel, F. J., and Eastman, D. E. (1979) *Phys. Rev. Lett.* **43**, 1829.

Jackson, J. D. (1975) *Classical Electrodynamics*, 2nd edn. (John Wiley and Sons, New York) Ch. 14, p. 654.

Jacob, W., Dose, V., Kolac, U., Fauster, T., and Goldmann, A. (1986) *Z. Phys. B: Condens. Matter* **63**, 459.

Jacobi, K. and Astaldi, C. (1988) *Surf. Sci.* **206**, L829.

Jacobi K., Scheffler, M., Kambe, K., and Forstmann, F. (1977) *Solid State Commun.* **22**, 17.

Janssen, G. J. M. and Nieuwpoort, W.C. (1988) *Phys. Rev.* **B38**, 3449.

Jensen, E. and Plummer, E. W. (1985) *Phys. Rev. Lett.* **55**, 1912.

Jo, T., Kotani, A., Parlebas, J. C., and Kanamori, J. (1983) *J. Phys. Soc. Jpn.* **52**, 2581.

Johnson, K. H. (1990) *Phys. Rev.* **B42**, 4783.

Johnson, P. D. (1992) in *Angle-resolved Photoemission – Theory and Current Applications*, ed. S. D. Kevan (Elsevier Science Publishers, Amsterdam) p. 509.

Johnson. P. D. and Davenport, J. W. (1985) *Phys. Rev.* **B31**, 7521.

Johnson, P. D., Hulbert, S. L., Garrett, R. F., and Howells, M. R. (1986) *Rev. Sci. Instr.* **57**, 1324.

Johnson, R. L. (1983) in *Handbook on Synchrotron Radiation*, Vol. 1, ed. E.-E. Koch (North-Holland, Amsterdam) p. 173.

Johnston, D. C., Chou, F. C., Borsa, F., Carretta, P., Lasciafari, A., Gooding, R. J., Salem, N. M., Vos, K. J. E., Cho, J. H., and Torgeson, D. R. (1996) *J. Supercond.* **9**, 337.

Jones, R. O. and Gunnarsson, O. (1989) *Rev. Mod. Phys.* **61**, 689.

Jordan, R. G. (1989) *J. Phys.: Condens. Matter* **1**, 9795.

Jorgensen, J. D., Veal, B. W., Paulikas, A. P., Nowicki, L. J., Crabtree, G. W., Claus, H., and Kwok, W. K. (1990) *Phys. Rev.* **B41**, 1864.

Joyce, J. J., del Giudice, M., and Weaver, J. H. (1989) *J. Electron Spectrosc. Relat. Phenom.* **49**, 31.

Jugnet, Y., Grenet, G., and Tran Min Duc (1987) in *Handbook on Synchrotron Radiation*, Vol. 2, ed. G. V. Marr (North-Holland, Amsterdam) p. 663.

Kampf, A. P. and Schrieffer, J. R. (1990) *Phys. Rev.* **B42**, 7967.

Kane, E. O. (1964) *Phys. Rev. Lett.* **12**, 97.

Kane, E. O. (1967) *Phys. Rev.* **159**, 624.

Kane, J. and Ng, K. W. (1996) *Phys. Rev.* **B53**, 2819.

Kanski, J., Nilsson, P. O., and Larsson, C. G. (1980) *Solid State Commun.* **35**, 397.

Karlsson, K., Gunnarsson, O., and Jepsen, O. (1992) *J. Phys.: Condens. Matter* **4**, 895.

Kelley, R. J., Ma, J., Margaritondo, G., and Onellion, M. (1993b) *Phys. Rev. Lett.* **71**, 4051.

Kelley, R. J., Ma, J., Onellion, M., Marsi, M., Alméras, P., Berger, H., and Margaritondo, G. (1993a) *Phys. Rev.* **B48**, 3534.

Kelley, R. J., Ma, J., Quitmann, C., Margaritondo, G., and Onellion, M. (1994) *Phys. Rev.* **B50**, 590.

Kelley, R. J., Quitmann, C., Onellion, M., Berger, H., Alméras, P., and Margaritondo, G. (1996) *Science* **271**, 1255.

Kemeny, P. C. and Shevchik, N. J. (1975) *Solid State Commun.* **17**, 255.

Kevan, S. D. (1983a) *Rev. Sci. Instr.* **54**, 1441.

Kevan, S. D. (1983b) *Phys. Rev. Lett.* **50**, 526.

Kevan, S. D. (1983c) *Phys. Rev.* **B28**, 4822.

Kevan, S. D. (1986) *Phys. Rev.* **B33**, 4364.

Kevan, S. D. (ed.) (1992) *Angle-resolved Photoemission – Theory and Current Applications*, (Elsevier Science Publishers, Amsterdam).

Kido, G., Komorita, K., Katayama-Yoshida, H., and Takahashi, T. (1991) *J. Phys. Chem. Solids* **52**, 1465.

Kim, C., Pianetta, P., and Kelly, M. A. (1995) *Rev. Sci. Instr.* **66**, 3159.

Kim, H.-D., Oh, S.-J., Yang, I.-S., and Hur, N. H. (1995) *Physica C* **253**, 351.

Kim, K. S. (1974) *J. Electron Spectrosc. Relat. Phenom.* **3**, 217.

Kimachi, Y., Hidaka, Y., Ohno, T. R., Kroll, G. H., and Weaver, J. H. (1991) *J. Appl. Phys.* **69**, 3176.

King, D. M., Shen, Z.-X., Dessau, D. S., Marshall, D. S., Park, C. H., Spicer, W. E., Peng, J. L., Li Z. Y., and Greene, R. L. (1994) *Phys. Rev. Lett.* **73**, 3298.

King, D. M., Shen, Z.-X., Dessau, D. S., Wells, B. O., Spicer, W. E., Arko, A. J., Marshall, D. S., DiCarlo, J., Loeser, A. G., Park, C. H., Ratner, E. R., Peng, J. L., Li, Z. Y., and Greene, R. L. (1993) *Phys. Rev. Lett.* **70**, 3159.

King, P. M., Borg, A., Kim, C., Pianetta, P., Lindau, I., Knapp, G. S., Keenlyside, M., and Browning, R. (1990) *Nucl. Instr. and Meth.* **A291**, 19.

Kitazawa, K. (1996) *Science* **271**, 313.

Kivelson, S. A., Emery, V. J., and Lin, H. Q. (1990) *Phys. Rev.* **B42**, 6523.

Klauder, M., Markl, J., Fink, C., Lunz, P., Saemann-Ischenko, G., Rau, F., Range, K.-J., Seemann, R., and Johnson, R. L. (1993) *Phys. Rev.* **B48**, 1217.

Klein, M. V. and Dierker, S. B. (1984) *Phys. Rev.* **B29**, 4976.

Kleinman, L. and Mednik, K. (1980) *Phys. Rev.* **B21**, 1549.

Kliewer, K. L. (1978) in *Photoemission and the Electronic Properties of Surfaces*, eds. B. Feuerbacher, B. Fitton, and R. F. Willis (John Wiley and Sons, New York) p. 45.

Knapp, J. A., Himpsel, F. J., and Eastman, D. E. (1979) *Phys. Rev.* **B19**, 4952.

Knapp, J. A., Lapeyre, G. J., Smith, N. V., and Traum, M. M. (1982) *Rev. Sci. Instr.* **53**, 781.

Knox, R. S. and Gold A. (1964) *Symmetry in the Solid State* (Benjamin, New York).

Koch, E.-E., Eastman, D. E., and Farge, Y. (1983) in *Handbook on Synchrotron Radiation*, Vol. 1, ed. E.-E. Koch (North-Holland, Amsterdam) p. 1

Kohn, W. and Sham, L. J. (1965) *Phys. Rev.* **140**, A1133.

Komeda, T., Waddill, G. D., Benning, P. J., and Weaver, J. H. (1991) *Phys. Rev.* **B43**, 8713.

Komolov, S. A. (1992) *Total Current Spectroscopy of Surfaces* (Gordon and Breach, Philadelphia).

Koster, G. F. (1957) in *Solid State Physics* **5**, ed. F. Seitz and D. Turnbull, 173.

Kotani, A. (1987) in *Handbook on Synchrotron Radiation*, Vol. 2, ed. G. V. Marr (North-Holland, Amsterdam) p. 611.

Kotani, A. and Toyozawa Y. (1973a) *J. Phys. Soc. Jpn.* **35**, 1073.

Kotani, A. and Toyozawa Y. (1973b) *J. Phys. Soc. Jpn.* **35**, 1082.

Kotani, A. and Toyozawa Y. (1973c) *J. Phys. Soc. Jpn.* **37**, 912.

Kotani, A. and Toyozawa Y. (1974) *J. Phys. Soc. Jpn.* **37**, 563.

Kotani, A. and Toyozawa, Y. (1979) in *Synchrotron Radiation – Techniques and Applications*, ed. C. Kunz (Springer-Verlag, Berlin) p. 169.

Kouznetsov, K. and Coffey, L. (1996) *Phys. Rev.* **B54**, 3617.

Kovacs, A., Nilsson, P. O. , and Kanski, J. (1982) *Physica Scripta* **25**, 791.

Koyama, R. Y. and Smith, N. V. (1970) *Phys. Rev.* **B2**, 3049.

Krakauer, H. and Pickett, W. E. (1988a) *Phys. Rev. Lett.* **60**, 1665.

Krakauer, H., Pickett, W. E., and Cohen, R. E. (1988b) *J. Supercond.* **1**, 111.

Kreutz, T. J., Aebi, P., and Osterwalder, J. (1995) *Solid State Commun.* **96**, 339.

Krinsky, S., Perlman, M. L., and Watson, R. E. (1983) in *Handbook on Synchrotron Radiation*, Vol. 1, ed. E.-E. Koch (North-Holland, Amsterdam) p. 65.

Kuhlenbeck, H., Odörfer, G., Jaeger, R. Illing, G., Menges, M., Mull, Th., Freund, H.-J., Pöhlchen, M., Staemmler, V., Witzel, S., Scharfschwerdt, C., Wennemann, K., Liedke, T., and Neumann, M. (1991) *Phys. Rev.* **B43**, 1969.

Kulkarni, P., Mahamuni, S., Kulkarni, S. K., and Nigavekar, A. S. (1990) *Physica C* **168**, 104.

Kunz, C. (1979a) in *Photoemission in Solids II*, eds. M. Cardona and L. Ley (Springer-Verlag, Berlin) p. 299.

Kunz, C. (1979b) in *Synchrotron Radiation – Techniques and Applications*, ed. C. Kunz (Springer-Verlag, Berlin) p. 1.

Kuper, C. G. (1968) *An Introduction to the Theory of Superconductivity* (Clarendon Press, Oxford).

Landau, L. D. and Lifshitz, E. M. (1965) *Electrodynamics of Continous Media* (Addison-Wesley, Reading MA)

Lang, J. K. and Baer, Y. (1979) *Rev. Sci. Instr.* **50**, 221.

Langer, M., Schmalian, J., Grabowski, S., and Bennemann, K. H. (1995) *Phys. Rev. Lett.* **75**, 4508.

Lapeyre, G. J., Anderson, J., Gobby, P. L., and Knapp, J. A. (1974b) *Phys. Rev. Lett.*. **33**, 1290.

Lapeyre, G. J., Baer, J. D., Hermanson, J., Anderson, J., Knapp, J. A., and Gobby, P. L. (1974a) *Solid State Commun.* **15**, 1601.

Law, D. S.-L., Mikheeva, M. N., Nazin, V. G., Svishchev, A. V., Tsetlin, M. B., Hayes, N., and Downes, S. (1996) *Physica C* **272**, 233.

Leckey, R. C. G. (1982) *Appl. of Surf. Sci.* **13**, 125.

Leckey, R. C. G. and Riley, J. D. (1985) *Appl. of Surf. Sci.* **22/23**, 196.

Leiro, J. A., Heinonen, M. H., and Elboussiri, K. (1995) *Phys. Rev.* **B52**, 82.

Lenard, P. (1900) *Ann. d. Physik* **2**, 359.

Levinson, H. J. and Plummer, E. W. (1981) *Phys. Rev.* **B24**, 628.

Levinson, H. J., Greuter, F., and Plummer, E. W. (1983) *Phys. Rev.* **B27**, 727.

Ley, L. and Cardona, M. (eds.) (1979) *Photoemission in Solids II*, (Springer-Verlag, Berlin).

Li, X., Zhang, Z., and Henrich, V. E. (1993) *J. Electron Spectrosc. Relat. Phenom.* **63**, 253.

Lieb, E. H. and Wu, F. Y. (1968) *Phys. Rev. Lett.* **20**, 1445.

Liebsch, A. (1976) *Phys. Rev.* **B13**, 544.

Liebsch, A. (1978) in *Electron and Ion Spectroscopy of Solids*, eds. L. Fiermans, J. Vennik, and W. Dekeyser (Plenum Press, New York) p. 93.

Liebsch, A. (1979a) in *Feskörperprobleme XIX*, ed. J. Treusch (Vieweg, Braunschweig) p. 209.

Liebsch, A. (1979b) *Phys. Rev. Lett.* **43**, 1431.

Liebsch, A. (1981) *Phys. Rev.* **B23**, 5203.

Liechtenstein, A. I. and Mazin, I. I. (1995) *Phys. Rev. Lett.* **74**, 1000.

Liechtenstein, A. I., Gunnarsson, O., Andersen, O. K., and Martin, R. M. (1996) *Phys. Rev.* **B54**, 12505.

Lindau, I. and Spicer, W. E. (1974) *J. Electron Spectrosc. Relat. Phenom.* **3**, 409.

Lindau, I., Pianetta, P., Yu, K. Y., and Spicer, W. E. (1976) *Phys. Rev.* **B13**, 492.

Lindberg, P. A. P., Shen, Z.-X., Dessau, D. S., Wells, B. O., Mitzi, D. B., Lindau, I., Spicer, W. E., and Kapitulnik, A. (1989a) *Phys. Rev.* **B40**, 5169.

Lindberg, P. A. P., Shen, Z.-X., Spicer, W. E., and Lindau, I. (1990a) *Surf. Sci. Rep.* **11**, 1.

Lindberg, P. A. P., Wells, B. O., Shen, Z.-X., Dessau, D. S. Lindau, I., Spicer, W. E., Mitzi, D. B., and Kapitulnik, A. (1989b) *J. Appl. Phys.* **67**, 2667.

Lindberg, P. A. P., Shen, Z.-X., Wells, B. O., Dessau, D. S., Mitzi, D. B., Lindau, I., Spicer, W. E., and Kapitulnik, A. (1989d) *Phys. Rev.* **B39**, 2890.

Lindberg, P. A. P., Wells, B. O., Shen, Z.-X., Dessau, D. S., Lindau, I., Spicer, W. E., Mitzi,

D. B., and Kapitulnik, A. (1990b) *Appl. Phys. Lett.* **67**, 2667.

Lindberg, P. A. P., Shen, Z.-X., Wells, B. O., Mitzi, D. B., Lindau, I., Spicer, W. E., and Kapitulnik, A. (1988) *Appl. Phys. Lett.* **53**, 2563.

Lindberg, P. A. P., Shen, Z.-X., Wells, B. O., Mitzi, D. B., Lindau, I., Spicer, W. E., and Kapitulnik, A. (1989c) *Phys. Rev.* **B40**, 8769.

Lindgren, S. A., Paul, J., and Walldén, L. (1982) *Solid State Commun.* **44**, 639.

Lindroos, M. and Bansil, A. (1991) *J. Phys. Chem. Solids* **52**, 1447.

Lindroos, M. and Bansil, A. (1995) *Phys. Rev. Lett.* **75**, 1182.

Lindroos, M. and Bansil, A. (1996) *Phys. Rev. Lett.* **77**, 2985.

Lindroos, M., Asonen, H., Pessa, M., and Smith, N. V. (1982) *Solid State Commun.* **39**, 285.

Lindroos, M., Bansil, A., Gofron, K., Campuzano, J. C., Ding, H., Liu, R., and Veal, B. W. (1993) *Physica C* **212**, 347.

List, R. S. (1990) (unpublished).

List, R. S., Arko, A. J., Bartlett, R. J., Olson, C. G., Yang, A.-B., Liu, R., Gu, C., Veal, B. W., Chang, Y., Jiang, P. Z., Vandevoort, K., Paulikas, A. P., and Campuzano, J. C. (1989) *Physica C* **159**, 439.

List, R. S., Arko, A. J., Fisk, Z., Cheong, S.-W., Conradson, S. D., Thompson, J. D., Pierce, C. B., Peterson, D. E., Bartlett, R. J., Shinn, N. D., Schirber, J. E., Veal, B. W., Paulikas, A. P., and Campuzano, J. C. (1988) *Phys. Rev.* **B38**, 11966.

Liu, L. Z., Anderson, R. O., and Allen, J. W. (1991) *J. Phys. Chem. Solids* **52**, 1473.

Liu, R., Olson, C. G., Yang, A.-B., Gu, C., Lynch, D. W., Arko, A. J., List, R. S., Bartlett, R. J., Veal, B. W., Liu, J. Z., Paulikas, A. P., and Vandervoort, K. (1989) *Phys. Rev.* **B40**, 2650.

Liu, R., Veal, B. W., Gu, C., Paulikas, A. P., Kostic, P., and Olson, C. G. (1995) *Phys. Rev.* **B52**, 553.

Liu, R., Veal, B. W., Paulikas, A. P., Downey, J. W., Kostic, P., Fleshler, S., Welp, U., Olson, C. G., Wu, X., Arko, A. J., and Joyce, J. J. (1992a) *Phys. Rev.* **B45**, 5614.

Liu, R., Veal, B. W., Paulikas, A. P., Downey, J. W., Kostic, P., Fleshler, S., Welp, U., Olson, C. G., Wu, X., Arko, A. J., and Joyce, J. J. (1992b) *Phys. Rev.* **B46**, 11056.

Liu, S. H. and Klemm, R. A. (1994) *Phys. Rev. Lett.* **73**, 1019.

Liu, S. H. and Klemm, R. A. (1995) *Phys. Rev.* **B52**, 9657.

Loeser, A. G., Shen, Z.-X., Dessau, D. S., Marshall, D. S., Park, C. H., Fournier, P., and Kapitulnik, A. (1996a) *Science* **273**, 325.

Loeser, A. G., Shen, Z.-X., and Dessau, D. S. (1996b) *J. Supercond.* **9**, 373.

Loeser, A. G., Shen, Z.-X., Dessau, D. S., and Spicer, W. E. (1994) *J. Electron Spectrosc. Relat. Phenom.* **66**, 359.

López, M., Gutierrez, A., Laubschat, C., and Kaindl, G. (1995) *J. Electron Spectrosc. Relat. Phenom.* **71**, 73.

López, M., Laubschat, C., Gutierrez, A., Höhr, A., Domke, M., Kaindl, G., and Abbate, M. (1994a) *Z. Phys. B* **95**, 9.

López, M., Laubschat, C., Gutierrez, A., Weschke, E., Höhr, A., Domke, M., and Kaindl, G. (1994b) *Surf. Sci.* **307/309**, 907.

Luo, Y. S., Yang, Y.-N., and Weaver, J. H. (1992) *Phys. Rev.* **B46**, 1114.

Luttinger, J. M. (1960) *Phys. Rev.* **119**, 1153.

Lynn, J. W. (ed.) (1990), *High-Temperature Superconductivity* (Springer-Verlag, New York).

Lyo, I.-W. and Plummer, E. W. (1988) *Phys. Rev. Lett.* **60**, 1668.

Ma, J., Alméras, P., Kelley, R. J., Berger, H., Margaritondo, G., Cai, X. Y., Feng, Y., and Onellion, M. (1995c) *Phys. Rev.* **B51**, 9271.

Ma, J., Quitmann, C., Kelley, R. J., Alméras, P., Berger, H., Margaritondo, G., and Onellion, M. (1995b) *Phys. Rev.* **B51**, 3832.

Ma, J., Quitmann, C., Kelley, R. J., Berger, H., Margaritondo, G., and Onellion, M. (1995a) *Science* **267**, 862.

Ma, J., Quitmann, C., Kelley, R. J., Margaritondo, G., and Onellion, M. (1995d) *Solid State Commun.* **94**, 27.

Ma, J., Quitmann, C., Onellion, M., Beschoten, B., Güntherodt, G., Auge, J., Kurz, H., Polyakov, S. N., and Kov'ev, E. K. (1995e) *Solid State Commun.* **95**, 85.

Ma, S.-K. and Shung, K. W.-K. (1994) *Phys. Rev.* **B49**, 10617.

Maetz, C. J., Gerhardt, U., Dietz, E., Ziegler, A., and Jelitto, R. J. (1982) *Phys. Rev. Lett.* **48**, 1686.

Maggio-Aprile, I., Renner, Ch., Erb, A., Walker, E., and Fischer, O. (1995) *Phys. Rev. Lett.* **75**, 2754.

Mahan, G. D. (1970) *Phys. Rev.* **B2**, 4334.

Mahan, G. D. (1978) in *Electron and Ion Spectroscopy of Solids,* eds. L. Fiermans, J. Vennik, and W. Dekeyser (Plenum Press, New York) p. 1.

Mahan, G. D. (1990) *Many-Particle Physics,* 2nd edn. (Plenum Press, New York) Chs. 3, 5.

Mahan, G. D. (1993) *Phys. Rev. Lett.* **71**, 4277.

Manghi, F., Calandra, C. and Ossicini, S. (1994) *Phys. Rev. Lett.* **73**, 3129.

Mann, A. and Linde, F. (1988) *J. Phys. E* **21**, 805.

Manson, S. T. (1978) in *Photoemission in Solids I*, eds. M. Cardona and L. Ley (Springer-Verlag, Berlin) p. 135.

Mante, G., Claessen, R., Buslaps, T., Harm, S., Manzke, R., Skibowski, M., and Fink, J. (1990) *Z. Phys. B* **80**, 181.

Mante, G., Claessen, R., Huss, A., Manzke, R., Skibowski, M., Wolf, Th., Knupfer, M., and Fink, J. (1991) *Phys. Rev.* **B44**, 9500.

Mante, G., Schmalz, T., Manzke, R., Skibowski, M, Alexander, M., and Fink, J. (1992) *Surf. Sci.* **269/270**, 1071.

Manuel, A. A., Shukla, A., Hoffmann, L., Jarlborg, T., Barbiellini, B., Massidda, S., Sadowski, W., Walker, E., Erb, A., and Peter, M. (1995) *J. Phys. Chem. Solids* **56**, 1951.

Manzke, R., Buslaps, T., Claessen, R., and Fink, J. (1989a) *Europhysics Lett.* **9**, 477.

Manzke, R., Buslaps, T., Claessen, R., Mante, G., and Zhao, Z. X. (1989c) *Solid State Commun.* **70**, 67.

Manzke, R., Buslaps, T., Claessen, R., Skibowski, M., and Fink, J. (1989b) *J. Phys. C* **162–164**, 1381.

Manzke, R., Mante, G., Claessen, R., Skibowski, M., and Fink, J. (1992) *Surf. Sci.* **269/270**, 1066.

Margaritondo, G. (1988) *Introduction to Synchrotron Radiation* (Oxford University Press, New York).

Margaritondo, G. and Cerrina, F. (1990) *Nucl. Instr. and Meth.* **A291**, 26.

Margaritondo, G. and Weaver, J. H. (1983) in *Methods of Experimental Physics: Surfaces*, eds. M. G. Legally and R. L. Park (Academic Press, New York) Ch. IV.

Margaritondo, G., Weaver, J. H., and Stoffel, N. G. (1979) *J. Phys. E* **12**, 662.

Markiewicz, R. S. (1997) *J. Phys. Chem. Solids* **58**, 1179.

Marksteiner, P., Yu, J., Massidda, S., Freeman, A. J., Redinger, J., and Weinberger, P. (1989) *Phys. Rev.* **B39**, 2894.

Marshall, D. S., Dessau, D. S., King, D. M., Park, C.-H., Matsuura, A. Y., Shen, Z.-X., Spicer, W. E., Eckstein, J. N., and Bozovic, I. (1995) *Phys. Rev.* **B52**, 12548.

Marshall, D. S., Dessau, D. S., Loeser, A.G., Park, C.-H., Matsuura, A. Y., Eckstein, J. N., Bozovic, I., Fournier, P., Kapitulnik, A., Spicer, W. E., and Shen, Z.-X. (1996) *Phys. Rev. Lett.* **76**, 4841.

Mårtensson, N., Baltzer, P., Brüweiler, P. A., Forsell, J.-O., Nilsson, A., Stenborg, A., and Wannberg, B. (1994) *J. Electron Spectrosc. Relat. Phenom.* **70**, 117.

Maruyama, T., Aiura, Y., Nishihara, Y., Oka, K., Ohashi, Y., Haruyama, Y., and Kato, H. (1994) *Physica C* **235–240**, 1049.

Massidda, S., Hamada, N., Yu, J., and Freeman, A. J. (1989) *Physica C* **157**, 571.

Massidda, S., Yu, J., and Freeman, A. J. (1988) *Physica C* **152**, 251.

Massidda, S., Yu, J., Freeman, A. J., and Koelling, D. D. (1987) *Phys. Lett. A* **122**, 198.

Massidda, S., Yu, J., Park, K. T., and Freeman, A. J. (1991) *Physica C* **176**, 159.

Matsui, T., Yamaguchi, D., and Kamijo, H. (1996) *Jpn. J. Appl. Phys.* **35**, L97.

Matzdorf, R., Goldmann, A., Braun, J., and Borstel, G. (1994) *Solid State Commun.* **91**, 163.

Matzdorf, R., Meister, G., and Goldmann, A. (1993a) *Surf. Sci.* **286**, 56.

Matzdorf, R., Meister, G., and Goldmann, A. (1993b) *Surf. Sci.* **296**, 241.

Mazin, I. I., Jepsen, O., Andersen, O. K., and Liechtenstein, A. I. (1992) *Phys. Rev. Lett.* **68**, 3936.

McDougall, B. A., Balasubramanian, T., and Jensen, E. (1995) *Phys. Rev.* **B51**, 13891.

McLean, A. B. (1995) *Surf. Sci.* **329**, L629.

McLean, A. B., Mitchell, C. E. J., and Hill, I. G. (1994) *Surf. Sci.* **314**, L925.

Mehta, M. and Fadley, C. S. (1977) *Phys. Rev. Lett.* **59**, 1569.

Mehta, M. and Fadley, C. S. (1979) *Phys. Rev.* **B20**, 2280.

Meinders, M. B. J., Eskes, H., and Sawatzky, G. A. (1993) *Phys. Rev.* **B48**, 3916.

Meyer, H. M. and Weaver, J. H. (1990) in *Physical Properties of High Temperature Superconductors*, Vol. 2, ed. D. M. Ginsberg (World Scientific, Singapore) p. 369.

Meyer, H. M., Hill, D. M., Anderson, S. G., Weaver, J. H., and Capone, D. W. (1987b) *Appl. Phys. Lett.* **51**, 1750.

Meyer, H. M., Hill, D. M., Wagener, T. J., Gao, Y., Weaver, J. H., Capone, D. W., and Goretta, K. C. (1988a) *Phys. Rev.* **B38**, 6500.

Meyer, H. M., Hill, D. M., Weaver, J. H., Nelson, D. L., and Gallo, C. F. (1988b) *Phys. Rev.* **B38**, 7144.

Meyer, H. M., Hill, D. M., Weaver, J. H., Nelson, D. L., and Goretta, K. C. (1988c) *Appl. Phys. Lett.* **53**, 1004.

Meyer, H. M., Wagener, T. J., Hill, D. M., Gao. Y., Anderson, S. G., Krahn, S.D., and Weaver, J. H. (1987a) *Appl. Phys. Lett.* **51**, 1118.

Meyer, H. M., Wagener, T. J., Weaver, J. H., and Ginley, D. S. (1989) *Phys. Rev.* **B39**, 7343.

Meyer, H. M., Weaver, J. H., and Goretta, K. C. (1990) *J. Appl. Phys.* **67**, 1995.

Meyer, M., Prescher, T., von Raven, E., Richter, M., Schmidt. E., Sonntag, B., and Wetzel, H.-E. (1986) *Z. Phys. D* **3**, 347.

Mikheeva, M. N., Nazin, V. G., Svishchev, A. V., Law, D. S.-L., Turner, T. S., Bailey, P., Teehan, D., Barilo, S. N., and Gritskov, P. V. (1993) *Surf. Sci.* **287/288**, 662.

Miller, T., McMahon, W. E., and Chiang, T.-C. (1996) *Phys. Rev. Lett.* **77**, 1167.

Minami, F., Kimura, T., and Takekawa, S. (1989) *Phys. Rev.* **B39**, 4788.

Miyake, K. and Narikiyo, O. (1995) *J. Phys. Soc. Jpn.* **64**, 1040.

Mizokawa, T. and Fujimori, A. (1996) *Phys. Rev.* **B53**, R4201.

Mogilevsky, R., Levi-Setti R., Pashmakov, B., Liu, L., Zhang, K., Jaeger, H. M., Buchholz, D.B., Chang., R. P. H., and Veal, B. W. (1994) *Phys. Rev.* **B49**, 6420.

Molodtsov, S. L., Richter, M., Danzenbächer, S., Wieling, S., Steinbeck, L., and Laubschat, C. (1997) *Phys. Rev. Lett.* **78**, 142.

Monthoux, P. (1997) *Phys. Rev.* **B55**, 11111.

Moore, J. H., Davis, C. C., and Coplan, M. A. (1983) *Building Scientific Apparatus* (Addison-Wesley Publishing Co., Reading) Ch. 5, p. 287.

Mosser, A., Romeo, M. A., Parlebas, J. C., Okada, K., and Kotani, A. (1991) *Solid State Commun.* **79**, 641.

Mueller, D. R., Wallace, J. S., Jia, J. J., O'Brien, W. L., Dong, Q.-Y., Callcott, T. A., Miyano, K. E., and Ederer, D. L. (1994) *Phys. Rev.* **B52**, 9702.

Mueller, F. M., Fowler, C. M., Freeman, B. J., Hults, W. L., King, J. C., and Smith, J. L. (1992) *Phys. Rev. Lett.* **68**, 3937.

Musket, R. G., McLean, W., Colmenares, C. A., Makowiecki, D. M., and Siekhaus, W. J. (1982) *Appl. of Surf. Sci.* **10**, 143.

Nagoshi, M., Fukuda, Y., and Suzuki, T. (1994) *J. Electron Spectrosc. Relat. Phenom.* **66**, 257.

Nagoshi, M., Fukuda, Y., Sanada, N., Syono, Y., Tokiwa-Yamamoto, A, and Takichi, M. (1993) *J. Electron Spectrosc. Relat. Phenom.* **61**, 309.

Nagoshi, M., Syono, Y., Tachiki, M., and Fukuda, Y. (1995) *Phys. Rev.* **B51**, 9352.

Nagoshi, M., Syono, Y., Tachiki, M., and Fukuda, Y. (1996) *Physica C* **263**, 294.

Namatame, H., Fujimori, A., Tokura, Y., Nakamura, M., Yamaguchi, K., Misu, A., Matsubara, H., Suga, S., Eisaki, H., Ito, T., Tagaki, H. and Uchida, S. (1990) *Phys. Rev.* **B41**, 7205.

Naumović, D., Stuck, A., Greber, T., Osterwalder, J., and Schlapbach, L. (1993) *Phys. Rev.* **B47**, 7462.

Nemoshkalenko, V. V. and Aleshin, V. G. (1979) *Electron Spectroscopy of Crystals*, (Plenum Press, New York) (Russian original published in 1976).

Nicol, E. J., Carbotte, J. P., and Timusk, T. (1991) *Phys. Rev.* **B43**, 473.

Niles, D. W., Rioux, D., and Höchst, H. (1992) *Phys. Rev.* **B46**, 12547.

Nilsson, A., Stenborg, A., Tillborg, H., Gunnelin, K., and Mårtensson, N. (1993) *Phys. Rev.* **B47**, 13590.

Nilsson, P. O. and Dahlbäck, N. (1979) *Solid State Commun.* **29**, 303.

Nilsson, P. O. and Kovacs, A. (1983) *Physica Scripta* **T4**, 61.

Nilsson, P. O. and Larsson, C. G. (1983) *Phys. Rev.* **B27**, 6143.

Norman, M. R. (1994) *Phys. Rev. Lett.* **73**, 3044.

Norman, M. R. and Freeman, A. J. (1986) *Phys. Rev.* **B33**, 8896.

Norman, M. R., Ding, H., Campuzano, J. C., Takeuchi, T., Randeria, M., Yokoya, T., Takahashi, T., Mochiku, T., and Kadowaki, K. (1997a) *Phys. Rev. Lett.* **79**, 3506.

Norman, M. R., Ding, H., Randeria, M., Campuzano, J C., Yokoya, T., Takeuchi, T., Takahashi, T., Mochiku, T., Kadowaki, K., Guptasarma, P., and Hinks, D. G. (1998) *Nature* **392**, 157.

Norman, M. R., Randeria, M., Ding, H., and Campuzano, J. C. (1997b) preprint.

Norman, M. R., Randeria, M., Ding, H., and Campuzano, J. C. (1995a) *Phys. Rev.* **B52**, 615.

Norman, M. R., Randeria, M., Ding, H., Campuzano, J. C., and Bellman, A. F. (1995b) *Phys. Rev.* **B52**, 15107.

Northrup, J. E., Hybertsen, M. S., and Louie, S. G. (1987) *Phys. Rev. Lett.* **59**, 819.

Northrup, J. E., Hybertsen, M. S., and Louie, S. G. (1989) *Phys. Rev.* **B39**, 8198.

Novikov, D. L. and Freeman, A. J. (1993) *Physica C* **212**, 233.

Nücker, N., Pellegrin, E., Schweiss, P., Fink, J., Molodtsov, S. L., Simmons, C. T., Kaindl, G., Frentrup, W., Erb, A., and Müller-Vogt, G. (1995) *Phys. Rev.* **B51**, 8529.

Ogawa, K., Fujiwara, J., Takei, H., and Asaoka, H. (1991) *Physica C* **190**, 39.

Oh, S.-J., Allen, J. W., Lindau, I., and Mikkelson, J. C. (1982) *Phys. Rev.* **B26**, 4845.

Ohlin, P. (1942) *Ark. Mat. Astr. Fiz.* **29a**, 3.

Ohno, T. R., Patrin, J. C., Meyer, H. M., Weaver, J. H., Kimachi, Y., and Hidaka, Y. (1990) *Phys. Rev.* **B41**, 11677.

Ohno, T. R., Yang, Y.-N., Kroll, G. H., Krause, K., Schmidt. L. D., Weaver, J. H., Kimachi, Y.,

Hidaka, Y., Pan, S. H., and de Lozanne, A. L. (1991) *Phys. Rev.* **B43**, 7980.

Ohtaka, K. and Tanabe, Y. (1990) *Rev. Mod. Phys.* **62**, 929.

Okada, K. and Kotani, A. (1990) *J. Electron Spectrosc. Relat. Phenom.* **52**, 313.

Olson, C. G. and Lynch, D. W. (1982) *J. Opt. Soc. Amer.* **72**, 88.

Olson, C. G., Benning, P. J., Schmidt, M., Lynch, D. W., Canfield, P., and Wieliczka, D. M. (1996) *Phys. Rev. Lett.* **76**, 4265.

Olson, C. G., Liu, R., Lynch, D. W., List, R. S., Arko, A. J., Veal, B. W., Chang, Y. C., Jiang, P. Z., and Paulikas, A. P. (1990a) *Phys. Rev.* **B42**, 381.

Olson, C. G., Liu, R., Lynch, D. W., List, R. S., Arko, A. J., Veal, B. W., Chang, Y. C., Jiang, P. Z., and Paulikas, A. P. (1990b) *Solid State Commun.* **76**, 411.

Olson, C. G., Liu, R., Yang, A.-B., Lynch, D. W., Arko, A. J., List, R. S., Veal, B. W., Chang, Y. C., Jiang, P. Z., and Paulikas, A. P. (1989) *Science* **245**, 731.

Olson, C. G., Tobin, J. G., Waddill, G. D., Lynch, D. W., and Liu, J. Z. (1995) *J. Phys. Chem. Solids* **56**, 1879.

Ori, D. M., Goldoni, A., del Pennino, U., and Parmigiani, F. (1995) *Phys. Rev.* **B52**, 3727.

Osofsky, M. S., Qadri, S. B., Skelton, E. F., and Vanderah, T. A. (1996) in *Oxide Superconducting Physics and Nanoengineering II*, eds. I. Bozovic and D. Pavuna (SPIE, Bellingham) p. 9.

Osterwalder, J., Aebi, P., Schwaller, P., Schlapbach, L., Shimoda, M., Mochiku, T., and Kadowaki, K. (1995) *Appl. Phys. A* **60**, 247.

Osterwalder, J., Greber, T., Aebi, P., Fasel, R., and Schlapbach, L. (1996) *Phys. Rev.* **B53**, 10209.

Osterwalder, J., Greber, T., Hüfner, S., and Schlapbach, L. (1990) *Phys. Rev. Lett.* **64**, 2683.

Osterwalder, J., Greber, T., Stuck, A., and Schlapbach, L. (1991) *Phys. Rev.* **B44**, 13764.

Otto, A. and Reihl, B. (1990) *Phys. Rev.* **B41**, 9752.

Overhauser. A. W. (1985) *Phys. Rev. Lett.* **55**, 1916.

Ovrebo, G. K. and Erskine, J. L. (1981) *J. Electron Spectrosc. Relat. Phenom.* **24**, 189.

Paniago, R., Matzdorf, R., Meister, G., and Goldmann, A. (1995) *Surf. Sci.* **331–333**, 1233.

Park, W. G., Nepijko, S. A., Fanelsa, A., Kisker, E., Winkeler, L., and Güntherodt, G. (1994) *Solid State Commun.* **91**, 655.

Parlebas, J. C., Kotani, A., and Kanamori, J. (1982) *J. Phys. Soc. Jpn.* **51**, 124.

Parlebas, J. C., Khan, M. A., Uozumi, T., Okada, K., and Kotani, A. (1995) *J. Electron Spectrosc. Relat. Phenom.* **71**, 117.

Parmigiani, F. and Sangaletti, L. (1994) *J. Electron Spectrosc. Relat. Phenom.* **66**, 223.

Parmigiani, F., Depero, L. E., Minerva, T., and Torrance, J. B. (1992) *J. Electron Spectrosc. Relat. Phenom.* **58**, 315.

Parmigiani, F., Pacchioni, C., Brundle, C. R., Fowler, D. E., and Bagus, P. S. (1991b) *Phys. Rev.* **B43**, 3695.

Parmigiani, F., Shen, Z.-X., Mitzi, D. B., Lindau, I., Spicer, W. E., and Kapitulnik, A. (1991a) *Phys. Rev.* **B43**, 3085.

Pellegrin, E., Nücker, N., Fink, J., Molodtsov, S. L., Gutiérrez, A., Navas, E., Strebel, O., Hu, Z., Domke, M., Kaindl, G., Uchida, S., Nakamura, Y., Markl, J., Klauda, M., Seamann-Ischenko, G., Krol, A., Peng, J. L., Li, Z. Y., and Greene, R. L. (1993) *Phys. Rev.* **B47**, 3354.

Pendry, J. B. (1976) *Surf. Sci.* **57**, 679.

Pendry, J. B. (1980) *Phys. Rev. Lett.* **45**, 1356.

Pendry, J. B. (1981) *J. Phys. C* **14**, 1381.

Pendry, J. B. and Martín-Moreno, L. (1994) *Phys. Rev.* **B50**, 5062.

Penn, D. R. (1979) *Phys. Rev. Lett.* **42**, 921.

Pétroff, Y. and Thiry, P. (1980) *Appl. Opt.* **19**, 3957.

Pham, A. Q., Studer, F., Merrien, N., Maignan, A., Michel, C., and Raveau, B. (1993) *Phys. Rev.* **B48**, 1249.

Phelps, R. B., Akavoor, P., Kesmodel, L. L., Barr, A. L., Markert, J. T., Ma, J., Kelley, R. J., and Onellion, M. (1994) *Phys. Rev. Lett.* **50**, 6526.

Pickett, W. E. (1989) *Rev. Mod. Phys.* **61**, 433.

Pickett, W. E., Cohen, R. E., and Krakauer, H. (1990) *Phys. Rev.* **B42**, 8764.

Pickett, W. E., Krakauer, H., Cohen, R. E., and Singh, D. J. (1992) *Science* **255**, 46.

Plakida, N. M. (1995) *High-Temperature Superconductivity* (Springer-Verlag, Berlin).

Plummer, E. W. (1980) *Nucl. Instr. and Meth.* **177**, 179.

Plummer E. W. and Eberhardt, W. (1982) in *Advances in Chemical Physics*, Vol. XLIV, eds. I. Prigogine and S. A. Rice (John Wiley and Sons, New York) p. 533.

Poole, C. P., Farach, H. A., and Creswick, R. J. (1995) *Superconductivity* (Academic Press, San Diego).

Pothuizen, J. J. M. Eder, R., Hien, N. T., Matoba, M., Menovsky, A. A., and Sawatzky, G. A. (1997) *Phys. Rev. Lett.* **78**, 717.

Powell, C. J. (1974) *Surf. Sci.* **44**, 29.

Powell, C. J. (1985) *Surf. and Interface Anal.* **7**, 256.

Powell, C. J. (1988) *J. Electron Spectrosc. Relat. Phenom.* **47**, 197.

Przybylski, H., Baalmann, A., Borstel, G., and Neumann, M. (1983) *Phys. Rev.* **B27**, 6669.

Puppin, E. and Ragaini, E. (1995) *J. Electron Spectrosc. Relat. Phenom.* **73**, 53

Qadri, S. B., Osofsky, M. S., Browning, V. M., and Skelton, E. F. (1996) *Appl. Phys. Lett.* **68**, 2729.

Qu, Z., Goonewardene, A., Subramanian, K., Karunamuni, J., Mainkar, N., Ye, L., Stockbauer, R. L., and Kurtz, R. L. (1995) *Surf. Sci.* **324**, 133.

Quitmann, C., Alméras, P., Ma, J., Kelley, R. J., Berger, H., Margaritondo, G., and Onellion, M. (1995b) *J. Supercond.* **8**, 635.

Quitmann, C., Alméras, P., Ma, J., Kelley, R. J., Berger, H., Cai, X., Margaritondo, G., and Onellion, M. (1996) *Phys. Rev.* **B53**, 6819.

Quitmann, C., Andrich, D., Jarchow, C., Fleuster, M., Beschoten, B., Güntherodt, G., Moshchalkov, V. V., Mante, G., and Manzke, R. (1992) *Phys. Rev.* **B46**, 11813.

Quitmann, C., Beschoten, B., Kelley, R. J., Güntherodt, G., and Onellion, M. (1995a) *Phys. Rev.* **B51**, 11647.

Quitmann, C., Ma, J., Kelley, R. J., Margaritondo, G., and Onellion, M. (1994) *Physica C* **235–240**, 1019.

Qvarford, M., Andersen, J. N., Nyholm, R., van Acker, J. F., Lundgren E., Lindau, I., Söderholm, S., Bernhoff, H., Karlsson, U. O., and Flodström, S. A. (1992) *Phys. Rev.* **B46**, 14126.

Qvarford, M., Söderholm, S., Chiaia, G., Nyholm, R., Andersen, J., N., Lindau, I., Karlsson, U. O., Leonyuk, L., Nilsson, A., and Mårtensson, N. (1996) *Phys. Rev.* **B53**, R14753.

Qvarford, M., Söderholm, S., Tjernberg, O., Chiaia, G., Nylén, H., Nyholm, R., Lindau, I., Karlsson, U. O., and Bernhoff, H. (1996) *Physica C* **265**, 113.

Qvarford, M., van Acker, J. F., Andersen, J. N., Nyholm, R., Lindau, I., Chiaia, G., Lundgren, E., Söderholm, S., Karlsson, U. O., Flodström, S. A., and Leonyuk, L. (1995) *Phys. Rev.* **B51**, 410.

Raether, H. (1965) *Springer Tracts in Modern Physics* **38**, 85.

Randeria, M., Ding, H., Campuzano, J. C., Bellman, A. F., Jennings, G., Yokoya, T., Takahashi, T., Katayama-Yoshida, H., Mochiku, T., and Kadowaki, K. (1995) *Phys. Rev. Lett.* **74**, 4951.

Rao, C. N. R., Rao, G. R., Rajumon, M. K., and Sarma, D. D. (1990) *Phys. Rev.* **B42**, 1026.

Rao, C. N. R., Santra, A. K., and Sarma, D. D. (1992) *Phys. Rev.* **B45**, 10814.

Raoux, D. (1993) in *Hercules–Neutron and Synchrotron Radiation for Condensed Matter Studies*, eds. J. Baruchel, J.-L, Hodeau, M. S.

Lehmann, J.-R. Regnard, and C. Schlenker (Springer-Verlag, Berlin) p.79.

Rarback, H., Shu, D., Feng, S. C., Ade, H., Kirz, J., McNulty, I., Kern, D. P., Chang, T. H. P., Vladimirsky, Y., Iskander, N., Attwood, D., McQuaid, K., and Rothman, S. (1988) *Rev. Sci. Instr.* **59**, 52.

Ratner, E. R., Shen, Z.-X., Dessau, D. S., Wells, B. O., Marshall, D. S., King, D. M., Spicer, W. E., Peng, J. L., Li, Z. Y., and Greene, R. L. (1993) *Phys. Rev.* **B48**, 10482.

Ratz, S., Böttner, R., Freund, W., Schroeder, W., Marquardt, S., Weiss, S., Günther, M., Gerhardt, U., and Wolf, Th. (1996) *Physica C* **262**, 255.

Ratz, S., Schroeder, N., Böttner, R., Dietz, E., Gerhardt, U., and Wolf, Th. (1992) *Solid State Commun.* **82**, 245.

Rehfeld, N., Gerhardt, U. and Dietz, E. (1973) *Appl. Phys.* **1**, 229.

Reihl, B., Maeno, Y., Mangelschots, I., Magnusson, K. O., and Rossel, C. (1990) *Solid State Commun.* **74**, 31.

Reimer, A., Schirmer, J., Feldhaus, J., Bradshaw, A. M., Becker, U., Kerkhoff, H. G., Langer, B., Szostak, D., Wehlitz, R., and Braun, W. (1986) *Phys. Rev. Lett.* **57**, 1707.

Reinert, F., Steiner, P., Hüfner, S., Schmitt, H., Fink, J., Knupfer, M., Sandl, P., and Bertel, E. (1995) *Z. Phys. B* **97**, 83.

Renner, Ch. and Fischer, Ø. (1995) *Phys. Rev.* **B51**, 9208.

Reyes-Gasga, J., Krekels, T., Van Tendeloo, G., Van Landuyt, J., Amelinckx, S, Briggink, W. H. M., and Verweij, H. (1989) *Physica C* **159**, 831.

Richter, M., Meyer, M., Pahler, M., Prescher, T., von Raven, E., Sonntag, B., and Wetzel, H. E. (1988) *Phys. Rev.* **A39**, 5666.

Rietveld, G., Collocott, S. J., Driver, R., and van der Marel, D. (1995) *Physica C* **241**, 273.

Rietveld, G., Glastra, M., and van der Marel, D. (1993) *Physica C* **241**, 257.

Ritchie, R. H. (1957) *Phys. Rev.* **106**, 874.

Rivière, J. C. (1983) in *Practical Surface Analysis by Auger and X-ray Photoelectron Spectroscopy*, eds. D. Briggs and M. P. Seah (John Wiley and Sons, New York) p. 17.

Robertson, J. (1983) *Phys. Rev.* **B28**, 3378.

Rocker, G. and Göpel, W. (1987) *Surf. Sci.* **181**, 530.

Rose-Innes, A. C. and Rhoderick, E. H. (1969) *Introduction to Superconductivity* (Pergamon Press, Oxford).

Rosei, R., Lässer, R., Smith, N. V., and Benbow, R. L. (1980) *Solid State Commun.* **35**, 979.

Rosenberg, R. A. and Wen, C.-R. (1988a) *Phys. Rev.* **B37**, 5841.

Rosenberg, R. A. and Wen, C.-R. (1988b) *Phys. Rev.* **B37**, 9852.

Rossi, G., Lindau, I., Braicovich, L., and Abbati, I. (1983) *Phys. Rev.* **B28**, 3031.

Rowe, E. M. (1979) in *Synchrotron Radiation – Techniques and Applications*, ed. C. Kunz (Springer-Verlag, Berlin) p. 25.

Roy, D. and Carette, D. J. (1977) in *Electron Spectroscopy for Surface Analysis*, ed. H. Ibach (Springer-Verlag, Berlin) p. 13.

Roy, R., Guo, R., Bhalla, A. S., and Cross, L. E. (1994) *J. Vac. Sci. Technol. A* **12**, 269.

Sachs, M. (1957) *Phys. Rev.* **107**, 437.

Saile, V., Schwentner, N., Koch, E.-E., Skibowski, M., Steinmann, W., Ophir, Z., Raz, B., and Jortner, J. (1974) in *Vacuum Ultraviolet Radiation Physics*, eds. E.-E. Koch, R. Haensel, and C. Kunz (Vieweg, Braunschweig) p. 352.

Sakisaka, Y. (1989) *Phys. Rev. Lett.* **63**, 2315.

Sakisaka, Y. (1994) *J. Electron Spectrosc. Relat. Phenom.* **66**, 387.

Sakisaka, Y., Komeda, T., Maruyama, T., Onchi, M., Kato, H., Aiura, Y., Yanashima, H., Terashima, T., Bando, Y., Iijima, K., Yamamoto, K., and Hirata, K.,(1989a) *Phys. Rev.* **B39**, 2304.

Sakisaka, Y., Komeda, T., Maruyama, T., Onchi, M., Kato, H., Aiura, Y., Yanashima, H., Terashima, T., Bando, Y., Iijima, K., Yamamoto, K., and Hirata, K. (1989b) *Phys. Rev.* **B39**, 9080.

Sakisaka, Y., Maruyama, T., Morikawa, Y., Kato, H., Edamoto, K., Okusawa, M., Aiura, Y., Yanashima, H., Terashima, T., Bando, Y., Iijima, K., Yamamoto, K., and Hirata, K. (1990a) *Solid State Commun.* **74**, 609.

Sakisaka, Y., Maruyama, T., Morikawa, Y., Kato, H., Edamoto, K., Okusawa, M., Aiura, Y., Yanashima, H., Terashima, T., Bando, Y., Iijima, K., Yamamoto, K., and Hirata, K. (1990b) *Phys. Rev.* **B42**, 4189.

Sakisaka, Y., Komeda, T., Onchi, M., Kato, H., Masuda, S., and Yagi, K. (1987) *Phys. Rev.* **B36**, 9383.

Salkola, M. I., Emery, V. J., and Kivelson, S. A. (1996a) *Phys. Rev. Lett.* **77**, 155.

Salkola, M. I., Emery, V. J., and Kivelson, S. A. (1996b) *J. Supercond.* **9**, 401.

Samson, J. A. R. (1967) *Techniques of Vacuum Ultraviolet Spectroscopy* (John Wiley and Sons, New York), reprinted in 1981 by Pied, Lincoln, NE.

Sanada, N., Shimomura, M., Suzuki, Y., Fukuda, Y., Nagoshi, M., Ogita, M., Syono, Y., and Takichi, M. (1994) *Phys. Rev.* **B49**, 13119.

Santoni, A. and Himpsel, F. J. (1991) *Phys. Rev.* **B43**, 1305.

Santoni, A., Terminello, L. J., Himpsel, F. J., and Takahashi, T. (1991) *Appl. Phys. A* **52**, 299.

Santoro, A. (1990) in *High-Temperature Superconductivity*, ed. J. W. Lynn (Springer-Verlag, New York) p. 84.

Santra, A. K., Sarma, D. D., and Rao, C. N. R. (1991) *Phys. Rev.* **B43**, 5612.

Sarma, D. D., Sen, P., Carbone, C., Cimino, R., and Gudat, W. (1989) *Phys. Rev.* **B39**, 123287.

Sawatzky, G. A. (1977) *Phys. Rev. Lett.* **39**, 504.

Sawatzky, G. A. (1988) in *Auger Electron Spectroscopy*, eds. C. L. Briant and R. P. Messmer (Academic Press, New York) p. 167.

Sawatzky, G. A. (1990) in *Earlier and Recent Aspects of Superconductivity*, eds. J. G. Bednorz and K. A. Müller (Springer-Verlag, Berlin) p. 345.

Sawatzky, G. A. (1991) in *High-Temperature Superconductivity*, ed. D. P. Tunstall and W. Barford (Adam Hilger, Bristol) p. 145.

Sawatzky, G. A. and Allen, J. W. (1984) *Phys. Rev. Lett.* **53**, 2339.

Sawatzky, G. A. and Lenselink, A. (1980) *Phys. Rev.* **B21**, 1790.

Scalapino, D. J. (1995) *Phys. Rep.* **250**, 329.

Schabel, M. C., Park, C.-H., Matsuura, A., Shen, Z. X., Bonn, D. A., Liang, R., and Hardy, W. N. (1997) *Phys. Rev.* **B55**, 2796.

Schäffer, I., Drube, W., Schlüter, M., Plagemann, G., and Skibowski, M. (1987) *Rev. Sci. Instr.* **58**, 710.

Schäffer, I., Schlüter, M., and Skibowski, M. (1987) *Phys. Rev.* **B35**, 7663.

Schaich, W. L. (1978) in *Photoemission in Solids I*, eds. M. Cardona and L. Ley (Springer-Verlag, Berlin) p. 105.

Schaich, W. L. and Ashcroft, N. W. (1971) *Phys. Rev.* **B3**, 2452.

Scheidt, H., Glöbl, M., and Dose, V. (1981) *Surf. Sci.* **112**, 97.

Schirmer, J., Braunstein, M., and McVoy, V. (1991) *Phys. Rev.* **A44**, 5762.

Schnatterly, S. E. (1970) in *Solid State Physics* **34**, eds. F. Seitz and D. Turnbull, 275.

Schneider, R. and Dose, V. (1992) in *Unoccupied Electronic States*, eds. J. C. Fuggle and J. E. Inglesfield (Springer-Verlag, Berlin) p. 277.

Schneider, R., Dürr, H., Fauster, T., and Dose, V. (1990) *Phys. Rev.* **B42**, 1638.

Schnell, R. D., Rieger, D., and Steinmann, W. (1984) *J. Phys. E* **17**, 221.

Schrieffer, J. R. (1964) *Theory of Superconductivity* (W. A. Benjamin, New York).

Schrieffer, J. R. and Kampf, A. P. (1995) *J. Phys. Chem. Solids* **56**, 1673.

Schroeder, N., Böttner, R., Ratz, S., Dietz, E., Gerhardt, U., and Wolf, Th. (1993b) *Phys. Rev.* **B47**, 5287.

Schroeder, N., Böttner, R., Ratz, S., Marquardt, S., Dietz, E., Gerhardt, U., and Wolf, Th. (1993c) *Solid State Commun.* **87**, 277.

Schroeder, N., Weiss, S., Böttner, R., Marquardt, S., Ratz, S., Dietz, E., Gerhardt, U., Ecke, G., Rössler, H., and Wolf, Th. (1993a) *Physica C* **217**, 220.

Schulz, K. H. and Cox, D. F. (1991) *Phys. Rev.* **B43**, 1610.

Schuppler, S., Fischer, N., Fauster, Th., and Steinmann, W. (1992) *Phys. Rev.* **B46**, 13539.

Schwaller, P., Aebi, P., Osterwalder, J., Schlapbach, L., Shimoda, M., Mochiku, T., and Kadowaki, K. (1993) *Phys. Rev.* **B48**, 6732.

Schwaller, P., Aebi, P., Berger, H., Beeli, C., Osterwalder, J., and Schlapbach, L. (1995) *J. Electron Spectrosc. Relat. Phenom.* **76**, 127.

Schwentner, N. (1974) thesis, Universität München. Quoted in Schwentner, N., Himpsel, F. J., Saile, V., Skibowski, M., Steinmann, W., and Koch, E.-E. (1975) *Phys. Rev. Lett.* **34**, 529.

Seah, M. P. (1995) *J. Electron Spectrosc. Relat. Phenom.* **71**, 191.

Seah, M. P. and Dench, W. A. (1979) *Surf. and Interface Anal.* **1**, 2.

Seino, Y., Okada, K., and Kotani, A. (1990) *J. Phys. Soc. Jpn.* **59**, 1384.

Sevier, K. D. (1972) *Low-Energy Electron Spectrometry* (John Wiley and Sons, New York).

Shaked, H., Keane, P. M., Rodriguez, J. C., Owen, F. F., Hitterman, R. L., and Jorgenson, J. D. (1994) *Crystal Structures of the High-T$_c$ Superconducting Copper Oxides* (Elsevier, Amsterdam).

Shen, Z.-X. (1992) *J. Phys. Chem. Solids* **53**, 1583.

Shen, Z.-X. and Dessau, D. S. (1995) *Phys. Rep.* **253**, 1.

Shen, Z.-X. and Schrieffer, J. R. (1997) *Phys. Rev. Lett.* **78**, 1771.

Shen, Z.-X., Allen, J. W., Yeh, J. J., Kang, J.-S., Ellis, W., Spicer, W., Lindau, I., Maple, M. B., Dalichaouch, Y. D., Torikachvili, M. S., Sun, J. Z., and Geballe, T. H. (1987) *Phys. Rev.* **B36**, 8414.

Shen, Z.-X., Dessau, D. S., Wells, B. O., and King, D. M. (1993a) *J. Phys. Chem. Solids*, **54**, 1169.

Shen, Z.-X., Dessau, D. S., Wells, B. O., King, D. M., Spicer, W. E., Arko, A. J., Marshall, D., Lombardo, L. W., Kapitulnik, A., Dickinson,

P., Doniach, S., DiCarlo, J., Loeser, A. G., and Park, C. H. (1993b) *Phys. Rev. Lett.* **70**, 1553.

Shen, Z.-X., Dessau, D. S., Wells, B. O., Olson, C. G., Mitzi, D. B., Lombardo, L., List, R. S., and Arko, A. J. (1991b) *Phys. Rev.* **B44**, 12098.

Shen, Z.-X., Lindberg, P. A. P., Lindau, I., Spicer, S. E., Eom, C. B., and Geballe, T. H. (1988) *Phys. Rev.* **B38**, 7152.

Shen, Z.-X., List, R. S., Dessau, D. S., Wells, B. O., Jepsen, O., Arko, A. J., Bartlett, R., Shih, C. K., Parmigiani, F., Huang, J. C., and Lindberg, P. A. P. (1991a) *Phys. Rev.* **B44**, 3604.

Shen, Z.-X., Spicer, W. E., King, D. M., Dessau, D. S., and Wells, B. O. (1995) *Science* **267**, 343.

Sherwood, P. M. A. (1983) in *Practical Surface Analysis by Auger and X-ray Photoelectron Spectroscopy*, eds. D. Briggs and M. P. Seah (John Wiley and Sons, New York) p. 465.

Shevchik, N. J. (1977) *Phys. Rev.* **B16**, 3428.

Shichi, Y., Inoue, Y., Munakata, F., and Yamanaka, M. (1990) *Phys. Rev.* **B42**, 939.

Shih, C. K., Feenstra, R. M., and Chandrashekhar, G. V. (1991) *Phys. Rev.* **B43**, 7913.

Shih, C. K., Feenstra, R. M., Kirtley, J. R., and Chandrashekhar, G. V. (1989) *Phys. Rev.* **B40**, 2682.

Shirley, D. A. (1972) *Phys. Rev.* **B5**, 4709.

Shirley, D. A. (1978) in *Photoemission in Solids I*, eds. M. Cardona and L. Ley (Springer-Verlag, Berlin) p. 165.

Shukla, A., Hoffmann, L., Manuel, A. A., Walker, Barbiellini, B., and Peter, M. (1995) *Phys. Rev.* **B51**, 6028.

Shung, K. W.-K. (1991) *Phys. Rev.* **B44**, 13112.

Shung, K. W.-K and Mahan, G. D. (1986) *Phys. Rev. Lett.* **57**, 1076.

Shung, K. W. K and Mahan, G. D. (1988) *Phys. Rev.* **B38**, 3856.

Shung, K. W.-K, Sernelius, B. E., and Mahan, G. D. (1987) *Phys. Rev.* **B36**, 4499.

Siegbahn, K., Nordling, C., Fahlman, A., Nordberg, R., Hamerin, K., Hedman, J., Johansson, G., Bergmark, T., Karlsson, S.-E., Lindgren, I., and Lindberg, B. (1967) *Electron Spectroscopy for Chemical Analysis–Atomic, Molecular, and Solid State Studies by Means of Electron Spectroscopy* (Almqvist and Wiksell, Stockholm); *Nova Acta Regiae Soc. Ups.* Ser. IV, p. 20.

Sieger, M. T., Miller, T., and Chiang, T.-C. (1995) *Phys. Rev. Lett.* **75**, 2043.

Simmons, C. T., Molodsov, S. L., Fedoseenko, S. I., Allman, J. C., Laubschat, C., Kaindl, G., Moshikin, S. V., Kuzmina, M. A., Vlasov, M. U., and Vyvenko, O. F. (1992) *Z. Phys. B* **87**, 271.

Singh, D. J. and Pickett, W. E. (1995) *Phys. Rev.* **B51**, 3128.

Skelton, E. F., Drews, A. R., Osofsky, M. S., Qadri, S. B., Hu, J. Z., Vanderah, T. A., Peng, J. L., and Greene, R. L. (1994) *Science* **263**, 1416.

Skibowski, M. and Kipp, L. (1994) *J. Electron Spectrosc. Relat. Phenom.* **68**, 77.

Slakey, F., Klein, M. V., Rice, J. P., and Ginsberg, D. M. (1990) *Phys. Rev.* **B42**, 2643.

Smith, G. C. (1994a) *Surface Analysis by Electron Spectroscopy* (Plenum Press, New York).

Smith, K. (1995) in *Atomic, Molecular, and Optical Physics: Charged Particles*, eds. F. B. Dunning and R. G. Hulet (Academic Press, San Diego) p. 253.

Smith, K. E. and Kevan, S. D. (1991) *Prog. Solid State Chem.* **21**, 49.

Smith, N. V. (1971) *C. R. C. Critical Reviews in Solid State Science* **2**, 45.

Smith, N. V. (1978) in *Photoemission in Solids I*, eds. M. Cardona and L. Ley (Springer-Verlag, Berlin) p. 237.

Smith, N. V. (1979) *Phys. Rev.* **B19**, 5019.

Smith, N. V. (1988) *Rep. Prog. Phys.* **51**, 1227.

Smith, N. V. (1994b) in *New Directions in Research with Third-Generation Soft X-Ray Synchrotron Radiation Sources,* eds. A. S. Schlachter and F. J. Wuilleumier (Kluwer Academic Publishers, Dordrecht) p. 203.

Smith, N. V. and Himpsel, F. J. (1983) in *Handbook on Synchrotron Radiation* Vol. 1b, ed. E.-E. Koch (North-Holland, Amsterdam) p. 905.

Smith, N. V. and Kevan, S. D. (1982) *Nucl. Instr. and Meth.* **195**, 309.

Smith, N. V. and Spicer, W. E. (1969) *Phys. Rev.* **188**, 593.

Smith, N. V. and Woodruff, N. V. (1986) *Prog. Surf. Sci.* **21**, 295.

Smith, N. V., Thiry, P., and Pétroff, Y. (1993) *Phys. Rev.* **B47**, 15476.

Smith, R. J., Anderson, J., Hermanson, J., and Lapeyre, G. J. (1977) *Solid State Commun.* **21**, 459.

Soda, K., Mori, T., Kitazawa, H., Hagiwara, M., Katsumata, K., Okusawa, M, and Ishiii, T. (1993) *J. Phys. Soc. Jpn.* **62**, 3590.

Söderholm, S., Qvarford, M., Bernhoff, H., Andersen, J. N., Lundgren, E., Nyholm, R., Karlsson, U. O., Lindau, I., and Flodström, S. A. (1996) *J. Phys: Condens. Matter* **8**, 1307.

Spanjaard, D., Guillot, C., Desjonquères, M.-C., Tréglia, G., and Lecante, J. (1985) *Surf. Sci. Rep.* **5**, 1.

Spencer, J. E. and Winick, H. (1980) in *Synchrotron Radiation Research,* eds. H. Winick and S. Doniach (Plenum Press, New York) p. 663.

Spiller, E. (1983) in *Handbook on Synchrotron Radiation,* Vol. 1, eds. E.-E. Koch (North-Holland, Amsterdam) p. 1091.

Srivastava, P., Sekhar, B. R., Saini, N. L., Sharma, S. K., Garg, K. B., Mercey, B., Lecoeur, Ph., and Murray, H. (1993) *Solid State Commun.* **88**, 105.

Stadlober, B., Krug, G., Nemetschek, R., Hackl, R., Cobb, J. T., and Markert, J. T. (1995) *Phys. Rev. Lett.* **74**, 4911.

Stampfl, A. P. J., Con Foo, J. A., Leckey, R. C. G., Riley, J. D., Denecke, R., and Ley, L. (1995) *Surf. Sci.* **331–333**, 1272.

Starace, A. F. (1972) *Phys. Rev.* **B5**, 1773.

Starnberg, H. (1995) *Surf. Sci.* **329**, L624.

Starnberg, H. I., Brauer, H. E., and Nilsson, P. O. (1993b) *Phys. Rev.* **B48**, 621.

Starnberg, H. I., Ilver, L., Nilsson, P. O., and Hughes, H. P. (1993a) *Phys. Rev.* **B47**, 4714.

Staufer, T., Nemetschek, R., Hackl, R., Müller, P., and Veith, H. (1992) *Phys. Rev. Lett.* **68**, 1069.

Steiner, P. and Hüfner, S. (1982) *Solid State Commun.* **44**, 619.

Steiner, P., Höchst, H., and Hüfner, S. (1979) in *Photoemission in Solids II*, eds. L. Ley and M. Cardona (Springer-Verlag, Berlin) p. 349.

Steiner, P., Hüfner, S., Kinsinger, V., Sandar, I., Singwart, B., Schmitt, H., Schultz, R., Junk, S., Schwitzgebel, G., Gold, A., Politis, C., Muller, H. P., Hoppe, R., Kemmler-Sack, S., and Kunz, C. (1988) *Z. Phys. B* **69**, 449.

Steinmann, W. (1989) *Appl. Phys. A* **49**, 365.

Steinmann, W. and Fauster, Th. (1996) in *Laser Spectroscopy and Photochemistry on Metal Surfaces, Part I*, eds. H.-L. Dai and W. Ho (World Scientific, Singapore), p. 184.

Stern, F. (1963) in *Solid State Physics* **15**, eds. F. Seitz and D. Turnbull, p. 299.

Stevens, H. A., Donoho, A. W., Turner, A. M., and Erskine, J. L. (1983) *J. Electron Spectrosc. Relat. Phenom.* **32**, 327.

Stoffel, N. G. and Johnson, P. D. (1984) *Nucl. Instr. and Meth.* **A234**, 230.

Stöhr, J. and Wu, Y. (1994) in *New Directions in Research with Third-Generation Soft X-Ray Synchrotron Radiation Sources,* eds. A. S. Schlachter and F. J. Wuilleumier (Kluwer Academic Publishers, Dordrecht) p. 221.

Stöhr, J., Jaeger, R., and Rehr, J. J. (1983) *Phys. Rev. Lett.* **51**, 821.

Straub, Th., Claessen, R., Steiner, P., Hüfner, S., Eyert, V., Freimelt, K., and Bucher, E. (1997) *Phys. Rev.* **B55**, 13473.

Strocov, V. N. and Starnberg, H. I. (1995) *Phys. Rev.* **B52**, 8759.

Strocov, V. N., Starnberg, H. I., and Nilsson, P. O. (1996) *J. Phys.: Condens. Matter* **8**, 7539

Sugar, J. (1972) *Phys. Rev.* **B5**, 1785.

Sugiyama, J., Itti, R., Yamauchi, H., Koshizuka, N., and Tanaka, S. (1992) *Phys. Rev.* **B45**, 4952.

Surh, M. P., Northrup, J. E., and Louie, S. G. (1988) *Phys. Rev.* **B38**, 5976.

Suzuki, T., Nagoshi, M., Fukuda, Y., Oh-ishii, K, Syono, Y., and Tachiki, M. (1990) *Phys. Rev.* **B42**, 4263.

Suzuki, T., Nagoshi, M., Fukuda, Y., Syono, Y., Kikuchi, M., Kobayashi, N., and Tachiki, M. (1989) *Phys. Rev.* **B40**, 5184.

Svane, A. and Gunnarsson, O. (1990) *Phys. Rev. Lett.* **65**, 1148.

Takahashi, M. and Igarashi, J. (1996a) *Phys. Rev.* **B54**, 13566.

Takahashi, M. and Igarashi, J. (1996b) *Ann. Physik.* **5**, 247.

Takahashi, T., Matsuyama, M., Katayama-Yoshida, H., Okabe, Y., Hosoya, S., Seki, K., Fujimoto, H., Sato, M., and Inokuchi, H. (1988) *Nature* (London) **334**, 691.

Takahashi, T., Matsuyama, H., Katayama-Yoshida, H., Okabe, Y., Hosoya, S., Seki, K., Fujimoto, H., Sato, M., and Inokuchi, H. (1989) *Phys. Rev.* **B39**, 6636.

Takahashi, T., Matsuyama, M., Katayama-Yoshida, H., Seki, K., Kamiya, K., and Inokuchi, H. (1990) *Physica C* **170**, 416.

Tanaka, A. and Jo, T. (1996) *J. Phys. Soc. Jpn.* **65**, 912.

Tanaka, S., Ueda, E., and Sato, M. (1994) *Physica C* **224**, 126.

Tanaka, S., Ueda, E., Sato, M., Tamasaku, K., and Uchida, S. (1995) *J. Phys. Soc. Jpn. Lett.* **64**, 181.

Tanner, D. B. and Timusk, T. (1992) in *Physical Properties of High-Temperature Superconductors*, Vol. 3, ed. D. M. Ginsberg (World Scientific, Singapore) p. 363.

Terada, N., Ahn, C. H., Lew, D., Suzuki, Y., Kihlstrom, K. E., Do, K. B., Arnason, S. B., Geballe, T. H., Hammond, R. H., and Beasley, M. R. (1994) *Appl. Phys. Lett.* **64**, 2581.

Terada, N., Ishibashi, S., Jo, M., Hirabayashi, M., Ihara, H., and Yamamoto, S. (1993) *Appl. Phys. Lett.* **63**, 2967.

Terakura, K., Oguchi, T., Williams, A. R., and Kübler, J. (1984) *Phys. Rev.* **B30**, 4734.

Tersoff, J. and Kevan, S. D. (1983) *Phys. Rev.* **B28**, 4267.

Tersoff, J., Falicov, L. M., and Penn, D. R. (1979) *Solid State Commun.* **32**, 1045.

Thiry, P., Chandesris, D., Le Cante, J., Guillot, C., Pincheaux, R. and Pétroff, Y. (1979) *Phys. Rev. Lett.* **43**, 82.

Thomann, U., Rangelov, G., and Fauster, T. (1995) *Surf. Sci.* **331–333**, 1283.

Thomsen, C., Cardona, M., Gegenheimer, B., Liu, R., and Simon, A. (1988) *Phys. Rev.* **B37**, 9860.

Thuler, M. R., Benbow, R. L. and Hurych, Z. (1982) *Phys. Rev.* **B26**, 669.

Thuler, M. R., Benbow, R. L. and Hurych, Z. (1983) *Phys. Rev.* **B27**, 2082.

Thurgate, S. M. (1996) *J. Electron Spectrosc. Relat. Phenom.* **81**, 1.

Thurgate, S. M., Lund, C. P., and Wedding, A. B. (1994) *Phys. Rev.* **B50**, 4810.

Tibbets, G. G. and Egelhoff, W. F. (1978) *Phys. Rev. Lett.* **71**, 188.

Timusk, T. and Tanner, D. B. (1989) in *Physical Properties of High-Temperature Superconductors*, Vol. 1, ed. D. M. Ginsberg (World Scientific, Singapore) p. 339.

Tinkham, M. (1964) *Group Theory and Quantum Mechanics* (McGraw-Hill, New York).

Tinkham, M. (1996) *Introduction to Superconductivity*, 2nd edn. (McGraw-Hill, New York).

Tjeng, L. H., Chen, C. T., and Cheong, S.-W. (1992) *Phys. Rev.* **B45**, 8205.

Tjeng, L. H., Chen, C. T., Ghijsen, J., Rudolf, P., and Sette, F. (1991) *Phys. Rev. Lett.* **67**, 501.

Tjeng, L. H., Sinkovic, B., Brookes, N. B., Goedkoop, J. B., Hesper, R., Pellegrin, E., de Groot, F. M. F., Altieri, S., Hulbert, S. L., Shekel, E., and Sawatzky, G. A. (1997) *Phys. Rev. Lett.* **78**, 1126.

Tjernberg, O., Nylén, H., Chiaia, G., Söderholm, S., Karlsson, U. O., Qvarford, M., Lindau, I., Puglia, C., Mårtensson, N., and Leonyuk, L. (1997) *Phys. Rev. Lett.* **79**, 499.

Tjernberg, O., Söderholm, S., Karlsson, U. O., Chiaia, G., Qvarford, M., Nylén, H., and Lindau, I. (1996) *Phys. Rev.* **B53**, 10372.

Tobin, J. G., Olson, C. G., Gu, C., Liu, J. Z., Solal, F. R., Fluss, M. J., Howell, M. H., O'Brien, J. C., Radousky, H. B., and Sterne, P. A. (1992) *Phys. Rev.* **B45**, 5563.

Tonner, B. P. (1990) *Nucl. Instr. and Meth.* **A291**, 60.

Tonner, B. P. and Dunham, D. (1994) *Nucl. Instr. and Meth.* **A347**, 436.

Tonner, B. P. and Harp, G. R. (1988) *Rev. Sci. Instr.* **59**, 853.

Tonner, B. P. and Harp, G. R. (1989) *J. Vac. Sci. Technol. A* **7**, 1.

Tonner, B. P., Dunham, D., Droubay, T., Kikuma, J., and Denlinger, J. (1996) *J. Electron Spectrosc. Relat. Phenom.* **78**, 13.

Tougaard, S. (1990) *J. Electron Spectrosc. Relat. Phenom.* **52**, 243.

Tougaard, S. and Kraaer, J. (1991) *Phys. Rev.* **B43**, 1651.

Tougaard, S. and Sigmund, P. (1982) *Phys. Rev.* **B25**, 4452.

Towler, M. D., Allan, N. L., Harrison, N. M., Saunders, V. R., Mackrodt, W. C., and Aprà, E. (1994) *Phys. Rev.* **B50**, 504.

Tranquada, J. M., Axe, J. D., Ichikawa, N., Moodenbaugh, A. R., Nakamura, Y., and Uchida, S. (1997) *Phys. Rev. Lett.* **78**, 338.

Tranquada, J. M., Axe, J. D., Ichikawa, N., Nakamura, Y., Uchida, S., and Nachumi, B. (1996) *Phys. Rev.* **B54**, 7489.

Tréglia, G., Desjonquères, M. C., Spanjaard, D., Sébilleau, D., Guillot, C., Chauveau, D., and Lecante, J. (1988) *J. Phys.: Condens. Matter* 1, 1879; (E) 5989.

Tréglia, G., Ducastelle, F., and Spanjaard, D. (1980) *Phys. Rev.* **B21**, 3729.

Ueba, H. (1995) *Surf. Sci.* **334**, L719.

Uhlenbrock, St., Scharfschwerdt, Chr., Neumann, M., Illing, G., and Freund, H. J. (1992) *J. Phys.: Condens. Matter* 4, 7973.

Unger, P., Igarashi, J., and Fulde, P. (1994) *Phys. Rev.* **B50**, 10485.

Ungier, L. and Thomas, T. D. (1984) *Phys. Rev. Lett.* **53**, 435.

van Bentum, P. J. M. and van Kampen, H. (1992) in *Scanning Tunneling Microscopy I*, eds. H.-J. Güntherodt and R. Wiesendanger (Springer-Verlag, Berlin) p. 207

van der Laan, G., Surman, M., Hoyland, M. A., Flipse, C. F. J., Thole, B. T., Seino, Y., Ogasawara, H., and Kotani, A. (1992) *Phys. Rev.* **B46**, 9336.

van Veenendaal, M. A. and Sawatzky, G. A. (1993b) *Phys. Rev. Lett.* **70**, 2459.

van Veenendaal, M. A. and Sawatzky, G. A. (1994b) *Phys. Rev.* **B49**, 3473.

van Veenendaal, M. A., Eskes, H., and Sawatzky, G. A. (1993c) *Phys. Rev.* **B47**, 11462.

van Veenendaal, M. A., Sawatzky, G. A., and Groen, W. A. (1994a) *Phys. Rev.* **B49**, 1407.

van Veenendaal, Schlatmann, R., Sawatzky, G. A., and Groen, W. A. (1993a) *Phys. Rev.* **B47**, 446.

Varma, C. M., Littlewood, P. B., Schmitt-Rink, S., Abrahams, E., and Ruckenstein, A. E. (1989) *Phys. Rev. Lett.* **63**, 1996

Vasquez, R. P. (1994a) *J. Electron Spectrosc. Relat. Phenom.* **66**, 209.

Vasquez, R. P. (1994b) *J. Electron Spectrosc. Relat. Phenom.* **66**, 241.

Vasquez, R. P. and Olson, W. L. (1991) *Physica C* **177**, 223.

Vasquez, R. P., Gupta, A., and Kussmaul, A. (1991) *Solid State Commun.* **78**, 303.

Vasquez, R. P., Ren, Z. F., and Wang, J. H. (1996) *Phys. Rev.* **B54**, 6115.

Vasquez, R. P., Rupp, M., Gupta, A., and Tsuei, C. C. (1995) *Phys. Rev.* **B51**, 15657.

Veal, B. W. and Gu, C. (1994) *J. Electron Spectrosc. Relat. Phenom.* **66**, 321.

Veal, B. W., Liu, J. Z., Paulikas, A. P., Vandervoort, K., Claus, H., Campuzano, J. C., Olson, C. G., Yang, A.-B., Liu, R., Gu, C., List, R. S., Arko, A. J., and Bartlett, R. J. (1989) *Physica C* **158**, 276.

Veal, B. W., Liu, R., Paulikas, A. P., Koelling, D. D., Shi, H., Downey, J. W., Olson, C. G., Arko, A. J., Joyce, J. J., and Blythe, R. (1990b) *Surf. Sci. Rep.* **19**, 121.

Veal, B. W., Paulikas, A. P., You, H., Shi, H., Fang, Y., and Downey, J. W. (1990a) *Phys. Rev.* **B42**, 4770.

Vicente Alvarez, M. A., Ascolani, H., and Zampieri, G. (1996) *Phys. Rev.* **B54**, 14703.

Viescas, A. J., Tranquada, J. M., Moodenbaugh, A. M., and Johnson, P. D. (1988) *Phys. Rev.* **B37**, 3738.

Vilk, Y. M. (1997) *Phys. Rev.* **B55**, 3870.

von Stetten, E. C., Berko, S., Lee, X. S., Lee, R. R., Brynesad, J., Singh, D., Krakauer, H., Pickett, W. E., and Cohen, R. E. (1988) *Phys. Rev. Lett.* **60**, 2198.

Voss, J., Kunz, C., Moewes, A., and Storjohann, I. (1992) *Rev. Sci. Instr.* **63**, 569.

Waddill, G. D., Komeda, T., Benning, P. J., Weaver, J. H., and Knapp, G. S. (1991) *J. Vac. Sci. Technol. A* **9**, 1634.

Waddill, G. D., Vitomirov, I. M., Aldao, C. M., Anderson, S. G., Capasso, C., Weaver, J. H., and Liliental-Weber, Z. (1990) *Phys. Rev.* **B41**, 5293.

Wagener, T. J., Hu. Y., Gao, Y.-J., Jost, M. B., Weaver, J. H., Spencer, M. D., and Goretta, K. C. (1989) *Phys. Rev.* **B39**, 2928.

Wagener, T. J., Hu, Y.-J., Jost, M. B., and Weaver, J. H. (1990a) *Phys. Rev.* **B42**, 1041.

Wagener, T. J., Hu. Y.-J., Jost, M. B., Weaver, J. H., Yan, Y. F., Chu, X., and Zhao, Z. X. (1990b) *Phys. Rev.* **B42**, 6317.

Wagener, T. J., Meyer, H. M., Ju, Y., Jost, M. B., Weaver, J. H., and Goretta, K. C. (1990c) *Phys. Rev.* **B41**, 4201.

Wagner, L. F., Hussain, Z., Fadley, C. S. and Baird, R. J. (1977) *Solid State Commun.* **21**, 453.

Waldram, J. R. (1996) *Superconductivity of Metals and Cuprates* (Institute of Physics Publishing, Bristol)

Wang, C. S. (1990) in *High-Temperature Superconductivity*, ed. J. W. Lynn (Springer-Verlag, New York) p. 122.

Wang, C. S. and Callaway, J. (1977) *Phys. Rev.* **B15**, 298.

Wang, D. S. and Freeman, A. J. (unpublished).

Wang, N. L., Ruan, K. Q., Chong, Y., Deng, M., Wang, C. Y., Chen, Z. J., Cao, L. Z., Wu, J. X., and Zhu, J. S. (1995) *Phys. Rev.* **B51**, 3791.

Wannberg, B. (1985) *Nucl. Instr. and Meth.* **A239**, 269.

Watanabe, T., Takahashi, T., Suzuki, S., Sato, S., Katayama-Yoshida, H., Yamanaka, A., Minami, F., and Takekawa, S. (1991) *Physica C* **176**, 274.

Weaver, J. H. (1994) in *Interfaces in High-T_c Superconducting Systems*, eds. S. L. Shindé and D. A. Rudman (Springer-Verlag, Berlin) p. 210.

Weaver, J. H., Gao, Y., Wagener, T. J., Flandermeyer, B., and Capone, D. W. (1987) *Phys. Rev.* **B36**, 3975.

Weaver, J. H., Meyer, H. M., Wagener, T. J., Hill, D. M., Gao, Y., Peterson, D., Fisk, Z., and Arko, A. J. (1988) *Phys. Rev.* **B38**, 4668.

Wei, P. and Qi, Z. Q. (1994) *Phys. Rev.* **B49**, 10864.

Weightman, P. (1982) *Rep. Prog. Phys.* **45**, 753.

Weinelt, M., Nilsson, A., Magnuson M., Weill, T., Wassdahl, N., Karis, O., Föhlisch, A., Mårtensson, N., Stöhr, J., and Samant, M. (1997) *Phys. Rev. Lett.* **78**, 967.

Weissmann, R. and Müller, K. (1981) *Surf. Sci. Rep.* **105**, 251.

Wells, B. O., Lindberg, P. A. P., Shen, Z.-X., Dessau, D. S., Spicer, W. E., Lindau, I., Mitzi, D. B., and Kapitulnik, A. (1989) *Phys. Rev.* **B40**, 5259.

Wells, B. O., Shen, Z.-X, Dessau, D. S., Spicer, W. E., Mitzi, D. B., Lombardo, L., Kapitulnik, A., and Arko, A. J. (1992) *Phys. Rev.* **B46**, 11830.

Wells, B. O., Shen, Z.-X, Dessau, D. S., Spicer, W. E., Olson, C. G., Mitzi, D. B., Kapitulnik, A., List, R. S., and Arko, A. J. (1990b) *Phys. Rev. Lett.* **65**, 3056.

Wells, B. O., Shen, Z.-X., Harrison, W.A., and Spicer, W. E. (1990a) *Phys. Rev.* **B42**, 4785.

Wells, B. O., Shen, Z.-X., Matsuura, A., King, D. M., Kastner, M. A., Greven, M., Birgeneau, R. J. (1995) *Phys. Rev. Lett.* **74**, 964.

Wendin, G. (1981) *Breakdown of the One-electron Pictures in Photoelectron Spectroscopy*, Vol. 45 of 'Structure and Bonding' (Springer-Verlag, Berlin).

Wendin, G. (1982) in *New Trends in Atomic Physics*, Vol. II, eds. G. Grynberg and R. Stora (North-Holland, Amsterdam) p. 555.

Werner, W. S. M. (1995), *Phys. Rev.* **B52**, 2964.

Wertheim, G. (1978) in *Electron and Ion Spectroscopy of Solids*, eds. L. Fiermans, J. Vennik, and W. Dekeyser (Plenum Press, New York) p. 192.

Wertheim, G. K. and Citrin, P. H. (1978) in *Photoemission in Solids I*, eds. M. Cardona and L. Ley (Springer-Verlag, Berlin) p. 197.

Wertheim, G. K. and Dicenzo, S. B. (1985) *J. Electron Spectrosc. Relat. Phenom.* **37**, 57.

Wertheim, G. K. and Hüfner, S. (1972) *Phys. Rev. Lett.* **28**, 1028.

Wertheim, G. K. and Riffe, D. M. (1995) *Solid State Commun.* **96**, 645.

Wertheim, G. K. and Walker, L. R. (1976) *J. Phys. F* **6**, 2297.

West, J. B. and Padmore, H. A. (1987) in *Handbook on Synchrotron Radiation*, Vol. 2, ed. G. V. Marr (North-Holland, Amsterdam) p. 21.

White, P. J., Shen, Z.-X., Kim, C., Harris, J. M., Loeser, A. G., Fournier, P., and Kapitulnik, A. (1996) *Phys. Rev.* **B54**, R15669.

White, R. C., Fadley, C. S., Sagurton, M., Roubin, P., Chandesris, D., Lecante, J., Guillot, C., and Hussain, Z. (1987) *Phys. Rev.* **B35**, 1147.

Wiedemann, H. (1991) in *Handbook on Synchrotron Radiation*, Vol. 3, eds. G. S. Brown and D. E. Moncton (North-Holland, Amsterdam) p. 1.

Williams, R. H., Srivastava, G. P. and McGovern, I. T. (1980) *Rep. Prog. Phys.* **43**, 87.

Williams, R. S., Wehner, P. S., Stöhr, J., and Shirley, D. A. (1977) *Phys. Rev. Lett.* **39**, 302.

Wilson, A. H. (1936) *Theory of Metals* (Cambridge University Press, Cambridge) p.133.

Wincott, P. L., Brookes, N. B., Law, D. S.-L., and Thornton, G. (1986) *Phys. Rev.* **B33**, 4373.

Winick, H. (1980) in *Synchrotron Radiation Research*, eds. H. Winick and S. Doniach (Plenum Press, New York) p. 11.

Winick, H. (1995) *Synchrotron Radiation Sources, a Primer* (World Scientific Publishing Co., Singapore)

Wöhlecke, M., Baalmann, A., and Neumann, M. (1984) *Solid State Commun.* **49**, 217.

Wolf, E. L. (1985) *Principles of Electron Tunneling Spectroscopy* (Oxford University Press, New York).

Wolter, H. (1952) *Ann. Phys. (10)* **94**, 286.

Woodruff, D. P. and Bradshaw, A. M. (1994) *Rept. Prog. Phys.* **57**, 1029.

Woodruff, D. P. and Delchar, T. A. (1994), *Modern Techniques of Surface Science*, 2nd edn. (Cambridge University Press, Cambridge).

Woodruff, D. P. and Smith, N. V. (1990) *Phys. Rev.* **B41**, 8150.

Woodruff, D. P., Smith, N. V., Johnson, P. D., and Royer, W. A. (1982) *Phys. Rev.* **B26**, 2943.

Wooten, F. (1972) *Optical Properties of Solids* (Academic Press, New York).

Xiao, M.-W. and Li, Z.-Z. (1994) *Phys. Rev.* **B49**, 13160.

Xu, S., Cao, J., Miller, C. C., Mantell, D. A., Miller, R. J. D., and Gao, Y. (1996) *Phys. Rev. Lett.* **76**, 483.

Yamamoto, A., Onoda, M., Takayama-Muromachi, E., Izumi, F., Ishigaki, T., and Asano, H. (1990) *Phys. Rev.* **B42**, 4228.

Yang, J. and Hu, C.-R. (1994) *Phys. Rev.* **B50**, 16766.

Yarmoff, J. A., Clarke, D. R., Drube, W., Karlsson, U. O., Taleb-Ibrahami, A., and Himpsel, F.J. (1987) *Phys. Rev.* **B36**, 3967.

Yeh, J. J. (1992) *Phys. Rev.* **B45**, 10816.

Yeh, J. J. and Lindau, I. (1985) *Atomic Data and Nuclear Data Tables* **32**, 1.

Yeh, J.-J., Lindau, I., Sun, J.-Z., Char, K., Missert, N., Kapitulnik, A., Geballe, T. H., and Beasley, M. R. (1990) *Phys. Rev.* **B42**, 8044.

Yokoya, T., Takahashi, T., Fujiwara, A., and Koike, Y. (1993) *Solid State Commun.* **87**, 553.

Yokoya, T., Takahashi, T., Mochiku, T., and Kadowaki, K. (1994) *Phys. Rev.* **B50**, 10225.

Yokoya, T., Takahashi, T., Mochiku, T., and Kadowaki, K. (1995) *Phys. Rev.* **B51**, 3945.

Yokoya, T., Takahashi, T., Mochiku, T., and Kadowaki, K. (1996) *Phys. Rev.* **B53**, 14055.

Yu, J., and Freeman, A. J. (1994) *J. Electron Spectrosc. Relat. Phenom.* **66**, 281.

Yu, J., Massidda, S., Freeman, A. J., and Koelling, D. D. (1987) *Phys. Lett. A* **122**, 203.

Yu, J., Park, K. T., and Freeman, A. J. (1991) *Physica C* **172**, 467.

Yu, W. J., Mao, Z. Q., Liu, X. M., Tian, M. L., Zhou, G. E., and Zhang, Y. H. (1996) *Physica C* **261**, 27.

Zaanen, J. (1992) in *Unoccupied Electronic States,* eds. J. C. Fuggle and J. E. Inglesfield (Springer-Verlag, Berlin) p. 89.

Zaanen, J., Sawatzky, G. A., and Allen, J. W. (1985) *Phys. Rev. Lett.* **55**, 418.

Zaanen, J., Westra, C., and Sawatzky, G. A. (1986) *Phys. Rev.* **B33**, 8060.

Zangwill, A. (1982) in *Atomic Physics* Vol. 8, eds. I. Lindgren, A. Rosen, and S. Svanberg (Plenum Press, New York) p. 339.

Zangwill, A. (1988) *Physics at Surfaces* (Cambridge University Press, Cambridge).

Zangwill, A. and Soven, P. (1980) *Phys. Rev. Lett.* **45**, 204.

Zasadzinski, J., Ozyuzer, L., Yusof, Z., Chen, J., Gray, K. E., Mogilevsky, R., Hinks, D. G., Cobb, J. L., and Markert, J. T. (1996) in *Spectroscopic Studies of Superconductors,* eds. I. Bozovic and D. van der Marel (SPIE, Bellingham) p. 338.

Zhang, F. C. and Rice, T. M. (1988) *Phys. Rev.* **B37**, 3759.

Zheng, Z. and Chen, C. (1995) *Phys. Rev.* **B51**, 14092.

Zhou, W. (1996) *J. Supercond.* **9**, 311.

Ziegler, Ch., Frank, G., and Göpel, W. (1990) *Z. Phys. B* **81**, 349.

Zunger, A. and Freeman, A. J. (1977) *Phys. Rev.* **B16**, 2901.

Index